Third Edition

HEAT EXCHANGERS
Selection, Rating, and Thermal Design

Third Edition

HEAT EXCHANGERS
Selection, Rating, and Thermal Design

Sadık Kakaç
Hongtan Liu
Anchasa Pramuanjaroenkij

CRC Press
Taylor & Francis Group
Boca Raton London New York

CRC Press is an imprint of the
Taylor & Francis Group, an **informa** business

GEN
TJ263
.K25
2012

CRC Press
Taylor & Francis Group
6000 Broken Sound Parkway NW, Suite 300
Boca Raton, FL 33487-2742

© 2012 by Taylor & Francis Group, LLC
CRC Press is an imprint of Taylor & Francis Group, an Informa business

No claim to original U.S. Government works

International Standard Book Number: 978-1-4398-4990-3 (Hardback)

This book contains information obtained from authentic and highly regarded sources. Reasonable efforts have been made to publish reliable data and information, but the author and publisher cannot assume responsibility for the validity of all materials or the consequences of their use. The authors and publishers have attempted to trace the copyright holders of all material reproduced in this publication and apologize to copyright holders if permission to publish in this form has not been obtained. If any copyright material has not been acknowledged please write and let us know so we may rectify in any future reprint.

Except as permitted under U.S. Copyright Law, no part of this book may be reprinted, reproduced, transmitted, or utilized in any form by any electronic, mechanical, or other means, now known or hereafter invented, including photocopying, microfilming, and recording, or in any information storage or retrieval system, without written permission from the publishers.

For permission to photocopy or use material electronically from this work, please access www.copyright.com (http://www.copyright.com/) or contact the Copyright Clearance Center, Inc. (CCC), 222 Rosewood Drive, Danvers, MA 01923, 978-750-8400. CCC is a not-for-profit organization that provides licenses and registration for a variety of users. For organizations that have been granted a photocopy license by the CCC, a separate system of payment has been arranged.

Trademark Notice: Product or corporate names may be trademarks or registered trademarks, and are used only for identification and explanation without intent to infringe.

Visit the Taylor & Francis Web site at
http://www.taylorandfrancis.com

and the CRC Press Web site at
http://www.crcpress.com

Printed and bound in Great Britain by
TJ International Ltd, Padstow, Cornwall

Contents

Preface xiii

1. **Classification of Heat Exchangers** 1
   1.1   Introduction 1
   1.2   Recuperation and Regeneration 1
   1.3   Transfer Processes 6
   1.4   Geometry of Construction 8
         1.4.1   Tubular Heat Exchangers 8
                 1.4.1.1   Double-Pipe Heat Exchangers 8
                 1.4.1.2   Shell-and-Tube Heat Exchangers 9
                 1.4.1.3   Spiral-Tube-Type Heat Exchangers 12
         1.4.2   Plate Heat Exchangers 12
                 1.4.2.1   Gasketed Plate Heat Exchangers 12
                 1.4.2.2   Spiral Plate Heat Exchangers 14
                 1.4.2.3   Lamella Heat Exchangers 15
         1.4.3   Extended Surface Heat Exchangers 17
                 1.4.3.1   Plate-Fin Heat Exchanger 17
                 1.4.3.2   Tubular-Fin Heat Exchangers 18
   1.5   Heat Transfer Mechanisms 23
   1.6   Flow Arrangements 24
   1.7   Applications 25
   1.8   Selection of Heat Exchangers 26
   References 30

2. **Basic Design Methods of Heat Exchangers** 33
   2.1   Introduction 33
   2.2   Arrangement of Flow Paths in Heat Exchangers 33
   2.3   Basic Equations in Design 35
   2.4   Overall Heat Transfer Coefficient 37
   2.5   LMTD Method for Heat Exchanger Analysis 43
         2.5.1   Parallel- and Counterflow Heat Exchangers 43
         2.5.2   Multipass and Crossflow Heat Exchangers 47
   2.6   The ε-NTU Method for Heat Exchanger Analysis 56
   2.7   Heat Exchanger Design Calculation 66
   2.8   Variable Overall Heat Transfer Coefficient 67
   2.9   Heat Exchanger Design Methodology 70
   Nomenclature 73
   References 78

v

3. Forced Convection Correlations for the Single-Phase Side of Heat Exchangers 81
- 3.1 Introduction 81
- 3.2 Laminar Forced Convection 84
  - 3.2.1 Hydrodynamically Developed and Thermally Developing Laminar Flow in Smooth Circular Ducts 84
  - 3.2.2 Simultaneously Developing Laminar Flow in Smooth Ducts 85
  - 3.2.3 Laminar Flow through Concentric Annular Smooth Ducts 86
- 3.3 Effect of Variable Physical Properties 88
  - 3.3.1 Laminar Flow of Liquids 90
  - 3.3.2 Laminar Flow of Gases 92
- 3.4 Turbulent Forced Convection 93
- 3.5 Turbulent Flow in Smooth Straight Noncircular Ducts 99
- 3.6 Effect of Variable Physical Properties in Turbulent Forced Convection 103
  - 3.6.1 Turbulent Liquid Flow in Ducts 103
  - 3.6.2 Turbulent Gas Flow in Ducts 104
- 3.7 Summary of Forced Convection in Straight Ducts 107
- 3.8 Heat Transfer from Smooth-Tube Bundles 111
- 3.9 Heat Transfer in Helical Coils and Spirals 114
  - 3.9.1 Nusselt Numbers of Helical Coils—Laminar Flow 116
  - 3.9.2 Nusselt Numbers for Spiral Coils—Laminar Flow 117
  - 3.9.3 Nusselt Numbers for Helical Coils—Turbulent Flow 117
- 3.10 Heat Transfer in Bends 118
  - 3.10.1 Heat Transfer in 90° Bends 118
  - 3.10.2 Heat Transfer in 180° Bends 119
- Nomenclature 120
- References 125

4. Heat Exchanger Pressure Drop and Pumping Power 129
- 4.1 Introduction 129
- 4.2 Tube-Side Pressure Drop 129
  - 4.2.1 Circular Cross-Sectional Tubes 129
  - 4.2.2 Noncircular Cross-Sectional Ducts 132
- 4.3 Pressure Drop in Tube Bundles in Crossflow 135
- 4.4 Pressure Drop in Helical and Spiral Coils 137
  - 4.4.1 Helical Coils—Laminar Flow 138
  - 4.4.2 Spiral Coils—Laminar Flow 138
  - 4.4.3 Helical Coils—Turbulent Flow 139
  - 4.4.4 Spiral Coils—Turbulent Flow 139
- 4.5 Pressure Drop in Bends and Fittings 140
  - 4.5.1 Pressure Drop in Bends 140
  - 4.5.2 Pressure Drop in Fittings 142

4.6	Pressure Drop for Abrupt Contraction, Expansion, and Momentum Change	147
4.7	Heat Transfer and Pumping Power Relationship	148

Nomenclature 150
References 155

5. Micro/Nano Heat Transfer 157
   5.1 PART A—Heat Transfer for Gaseous and Liquid Flow in Microchannels 157
      5.1.1 Introduction of Heat Transfer in Microchannels 157
      5.1.2 Fundamentals of Gaseous Flow in Microchannels 158
            5.1.2.1 Knudsen Number 158
            5.1.2.2 Velocity Slip 160
            5.1.2.3 Temperature Jump 160
            5.1.2.4 Brinkman Number 161
      5.1.3 Engineering Applications for Gas Flow 163
            5.1.3.1 Heat Transfer in Gas Flow 165
            5.1.3.2 Friction Factor 169
            5.1.3.3 Laminar to Turbulent Transition Regime 173
      5.1.4 Engineering Applications of Single-Phase Liquid Flow in Microchannels 177
            5.1.4.1 Nusselt Number and Friction Factor Correlations for Single-Phase Liquid Flow 179
            5.1.4.2 Roughness Effect on Friction Factor 185
   5.2 PART B—Single-Phase Convective Heat Transfer with Nanofluids 186
      5.2.1 Introduction of Convective Heat Transfer with Nanofluids 186
            5.2.1.1 Particle Materials and Base Fluids 187
            5.2.1.2 Particle Size and Shape 187
            5.2.1.3 Nanofluid Preparation Methods 188
      5.2.2 Thermal Conductivity of Nanofluids 188
            5.2.2.1 Classical Models 189
            5.2.2.2 Brownian Motion of Nanoparticles 191
            5.2.2.3 Clustering of Nanoparticles 193
            5.2.2.4 Liquid Layering around Nanoparticles 196
      5.2.3 Thermal Conductivity Experimental Studies of Nanofluids 203
      5.2.4 Convective Heat Trasfer of Nanofluids 207
      5.2.5 Analysis of Convective Heat Transfer of Nanofluids 212
            5.2.5.1 Constant Wall Heat Flux Boundary Condition ... 212
            5.2.5.2 Constant Wall Temperature Boundary Condition 214
      5.2.6 Experimental Correlations of Convective Heat Transfer of Nanofluids 216

Nomenclature 224
References 228

6. Fouling of Heat Exchangers 237
6.1 Introduction 237
6.2 Basic Considerations 237
6.3 Effects of Fouling 239
    6.3.1 Effect of Fouling on Heat Transfer 240
    6.3.2 Effect of Fouling on Pressure Drop 241
    6.3.3 Cost of Fouling 243
6.4 Aspects of Fouling 244
    6.4.1 Categories of Fouling 244
        6.4.1.1 Particulate Fouling 244
        6.4.1.2 Crystallization Fouling 245
        6.4.1.3 Corrosion Fouling 245
        6.4.1.4 Biofouling 245
        6.4.1.5 Chemical Reaction Fouling 246
    6.4.2 Fundamental Processes of Fouling 246
        6.4.2.1 Initiation 246
        6.4.2.2 Transport 246
        6.4.2.3 Attachment 247
        6.4.2.4 Removal 247
        6.4.2.5 Aging 248
    6.4.3 Prediction of Fouling 248
6.5 Design of Heat Exchangers Subject to Fouling 250
    6.5.1 Fouling Resistance 250
    6.5.2 Cleanliness Factor 256
    6.5.3 Percent over Surface 257
        6.5.3.1 Cleanliness Factor 260
        6.5.3.2 Percent over Surface 260
6.6 Operations of Heat Exchangers Subject to Fouling 262
6.7 Techniques to Control Fouling 264
    6.7.1 Surface Cleaning Techniques 264
        6.7.1.1 Continuous Cleaning 264
        6.7.1.2 Periodic Cleaning 264
    6.7.2 Additives 265
        6.7.2.1 Crystallization Fouling 265
        6.7.2.2 Particulate Fouling 266
        6.7.2.3 Biological Fouling 266
        6.7.2.4 Corrosion Fouling 266
Nomenclature 266
References 270

7. Double-Pipe Heat Exchangers 273
7.1 Introduction 273

	7.2	Thermal and Hydraulic Design of Inner Tube	276

- 7.2 Thermal and Hydraulic Design of Inner Tube ... 276
- 7.3 Thermal and Hydraulic Analysis of Annulus ... 278
  - 7.3.1 Hairpin Heat Exchanger with Bare Inner Tube ... 278
  - 7.3.2 Hairpin Heat Exchangers with Multitube Finned Inner Tubes ... 283
- 7.4 Parallel–Series Arrangements of Hairpins ... 291
- 7.5 Total Pressure Drop ... 294
- 7.6 Design and Operational Features ... 295
- Nomenclature ... 297
- References ... 304

8. Design Correlations for Condensers and Evaporators ... 307
- 8.1 Introduction ... 307
- 8.2 Condensation ... 307
- 8.3 Film Condensation on a Single Horizontal Tube ... 308
  - 8.3.1 Laminar Film Condensation ... 308
  - 8.3.2 Forced Convection ... 309
- 8.4 Film Condensation in Tube Bundles ... 312
  - 8.4.1 Effect of Condensate Inundation ... 313
  - 8.4.2 Effect of Vapor Shear ... 317
  - 8.4.3 Combined Effects of Inundation and Vapor Shear ... 317
- 8.5 Condensation inside Tubes ... 322
  - 8.5.1 Condensation inside Horizontal Tubes ... 322
  - 8.5.2 Condensation inside Vertical Tubes ... 327
- 8.6 Flow Boiling ... 329
  - 8.6.1 Subcooled Boiling ... 329
  - 8.6.2 Flow Pattern ... 331
  - 8.6.3 Flow Boiling Correlations ... 334
- Nomenclature ... 353
- References ... 356

9. Shell-and-Tube Heat Exchangers ... 361
- 9.1 Introduction ... 361
- 9.2 Basic Components ... 361
  - 9.2.1 Shell Types ... 361
  - 9.2.2 Tube Bundle Types ... 364
  - 9.2.3 Tubes and Tube Passes ... 366
  - 9.2.4 Tube Layout ... 368
  - 9.2.5 Baffle Type and Geometry ... 371
  - 9.2.6 Allocation of Streams ... 376
- 9.3 Basic Design Procedure of a Heat Exchanger ... 378
  - 9.3.1 Preliminary Estimation of Unit Size ... 380
  - 9.3.2 Rating of the Preliminary Design ... 386
- 9.4 Shell-Side Heat Transfer and Pressure Drop ... 387
  - 9.4.1 Shell-Side Heat Transfer Coefficient ... 387

	9.4.2	Shell-Side Pressure Drop	389
	9.4.3	Tube-Side Pressure Drop	390
	9.4.4	Bell–Delaware Method	395
		9.4.4.1 Shell-Side Heat Transfer Coefficient	396
		9.4.4.2 Shell-Side Pressure Drop	407
Nomenclature			419
References			425

10. Compact Heat Exchangers ... 427

10.1	Introduction	427
	10.1.1 Heat Transfer Enhancement	427
	10.1.2 Plate-Fin Heat Exchangers	431
	10.1.3 Tube-Fin Heat Exchangers	431
10.2	Heat Transfer and Pressure Drop	433
	10.2.1 Heat Transfer	433
	10.2.2 Pressure Drop for Finned-Tube Exchangers	441
	10.2.3 Pressure Drop for Plate-Fin Exchangers	441
Nomenclature		446
References		449

11. Gasketed-Plate Heat Exchangers ... 451

11.1	Introduction	451
11.2	Mechanical Features	451
	11.2.1 Plate Pack and the Frame	453
	11.2.2 Plate Types	455
11.3	Operational Characteristics	457
	11.3.1 Main Advantages	457
	11.3.2 Performance Limits	459
11.4	Passes and Flow Arrangements	460
11.5	Applications	461
	11.5.1 Corrosion	462
	11.5.2 Maintenance	465
11.6	Heat Transfer and Pressure Drop Calculations	466
	11.6.1 Heat Transfer Area	466
	11.6.2 Mean Flow Channel Gap	467
	11.6.3 Channel Hydraulic Diameter	468
	11.6.4 Heat Transfer Coefficient	468
	11.6.5 Channel Pressure Drop	474
	11.6.6 Port Pressure Drop	474
	11.6.7 Overall Heat Transfer Coefficient	475
	11.6.8 Heat Transfer Surface Area	475
	11.6.9 Performance Analysis	476
11.7	Thermal Performance	481
Nomenclature		484
References		488

12. Condensers and Evaporators ... 491
- 12.1 Introduction ... 491
- 12.2 Shell and Tube Condensers ... 492
  - 12.2.1 Horizontal Shell-Side Condensers ... 492
  - 12.2.2 Vertical Shell-Side Condensers ... 495
  - 12.2.3 Vertical Tube-Side Condensers ... 495
  - 12.2.4 Horizontal in-Tube Condensers ... 497
- 12.3 Steam Turbine Exhaust Condensers ... 500
- 12.4 Plate Condensers ... 501
- 12.5 Air-Cooled Condensers ... 502
- 12.6 Direct Contact Condensers ... 503
- 12.7 Thermal Design of Shell-and-Tube Condensers ... 504
- 12.8 Design and Operational Considerations ... 515
- 12.9 Condensers for Refrigeration and Air-Conditioning ... 516
  - 12.9.1 Water-Cooled Condensers ... 518
  - 12.9.2 Air-Cooled Condensers ... 519
  - 12.9.3 Evaporative Condensers ... 519
- 12.10 Evaporators for Refrigeration and Air-Conditioning ... 522
  - 12.10.1 Water-Cooling Evaporators (Chillers) ... 522
  - 12.10.2 Air-Cooling Evaporators (Air Coolers) ... 523
- 12.11 Thermal Analysis ... 525
  - 12.11.1 Shah Correlation ... 526
  - 12.11.2 Kandlikar Correlation ... 528
  - 12.11.3 Güngör and Winterton Correlation ... 529
- 12.12 Standards for Evaporators and Condensers ... 531
- Nomenclature ... 536
- References ... 540

13. Polymer Heat Exchangers ... 543
- 13.1 Introduction ... 543
- 13.2 Polymer Matrix Composite Materials (PMC) ... 547
- 13.3 Nanocomposites ... 551
- 13.4 Application of Polymers in Heat Exchangers ... 552
- 13.5 Polymer Compact Heat Exchangers ... 563
- 13.6 Potential Applications for Polymer Film Compact Heat Exchangers ... 567
- 13.7 Thermal Design of Polymer Heat Exchangers ... 570
- References ... 573

Appendix A ... 577

Appendix B ... 583

Index ... 607

Preface

This third edition of *Heat Exchangers: Selection, Rating, and Thermal Design* has retained the basic objectives and level of the second edition to present a systematic treatment of the selections, thermal–hydraulic designs, and ratings of the various types of heat exchanging equipment. All the popular features of the second edition are retained while new ones are added. In this edition, modifications have been made throughout the book in response to users' suggestions and input from students who heard lectures based on the second edition of this book.

Included are 58 solved examples to demonstrate thermal–hydraulic designs and ratings of heat exchangers; these examples have been extensively revised in the third edition. A complete solutions manual is now also available, which provides guidance for approaching the thermal design problems of heat exchangers and for the design project topics suggested at the end of each chapter.

Heat exchangers are vital in power producing plants, process and chemical industries, and in heating, ventilating, air-conditioning, refrigeration systems, and cooling of electronic systems. A large number of industries are engaged in designing various types of heat exchange equipment. Courses are offered at many colleges and universities on thermal design under various titles.

There is extensive literature on this subject; however, the information has been widely scattered. This book provides a systematic approach and should be used as an up-to-date textbook based on scattered literature for senior undergraduate and first-year graduate students in mechanical, nuclear, aerospace, and chemical engineering programs who have taken introductory courses in thermodynamics, heat transfer, and fluid mechanics. This systematic approach is also essential for beginners who are interested in industrial applications of thermodynamics, heat transfer, and fluid mechanics, and for the designers and the operators of heat exchange equipment. This book focuses on the selections, thermohydraulic designs, design processes, ratings, and operational problems of various types of heat exchangers.

One of the main objectives of this textbook is to introduce thermal design by describing various types of single- and two-phase flow heat exchangers, detailing their specific fields of application, selection, and thermohydraulic design and rating, and showing thermal design and rating processes with worked examples and end-of-chapter problems including student design projects.

Much of this text is devoted to double-pipe, shell-and-tube, compact, gasketed-plate heat exchanger types, condensers, and evaporators. Their design processes are described and thermal–hydraulic design examples are presented. Some other types, mainly specialized ones, are briefly described without design examples. Thermal design factors and methods are common, however, to all heat exchangers, regardless of their function.

This book begins in Chapter 1 with the classification of heat exchangers according to different criteria. Chapter 2 provides the basic design methods for sizing and rating heat exchangers. Chapter 3 is a review of single-phase forced convection correlations in channels. A large number of experimental and analytical correlations are available for the heat transfer coefficient and flow friction factor for laminar and turbulent flow through ducts. Thus, it is often a difficult and confusing task for a student, and even a designer, to choose appropriate correlations. In Chapter 3, recommended correlations for the single-phase side of heat exchangers are given with worked examples. Chapter 4 discusses pressure drop and pumping power for heat exchangers and their piping circuit analysis. The thermal design fundamentals for microscale heat exchangers and the enhancement heat transfer for the applications to the heat exchanger design with nanofluids are provided in Chapter 5. Also presented in Chapter 5 are single-phase forced convection correlations and flow friction factors for microchannel flows for heat transfer and pumping power calculations. One of the major unresolved problems in heat exchanger equipment is fouling; the design of heat exchangers subject to fouling is presented in Chapter 6. Double-pipe heat exchanger design methods are presented in Chapter 7. The important design correlations for the design of two-phase flow heat exchangers are given in Chapter 8. The thermal design methods and processes for shell-and-tube, compact, and gasketed-plate heat exchangers are presented in Chapters 9, 10, and 11 for single-phase duties, respectively. Chapter 10 deals with the gasketed-plate heat exchangers, and has been revised with new correlations to calculate heat transfer and friction coefficients for chevron-type plates provided; solved examples in Chapter 10 and throughout the book have been modified. With this arrangement, both advanced students and beginners will achieve a better understanding of thermal design and will be better prepared to specifically understand the thermal design of condensers and evaporators that is introduced in Chapter 12. An overview of polymer heat exchangers is introduced in Chapter 13 as a new chapter; in some applications, the operating limitations of metallic heat exchangers have created the need to develop alternative designs using other materials, such as polymers, which have the ability to resist fouling and corrosion. Besides, the use of polymers offers substantial reductions in weight, cost, water consumption, volume, and space, which can make these heat exchangers more competitive over metallic heat exchangers.

The appendices provide the thermophysical properties of various fluids, including the new refrigerants.

In every chapter, examples illustrating the relevant thermal design methods and procedures are given. Although the use of computer programs is essential for the thermal design and rating of these exchangers, for all advanced students as well as beginners, manual thermal design analysis is essential during the initial learning period. Fundamental design knowledge is needed before one can correctly use computer design software and develop new reliable and sophisticated computer software for rating to obtain an optimum solution. Therefore, one of the primary goals of this book is to encourage students and practicing engineers to develop a systematic approach to the thermal–hydraulic design and rating of heat exchangers. A solution manual accompanies the text. Additional problems are added to the solution manual, which may be helpful to instructors.

Design of heat exchange equipment requires explicit consideration of mechanical design, economics, optimization techniques, and environmental considerations. Information on these topics is available in various standard references and handbooks and from manufacturers.

Several individuals have made very valuable contributions to this book. E.M. Sparrow and A. Bejan reviewed the manuscript of the second edition and provided very helpful suggestions. We gratefully appreciate their support. The first author has edited several books on the fundamentals and design of heat exchangers, to which many leading scientists and experts have made invaluable contributions, and that author is very thankful to them. The authors are especially indebted to the following individuals whose contributions to the field of heat exchangers made this book possible: Kenneth J. Bell, David Butterworth, John Collier, Franz Mayinger, Paul J. Marto, Mike B. Pate, Ramesh K. Shah, J. Taborek, R. M. Manglik, Bengt Sunden, Almıla G. Yazıcıoğlu, and Selin Aradağ.

The authors express their sincere appreciation to their students, who contributed to the improvement of the manuscript by their critical questions. The authors wish to thank Büryan Apaçoğlu, Fatih Aktürk, Gizem Gülben, Amarin Tongkratoke, Sezer Özerinç, and our students in heat exchanger courses for their valuable assistance during classroom teaching, various stages of this project, their assistance in the preparation of the third edition and the solution manual. Thanks are also due to Amy Blalock, Project Coordinator, Editorial Project Development, and Jonathan W. Plant, Executive Editor—Mechanical, Aerospace, Nuclear & Energy Engineering, and other individuals at CRC Press LLC, who initiated the first edition and contributed their talents and energy to the second and third editions.

Finally, we wish to acknowledge the encouragement and support of our lovely families who made many sacrifices during the preparation of this text.

Sadik Kakaç
sadikkakac@yahoo.com

Hongtan Liu
hliu@miami.edu

Anchasa Pramuanjaroenkij
anchasa@gmail.com

1

Classification of Heat Exchangers

1.1 Introduction

Heat exchangers are devices that provide the transfer of thermal energy between two or more fluids at different temperatures. Heat exchangers are used in a wide variety of applications, such as power production, process, chemical and food industries, electronics, environmental engineering, waste heat recovery, manufacturing industry, air-conditioning, refrigeration, space applications, etc. Heat exchangers may be classified according to the following main criteria[1,2]:

1. Recuperators/regenerators
2. Transfer processes: direct contact and indirect contact
3. Geometry of construction: tubes, plates, and extended surfaces
4. Heat transfer mechanisms: single phase and two phase
5. Flow arrangements: parallel flows, counter flows, and cross flows

The preceding five criteria are illustrated in Figure 1.1.[1]

1.2 Recuperation and Regeneration

The conventional heat exchanger, shown diagrammatically in Figure 1.1a with heat transfer between two fluids, is called a recuperator because the hot stream A recovers (recuperates) some of the heat from stream B. The heat transfer occurs through a separating wall or through the interface between the streams as in the case of the direct-contact-type heat exchangers (Figure 1.1c). Some examples of the recuperative type of exchangers are shown in Figure 1.2.

In regenerators or storage-type heat exchangers, the same flow passage (matrix) is alternately occupied by one of the two fluids. The hot fluid stores the thermal energy in the matrix; during the cold fluid flow through the

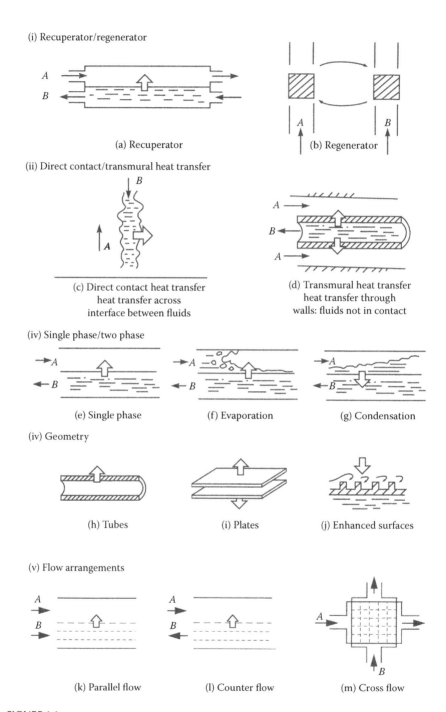

FIGURE 1.1
Criteria used in the classification of heat exchangers. (From Hewitt, G. F., Shires, G. L., and Bott, T. R., *Process Heat Transfer*, CRC Press, Boca Raton, FL, 1994. With permission.)

Classification of Heat Exchangers

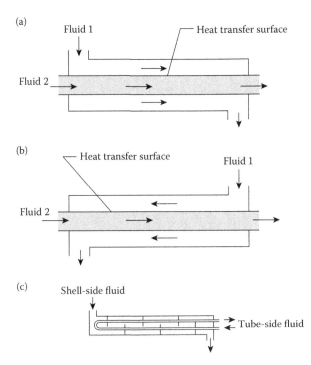

FIGURE 1.2
Indirect-contact-type heat exchangers: (a) double-pipe heat exchanger; (b) shell-and-tube-type heat exchanger.

same passage at a later time, stored energy is extracted from the matrix. Therefore, thermal energy is not transferred through the wall as in a direct-transfer-type heat exchanger (recuperator). This cyclic principle is illustrated in Figure 1.1b. While the solid is in the cold stream A, it loses heat; while it is in the hot stream B, it gains heat, i.e., the heat is regenerated. Some examples of storage-type heat exchangers are a rotary regenerator for preheating the air in a large coal-fired steam power plant, a gas turbine rotary regenerator, fixed-matrix air preheaters for blast furnace stoves, steel furnaces, open-hearth steel melting furnaces, and glass furnaces.

Regenerators can be classified as follows: (1) rotary regenerators and (2) fixed matrix regenerators. Rotary regenerators can be further subclassified as follows: (a) disk-type regenerators and (b) drum-type regenerators, both are shown schematically in Figure 1.3. In a disk-type regenerator, heat transfer surface is in a disk form and fluids flow axially. In a drum-type regenerator, the matrix is in a hollow drum form and fluids flow radially.

These regenerators are periodic flow heat exchangers. In rotary regenerators, the operation is continuous. To achieve this, the matrix moves periodically in and out of the fixed stream of gases. A rotary regenerator for air heating is illustrated in Figure 1.4. There are two kinds of regenerative

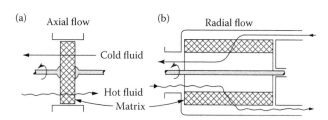

FIGURE 1.3
Rotary regenerators: (a) disk-type; (b) drum-type. (From Shah, R. K., *Heat Exchangers: Thermal Hydraulic Fundamentals and Design*, Hemisphere, Washington, DC, pp. 455–459, 1981. With permission.)

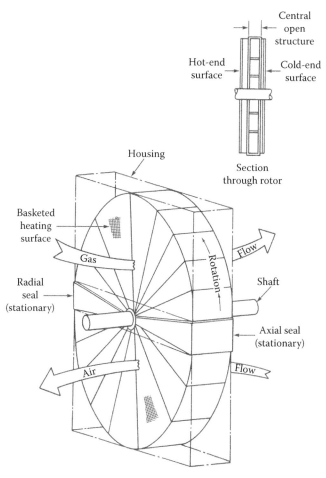

FIGURE 1.4
Rotary storage-type heat exchanger.

Classification of Heat Exchangers

air preheaters used in conventional power plants[3]: the rotating-plate type (Figures 1.4 and 1.5) and the stationary-plate type (Figure 1.6). The rotor of the rotating-plate air heater is mounted within a box housing and is installed with the heating surface in the form of plates, as shown in Figure 1.5. As the rotor rotates slowly, the heating surface is exposed alternately to flue gases and to the entering air. When the heating surface is placed in the flue gas stream, the heating surface is heated, and then when it is rotated by mechanical devices into the air stream, the stored heat is released to the entering air; thus, the air stream is heated. In the stationary-plate air heater (Figure 1.6), the heating plates are stationary, while cold-air hoods—top and bottom—are

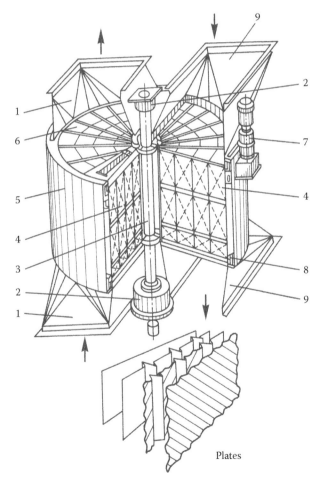

FIGURE 1.5
Rotating-plate regenerative air preheater in a large coal-fired steam power plant: (1) air ducts; (2) bearings; (3) shaft; (4) plates; (5) outer case; (6) rotor; (7) motor; (8) sealings; (9) flue gas ducts. (From Lin, Z. H., *Boilers, Evaporators and Condensers*, John Wiley & Sons, New York, 1991. With permission.)

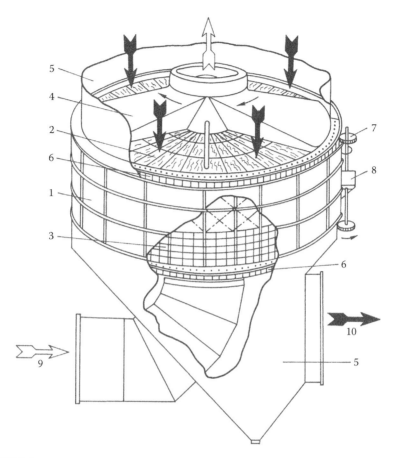

FIGURE 1.6
Stationary-plate regenerative air preheater: (1) outer case; (2) plates; (3) plates in the lower temperature region; (4) rotating air ducts; (5) flue gas ducts; (6,7) drive; (8) motor and drive-down devices; (9) air inlet; (10) gas exit. (From Lin, Z. H., *Boilers, Evaporators and Condensers*, John Wiley & Sons, New York, 1991. With permission.)

rotated across the heating plates. The heat transfer principles are the same as those of the rotating-plate regenerative air heater. In a fixed-matrix regenerator, the gas flow must be diverted to and from the fixed matrices. Regenerators are compact heat exchangers and are designed for surface area densities of up to approximately 6,600 m^2/m^3.

1.3 Transfer Processes

According to transfer processes, heat exchangers are classified as direct contact type and indirect contact type (transmural heat transfer). In

Classification of Heat Exchangers

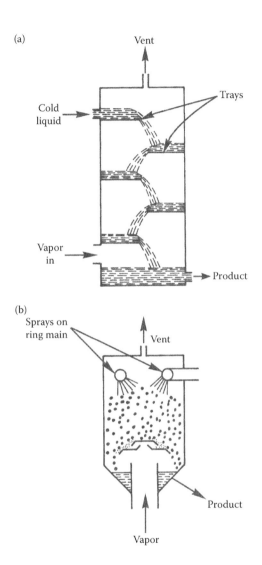

FIGURE 1.7
Direct-contact-type heat exchangers: (a) tray condenser; (b) spray condenser. (Adapted from Butterworth, D., *Two-Phase Flow Heat Exchangers: Thermal-Hydraulic Fundamentals and Design*, Kluwer, Dordrecht, The Netherlands, 1988.)

direct-contact-type heat exchangers, heat is transferred between the cold and hot fluids through direct contact between these fluids. There is no wall between the hot and cold streams, and the heat transfer occurs through the interface between the two streams, as illustrated in Figure 1.1c. In direct-contact-type heat exchangers, the streams are two immiscible liquids, a gas–liquid pair, or a solid particle–fluid combination. Spray and tray condensers (Figure 1.7) and cooling towers are good examples of such heat exchangers.[4,5]

Very often in such exchangers, heat and mass transfer occur simultaneously. In a cooling tower, a spray of water falling from the top of the tower is directly contacted and cooled by a stream of air flowing upward (see Figures 11.14 and 11.15).

In an indirect contact heat exchanger, the thermal energy is exchanged between hot and cold fluids through a heat transfer surface, i.e., a wall separating the two fluids. The cold and hot fluids flow simultaneously while the thermal energy is transferred through a separating wall, as illustrated in Figure 1.1d. The fluids are not mixed. Examples of this type of heat exchangers are shown in Figure 1.2.

Indirect contact and direct transfer heat exchangers are also called recuperators, as discussed in Section 1.2. Tubular (double-pipe or shell-and-tube), plate, and extended surface heat exchangers, cooling towers, and tray condensers are examples of recuperators.

1.4 Geometry of Construction

Direct-transfer-type heat exchangers (transmural heat exchangers) are often described in terms of their construction features. The major construction types are tubular, plate, and extended surface heat exchangers.

1.4.1 Tubular Heat Exchangers

Tubular heat exchangers are built of circular tubes. One fluid flows inside the tubes and the other flows on the outside of the tubes. Tube diameter, the number of tubes, the tube length, the pitch of the tubes, and the tube arrangement can be changed. Therefore, there is considerable flexibility in their design.

Tubular heat exchangers can be further classified as follows:

1. Double-pipe heat exchangers
2. Shell-and-tube heat exchangers
3. Spiral-tube-type heat exchangers

1.4.1.1 Double-Pipe Heat Exchangers

A typical double-pipe heat exchanger consists of one pipe placed concentrically inside another pipe of larger diameter with appropriate fittings to direct the flow from one section to the next, as shown in Figures 1.2 and 1.8. Double-pipe heat exchangers can be arranged in various series and parallel arrangements to meet pressure drop and mean temperature difference

Classification of Heat Exchangers

Cross section view of fintube inside shell

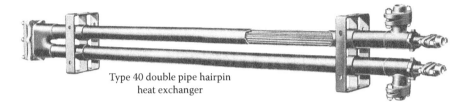
Type 40 double pipe hairpin heat exchanger

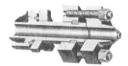

Return bend housing and cover plate (500 psig)   500 psig pressure shell to tube closure and tubeside joint   500 psig pressure shell to tube closure and high-pressure tubeside joint

FIGURE 1.8
Doublepipe hair-pin heat exchanger with cross-section view and return bend housing. (Courtesy of Brown Fintube.)

requirements. The major use of double-pipe exchangers is for sensible heating or cooling of process fluids where small heat transfer areas (to 50 m^2) are required. This configuration is also very suitable when one or both fluids are at high pressure. The major disadvantage is that double-pipe heat exchangers are bulky and expensive per unit transfer surface. Inner tubing may be single tube or multitubes (Figure 1.8). If the heat-transfer coefficient is poor in the annulus, axially finned inner tube (or tubes) can be used. Double-pipe heat exchangers are built in modular concept, i.e., in the form of hairpins.

1.4.1.2 Shell-and-Tube Heat Exchangers

Shell-and-tube heat exchangers are built of round tubes mounted in large cylindrical shells with the tube axis parallel to that of the shell. They are widely used as oil coolers, power condensers, preheaters in power plants, steam generators in nuclear power plants, in process applications, and in chemical industry. The simplest form of a horizontal shell-and-tube type condenser with various components is shown in Figure 1.9. One fluid stream flows through the tubes while the other flows on the shell side, across or

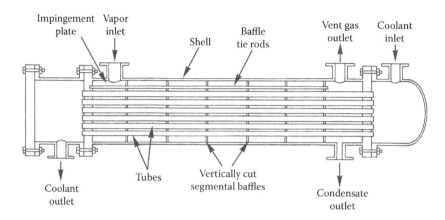

FIGURE 1.9
Shell-and-tube heat exchanger as a shell-side condenser: TEMA E-type shell with single tube-side. (Adapted from Butterworth, D., *Two-Phase Flow Heat Exchangers: Thermal-Hydraulic Fundamentals and Design*, Kluwer, Dordrecht, The Netherlands, 1988.)

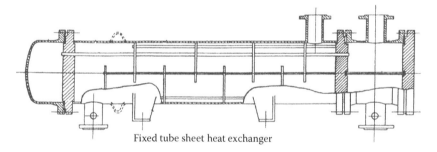

FIGURE 1.10
A two-pass tube, baffled single-pass shell, shell-and-tube heat exchanger. (From *Standards of the Tubular Exchanger Manufacturers Association*, 1988. With permission. ©1988 by Tubular Exchanger Manufacturers Association.)

along the tubes. In a baffled shell-and-tube heat exchanger, the shell-side stream flows across pairs of baffles and then flows parallel to the tubes as it flows from one baffle compartment to the next. There are wide differences between shell-and-tube heat exchangers depending on the application.

The most representative tube bundle types used in shell-and-tube heat exchangers are shown in Figures 1.10–1.12. The main design objectives here are to accommodate thermal expansion, to provide ease of cleaning, or to achieve the least expensive construction if other features are of no importance.[6] In a shell-and-tube heat exchanger with fixed tube sheets, the shell is welded to the tube sheets, and there is no access to the outside of the tube bundle for cleaning. This low-cost option has only limited thermal expansion, which can be somewhat increased by expansion bellows. Cleaning of the tube is easy (Figure 1.10).

Classification of Heat Exchangers

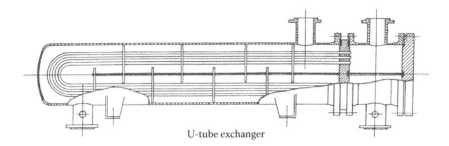

FIGURE 1.11
A U-tube, baffled single-pass shell, shell-and-tube heat exchanger. (From *Standards of the Tubular Exchanger Manufacturers Association*, 1988. With permission. ©1988 by Tubular Exchanger Manufacturers Association.)

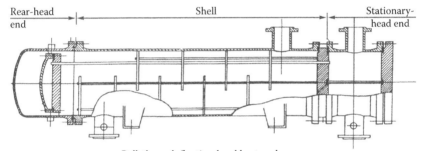

FIGURE 1.12
A heat exchanger similar to that of Figure 1.11, but with a pull-through floating-head shell-and-tube exchanger. (From *Standards of the Tubular Exchanger Manufacturers Association*, 1988. With permission. ©1988 by Tubular Exchanger Manufacturers Association.)

The U-tube is the least expensive construction because only one tube sheet is needed. The tube side cannot be cleaned by mechanical means because of the sharp U-bend. Only even number of tube passes can be accommodated, but thermal expansion is unlimited (Figure 1.11). Several designs have been developed that permit the tube sheet to "float," i.e., to move with thermal expansion. Figure 1.12 shows the classic type of pull-through floating head, which permits tube bundle removal with minimum disassembly, as required for heavily fouling units. The cost is high.

A number of shell- and tube-side flow arrangements are used in shell-and-tube heat exchangers depending on heat duty, pressure drop, pressure level, fouling, manufacturing techniques, cost, corrosion control, and cleaning problems. The baffles are used in shell-and-tube heat exchangers to promote a better heat-transfer coefficient on the shell side and to support the tubes. Shell-and-tube heat exchangers are designed on a custom basis for any capacity and operating conditions. This is contrary to many other types of heat exchangers.

1.4.1.3 Spiral-Tube-Type Heat Exchangers

This consists of spirally wound coils placed in a shell or designed as coaxial condensers and coaxial evaporators that are used in refrigeration systems. The heat-transfer coefficient is higher in a spiral tube than in a straight tube. Spiral-tube heat exchangers are suitable for thermal expansion and clean fluids, since cleaning is almost impossible.

1.4.2 Plate Heat Exchangers

Plate heat exchangers are built of thin plates forming flow channels. The fluid streams are separated by flat plates which are smooth or between which lie corrugated fins. Plate heat exchangers are used for transferring heat for any combination of gas, liquid, and two-phase streams. These heat exchangers can further be classified as gasketed plate, spiral plate, or lamella.

1.4.2.1 Gasketed Plate Heat Exchangers

A typical gasketed plate heat exchanger is shown in Figures 1.13 and 1.14.[7] A gasketed plate heat exchanger consists of a series of thin plates with corrugation or wavy surfaces that separate the fluids. The plates come with corner parts are so arranged that the two media between which heat is to be

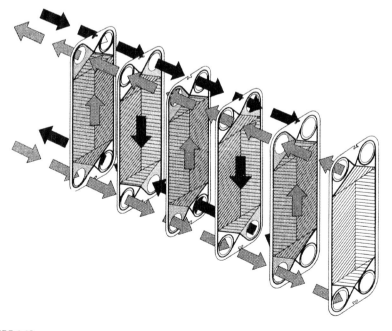

FIGURE 1.13
A diagram showing the flow paths in a gasketed plate heat exchanger. (Courtesy of Alfa Laval Thermal AB.)

Classification of Heat Exchangers

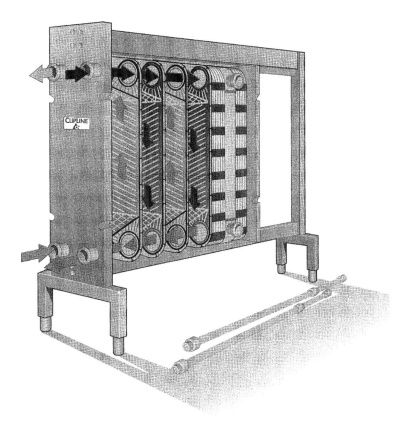

FIGURE 1.14
A gasketed plate heat exchanger showing flow paths and the constructions. (Courtesy of Alfa Laval Thermal AB.)

exchanged flow through alternate interplate spaces. Appropriate design and gasketing permit a stack of plates to be held together by compression bolts joining the end plates. Gaskets prevent intermixing of the two fluids and fluid leaking to the outside, as well as directing the fluids in the plates as desired. The flow pattern is generally chosen so that the media flow countercurrent to each other. Plate heat exchangers are usually limited to fluid streams with pressures below 25 bars and temperature below about 250°C. Since the flow passages are quite small, strong eddying gives high heat-transfer coefficients, high pressure drops, and high local shear, which minimizes fouling. These exchangers provide a relatively compact and lightweight heat transfer surface. They are temperature and pressure limited due to the construction details and gasket materials. Gasketed plate heat exchangers are typically used for heat exchange between two liquid streams. They are easily cleaned and sterilized because they can be completely disassembled, so they have a wide application in the food processing industry.

1.4.2.2 Spiral Plate Heat Exchangers

Spiral plate heat exchangers are formed by rolling two long, parallel plates into a spiral using a mandrel and welding the edges of adjacent plates to form channels (Figure 1.15). The distance between the metal surfaces in both spiral channels is maintained by means of distance pins welded to the metal sheet. The length of the distance pins may vary between 5 and 20 mm. It is therefore possible to choose different channel spacing according to the flow rate. This means that the ideal flow conditions, and therefore the smallest possible heating surfaces, can be obtained.

The two spiral paths introduce a secondary flow, increasing the heat transfer and reducing fouling deposits. These heat exchangers are quite compact but are relatively expensive due to their specialized fabrication. Sizes range from 0.5 to 500 m^2 of heat transfer surface in one single spiral body. The maximum operating pressure (up to 15 bar) and operating temperature (up to 500°C) are limited. The spiral heat exchanger is particularly effective in handling sludge, viscous liquids, and liquids with solids in suspension including slurries.

The spiral heat exchanger is made in three main types that differ in the connections and flow arrangements. Type I has flat covers over the spiral channels. The media flow countercurrent through the channels via the connections in the center and at the periphery. This type is used to exchange heat between media without phase changes, such as liquid–liquid, gas–liquid, or gas–gas. One stream enters at the center of the unit and flows from inside outward. The other stream enters at the periphery and flows toward the center. Thus, true counterflow is achieved (Figure 1.15a).

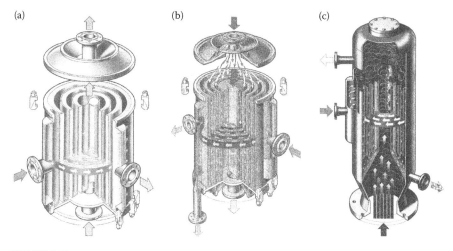

FIGURE 1.15
Spiral plate heat exchangers: (a) Type I; (b) Type II; (c) Type G. (Courtesy of Alfa Laval Thermal AB.)

Classification of Heat Exchangers

Type II is designed for crossflow operation (Figure 1.15b). One channel is completely seal-welded, while the other is open along both sheet metal edges. Therefore, this type has one medium in spiral flow and the other in crossflow. The passage with the medium in spiral flow is welded shut on each side, and the medium in crossflow passes through the open spiral annulus. It is highly effective as a vaporizer. Two spiral bodies are often built into the same jacket and are mounted below each other.

Type III, the third standard type, is, in principle, similar to type I with alternately welded up channels, but type III is provided with a specially designed top-cover. This type of heat exchanger is mainly intended for condensing vapors with subcooling of condensate and noncondensable gases. The top cover, therefore, has a special distribution cone where the vapor is distributed to the uncovered spiral turns in order to maintain a constant vapor velocity along the channel opening.

For subcooling, the two to three outer turns of the vapor channel are usually covered, which means that a spiral flow path in countercurrent to the cooling medium is obtained. The vapor–gas mixture and the condensate are separated, and the condensate then flows out through a downward connection to the periphery box while the gas through an upward connection.

A spiral heat exchanger type G,[7] also shown in Figure 1.15c, is used as a condenser. The vapor enters the open center tube, reverses flow direction in the upper shell extension, and is condensed in downward crossflow in the spiral element.

1.4.2.3 Lamella Heat Exchangers

The lamella (Ramen) type of heat exchanger consists of a set of parallel, welded, thin plate channels or lamellae (flat tubes or rectangular channels) placed longitudinally in a shell (Figure 1.16).[7] It is a modification of the floating-head type shell-and-tube heat exchanger. These flattened tubes, called lamellae (lamellas), are made up of two strips of plates, profiled and spot- or seam-welded together in a continuous operation. The forming of the strips creates space inside the lamellae and bosses acting as spacers for the flow sections outside the lamellae on the shell side. The lamellae are welded together at both ends by joining the ends with steel bars in between, depending on the space required between lamellae. Both ends of the lamella bundle are joined by peripheral welds to the channel cover which, at the outer ends, is welded to the inlet and outlet nozzle. The lamella side is thus completely sealed in by welds. At the fixed end, the channel cover is equipped with an outside flange ring, which is bolted to the shell flange. The flanges are of spigot and recess type, where the spigot is an extension of the shell. The difference in expansion between the heating surface and the shell is well taken care of by a box in the floating end; this design improves the reliability and protects the lamella bundle against failure caused by thermal stresses and

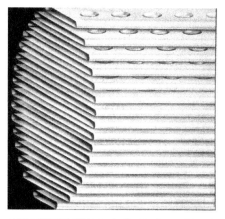
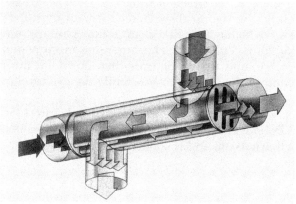

FIGURE 1.16
Lamella heat exchanger. (Courtesy of Alfa Laval Thermal AB.)

strain from external forces. The end connection is designed with a removable flange. By removing this flange and loosening the fixed end shell flanges, the lamella bundle can be pulled out of the shell. The surfaces inside the lamellae are suitable for chemical cleaning. Therefore, fouling fluids should flow through the shell side. The channel walls are either plain or have dimples. The channels are welded into headers at each end of the plate bundle that are allowed to expand and contract independently of the shell by the use of a packing gland at the lower end. The shell-side flow is typically a single pass around the plates and flows longitudinally in the spaces between channels. There are no shell-side baffles, and therefore, lamella heat exchangers can be arranged for true countercurrent flow. Because of the high turbulence, uniformly distributed flow, and smooth surfaces, the lamellae do not foul easily. The plate bundle can be easily removed for inspection and cleaning. This design is capable of pressure up to 35 bar and temperature of 200°C for Teflon® gaskets and 500°C for asbestos gaskets.

1.4.3 Extended Surface Heat Exchangers

Extended surface heat exchangers are devices with fins or appendages on the primary heat transfer surface (tubular or plate) with the object of increasing heat transfer area. As it is well known that the heat-transfer coefficient on the gas side is much lower than those on the liquid side, finned heat transfer surfaces are used on the gas side to increase the heat transfer area. Fins are widely used in gas-to-gas and gas-to-liquid heat exchangers whenever the heat-transfer coefficient on one or both sides is low and there is a need for a compact heat exchanger. The two most common types of extended surface heat exchangers are plate-fin heat exchangers and tube-fin heat exchangers.

1.4.3.1 Plate-Fin Heat Exchanger

The plate-fin heat exchangers are primarily used for gas-to-gas applications and tube-fin exchangers for liquid-to-gas heat exchangers. In most of the applications (i.e., in trucks, cars, and airplanes), mass and volume reductions are particularly important. Because of this gain in volume and mass, compact heat exchangers are also widely used in cryogenic, energy recovery, process industry, and refrigeration and air conditioning systems. Figure 1.17 shows the general form of a plate-fin heat exchanger. The fluid streams are separated by flat plates, between which are sandwiched corrugated fins. The plates can be arranged into a variety of configurations with respect to the fluid streams. Figure 1.17 shows the arrangement for parallel or counterflow

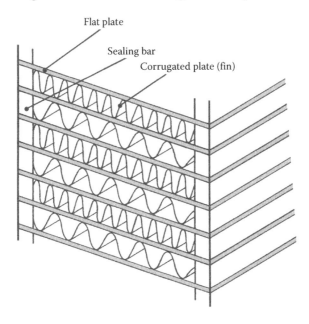

FIGURE 1.17
Basic construction of a plate-fin exchanger.

and crossflow between the streams. These heat exchangers are very compact units with a heat transfer area per unit volume of around 2,000 m²/m³. Special manifold devices are provided at the inlet to these exchangers to provide good flow distributions across the plates and from plate to plate. The plates are typically 0.5–1.0 mm thick, and the fins are 0.15–0.75 mm thick. The whole exchanger is made of aluminum alloy, and the various components are brazed together in a salt bath or in a vacuum furnace.

The corrugated sheets that are sandwiched between the plates serve both to give extra heat transfer area and to give structural support to the flat plates. There are many different forms of corrugated sheets used in these exchangers, but the most common types are as follows:

1. Plain fin
2. Plain-perforated fin
3. Serrated fin (also called "lanced," "interrupted," "louver," or "multientry")
4. Herringbone or wavy fin

By the use of fins, discontinuous in the flow direction, the boundary layers can be completely disrupted; if the surface is wavy in the flow direction, the boundary layers are either thinned or interrupted, which results in higher heat-transfer coefficients and a higher pressure drop.

Figure 1.18 shows these four types.[4] The perforated type is essentially the same as the plain type except that it has been formed from a flat sheet with small holes in it. Many variations of interrupted fins have been used in the industry.

The flow channels in plate-fin exchangers are small, which means that the mass velocity also has to be small [10–300 kg/(m² s)] to avoid excessive pressure drops. This may make the channel prone to fouling which, when combined with the fact that they cannot be mechanically cleaned, means that plate-fin exchangers are restricted to clean fluids. Plate-fin exchangers are frequently used for condensation duties in air liquefaction plants. Further information on these exchangers is given by HTFS.[8]

Plate-fin heat exchangers have been established for use in gas turbines, conventional and nuclear power plants, propulsion engineering (airplanes, trucks, and automobiles), refrigeration, heating, ventilating, and air conditioning, waste heat recovery systems, in chemical industry, and for the cooling of electronic devices.

1.4.3.2 Tubular-Fin Heat Exchangers

These heat exchangers are used in gas-to-liquid heat exchanges. The heat-transfer coefficients on the gas side are generally much lower than those on the liquid side and fins are required on the gas side. A tubular-fin (or tube-fin) heat exchanger consists of an array of tubes with fins fixed on the outside

Classification of Heat Exchangers

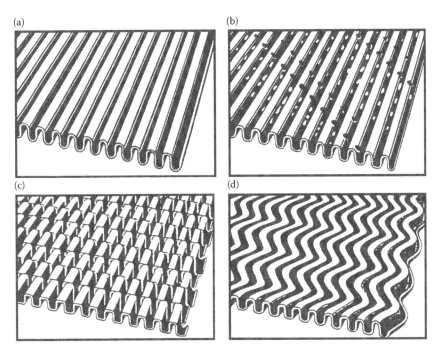

FIGURE 1.18
Fin types in plate-fin exchangers: (a) plain; (b) perforated; (c) serrated; (d) herringbone. (Adapted from Butterworth, D., *Boilers, Evaporators and Condensers*, John Wiley & Sons, New York, 1991.)

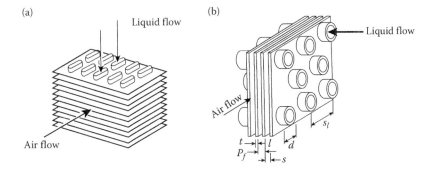

FIGURE 1.19
Tube-fin heat exchanger: (a) flattened tube-fin; (b) round tube-fin.

(Figures 1.19–1.21). The fins on the outside of the tubes may be normal on individual tubes and may also be transverse, helical, or longitudinal (or axial), as shown in Figure 1.20. Longitudinal fins are commonly used in double-pipe or shell-and-tube heat exchangers with no baffles. The fluids may be gases or viscous liquids (oil coolers). Alternately, continuous plate-fin sheets may be fixed on the array of tubes. Examples are shown in Figure 1.19. As can

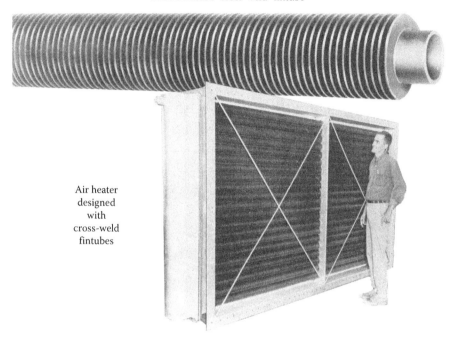

FIGURE 1.20
Fin-tube air heater. (Courtesy of Brown Fintube.)

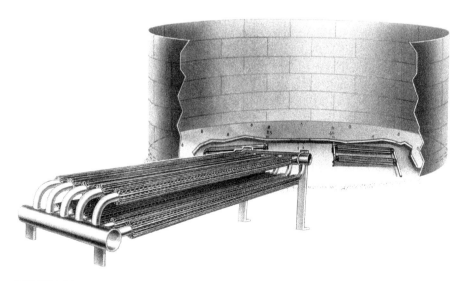

FIGURE 1.21
Longitudinally finned tubes. (Courtesy of Brown Fintube.)

be seen from that figure, in tube-fin exchangers, tubes of round, rectangular, or elliptical shapes are generally used. Fins are attached to the tubes by soldering, brazing, welding, extrusion, mechanical fit, or tension wound, etc. Plate-fin-tube heat exchangers are commonly used in heating, ventilating, refrigeration, and air conditioning systems.

Some of the extended surface heat exchangers are compact heat exchangers. A heat exchanger having a surface area density on at least one side of the heat transfer surface greater than 700 m^2/m^3 is arbitrarily referred to as a compact heat exchanger. These heat exchangers are generally used for applications where gas flows on at least one side of the heat transfer surface. These heat exchangers are generally plate-fin, tube-fin, and regenerative. Extremely high heat-transfer coefficients are achievable with small hydraulic diameter flow passages with gases.

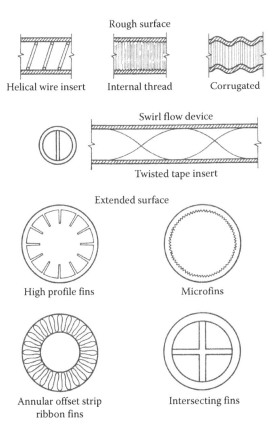

FIGURE 1.22
Examples of in-tube enhancement techniques for evaporating (and condensing) refrigerants. (From Pate, M. B., *Boilers, Evaporators and Condensers*, John Wiley & Sons, New York, 1991. With permission.)

Extended surfaces on the insides of the tubes are very commonly used in the condensers and evaporators of refrigeration systems. Figure 1.22 shows examples of in-tube enhancement techniques for evaporating and condensing refrigerants.[9,10]

Air-cooled condensers and waste heat boilers are typically tube-fin exchangers, which consist of horizontal bundle of tubes with air/gas being blown across the tubes on the outside and condensation/boiling occurring inside the tubes (Figures 1.23 and 1.24).[11]

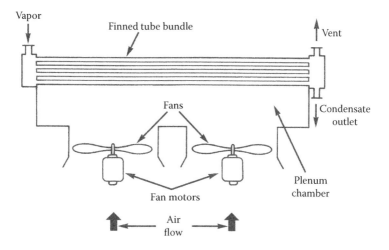

FIGURE 1.23
Forced-draft, air-cooled exchanger used as a condenser. (Adapted from Butterworth, D., *Boilers, Evaporators and Condensers*, John Wiley & Sons, New York, 1991.)

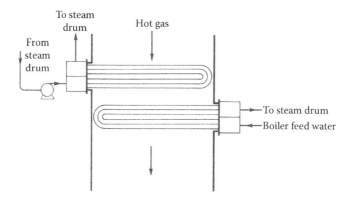

FIGURE 1.24
Horizontal U-tube waste heat boiler. (Adapted from Collier, J. G., *Two-Phase Flow Heat Exchangers: Thermal-Hydraulic Fundamentals and Design*, Kluwer, Boston, MA, 1988.)

An alternative design for air-cooled condensers is the induced-draft unit that has fans on top to suck the air over the tubes. The tubes are finned with transverse fins on the outside to overcome the effects of low air-side coefficients. There would normally be a few tube rows, and the process stream may take one or more passes through the unit. With multipass condensers, the problem arises with redistributing the two-phase mixture upon entry to the next pass. This can be overcome in some cases by using U-tubes or by having separate passes just for subcooling or desuperheating duties. In multipass condensers, it is important to have each successive pass occurs below the previous one to enable the condensate to continue downwards. Further information on air-cooled heat exchangers is given by the American Petroleum Institute.[12]

1.5 Heat Transfer Mechanisms

Heat exchanger equipment can also be classified according to the heat transfer mechanisms as

1. Single-phase convection on both sides
2. Single-phase convection on one side, two-phase convection on other side
3. Two-phase convection on both sides

The principles of these types are illustrated in Figure 1.1e–g. Figure 1.1f and g illustrate two possible modes of two-phase flow heat exchangers. In Figure 1.1f, fluid A is being evaporated, receiving heat from fluid B, and in Figure 1.1g, fluid A is being condensed, giving up heat to fluid B.

In heat exchangers, such as economizers and air heaters in boilers, compressor intercoolers, automotive radiators, regenerators, oil coolers, space heaters, etc., single-phase convection occurs on both sides. Condensers, boilers, steam generators used in pressurized water reactors (PWR) and power plants, evaporators, and radiators used in air conditioning and space heating include the mechanisms of condensation, boiling, and radiation on one of the surfaces of the heat exchanger. Two-phase heat transfer could also occur on both sides of the heat exchanger such as condensing on one side and boiling on the other side of the heat transfer surface. However, without phase change, we may also have a two-phase flow heat transfer mode as in the case of fluidized beds where a mixture of gas and solid particles are transporting heat to or from a heat transfer surface.

1.6 Flow Arrangements

Heat exchangers may be classified according to the fluid-flow path through the heat exchanger. The three basic flow configurations are as follows:

1. Parallel-flow
2. Counter-flow
3. Cross-flow

In parallel-flow heat exchangers, the two fluid streams enter together at one end, flow through in the same direction, and leave together at the other end (Figure 1.25a). In counter-flow heat exchangers, two fluid streams flow in opposite directions (Figure 1.25b). In a single cross-flow heat exchanger, one fluid flows through the heat transfer surface at right angles to the flow path of the other fluid. Cross-flow arrangements with both fluids unmixed or

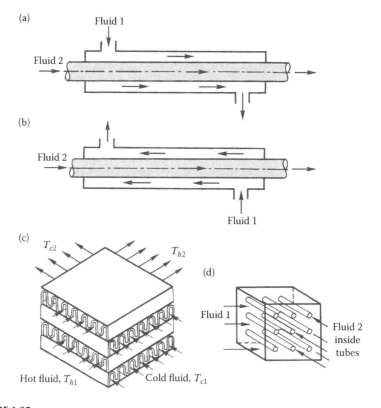

FIGURE 1.25
Heat exchanger classification according to flow arrangement: (a) parallel-flow; (b) counterflow; (c) crossflow, both fluids unmixed; (d) crossflow, fluid 1 mixed, fluid 2 unmixed.

one fluid mixed and the other unmixed are illustrated in Figure 1.25c and d. Multipass cross-flow configurations can also be organized by having the basic arrangements in a series. For example, in a U-tube baffled single-pass shell-and-tube heat exchanger, one fluid flows through the U-tube while the other fluid flows first downward and then upward, across the flow paths of the other fluid stream, which is also referred to as cross-counter, cross-parallel flow arrangements.

Multipass flow arrangements are frequently used in shell-and-tube heat exchangers with baffles (Figures 1.10–1.12). The main difference between the flow arrangements lies in the temperature distribution along the length of the heat exchanger and the relative amounts of heat transfer under given temperature specifications for specified heat exchanger surfaces (as will be shown later for given flow and specified temperatures, a counter-flow heat exchanger requires minimum area, a parallel-flow heat exchanger requires maximum area, and a cross-flow heat exchanger requires an area in between).

In the cross-flow arrangement, the flow may be called mixed or unmixed, depending on the design. Figure 1.25c shows an arrangement in which both hot and cold fluids flow through individual flow channels with no fluid mixing between adjacent flow channels. In this case, each fluid stream is said to be unmixed.

In the flow arrangements shown in Figure 1.25d, fluid 2 flows inside the tubes and is therefore not free to move in the transverse direction; fluid 2 is therefore considered unmixed, while fluid 1 is free to move in the transverse direction and mix itself. Consequently, this heat exchanger is called an unmixed–mixed cross-flow heat exchanger. For extended surface heat exchangers, it is also possible to have the basic cross-flow arrangements in a series to form multipass arrangements as cross-counter-flow and cross-parallel flow. This usually helps to increase the overall effectiveness of the heat exchanger. In a gasketed plate heat exchanger, it is also possible to have more than one pass simply by properly gasketing around the parts in the plates.

The detailed information on the classification of heat exchangers is also given in Shah and Sekulić.[13]

1.7 Applications

The most common heat exchangers are two-fluid heat exchangers. Three-fluid heat exchangers are widely used in cryogenics. They are also used in chemical and process industries such as air separation systems, purification and liquefaction of hydrogen, ammonia gas synthesis, etc. Heat exchangers with three or more components are very complex. They may also include multicomponent two-phase convection as in the condensation of mixed vapors in the distillation of hydrocarbons.

Heat exchangers are used in a wide variety of applications as in the process, power, air conditioning, refrigeration, cryogenics, heat recovery, and manufacturing industries.[11] In the power industry, various kinds of fossil boilers, nuclear steam generators, steam condensers, regenerators, and cooling towers are used. In the process industry, two-phase flow heat exchangers are used for vaporizing, condensing, freezing in crystallization, and as fluidized beds with catalytic reactions. The air conditioning and refrigeration industries need large amount of condensers and evaporators.

Energy can be saved by direct contact condensation. By direct contact condensation of a vapor in a liquid of the same substance under high pressure, thermal energy can be stored in a storage tank. When the energy is needed again, the liquid is depressurized and flashing occurs, which results in the production of vapor. The vapor can then be used for heating or as a working fluid for an engine.

There have been considerable developments in heat exchanger applications.[12,14] One of the main stages in the early development of boilers was the introduction of the water-tube boilers. The demand for more powerful engines created a need for boilers that operated at higher pressures, and, as a result, individual boilers were built larger and larger. The boiler units used in modern power plants for steam pressures above 1,200 PSI (80 bar) consist of furnace water-wall tubes, super heaters, and such heat recovery accessories as economizers and air heaters, as illustrated in Figure 1.26.[15,16] The development of modern boilers and more efficient condensers for the power industry has represented a major milestone in engineering. In the PWR, large size inverted U-tube type steam generators, each providing steam of 300–400 MW electrical power, are constructed (Figure 1.27).[11] In the process industry, engineers are concerned with designing equipment to vaporize a liquid. In the chemical industry, the function of an evaporator is to vaporize a liquid (vaporizer) or to concentrate a solution by vaporizing part of the solvent. Vaporizers may also be used in the crystallization process. Often the solvent is water, but in many cases, the solvent is valuable and recovered for reuse. The vaporizers used in the process chemical industry cover a wide range of sizes and applications.[9,14,17,18] The wide variety of applications of heat exchanger equipment is shown in Figure 1.28.

1.8 Selection of Heat Exchangers

The basic criteria for heat exchanger selection from various available types are as follows[19]:

- The heat exchanger must satisfy the process specifications; it must be able to function properly to the next scheduled shut down of the plant for maintenance.

Classification of Heat Exchangers

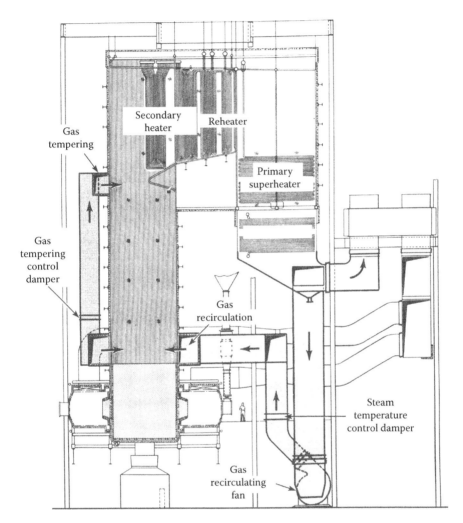

FIGURE 1.26
Boiler with water-cooled furnace with heat recovery units. (From Kitto, J. B., Jr. and Albrecht, M. J., *Boilers, Evaporators and Condensers*, John Wiley & Sons, New York, 1991. With permission.)

- The heat exchanger must withstand the service conditions of the plant environment. It must also resist corrosion by the process and service streams as well as the environment. The heat exchanger should also resist fouling.
- The exchanger must be maintainable, which usually implies choosing a configuration that permits cleaning and the replacement of any components that may be especially vulnerable to corrosion, erosion, or vibration. This requirement will dictate the positioning of the exchanger and the space requirement around it.

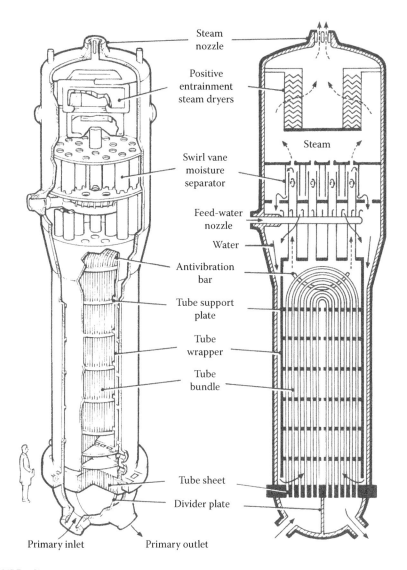

FIGURE 1.27
Inverted U-tube type steam generators. (Adapted from Collier, J. G., *Two-Phase Flow Heat Exchangers: Thermal-Hydraulic Fundamentals and Design*, Kluwer, Boston, MA, 1988.)

- The heat exchanger should be cost effective. The installed operating and maintenance costs, including the loss of production due to exchanger unavailability, must be calculated and the exchanger should cost as little as possible.
- There may be limitations on exchanger diameter, length, weight, and tube configurations due to site requirements, lifting and servicing capabilities, or inventory considerations.[19]

Classification of Heat Exchangers

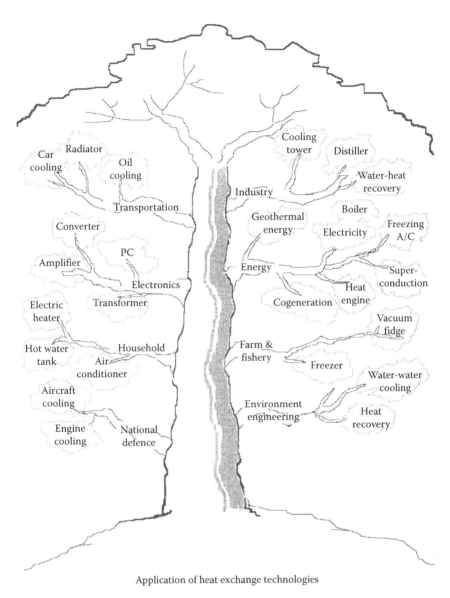

Application of heat exchange technologies

FIGURE 1.28
Application of heat exchangers.

PROBLEMS

1.1. What are the principal types of shell-and-tube construction?
1.2. What are the advantages and disadvantages of the principal types of shell-and-tube construction?
1.3. What are the different kinds of tubes used in heat exchangers?
1.4. Why are baffles used in shell-and-tube heat exchangers?

1.5. Where are fins (extended surfaces) used?
1.6. What are the types of fins that are used in heat exchangers?
1.7. What is fin efficiency? What does it depend upon?
1.8. What are enhanced surfaces and what are the advantages of enhanced surfaces?
1.9. What characterizes a unit as compact?
1.10. What are the types of heat exchangers used in a conventional plant? Classify them and discuss.
1.11. What are the advantages and the limitations of gasketed plate heat exchangers?
1.12. What are the different kinds of spiral heat exchangers and what are their limitations?
1.13. Name the specific exchanger construction type that may be used in each of the following applications:
   a. Milk pasteurizing
   b. Power condenser
   c. Automotive radiator
   d. Marine oil cooler
   e. Air-cooled condenser
1.14. What are the main selection criteria of a heat exchanger?
1.15. Select one type of heat exchanger and provide a brief report on its characteristics, typical areas of applications, advantages, limitations, and current status.

References

1. Hewitt, G. F., Shires, G. L., and Bott, T. R., *Process Heat Transfer*, CRC Press, Boca Raton, FL, 1994.
2. Shah, R. K., Classification of heat exchangers, in *Heat Exchangers—Thermo-Hydraulic Fundamentals and Design*, Kakaç, S., Bergles, A. E., and Mayinger, F., Eds., John Wiley & Sons, New York, 1981.
3. Lin, Z. H., Thermohydraulic design of fossil-fuel boiler components, in *Boilers, Evaporators and Condensers*, Kakaç, S., Ed., John Wiley & Sons, New York, 1991.
4. Butterworth, D., Steam power plant and process condensers, in *Boilers, Evaporators and Condensers*, Kakaç, S., Ed., John Wiley & Sons, New York, 1991.
5. Butterworth, D., Condensers and their design, in *Two-Phase Flow Heat Exchangers: Thermal-Hydraulic Fundamentals and Design*, Kakaç, S., Bergles, A. E., and Fernandes, E. O., Eds., Kluwer Publishers, The Netherlands, 1988.
6. Tubular Exchanger Manufacturers Association (TEMA), *Standards of the Tubular Exchanger Manufacturers Association*, 7th ed., New York, 1988.
7. Alfa Laval, *Heat Exchanger Guide*, S. 22101, Lund 1, Sweden, 1980.

8. HTFS Plate-Fin Study Group, *Plate-Fin Heat Exchangers—Guide to Their Specifications and Use*, Taylor, M. A., Ed., HTFS, Oxfordshire, UK, 1987.
9. Pate, M. B., Evaporators and condensers for refrigeration and air-conditioning systems, in *Boilers, Evaporators and Condensers*, Kakaç, S., Ed., John Wiley & Sons, New York, 1991.
10. Pate, M. B., Design considerations for air-conditioning evaporators and condenser coils, in *Two-Phase Flow Heat Exchangers: Thermal-Hydraulic Fundamentals and Design*, Kakaç, S., Bergles, A. E., and Fernandes, E. O., Eds., Kluwer Publishers, The Netherlands, 1988.
11. Collier, J. G., Evaporators, in *Two-Phase Flow Heat Exchangers: Thermal-Hydraulic Fundamentals and Design*, Kakaç, S., Bergles, A. E., and Fernandes, E. O., Eds., Kluwer Publishers, The Netherlands, 1988.
12. American Petroleum Institute, *Air Cooled Heat Exchangers for General Refinery Service*, API Standard 661, Washington, DC, 1986.
13. Shah, R. K. and Sekulić, D. P., *Fundamentals of Heat Exchanger Design*, John Wiley & Sons, New York, 2003.
14. Mayinger, F., Classification and applications of two-phase flow heat exchangers, in *Two-Phase Flow Heat Exchangers: Thermal-Hydraulic Fundamentals and Design*, Kakaç, S., Bergles, A. E., and Fernandes, E. O., Eds, Kluwer Publishers, The Netherlands, 1988.
15. Kitto, J. B., Jr. and Albrecht, M. J., Fossil-fuel-fired boilers: fundamentals and elements, in *Boilers, Evaporators and Condensers*, Kakaç, S., Ed., John Wiley & Sons, New York, 1991.
16. Stultz, S. C. and Kitto, J. B., *Steam, its Generation and Use*, Babcock and Wilcox, New York, 1992.
17. Sclünder, E. U. et al., Eds., *Heat Exchanger Design Handbook*, Hemisphere, Washington, DC, 1988.
18. Smith, R. A., *Vaporizers—Selection, Design and Operation*, John Wiley & Sons, New York, 1986.
19. Bell, K. J., Preliminary design of shell and tube heat exchangers, in *Heat Exchangers: Thermal-Hydraulic Fundamentals and Design*, Kakaç, S., Bergles, A. E., and Mayinger, F., Eds., Hemisphere, Washington, DC, 1981.

2

Basic Design Methods of Heat Exchangers

2.1 Introduction

The most common tasks in heat exchanger design are rating and sizing. The rating problem is concerned with the determination of the heat transfer rate and the fluid outlet temperatures for prescribed fluid flow rates, inlet temperatures, and allowable pressure drop for an existing heat exchanger; hence, the heat transfer surface area and the flow passage dimensions are available. The sizing problem, on the other hand, involves determination of the dimensions of the heat exchanger, that is, selecting an appropriate heat exchanger type and determining the size to meet the requirements of specified hot- and cold-fluid inlet and outlet temperatures, flow rates, and pressure drops. This chapter discusses the basic design methods, both rating and sizing, for two-fluid direct-transfer heat exchangers (recuperators).

2.2 Arrangement of Flow Paths in Heat Exchangers

A recuperator-type heat exchanger is classified according to the flow direction of the hot- and cold-fluid streams and the number of passes made by each fluid as it travels through the heat exchanger, as discussed in Chapter 1. Therefore, heat exchangers may have the following patterns of flow: (1) parallel flow, with two fluids flowing in the same direction (Figure 2.1a), (2) counterflow, with two fluids flowing parallel to one another but in opposite directions (Figure 2.1b), (3) crossflow, with two fluids crossing each other (Figure 2.1c and d), and (4) mixed flow where both fluids are simultaneously in parallel flow, in counterflow (Figure 2.2a and b), and in multipass crossflow (Figure 2.2c). Applications include various shell-and-tube heat exchangers.[1]

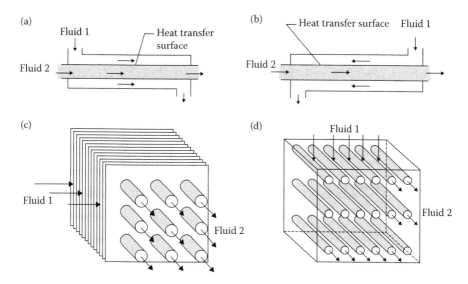

FIGURE 2.1
Heat exchanger classification according to flow arrangements.

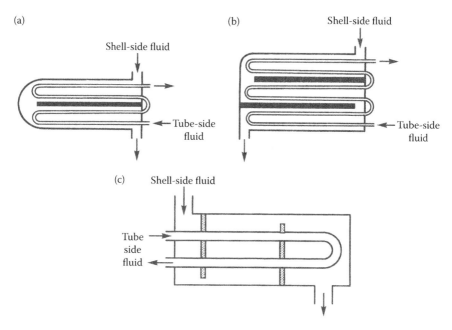

FIGURE 2.2
Multipass and multipass crossflow arrangements.

2.3 Basic Equations in Design

The term heat exchanger, although applicable to all four categories listed earlier, will be used in this chapter to designate a recuperator in which heat transfer occurs between two fluid streams that are separated by a heat transfer surface. Basic heat transfer equations will be outlined for the thermal analysis (sizing and rating calculations) of such heat exchangers. Although complete design of a heat exchanger requires structural and economical considerations in addition to these basic equations, the purpose of the thermal analysis given here will be to determine the heat transfer surface area of the heat exchanger (sizing problem). Performance calculations of a heat exchanger (rating problem) are carried out when the heat exchanger is available, but it is necessary to determine the amount of heat transferred, pressure losses, and outlet temperatures of both fluids.

The temperature variations in usual fluid-to-fluid heat transfer processes, depending on the flow path arrangement, are shown in Figure 2.3, in which the heat transfer surface area A is plotted along the x-axis and the temperature of the fluids is plotted along the y-axis. Figure 2.3a shows counterflow heat transfer with the two fluids flowing in opposite directions. Parallel-flow heat transfer with the two fluids flowing in the same direction is shown in Figure 2.3b. Heat transfer with the cold fluid at constant temperature (evaporator) is shown in Figure 2.3c. Figure 2.3d shows heat transfer with the hot fluid at constant temperature (condenser). The nature of the temperature profiles also depends on the heat capacity ratios ($\dot{m}c_p$) of the fluids and is shown later.

From the first law of thermodynamics for an open system, under steady-state, steady flow conditions, with negligible potential and kinetic energy changes, the change of enthalpy of one of the fluid streams is (Figure 2.4)

$$\delta Q = \dot{m} di \qquad (2.1)$$

where $\dot{m}$ is the rate of mass flow, i is the specific enthalpy, and δQ is the heat transfer rate to the fluid concerned associated with the infinitesimal state change. Integration of Equation 2.1 gives

$$Q = \dot{m}(i_2 - i_1) \qquad (2.2)$$

where i_1 and i_2 represent the inlet and outlet enthalpies of the fluid stream. Equation 2.2 holds true for all processes of Figure 2.3. Note that δQ is negative for the hot fluid. If there is negligible heat transfer between the exchanger and its surroundings (adiabatic process), integration of Equation 2.1 for hot and cold fluids gives

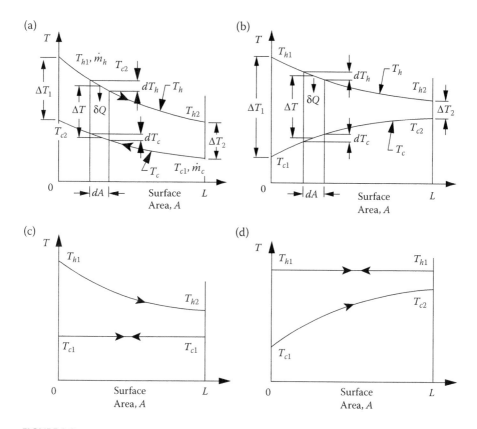

FIGURE 2.3
Fluid temperature variation in parallel-flow, counterflow, evaporator, and condenser heat exchangers: (a) counterflow; (b) parallel flow; (c) cold fluid evaporating at constant temperature; (d) hot fluid condensing at constant temperature.

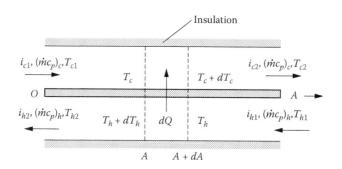

FIGURE 2.4
Overall energy balance for the hot and cold fluids of a two-fluid heat exchanger.

Basic Design Methods of Heat Exchangers

$$Q = \dot{m}_h(i_{h_1} - i_{h_2}) \tag{2.3}$$

and

$$Q = \dot{m}_c(i_{c_2} - i_{c_1}) \tag{2.4}$$

The subscripts h and c refer to the hot and cold fluids, and the numbers 1 and 2 designate the fluid inlet and outlet conditions. If the fluids do not undergo a phase change and have constant specific heats with $di = c_p\, dT$, then Equations 2.3 and 2.4 can be written as

$$Q = (\dot{m} c_p)_h (T_{h_1} - T_{h_2}) \tag{2.5}$$

and

$$Q = (\dot{m} c_p)_c (T_{c_2} - T_{c_1}) \tag{2.6}$$

As can be seen from Figure 2.3, the temperature difference between the hot and cold fluids ($\Delta T = T_h - T_c$) varies with position in the heat exchanger. Therefore, in the heat transfer analysis of heat exchangers, it is convenient to establish an appropriate mean value of the temperature difference between the hot and cold fluids such that the total heat transfer rate Q between the fluids can be determined from

$$Q = UA\,\Delta T_m \tag{2.7}$$

where A is the total hot-side or cold-side heat transfer area, and U is the average overall heat transfer coefficient based on that area. ΔT_m is a function of T_{h1}, T_{h2}, T_{c1}, and T_{c2}. Therefore, a specific form of ΔT_m must be obtained.

Equations 2.5–2.7 are the basic equations for the thermal analysis of a heat exchanger under steady-state conditions. If Q, the total heat transfer rate, is known from Equation 2.5 or 2.6, then Equation 2.7 is used to calculate the heat transfer surface area A. Therefore, it is clear that the problem of calculating the heat transfer area comes down to determining the overall heat transfer coefficient and the mean temperature difference ΔT_m.

2.4 Overall Heat Transfer Coefficient

Heat exchanger walls are usually made of a single material, although a wall may sometimes be bimetallic (steel with aluminum cladding) or coated with a plastic as a protection against corrosion. Most heat exchanger surfaces tend to acquire an additional heat transfer resistance that increases with time. This may either be a very thin layer of oxidation, or, at the other extreme, it may be a thick crust deposit, such as that which results from a salt-water coolant in steam condensers. This fouling effect can be taken into consideration by introducing an additional thermal resistance, termed the fouling resistance R_s. Its value depends on the type of fluid, fluid velocity, type of surface, and length of service of the heat exchanger.[2-4] Fouling will be discussed in Chapter 6.

In addition, fins are often added to the surfaces exposed to either or both fluids, and by increasing the surface area, they reduce the resistance to convection heat transfer. The overall heat transfer coefficient for a single smooth and clean plane wall can be calculated from

$$UA = \frac{1}{R_t} = \frac{1}{\frac{1}{h_i A} + \frac{t}{kA} + \frac{1}{h_o A}} \qquad (2.8)$$

where R_t is the total thermal resistance to heat flow across the surface between the inside and outside flow, t is the thickness of the wall, and h_i and h_o are heat transfer coefficients for inside and outside flows, respectively.

For the unfinned and clean tubular heat exchanger, the overall heat transfer coefficient is given by

$$U_o A_o = U_i A_i = \frac{1}{R_t} = \frac{1}{\frac{1}{h_i A_i} + \frac{\ln(r_o/r_i)}{2\pi kL} + \frac{1}{h_o A_o}} \qquad (2.9)$$

If the heat transfer surface is fouled with the accumulation of deposits, this in turn introduces an additional thermal resistance in the path of heat flow.[2] We define a scale coefficient of heat transfer h_s in terms of the thermal resistance R_s of this scale as

$$\frac{\Delta T_s}{Q} = R_s = \frac{1}{A h_s} \qquad (2.10)$$

where the area A is the original heat transfer area of the surface before scaling and ΔT_s is the temperature drop across the scale. $R_f = 1/h_s$ is termed the fouling factor (i.e., unit fouling resistance), which has the unit of m² K/W. This is discussed in detail in the following chapters, and tables are provided for the values of R_f.

Basic Design Methods of Heat Exchangers

We now consider heat transfer across a heat exchanger wall fouled by deposit formation on both the inside and outside surfaces. The total thermal resistance R_t can be expressed as

$$R_t = \frac{1}{UA} = \frac{1}{U_o A_o} = \frac{1}{U_i A_i} = \frac{1}{h_i A_i} + R_w + \frac{R_{fi}}{A_i} + \frac{R_{fo}}{A_o} + \frac{1}{A_o h_o} \quad (2.11)$$

The calculation of an overall heat transfer coefficient depends upon whether it is based on the cold- or hot-side surface area, since $U_o \neq U_i$ if $A_o \neq A_i$. The wall resistance R_w is obtained from the following equations:

$$R_w = \frac{t}{kA} \quad \text{(for a bare plane wall)} \quad (2.12a)$$

$$R_w = \frac{\ln(r_o/r_i)}{2\pi L k} \quad \text{(for a bare tube wall)} \quad (2.12b)$$

A separating wall may be finned differently on each side (Figure 2.5). On either side, heat transfer takes place from the fins (subscript f) as well as from the unfinned portion of the wall (subscript u). Introducing the fin efficiency η_f, the total heat transfer can be expressed as

$$Q = \left(\eta_f A_f h_f + A_u h_u\right) \Delta T \quad (2.13)$$

where ΔT is either $(T_h - T_{w1})$ or $(T_{w2} - T_c)$. The subscripts h and c refer to the hot and cold fluids, respectively (see Figure 2.5).

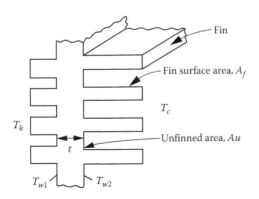

FIGURE 2.5
Finned wall.

By taking $h_u = h_f = h$ and rearranging the right-hand side of Equation 2.13, we get

$$Q = hA\left[1 - \frac{A_f}{A}(1 - \eta_f)\right]\Delta T \tag{2.14}$$

or

$$Q = \eta h A \Delta T \tag{2.15}$$

where $\eta = [1 - (1 - \eta_f)A_f/A]$ is called the *overall surface efficiency*, ΔT is the temperature difference between the fluid stream and the base temperature, and the total surface area on one side is $A = A_u + A_f$.

As can be seen from Equation 2.15, there will be additional thermal resistances for finned surfaces as $1/\eta h A$ on both sides of the finned wall; this is the combined effective surface resistance which accounts for parallel heat flow paths by conduction and convection in the fins and by convection from the prime surface. Therefore, an overall thermal resistance for the entire wall is then given by

$$R_t = \frac{1}{UA} = \frac{1}{U_o A_o} = \frac{1}{U_i A_i} = \frac{1}{\eta_i h_i A_i} + \frac{R_{fi}}{\eta_i A_i} + R_w + \frac{R_{fo}}{\eta_o A_o} + \frac{1}{\eta_o h_o A_o} \tag{2.16}$$

It should be noted that for finned surfaces, A_o and A_i represent the total surface area of the outer and inner surfaces, respectively.

Contact resistance may be finite between a tube or a plate and the fin. In this case, the contact resistance terms on the hot and cold sides are added to Equation 2.16.

In heat exchanger applications, the overall heat transfer coefficient is usually based on the outer area (cold side or hot side). Then, Equation 2.16 can be represented in terms of the overall heat transfer coefficient based on the outside surface area of the wall as

$$U_o = \frac{1}{\dfrac{A_o}{A_i}\dfrac{1}{\eta_i h_i} + \dfrac{A_o R_{fi}}{\eta_i A_i} + A_o R_w + \dfrac{R_{fo}}{\eta_o} + \dfrac{1}{\eta_o h_o}} \tag{2.17}$$

The expressions of magnitude of η_f for a variety of fin configurations are available in the literature.[5] If a straight or pin fin of length L and uniform cross section is used and an adiabatic tip is assumed, then the fin efficiency is given by

$$\eta_f = \frac{\tanh(mL)}{mL} \tag{2.18}$$

where

$$m = \sqrt{\frac{2h}{\delta k_f}} \qquad (2.19)$$

where δ is the fin thickness and L is the fin length.

For the unfinned tubular heat exchangers shown in Figures 2.1a and b and 2.2, Equation 2.17 reduces to

$$U_o = \frac{1}{\frac{r_o}{r_i}\frac{1}{h_i} + \frac{r_o}{r_i}R_{fi} + \frac{r_o \ln(r_o/r_i)}{k} + R_{fo} + \frac{1}{h_o}} \qquad (2.20)$$

The overall heat transfer coefficient can be determined from knowledge of the inside and outside heat transfer coefficients, fouling factors, and appropriate geometrical parameters. For a plane wall of thickness t and h_i and h_o on either side with fouling only on one side, Equation 2.17 becomes

$$\frac{1}{U} = \frac{1}{h_i} + R_{fi} + \frac{t}{k} + \frac{1}{h_o} \qquad (2.21)$$

The order of magnitude and range of h for various conditions are given in Table 2.1.

TABLE 2.1

Order of Magnitude of Heat Transfer Coefficient

Fluid	h, W/(m²·K)
Gases (natural convection)	3–25
Engine oil (natural convection)	30–60
Flowing liquids (nonmetal)	100–10,000
Flowing liquid metals	5,000–250,000
Boiling heat transfer:	
Water, pressure < 5 bars abs, $\Delta T = 25$ K	5,000–10,000
Water, pressure 5–100 bars abs, $\Delta T = 20$ K	4,000–15,000
Film boiling	300–400
Condensing heat transfer at 1 atm:	
Film condensation on horizontal tubes	9,000–25,000
Film condensation on vertical surfaces	4,000–11,000
Dropwise condensation	60,000–120,000

Example 2.1

Determine the overall heat transfer coefficient U for liquid-to-liquid heat transfer through a 0.003-m-thick steel plate [$k = 50$ W/m·K] for the following heat transfer coefficients and fouling factor on one side:

$$h_i = 2{,}500 \text{ W/m}^2 \text{ K}$$

$$h_o = 1{,}800 \text{ W/m}^2 \text{ K}$$

$$R_{fi} = 0.0002 \text{ m}^2 \text{ K/W}$$

Solution

Substituting h_i, h_i, R_{fi}, t, and k into Equation 2.21, we get

$$\frac{1}{U} = \frac{1}{2500} + 0.0002 + \frac{0.003}{50} + \frac{1}{1800}$$
$$= 0.0004 + 0.0002 + 0.00006 + 0.00056 = 0.00122 \text{ m}^2 \cdot \text{K/W}$$
$$U \approx 820 \text{ W/m}^2 \cdot \text{K}$$

In this case, no resistances are negligible.

Example 2.2

Replace one of the flowing liquids in Example 2.1 with a flowing gas [$h_o = 50$ W/m² K]:

Solution

$$\frac{1}{U} = \frac{1}{2500} + 0.0002 + \frac{0.003}{50} + \frac{1}{50}$$
$$= 0.0004 + 0.0002 + 0.00006 + 0.02 = 0.02066 \text{ m}^2 \cdot \text{K/W}$$
$$U \approx 48 \text{ W/m}^2 \cdot \text{K}$$

In this case, only the gas side resistance is significant.

Example 2.3

Replace the remaining flowing liquid in Example 2.2 with another flowing gas ($h_i = 20$ W/m² K):

Solution

$$\frac{1}{U} = \frac{1}{20} + 0.0002 + \frac{0.003}{50} + \frac{1}{50}$$

$$= 0.05 + 0.0002 + 0.00006 + 0.02 = 0.07026 \text{ m}^2 \cdot \text{K/W}$$
$$U \approx 14 \text{ W/m}^2 \cdot \text{K}$$

Here, the wall and scale resistances are negligible.

2.5 LMTD Method for Heat Exchanger Analysis

2.5.1 Parallel- and Counterflow Heat Exchangers

In the heat transfer analysis of heat exchangers, the total heat transfer rate Q through the heat exchanger is the quantity of primary interest. Let us consider a simple counterflow or parallel-flow heat exchanger (Figure 2.3a and b). The form of ΔT_m in Equation 2.7 may be determined by applying an energy balance to a differential area element dA in the hot and cold fluids. The temperature of the hot fluid will drop by dT_h. The temperature of the cold fluid will also drop by dT_c over the element dA for counterflow, but it will increase by dT_c for parallel flow if the hot-fluid direction is taken as positive. Consequently, from the differential forms of Equations 2.5 and 2.6 or from Equation 2.1 for adiabatic, steady-state, steady flow, the energy balance yields

$$\delta Q = -(\dot{m} c_p)_h dT_h = \pm (\dot{m} c_p)_c dT_c \tag{2.22a}$$

or

$$\delta Q = -C_h dT_h = \pm C_h dT_c \tag{2.22b}$$

where C_h and C_c are the hot- and cold-fluid heat capacity rates, respectively, and the + and – signs correspond to parallel- and counterflow, respectively. The rate of heat transfer δQ from the hot fluid to the cold fluid across the heat transfer area dA may also be expressed as

$$\delta Q = U(T_h - T_c) dA \tag{2.23}$$

From Equation 2.22b for counterflow, we get

$$d(T_h - T_c) = dT_h - dT_c = \delta Q \left(\frac{1}{C_c} - \frac{1}{C_h} \right) \tag{2.24}$$

By substituting the value of δQ from Equation 2.23 into Equation 2.24, we obtain

$$\frac{d(T_h - T_c)}{(T_h - T_c)} = U \left(\frac{1}{C_c} - \frac{1}{C_h} \right) dA \tag{2.25}$$

which, when integrated with constant values of U, C_h, and C_c over the entire length of the heat exchangers, results in

$$\ln\frac{T_{h2}-T_{c1}}{T_{h1}-T_{c2}} = UA\left(\frac{1}{C_c}-\frac{1}{C_h}\right) \quad (2.26a)$$

or

$$T_{h2}-T_{c1} = (T_{h1}-T_{c2})\exp\left[UA\left(\frac{1}{C_c}-\frac{1}{C_h}\right)\right] \quad (2.26b)$$

It can be shown that for a parallel-flow heat exchanger, Equation 2.26b becomes

$$T_{h2}-T_{c2} = (T_{h1}-T_{c1})\exp\left[-UA\left(\frac{1}{C_c}+\frac{1}{C_h}\right)\right] \quad (2.26c)$$

Equations 2.25 and 2.26 demonstrate that the temperature difference along the heat exchanger is an exponential function of A. Hence, in a counterflow heat exchanger, the temperature difference $(T_h - T_c)$ increases in the direction of flow of the hot fluid, if $C_h > C_c$ (Figure 2.6a). It can be shown that both temperature curves are concave (see Problem 2.3).[6] If the length of the heat exchanger is infinite ($A = \infty$), the cold-fluid outlet temperature becomes equal to the inlet temperature of the hot fluid. If $C_h < C_c$, both curves are convex and $(T_h - T_c)$ decreases in the direction of flow of the hot fluid. If the length is infinite ($A = \infty$), the hot-fluid exit temperature becomes equal to the inlet temperature of the cold fluid (Figure 2.6b). The expressions for

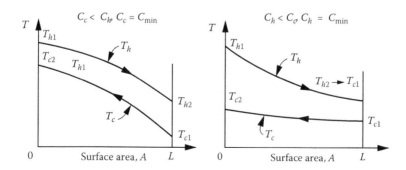

FIGURE 2.6
Temperature variation for a counterflow heat exchanger.

Basic Design Methods of Heat Exchangers

C_c and C_h can now be obtained from Equations 2.5 and 2.6. Substituting C_c and C_h into Equation 2.26a and solving for Q, we obtain

$$Q = UA \frac{(T_{h1} - T_{c2}) - (T_{h2} - T_{c1})}{\ln\left(\dfrac{T_{h1} - T_{c2}}{T_{h2} - T_{c1}}\right)} \tag{2.27a}$$

or

$$Q = UA \frac{\Delta T_1 - \Delta T_2}{\ln(\Delta T_1 / \Delta T_2)} \tag{2.27b}$$

where ΔT_1 is the temperature difference between the two fluids at one end of the heat exchanger and ΔT_2 is the temperature difference of the fluids at the other end of the heat exchanger.

Comparison of the preceding expression with Equation 2.7 reveals that the appropriate average temperature difference between the hot and cold fluids, over the entire length of the heat exchanger, is given by

$$\Delta T_{lm} = \frac{\Delta T_1 - \Delta T_2}{\ln(\Delta T_1 / \Delta T_2)} \tag{2.28}$$

which is called the LMTD (log-mean temperature difference). Accordingly, we may write the total heat transfer rate between the hot and cold fluids in a counterflow arrangement as

$$Q = AU\,\Delta T_{lm} \tag{2.29}$$

In the case of counterflow with $(\dot{m}c_p)_h = (\dot{m}c_p)_c$, the quantity ΔT_{lm} is indeterminate because

$$(T_{h1} - T_{h2}) = (T_{c2} - T_{c1}) \text{ and } \Delta T_1 = \Delta T_2 \tag{2.30}$$

In this case, it can be shown using L'Hospital's rule that $\Delta T_{lm} = \Delta T_1 = \Delta T_2$ and, therefore,

$$Q = UA(T_h - T_c) \text{ with } (T_h - T_c) = \Delta T_1 = \Delta T_2 \tag{2.31}$$

Starting with Equation 2.22 for a parallel-flow arrangement, it can be shown that Equation 2.28 is also applicable. However, for a parallel-flow heat exchanger, the end point temperature differences must now be defined as $\Delta T_1 = (T_{h1} - T_{c1})$ and $\Delta T_2 = (T_{h2} - T_{c2})$. Moreover, the total heat transfer rate between hot and cold fluids for all single-pass flow arrangements shown in Figure 2.3 is determined from Equation 2.29.

Note that, for the same inlet and outlet temperatures, the LMTD for counterflow exceeds that for parallel flow, $\Delta T_{lm,cf} > \Delta T_{lm,pf}$; that is, LMTD represents the maximum temperature potential for heat transfer that can only be obtained in a counterflow exchanger. Hence, the surface area required to affect a prescribed heat transfer rate Q is smaller for a counterflow arrangement than that required for a parallel-flow arrangement, assuming the same value of U. Also note that T_{c2} can exceed T_{h2} for counterflow but not for parallel flow.

Example 2.4

Consider a shell-and-tube heat exchanger constructed from a 0.0254 m OD tube to cool 6.93 kg/s of a 95% ethyl alcohol solution (c_p = 3810 J/kg · K) from 66°C to 42°C, using water available at 10°C (c_p = 4,187 J/kg · K) at a flow rate of 6.30 kg/s. In the heat exchanger, 72 tubes will be used. Assume that the overall coefficient of heat transfer based on the outer-tube area is 568 W/m² · K. Calculate the surface area and the length of the heat exchanger for each of the following arrangements:

1. Parallel-flow shell-and-tube heat exchanger
2. Counterflow shell-and-tube heat exchanger

Solution

The heat transfer rate may be obtained from the overall energy balance for the hot fluid, Equation 2.5

$$Q = (\dot{m} c_p)_h (T_{h1} - T_{h2})$$
$$Q = 6.93 \times 3810 (66 - 42) = 633.7 \times 10^3 \text{ W}$$

By applying Equation 2.6, the water outlet temperature is

$$T_{c2} = \frac{Q}{(\dot{m} c_p)_c} + T_{c1}$$

$$T_{c2} = \frac{633.7 \times 10^3}{6.30 \times 4187} + 10 = 34°C$$

The LMTD can be obtained from Equation 2.28

$$\Delta T_{lm,pf} = \frac{\Delta T_1 - \Delta T_2}{\ln(\Delta T_1 / \Delta T_2)} = \frac{56 - 8}{\ln(56/8)} = 24.67°C$$

The outside surface area can be calculated from Equation 2.7

$$Q = U_o A_o \Delta T_{lm,pf}$$

$$A_o = \frac{Q}{U_o \Delta T_{lm,pf}} = \frac{633.7 \times 10^3}{568 \times 24.67} = 45.2 \text{ m}^2$$

Basic Design Methods of Heat Exchangers

Because $A_o = \pi d_o N_T L$, the required heat exchanger length may now be obtained as

$$L = \frac{A_o}{d_o \pi N_T} = \frac{45.2}{0.0254 \times \pi \times 72} = 7.87 \text{ m}$$

For a counterflow arrangement, we have the case given by Equation 2.30

$$\Delta T_{lm,cf} = \Delta T_1 = \Delta T_2 = 32°C$$

$$A = \frac{Q}{U \Delta T_{lm,cf}} = \frac{633.7 \times 10^3}{568 \times 32} = 34.9 \text{ m}^2$$

$$L = \frac{A}{d_o \pi N_T} = \frac{34.9}{0.0254 \times \pi \times 72} = 6.07 \text{ m}$$

Therefore, the surface area required under the same condition of a prescribed heat transfer rate Q is smaller for a counterflow arrangement than that required for a parallel-flow arrangement.

2.5.2 Multipass and Crossflow Heat Exchangers

The LMTD developed previously is not applicable for the heat transfer analysis of crossflow and multipass flow heat exchangers. The integration of Equation 2.23 for these flow arrangements results in a form of an integrated mean temperature difference ΔT_m such that

$$Q = UA \Delta T_m \tag{2.32}$$

where ΔT_m is the true (or effective) mean temperature difference and is a complex function of T_{h1}, T_{h2}, T_{c1}, and T_{c2}. Generally, this function ΔT_m can be determined analytically in terms of the following quantities[7]:

$$\Delta T_{lm,cf} = \frac{(T_{h2} - T_{c1}) - (T_{h1} - T_{c2})}{\ln\left[(T_{h2} - T_{c1})/(T_{h1} - T_{c2})\right]} \tag{2.33}$$

$$P = \frac{T_{c2} - T_{c1}}{T_{h1} - T_{c1}} = \frac{\Delta T_c}{\Delta T_{max}} \tag{2.34}$$

and

$$R = \frac{C_c}{C_h} = \frac{T_{h1} - T_{h2}}{T_{c2} - T_{c1}} \tag{2.35}$$

where $\Delta T_{lm,cf}$ is the LMTD for a counterflow arrangement with the same fluid inlet and outlet temperatures. P is a measure of the ratio of the heat actually transferred to the heat which would be transferred if the same cold fluid temperature was raised to the hot-fluid inlet temperature; therefore, P is the temperature effectiveness of the heat exchanger on the cold-fluid side. R is the ratio of the $(m^u c_p)$ value of the cold fluid to that of the hot fluid and it is called the heat capacity rate ratio (regardless of which fluid is in the tube side or shell side in a shell-and-tube heat exchanger).

For design purposes, Equation 2.29 can also be used for multipass and cross-flow heat exchangers by multiplying ΔT_{lm}, which would be computed under the assumption of a counterflow arrangement with a correction factor F

$$Q = UA\Delta T_{lm} = UAF\Delta T_{lm,cf} \tag{2.36}$$

F is nondimensional; it depends on the temperature effectiveness P, the heat capacity rate ratio R, and the flow arrangement

$$F = \phi(P, R, \text{flow arrangement}) \tag{2.37}$$

The correction factors are available in chart form as prepared by Bowman et al.[8] for practical use for all common multipass shell-and-tube and cross-flow heat exchangers, and selected results are presented in Figures 2.7–2.11.

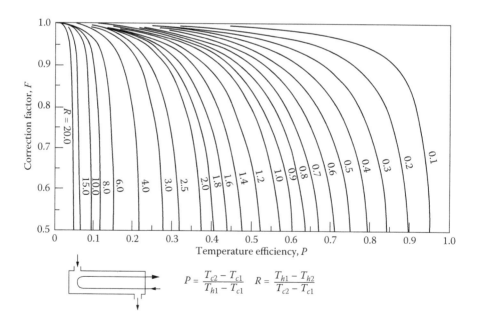

FIGURE 2.7
LMTD correction factor F for a shell-and-tube heat exchanger with one shell pass and two or a multiple of two tube passes. (From Standards of the Tubular Exchanger Manufacturers Association, 1988. With permission. ©1988 by Tubular Exchanger Manufacturers Association.)

Basic Design Methods of Heat Exchangers

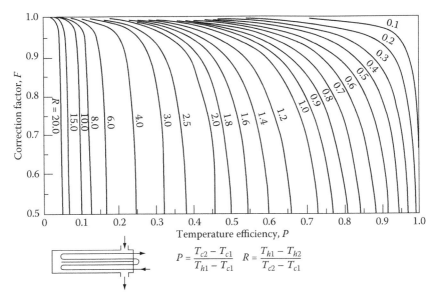

FIGURE 2.8
LMTD correction factor F for a shell-and-tube heat exchanger with two shell passes and four or a multiple of four tube passes. (From *Standards of the Tubular Exchanger Manufacturers Association*, 1988. With permission. ©1988 by Tubular Exchanger Manufacturers Association.)

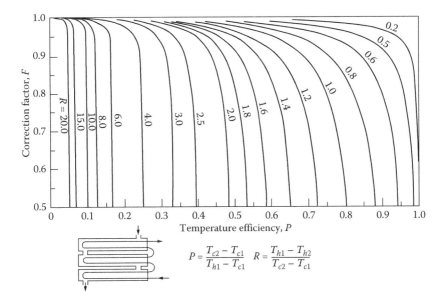

FIGURE 2.9
LMTD correction factor F for a shell-and-tube heat exchanger with three two-shell passes and six or more even number tube passes. (From *Standards of the Tubular Exchanger Manufacturers Association*, 1988. With permission. ©1988 by Tubular Exchanger Manufacturers Association.)

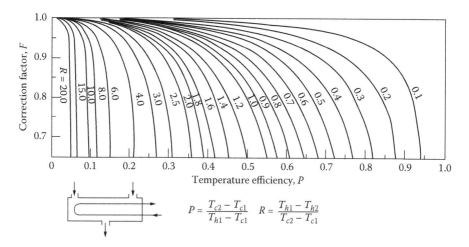

FIGURE 2.10
LMTD correction factor F for a divided-flow shell-type heat exchanger with one divided-flow shell pass and an even number of tube passes. (From *Standards of the Tubular Exchanger Manufacturers Association*, 1988. With permission. ©1988 by Tubular Exchanger Manufacturers Association.)

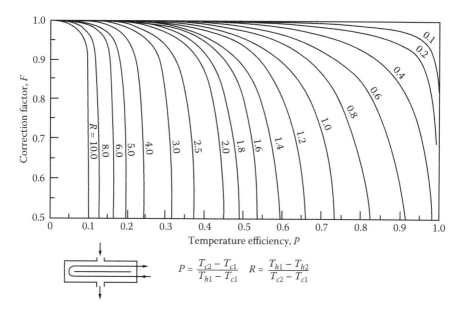

FIGURE 2.11
LMTD correction factor F for a split-flow shell-type heat exchanger with one split-flow shell pass and two-tube passes. (From *Standards of the Tubular Exchanger Manufacturers Association*, 1988. With permission. ©1988 by Tubular Exchanger Manufacturers Association.)

Basic Design Methods of Heat Exchangers 51

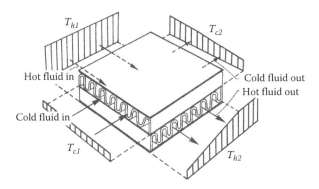

FIGURE 2.12
Temperature distribution in a crossflow heat exchanger.

In calculating P and R to determine F, it is immaterial whether the colder fluid flows through the shell or inside the tubes.

The correction factor F is less than 1 for crossflow and multipass arrangements; it is 1 for a true counterflow heat exchanger. It represents the degree of departure of the true mean temperature difference from the LMTD for a counterflow arrangement.

In a multipass or a crossflow arrangement, the fluid temperature may not be uniform at a particular location in the exchanger unless the fluid is well mixed along the path length. For example, in crossflow (Figure 2.12), the hot and cold fluids may enter at uniform temperatures, but if there are channels in the flow path (with or without corrugated spacers), to prevent mixing, the exit temperature distributions will be as shown in Figure 2.12. If such channels are not present, the fluids may be well mixed along the path length and the exit temperatures are more nearly uniform as in the flow normal to the tube bank in Figure 2.1d. A similar stratification of temperatures occurs in the shell-and-tube multipass exchanger. A series of baffles may be required if mixing of the shell fluid is to be obtained. Figures 2.13 and 2.14 present charts for both mixed and unmixed fluids.

Example 2.5

Repeat Example 2.4 for a shell-and-tube heat exchanger with one shell pass and multiples of two tube passes.

Solution

For a multipass arrangement, from Equation 2.36 we have

$$Q = UAF\Delta T_{lm,cf}$$

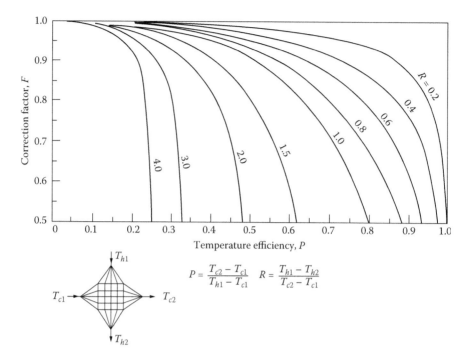

FIGURE 2.13
LMTD correction factor F for a crossflow heat exchanger with both fluids unmixed. (From Bowman, R. A., Mueller, A. C., and Nagle, W. M., *Trans. ASME*, 62, 283, 1940. With permission.)

P and R can be obtained from Equations 2.34 and 2.35

$$P = \frac{T_{c2} - T_{c1}}{T_{h1} - T_{c1}} = \frac{24}{56} = 0.43$$

$$R = \frac{T_{h1} - T_{h2}}{T_{c2} - T_{c1}} = \frac{24}{24} = 1.0$$

From Figure 2.7, we get $F = 0.89$, then

$$A = \frac{Q}{UF\Delta T_{lm,cf}} = \frac{633.68 \times 10^3}{568 \times 32 \times 0.89} = 39.17 \text{ m}^2$$

$$L = \frac{A}{\pi d_o N_T} = \frac{39.17}{\pi \times 0.0254 \times 72} = 6.19 \text{ m}$$

Therefore, under the same conditions, the surface area required for a multipass arrangement is between counterflow and parallel-flow arrangements.

The preceding analysis assumed U to be uniform throughout the heat exchanger. If U is not uniform, the heat exchanger calculations may be

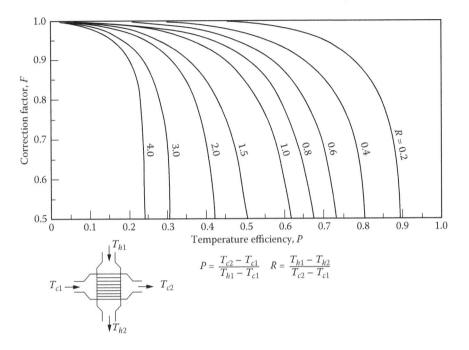

FIGURE 2.14
LMTD correction factor F for a single-pass crossflow heat exchanger with one fluid mixed and the other unmixed. (From Bowman, R. A., Mueller, A. C., and Nagle, W. M., *Trans. ASME*, 62, 283, 1940. With permission.)

made by subdividing the heat exchanger into sections over which U is nearly uniform and by applying the previously developed relations to each subdivision (see Section 2.10).

One important implication of Figures 2.7–2.11, 2.13, and 2.14 is that if the temperature change of one fluid is negligible, either P or R is zero and F is 1. Hence, heat exchanger behavior is independent of specific configuration. Such will be the case if one of the fluids undergoes a phase change. We note from Figures 2.7–2.11, 2.13, and 2.14 that the value of temperature effectiveness P ranges from zero to 1. The value of R ranges from zero to ∞, with zero corresponding to pure vapor condensation and ∞ corresponding to evaporation. It should be noted that a value of F close to 1 does not mean a highly efficient heat exchanger; rather, it means a close approach to the counterflow behavior for comparable operating conditions of flow rates and inlet fluid temperatures.

Example 2.6

A heat exchanger with two shell passes and four tube passes is used to heat water with hot exhaust gas. Water enters the tubes at 50°C and leaves at 125°C with a flow rate of 10 kg/s, while the hot exhaust gas enters the shell side at 300°C and leaves at 125°C. The total heat transfer surface is 800 m². Calculate the overall heat transfer coefficient.

Solution

The overall heat transfer coefficient can be determined from Equation 2.36

$$Q = UAF\Delta T_{lm,cf}$$

with

$$P = \frac{\Delta T_c}{T_{h1} - T_{c1}} = \frac{75}{300 - 50} = 0.3$$

$$R = \frac{T_{h1} - T_{h2}}{T_{c2} - T_{c1}} = \frac{300 - 125}{125 - 50} = 2.3$$

$$\Delta T_{lm,cf} = \frac{\Delta T_1 - \Delta T_2}{\ln(\Delta T_1/\Delta T_2)} = \frac{175 - 75}{\ln(175/75)} = 118°C$$

It follows from Figure 2.8 that $F \approx 0.96$. For water at 87.5°C, $c_p = 4203$ J/kg·

$$U = \frac{(\dot{m}c_p)_c (T_{c2} - T_{c1})}{AF\Delta T_{lm,cf}} = \frac{10 \times 4203(125 - 50)}{800 \times 0.96 \times 118} = 34.8 \text{ W/m}^2\text{K}$$

Example 2.7

A finned-tube, single-pass, crossflow heat exchanger with both fluids unmixed, as shown in Figure 2.13, is used to heat air with hot water. The total heat transfer rate is 200 kW. The water enters the tube at 85°C and leaves at 40°C, while the air enters the finned side at 15°C and leaves at 50°C. The overall heat transfer coefficient is 75 W/m²·K. Calculate the surface area required.

Solution

The desired heat transfer can be determined from Equation 2.36

$$A = \frac{Q}{UF\Delta T_{lm,cf}}$$

with

$$P = \frac{\Delta T_c}{T_{h1} - T_{c1}} = \frac{50 - 15}{85 - 15} = 0.5, \quad R = \frac{T_{h1} - T_{h2}}{T_{c2} - T_{c1}} = \frac{85 - 40}{50 - 15} = 1.29$$

It follows from Figure 2.13 that $F \approx 0.7$. From Equation 2.28 for counterflow,

$$\Delta T_{lm,cf} = \frac{\Delta T_1 - \Delta T_2}{\ln(\Delta T_1/\Delta T_2)} = \frac{35 - 25}{\ln(35/25)} = 29.7°C$$

$$A = \frac{200 \times 10^3}{75 \times 0.7 \times 29.7} = 128.3 \text{ m}^2$$

Basic Design Methods of Heat Exchangers

Example 2.8

Air flowing at a rate of 5 kg/s is to be heated in a shell-and-tube heat exchanger from 20°C to 50°C with hot water entering at 90°C and exiting at 60°C. The overall heat transfer coefficient is 400 W/m² · K. The length of the heat exchanger is 2 m. Determine the surface area of the heat exchanger and the number of tubes required by using

1. 1 to 2 shell-and-tube type heat exchanger
2. 2 to 4 shell-and-tube type heat exchanger

The tubes are schedule 40, 3/4″ nominal diameter.

Solution

For air at 35°C, $c_p = 1,005$ J/kg · K. From the overall energy balance, Equation 2.6, the heat transfer required of the exchanger is

$$Q = (\dot{m}c_p)_c (T_{c2} - T_{c1}) = 1005 \times 5(50 - 20) = 150.75 \text{ kW}$$

(1) For 1 to 2 shell-and-tube heat exchangers, the desired heat transfer area may be determined by using the LMTD method from Equation 2.36 with

$$P = \frac{T_{c2} - T_{c1}}{T_{h1} - T_{c1}} = \frac{50 - 20}{90 - 20} = 0.43, \quad R = \frac{T_{h1} - T_{h2}}{T_{c2} - T_{c1}} = \frac{90 - 60}{50 - 20} = 1$$

It follows from Figure 2.7 that $F \approx 0.90$. From Equation 2.30,

$$\Delta T_{lm, cf} = \Delta T_1 = \Delta T_2 = 40°C$$

The heat transfer area of the heat exchanger can be obtained from Equation 2.36

$$A = \frac{Q}{UF\Delta T_{lm, cf}} = \frac{150750}{400 \times 0.9 \times 40}$$

$$A = 10.47 \text{ m}^2$$

The heat transfer surface area of one tube (see Chapter 8, Table 8.2) is

$$A_t = \pi d_o L = \pi \times 0.02667 \times 2$$

$$A_t = 0.1676 \text{ m}^2$$

The number of tubes of the heat exchanger is

$$N_T = \frac{A}{A_t} = \frac{10.47}{0.1676} \approx 63 \text{ tubes}$$

(2) For 2 to 4 shell-and-tube heat exchangers from Figure 2.8, $F \approx 0.975$. The number of tubes

$$N_T = \frac{Q}{UF\Delta T_{lm,cf}(\pi d_o L)} = \frac{150750}{400 \times 0.975 \times 40(\pi \times 0.02667 \times 2)}$$

$$N_T \cong 58 \text{ tubes}$$

2.6 The ε-NTU Method for Heat Exchanger Analysis

When the inlet or outlet temperatures of the fluid streams are not known, a trial-and-error procedure could be applied for using the LMTD method in the thermal analysis of heat exchangers. The converged value of the LMTD will satisfy the requirement that the heat transferred in the heat exchanger (Equation 2.7) must be equal to the heat convected to the fluid (Equations 2.5 or 2.6). In these cases, to avoid a trial-and-error procedure, the method of the number of transfer units (NTU) based on the concept of heat exchanger effectiveness may be used. This method is based on the fact that the inlet or exit temperature differences of a heat exchanger are a function of UA/C_c and C_c/C_h (see Equations 2.26b and 2.26c).

Heat exchanger heat transfer equations such as Equations 2.5, 2.6, and 2.26 may be written in dimensionless form, resulting in the following dimensionless groups[8]:

1. Capacity rate ratio:

$$C^* = \frac{C_{min}}{C_{max}} \tag{2.38}$$

where C_{min} and C_{max} are the smaller and larger of the two magnitudes of C_h and C_c, respectively, and $C^* \leq 1$. $C^* = 0$ corresponds to a finite C_{min} and C_{max} approaching ∞ (a condensing or evaporating fluid).

2. Exchanger heat transfer effectiveness:

$$\varepsilon = \frac{Q}{Q_{max}} \tag{2.39}$$

which is the ratio of the actual heat transfer rate in a heat exchanger to the thermodynamically limited maximum possible heat transfer rate if an infinite heat transfer surface area were available in a counterflow heat exchanger.

The actual heat transfer is obtained by either the energy given off by the hot fluid or the energy received by the cold fluid, from Equations 2.5 and 2.6

$$Q = (\dot{m}c_p)_h (T_{h1} - T_{h2}) = (\dot{m}c_p)_c (T_{c2} - T_{c1})$$

If $C_h > C_c$, then $(T_{h1} - T_{h2}) < (T_{c2} - T_{c1})$ (2.40)

If $C_h < C_c$, then $(T_{h1} - T_{h2}) > (T_{c2} - T_{c1})$

The fluid that might undergo the maximum temperature difference is the fluid with the minimum heat capacity rate C_{min}. Therefore, the maximum possible heat transfer is expressed as

$$Q_{max} = (\dot{m}c_p)_c (T_{h1} - T_{c1}) \quad \text{if} \quad C_c < C_h \quad (2.41a)$$

or

$$Q_{max} = (\dot{m}c_p)_h (T_{h1} - T_{c1}) \quad \text{if} \quad C_h < C_c \quad (2.41b)$$

which can be obtained with a counterflow heat exchanger if an infinite heat transfer surface area were available (Figure 2.6). Heat exchanger effectiveness, ε, is therefore written as

$$\varepsilon = \frac{C_h (T_{h1} - T_{h2})}{C_{min} (T_{h1} - T_{c1})} = \frac{C_c (T_{c2} - T_{c1})}{C_{min} (T_{h1} - T_{c1})} \quad (2.42)$$

The first definition is for $C_h = C_{min}$, and the second is for $C_c = C_{min}$. Equation 2.42 is valid for all heat exchanger flow arrangements. The value of ε ranges between zero and 1. When the cold fluid has a minimum heat capacity rate, then temperature effectiveness, P, given by Equation 2.34, will be equal to heat exchanger effectiveness, ε. For given ε and Q_{max}, the actual heat transfer rate Q from Equation 2.39 is

$$Q = \varepsilon \left(\dot{m} c_p\right)_{min} \left(T_{h1} - T_{c1}\right). \tag{2.43}$$

If the effectiveness ε of the exchanger is known, Equation 2.43 provides an explicit expression for the determination of Q.

3. Number of transfer unit:

$$\mathrm{NTU} = \frac{AU}{C_{min}} = \frac{1}{C_{min}} \int_A U\, dA \tag{2.44}$$

If U is not constant, the definition of second equality applies. NTU designates the nondimensional heat transfer size of the heat exchanger.

Let us consider a single-pass heat exchanger, assuming $C_c > C_h$, so that $C_h = C_{min}$ and $C_c = C_{max}$. With Equation 2.44, Equation 2.26b may be written as

$$T_{h2} - T_{c1} = (T_{h1} - T_{c2}) \exp\left[-\mathrm{NTU}\left(\pm 1 - \frac{C_{min}}{C_{max}}\right)\right] \tag{2.45}$$

where the + is for counterflow and the − is for parallel flow. With Equations 2.5, 2.6, and 2.42, T_{h2} and T_{c2} in Equation 2.45 can be eliminated and the following expression is obtained for ε for counterflow:

$$\varepsilon = \frac{1 - \exp\left[-\mathrm{NTU}(1 - C_{min}/C_{max})\right]}{1 - (C_{min}/C_{max}) \exp\left[-\mathrm{NTU}(1 - C_{min}/C_{max})\right]} \tag{2.46}$$

if $C_c < C_h$ ($C_c = C_{min}$, $C_h = C_{max}$), the result will be the same.

In the case of parallel flow, a similar analysis may be applied to obtain the following expression:

$$\varepsilon = \frac{1 - \exp\left[-\mathrm{NTU}(1 + C_{min}/C_{max})\right]}{1 + (C_{min}/C_{max})} \tag{2.47}$$

Two limiting cases are of interest: C_{min}/C_{max} equal to 1 and zero. For $C_{min}/C_{max} = 1$, Equation 2.46 is indeterminate, but by applying L'Hospital's rule to Equation 2.46, the following result is obtained for counterflow:

Basic Design Methods of Heat Exchangers

$$\varepsilon = \frac{NTU}{1+NTU} \qquad (2.48)$$

For parallel flow, Equation 2.47 gives

$$\varepsilon = \frac{1}{2}\left(1 - e^{-2NTU}\right) \qquad (2.49)$$

For $C_{min}/C_{max} = 0$, as in boilers and condensers (Figures 2.3c and d) for parallel flow or counterflow, Equations 2.46 and 2.47 become

$$\varepsilon = 1 - e^{-NTU} \qquad (2.50)$$

It is noted from Equations 2.46 and 2.47 that

$$\varepsilon = \phi(NTU, C^*, \text{flow arrangement}) \qquad (2.51)$$

Similar expressions have been developed for heat exchangers having other flow arrangements such as crossflow, multipass, etc., and representative results are summarized in Table 2.2.[3,10,11]

Some ε-NTU relations are shown graphically in Figure 2.15.[9] The following observations may be made by reviewing this figure:

1. The heat exchanger thermal effectiveness ε increases with increasing values of NTU for a specified C^*.
2. The exchanger thermal effectiveness ε increases with decreasing values of C^* for a specified NTU.
3. For ε < 40%, the capacity rate ratio C^* does not have a significant influence on the exchanger effectiveness.

Because of the asymptotic nature of the ε-NTU curves, a significant increase in NTU and, hence, in the heat exchanger size, is required for a small increase in ε at high values of ε.

The counterflow exchanger has the highest exchanger effectiveness ε for specified NTU and C^* values compared to those for all other exchanger flow arrangements. Thus, for given NTU and C^*, a maximum heat transfer performance is achieved for counterflow; alternately, the heat transfer surface is utilized most efficiently in a counterflow arrangement as compared to all other flow arrangements.

Table 2.3 summarizes each of the methods discussed in the preceding sections.

TABLE 2.2
The ε-NTU Expressions

Type of heat exchanger exchange	ε (NTU, C*)	NTU(ε, C*)	Particular case C* = 0	Value of ε when NTU → ∞
Counter flow	$\varepsilon = \dfrac{1-\exp[-(1-C^*)NTU]}{1-C^*\exp[-(1-C^*)NTU]}$	$NTU = \dfrac{1}{1-C^*}\ln\left(\dfrac{1-\varepsilon C^*}{1-\varepsilon}\right)$	$\varepsilon = 1-\exp(-NTU)$	$\varepsilon = 1$ for all C^*
Parallel flow	$\varepsilon = \dfrac{1}{1+C^*}\{1-\exp[-(1+C^*)NTU]\}$	$NTU = -\dfrac{1}{1+C^*}\ln[1+\varepsilon(1+C^*)]$	$\varepsilon = 1-\exp(-NTU)$	$\varepsilon = \dfrac{1}{1+C^*}$
Cross flow, C_{min} mixed and C_{max} unmixed	$\varepsilon = 1-\exp\left[-\dfrac{1-\exp(-C^*NTU)}{C^*}\right]$	$NTU = -\dfrac{1}{C^*}\ln[1+C^*\ln(1-\varepsilon)]$	$\varepsilon = 1-\exp(-NTU)$	$\varepsilon = 1-\exp\left(-\dfrac{1}{C^*}\right)$
Cross flow, C_{max} mixed and C_{min} unmixed	$\varepsilon = \dfrac{1}{C^*}[1-\exp\{-C^*[1-\exp(-NTU)]\}]$	$NTU = -\ln\left[1+\dfrac{1}{C^*}\ln(1-\varepsilon C^*)\right]$	$\varepsilon = 1-\exp(-NTU)$	$\varepsilon = \dfrac{1}{C^*}[1-\exp(-C^*)]$
1-2 shell-and tube heat exchanger, TEMA E	$\varepsilon = \dfrac{2}{1+C^*+(1+C^{*2})^{1/2}\dfrac{1+\exp[-NTU(1+C^{*2})^{1/2}]}{1-\exp[-NTU(1+C^{*2})^{1/2}]}}$	$NTU = \dfrac{1}{(1+C^{*2})^{1/2}}\ln\dfrac{2-\varepsilon[1+C^*-(1+C^{*2})^{1/2}]}{2-\varepsilon[1+C^*+(1+C^{*2})^{1/2}]}$	$\varepsilon = 1-\exp(-NTU)$	$\varepsilon = \dfrac{2}{1+C^*+(1+C^{*2})^{1/2}}$

Basic Design Methods of Heat Exchangers

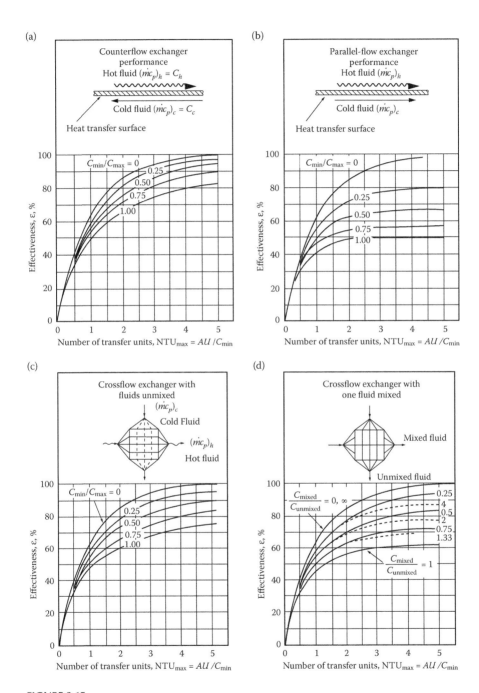

FIGURE 2.15
Effectiveness vs. NTU for various types of heat exchangers. (From Kays, W. M. and London, A. L., *Compact Heat Exchangers*, 3rd ed., McGraw-Hill, New York, 1984. With permission.)

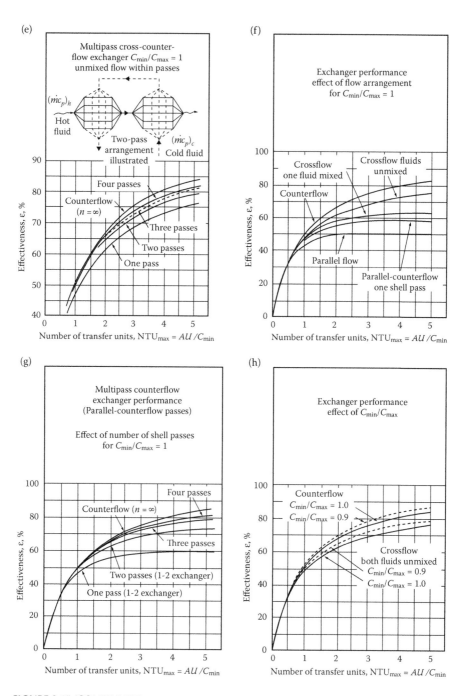

FIGURE 2.15 (CONTINUED)
Effectiveness vs. NTU for various types of heat exchangers. (From Kays, W. M. and London, A. L., *Compact Heat Exchangers*, 3rd ed., McGraw-Hill, New York, 1984. With permission.)

Basic Design Methods of Heat Exchangers

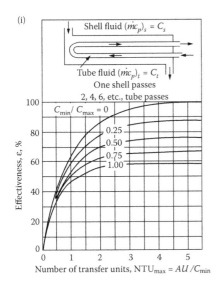
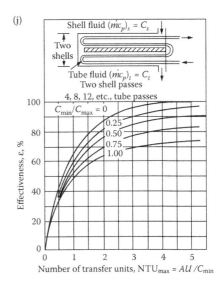

FIGURE 2.15 (CONTINUED)
Effectiveness vs. NTU for various types of heat exchangers. (From Kays, W. M. and London, A. L., *Compact Heat Exchangers*, 3rd ed., McGraw-Hill, New York, 1984. With permission.)

TABLE 2.3
Working Equation for the LMTD and ε-NTU Methods

LMTD	ε-NTU
$Q = UAF\Delta T_{lm,cf}$	$Q = \varepsilon (\dot{m}c_p)_{min}(T_{h1} - T_{c1})$
$LMTD = \Delta T_{lm,cf} = \dfrac{\Delta T_1 - \Delta T_2}{\ln(\Delta T_1/T_2)}$	$\varepsilon = \dfrac{C_h(T_{h1} - T_{h2})}{C_{min}(T_{h1} - T_{c1})} = \dfrac{C_c(T_{c2} - T_{c1})}{C_{min}(T_{h1} - T_{c1})}$
$\Delta T_1 = T_{h1} - T_{c2},\ \Delta T_2 = T_{h2} - T_{c1}$	$C^* = \dfrac{C_{min}}{C_{max}} = \dfrac{(\dot{m}c_p)_{min}}{(\dot{m}c_p)_{max}}$
$P = \dfrac{T_{c2} - T_{c1}}{T_{h1} - T_{c1}},\ R = \dfrac{T_{h1} - T_{h2}}{T_{c2} - T_{c1}}$	$NTU = \dfrac{AU}{C_{min}} = \dfrac{1}{C_{min}} \int_A U dA$
$F = \phi(P, R, \text{flow arrangement})$	$\varepsilon = \phi(NTU, C^*, \text{flow arrangement})$

Example 2.9

A shell-and-tube heat exchanger with two tube passes and baffled single shell pass is used as oil cooler. Cooling water at 20°C flows through the tubes at a flow rate of 4.082 kg/s. Engine oil enters the shell side at a flow rate of 10 kg/s. The inlet and outlet temperatures of oil are 90°C and 60°C, respectively. Determine the surface area of the heat exchanger by both the LMTD and ε-NTU methods if the overall heat transfer coefficient based on the outside tube area is 262 W/m² · K. The specific heats of water and oil are 4,179 J/kg · K and 2,118 J/kg · K, respectively.

Solution

First, using the LMTD method, we shall calculate the heat transfer rate Q and the LMTD for counterflow. Subsequently, P, R, and the correction factor F will be determined. Then, applying the heat transfer rate of Equation 2.36, the heat transfer surface area A will be determined.

The heat capacity rates for the shell fluid (oil) and the tube fluid (water) are

$$C_h = (\dot{m}c_p)_h = 10 \text{ kg/s} \times 2118 \text{ J/kg} \cdot \text{K} = 21180 \text{ W/K}$$
$$C_c = (\dot{m}c_p)_c = 4.082 \text{ kg/s} \times 4179 \text{ J/kg} \cdot \text{K} = 17058.7 \text{ W/K}$$

The heat transfer rate from the temperature drop of the oil is

$$Q = C_h(T_{h1} - T_{h2}) = 21180 \times (90 - 60) = 635400 \text{ W}$$

The water outlet temperature from the energy balance is

$$Q = C_c(T_{c2} - T_{c1}) = 635400 \text{ W}$$
$$T_{c2} = 635400/C_c + T_{c1} = 635400/17058.678 + 20 = 37.248 + 20 = 57.25°C$$

Now let us determine $\Delta T_{lm,cf}$. From the definition of $\Delta T_{lm,cf}$ in Equation 2.28 for a counterflow arrangement

$$\Delta T_{lm,cf} = \frac{32.75 - 40}{\ln(32.75/40)} = 36.4°C$$

Now the values of P and R from Equations 2.34 and 2.35 are

$$P = \frac{T_{c2} - T_{c1}}{T_{h1} - T_{c1}} = \frac{57.25 - 20}{90 - 20} = \frac{37.25}{70} = 0.532$$

$$R = \frac{T_{h1} - T_{h2}}{T_{c2} - T_{c1}} = \frac{C_c}{C_h} = \frac{17058.7}{21180} = 0.805$$

Therefore, from Figure 2.7, we get $F = 0.85$. Thus, the heat transfer area from Equation 2.36 is

$$A = \frac{Q}{UF\Delta T_{lm,cf}} = \frac{635400}{262 \times 0.85 \times 36.4} = 78.6 \text{ m}^2$$

Basic Design Methods of Heat Exchangers

Using the ε-NTU method, we will first determine ε and C^* and, subsequently, NTU and then A. In this problem, $C_h > C_c$ and, hence, $C_c = C_{min}$

$$C^* = \frac{C_{min}}{C_{max}} = \frac{17058.7}{21180} = 0.805$$

From the given temperature, for $C_c = C_{min}$, Equation 2.42 gives

$$\varepsilon = \frac{T_{c2} - T_{c1}}{T_{h1} - T_{c1}} = \frac{57.25 - 20}{90 - 20} = \frac{37.25}{70} = 0.532$$

Now calculate NTU either from the formula of Table 2.2 or from Figure 2.15 with a proper interpretation for ε, NTU, and C^*. From Table 2.2 for a 1 to 2 shell-and-tube heat exchanger, we have

$$NTU = \frac{1}{(1+C^{*2})^{1/2}} \ln\left[\frac{2-\varepsilon\left[1+C^*-(1+C^{*2})^{1/2}\right]}{2-\varepsilon\left[1+C^*+(1+C^{*2})^{1/2}\right]}\right]$$

$$= \frac{1}{(1+0.805^2)^{1/2}} \ln\left[\frac{2-0.532\left[1+0.805-(1+0.805^2)^{1/2}\right]}{2-0.532\left[1+0.805+(1+0.805^2)^{1/2}\right]}\right]$$

$$= 0.778965 \ln\left[\frac{1.7227}{0.35679}\right] = 1.226$$

Hence,

$$A = \frac{C_{min}}{U} NTU = \frac{17058.7}{262} \times 1.226 = 79.8 \text{ m}^2$$

In obtaining expressions for the basic design methods discussed in the preceding sections, Equations 2.22 and 2.23 are integrated across the surface area under the following assumptions:

1. The heat exchanger operates under steady-state, steady flow conditions.
2. Heat transfer to the surroundings is negligible.
3. There is no heat generation in the heat exchanger.
4. In counterflow and parallel-flow heat exchangers, the temperature of each fluid is uniform over every flow cross section; in crossflow heat exchangers, each fluid is considered mixed or unmixed at every cross section depending upon the specifications.
5. If there is a phase change in one of the fluid streams flowing through the heat exchanger, then that change occurs at a constant temperature for a single-component fluid at constant pressure.

6. The specific heat at constant pressure is constant for each fluid.
7. Longitudinal heat conduction in the fluid and the wall are negligible.
8. The overall heat transfer coefficient between the fluids is constant throughout the heat exchanger, including the case of phase change.

Assumption 5 is an idealization of a two-phase-flow heat exchanger. Specifically, for two-phase flows on both sides, many of the foregoing assumptions are not valid. The design theory of these types of heat exchangers is discussed and practical results are presented in the following chapters.

2.7 Heat Exchanger Design Calculation

Thus far, two methods for performing a heat exchanger thermal analysis have been discussed (Table 2.3). The rating and sizing of heat exchangers are two important problems encountered in the thermal analysis of heat exchangers.

For example, if inlet temperatures, one of the fluid outlet temperatures, and mass flow rates are known, then the unknown outlet temperature can be calculated from heat balances and the LMTD method can be used to solve this sizing problem with the following steps:

1. Calculate Q and the unknown outlet temperature from Equations 2.5 and 2.6.
2. Calculate ΔT_{lm} from Equation 2.28 and obtain the correction factor F, if necessary.
3. Calculate the overall heat transfer coefficient U.
4. Determine A from Equation 2.36.

The LMTD method may also be used for rating problems (performance analysis) for an available heat exchanger, but computation would be tedious, requiring iteration since the outlet temperatures are not known. In such situations, the analysis can be simplified by using the ε-NTU method. The rating analysis with the ε-NTU method is as follows:

1. Calculate the capacity rate ratio $C^* = C_{min}/C_{max}$ and $NTU = UA/C_{min}$ from the input data.
2. Determine the effectiveness ε from the appropriate charts or ε-NTU equations for the given heat exchanger and specified flow arrangement.

3. Knowing ε, calculate the total heat transfer rate from Equation 2.43.
4. Calculate the outlet temperatures from Equations 2.5 and 2.6.

The ε-NTU method may also be used for the sizing problem, and the procedure is as follows:

1. Knowing the outlet and inlet temperatures, calculate ε from Equation 2.42.
2. Calculate the capacity rate ratio $C^* = C_{min}/C_{max}$.
3. Calculate the overall heat transfer coefficient U.
4. Knowing ε, C^*, and the flow arrangement, determine NTU from charts or from ε-NTU relations.
5. Knowing NTU, calculate the heat transfer surface area A from Equation 2.44.

The use of the ε-NTU method is generally preferred in the design of compact heat exchangers for automotive, aircraft, air-conditioning, and other industrial applications where the inlet temperatures of the hot and cold fluids are specified and the heat transfer rates are to be determined. The LMTD method is traditionally used in the process, power, and petrochemical industries.

2.8 Variable Overall Heat Transfer Coefficient

In practical applications, the overall heat transfer coefficient varies along the heat exchanger and is strongly dependent on the flow Reynolds number, heat transfer surface geometry, and fluid physical properties. Methods to account for specific variations in U are given for counterflow, crossflow, and multipass shell-and-tube heat exchangers.

Figure 2.16 (Ref. 12) shows typical situations in which the variation of U within a heat exchanger might be substantially large. The case in which both fluids are changing phase is shown in Figure 2.16a, where there is no sensible heating and cooling; the temperatures simply remain constant throughout. The condenser shown in Figure 2.16b is perhaps more common than the condenser of Figure 2.3d. In the former, the fluid vapor enters at a temperature greater than the saturation temperature and the condensed liquid becomes subcooled before leaving the condenser. Figure 2.16c shows a corresponding situation, where the cold fluid enters as a subcooled liquid, is heated, evaporated, and then superheated. When the hot fluid consists of condensable vapor and noncondensable gases, the temperature distribution is more complex, as

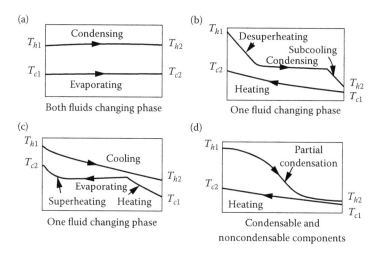

FIGURE 2.16
Typical cases of a heat exchanger with variable U. (From Walker, G., *Industrial Heat Exchanger. A Basic Guide.* Hemisphere, Washington, DC, 1990. With permission.)

represented in a general way in Figure 2.16d. The difficulty that one faces in designing such a heat exchanger is the continuous variation of U with position within the heat exchanger. If the three parts of the heat exchanger (Figures 2.16b and c) had constant values of U, then the heat exchanger could be treated as three different heat exchangers in series. For arbitrary variation of U through the heat exchanger, the exchanger is divided into many segments and a different value of U is then assigned to each segment. The analysis is best performed by a numerical or finite-difference method.

Let us consider a parallel-flow double-pipe heat exchanger (Figure 2.1a). The heat exchanger is divided into increments of surface area ΔA_i. For any incremental surface area, the hot- and cold-fluid temperatures are T_{hi} and T_{ci}, respectively, and it will be assumed that the overall heat transfer coefficient can be expressed as a function of these temperatures. Thus,

$$U_i = U_i(T_{hi}, T_{ci}) \tag{2.52}$$

The incremental heat transfer in ΔA_i can be calculated from Equation 2.22:

$$\Delta Q_i = -(\dot{m}c_p)_{hi}(T_{h(i+1)} - T_{hi}) = (\dot{m}c_p)_{ci}(T_{c(i+1)} - T_{ci}) \tag{2.53}$$

From Equation 2.23, δQ_i is also given by

$$\delta Q_i = U_i \Delta A_i (T_h - T_c)_i \tag{2.54}$$

where ΔA_i must be small enough to maintain the accuracy of the solution.

Equation 2.25 can be written for a parallel-flow arrangement as

$$\frac{d(T_h - T_c)}{(T_h - T_c)} = -U\left(\frac{1}{C_h} + \frac{1}{C_c}\right) dA \tag{2.55}$$

Equation 2.55 can be written in the finite-difference form as

$$\frac{(T_h - T_c)_{i+1} - (T_h - T_c)_i}{(T_h - T_c)_i} = -U_i\left(\frac{1}{C_{ci}} + \frac{1}{C_{hi}}\right) \Delta A_i \tag{2.56}$$

which can be solved for $(T_h - T_c)_{i+1}$

$$(T_h - T_c)_{i+1} = (T_h - T_c)_i (1 - M_i \Delta A_i) \tag{2.57}$$

where

$$M_i = U_i\left(\frac{1}{C_{ci}} + \frac{1}{C_{hi}}\right) \tag{2.58}$$

The numerical analysis can be carried out as follows:

1. Select a convenient value of ΔA_i for the analysis.
2. Calculate the inner and outer heat transfer coefficients and the value of U for the inlet conditions and through the initial ΔA increment.
3. Calculate the value of ΔQ_i for this increment from Equation 2.54.
4. Calculate the values of T_h, T_c, and $T_h - T_c$ for the next increment by the use of Equations 2.53 and 2.56.

The total heat transfer rate is then calculated from

$$Q_t = \sum_{i=1}^{n} \Delta Q_i \tag{2.59}$$

where n is the number of the increments in ΔA_i.

The treatment of counterflow shell-and-tube exchangers and the exchanger with one shell-side pass and two tube passes for the variable overall heat transfer coefficient is given by Butterworth.[13]

For the overall heat transfer coefficient U and ΔT varying linearly with Q, Colburn recommends the following expression to calculate Q[14]:

$$Q = U_m A \Delta T_{lm} = \frac{A(U_2 \Delta T_1 - U_1 \Delta T_2)}{\ln[(U_2 \Delta T_1)/(U_1 \Delta T_2)]} \quad (2.60)$$

where U_1 and U_2 are the values of the overall heat transfer coefficients on the ends of the exchanger having temperature differences of ΔT_1 and ΔT_2, respectively.

When both $1/U$ and ΔT vary linearly with Q, Butterworth[13] has shown that

$$Q = U_m A \Delta T_{lm} \quad (2.61)$$

where

$$\frac{1}{U_m} = \frac{1}{U_1}\left[\frac{\Delta T_{lm} - \Delta T_2}{\Delta T_1 - \Delta T_2}\right] = \frac{1}{U_2}\left[\frac{\Delta T_1 - \Delta T_{lm}}{\Delta T_1 - \Delta T_2}\right] \quad (2.62)$$

Equations 2.60 and 2.61 will not usually be valid over the whole heat exchanger but may apply to a small portion of it.

2.9 Heat Exchanger Design Methodology

The flow chart of heat exchanger design methodology is given in Figure 2.17.[15] The first criterion that a heat exchanger should satisfy is the fulfillment of the process requirements: the design specifications may contain all the necessary detailed information on flow rates of streams, operating pressures, pressure drop limitations for both streams, temperatures, size, length, and other design constraints such as cost, type of materials, heat exchanger type, and arrangements. The heat exchanger designer provides missing information based on his/her experiences, judgment, and the requirements of the customer.

Based on the problem specifications, the exchanger construction type, flow arrangement, surface or core geometry, and materials must be selected. In the selection of the type of heat exchanger, the operating pressure and temperature levels, maintenance requirements, reliability, safety, availability and manufacturability of surfaces, and cost must be considered.

As discussed in previous sections, heat exchanger thermal design may be classified as sizing (design problem) or rating (performance analysis). In the sizing problem, the surface area and heat exchanger dimensions are to

Basic Design Methods of Heat Exchangers

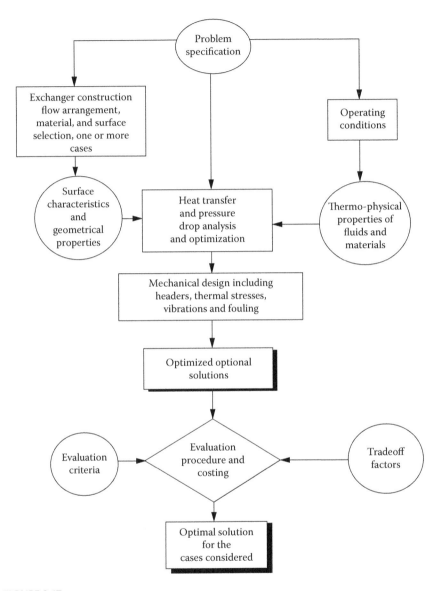

FIGURE 2.17
Heat exchanger design methodology. (From Shah, R. K., *Heat Exchangers: Thermal-Hydraulic Fundamentals and Design*, Hemisphere, Washington, DC, 455, 1981. With permission.)

be determined; inputs to the sizing problem are as follows: flow rates, inlet temperatures and one outlet temperature, surface geometries, pressure drop limitations, and thermophysical properties of streams and materials.

In the rating problem, the heat exchanger already exists or the heat exchanger configuration is selected by approximate sizing. Therefore, inputs to the rating problem are heat exchanger surface geometry and dimensions, fluid flow rates, inlet temperatures, and pressure drop limitations. The fluid outlet temperatures, total heat transferred, and pressure drop for both streams through the heat exchanger are to be calculated in the rating (performance) analysis. If the rating gives acceptable thermal performance with pressure drops in both streams below the maximum allowable ones, this heat exchanger configuration may be considered a solution to the problem. It is often possible to find a number of variant configurations that will meet these requirements; then the choice must be made on other criteria, usually the cost of the heat exchanger. If the heat exchanger selected for rating is found to be unsatisfactory, a new design must be chosen for the next rating. The process by which one moves from one design to a more satisfying design can be called design modification.[15] Rating is the computational process (hand method or computer method) by which one determines the thermal performance and pressure drops for two streams in a completely defined heat exchanger. The rating and sizing problems are discussed, and worked examples are given, in the various chapters of this book.

The selection criteria demand that the heat exchanger withstand the service conditions of the plant environment. Therefore, after thermal design analysis, the mechanical design is carried out, which includes the calculation of plate, tube, shell, and header thicknesses and arrangements. The heat exchanger must resist corrosion by the service and process streams and by the environment; this is mostly a matter of proper material selection. A proper design of inlet and outlet nozzles and connections, supporting materials, location of pressure and temperature measuring devices, and manifolds is to be made. Thermal stress calculations must be performed under steady-state and transit operating conditions. The additional important factors that should be considered and checked in the design are flow vibrations and the level of velocities to eliminate or minimize fouling and erosion.

Another criterion involves the maintenance requirements. The configuration and placement of the heat exchanger must be properly chosen to permit cleaning as required and replacement of tubes, gaskets, and any other components that are especially vulnerable to corrosion, erosion, vibration, or aging. Operating problems under severe climatic conditions (freezing) and the transportation of the unit should also be considered.

There may be limitations on exchanger diameter, length, weight, and/or tube matrix core specifications due to size requirements, lifting and servicing capabilities, and availability of replacement tubes and gaskets.

After the mechanical design is completed, the final cost analysis to arrive at an optimum solution must be considered. An overall optimum design, in general, is the one that meets the performance requirements at a minimum cost, which includes capital costs (the costs of materials, manufacturing, testing, shipment, and installation) and operating and maintenance costs (the costs of fluid pumping powers, repair, and cleaning). There are many interdependent factors that must be considered while designing and optimizing a heat exchanger.

The problem of heat exchanger design is very intricate. Only a part of the total design process consists of quantitative analytical evaluation. Because of a large number of qualitative judgments, trade-offs, and compromises, heat exchanger design is more of an art than a science at this stage. In general, no two engineers will come up with the same heat exchanger design for a given application.[16]

For heat exchangers design methodology, the readers may also refer to Shah and Sekulić.[17]

Nomenclature

A	total heat transfer area on one side of a recuperator, m²
A_c	cross-sectional area of a heat exchanger passage, m²
A_f	fin surface area on one side of a heat exchanger, m²
A_u	unfinned surface area on one side of a heat exchanger, m²
C	flow stream heat capacity rate, mc_p, W/K
C_{max}	maximum of C_c and C_h, W/K
C_{min}	minimum of C_c and C_h, W/K
C^*	heat capacity rate ratio, C_{min}/C_{max}
c_p	specific heat at constant pressure, J/kg·K
d_i	tube inside diameter, m
d_o	tube outside diameter, m
e	tube surface roughness, m
F	LMTD correction factor
i	specific enthalpy, J/kg
k	thermal conductivity, W/m·K
L	length of the heat exchanger, m
$\dot{m}$	fluid mass flow rate, kg/s
N_T	number of tubes
NTU	number of heat transfer units based on C_{min}, UA/C_{min}
P	temperature effectiveness, defined by Equation 2.34
Q	heat transfer rate, W
R	heat capacity rate ratio, defined by Equation 2.35
R	thermal resistance, m²·K/W
R_f	fouling factor, m²·K/W
r	tube radius, m
T	temperature, °C, K
T_c	cold-fluid temperature, °C, K
T_h	hot-fluid temperature, °C, K
ΔT	local temperature difference between two fluids, °C, K

ΔT_{lm} LMTD, defined by Equation 2.28, °C, K
ΔT_m true mean temperature difference, defined by Equation 2.7, °C, K
t wall thickness, m
U overall heat transfer coefficient, W/m²·K
u_m fluid mean velocity, m/s

Greek Symbols

Δ difference
δ fin thickness, m
ε heat exchanger effectiveness, defined by Equation 2.39
η_f fin efficiency
η extended surface efficiency, defined by Equation 2.15
μ dynamic viscosity, Pa·s
ν kinematic viscosity, m²/s
ρ fluid density, kg/m³
φ parameter, function of

Subscripts

c cold fluid
cf counterflow
f fin, finned, friction
h hot fluid, heat transfer
i inner, inside
m mean
max maximum
min minimum
o outer, outside, overall
u unfinned
w wall
x local
1 inlet
2 outlet

PROBLEMS

2.1. Starting from Equation 2.22, show that for a parallel-flow heat exchanger, Equation 2.26a becomes

$$\frac{T_{h_2} - T_{c_2}}{T_{h_1} - T_{c_1}} = \exp\left[-\left(\frac{1}{C_h} + \frac{1}{C_c}\right)UA\right]$$

2.2. Show that for a parallel-flow heat exchanger, the variation of the hot-fluid temperature along the heat exchanger is given by

$$\frac{T_h - T_{h_1}}{T_{h_1} - T_{c_1}} = -\frac{C_c}{C_h + C_c}\left\{1 - \exp\left[-\left(\frac{1}{C_h} + \frac{1}{C_c}\right)UA\right]\right\}.$$

Obtain a similar expression for the variation of the cold-fluid temperatures along the heat exchanger. Also show that for $A \to \infty$, the temperature will be equal to mixing cup temperatures of the fluids, given by

$$T_\infty = \frac{C_h T_{h_1} + C_c T_{c_1}}{C_c + C_h}$$

2.3. Show that the variations of the hot- and cold-fluid temperatures along a counterflow heat exchanger are given by

$$\frac{T_h - T_{h_1}}{T_{h_1} - T_{c_2}} = \frac{C_c}{C_c - C_h}\left\{\exp\left[\left(\frac{1}{C_c} - \frac{1}{C_h}\right)UA\right] - 1\right\}$$

and

$$\frac{T_c - T_{c_2}}{T_{h_1} - T_{c_2}} = \frac{C_h}{C_c - C_h}\left\{\exp\left[\left(\frac{1}{C_c} - \frac{1}{C_h}\right)UA\right] - 1\right\}$$

2.4. Problem 2.3 show that for the case $C_h < C_c$,

$$\frac{d^2 T_h}{dA^2} > 0 \text{ and } \frac{d^2 T_c}{dA^2} > 0$$

and, therefore, temperature curves are convex, and for the case $C_h > C_c$,

$$\frac{d^2 T_h}{dA^2} < 0 \text{ and } \frac{d^2 T_c}{dA^2} < 0$$

therefore, the temperature curves are concave (see Figure 2.6).

2.5. Show that for a counterflow heat exchanger, when the heat capacity rates of hot and cold fluids are equal ($C_c = C_h = C$), the variation of the hot and cold-fluid temperatures along the counterflow heat exchanger are linear with the surface area as

$$\frac{T_c - T_{c_2}}{T_{h_1} - T_{c_2}} = \frac{T_h - T_{h_1}}{T_{h_1} - T_{c_2}} = -\frac{UA}{C}$$

2.6. Assume that in a condenser, there will be no subcooling, and condensate leaves the condenser at saturation temperature, T_h. Show that variation of the coolant temperature along the condenser is given by

$$\frac{T_c - T_{c_1}}{T_h - T_{c_1}} = 1 - \exp\left(-\frac{UA}{C_c}\right).$$

2.7. In a boiler (evaporator), the temperature of hot gases decreases from T_{h1} to T_{h2}, while boiling occurs at a constant temperature T_s. Obtain an expression, as in Problem 2.6, for the variation of hot-fluid temperature with the surface area.

2.8. Show that Equation 2.46 is also applicable for $C_h > C_c$, that is, $C^* = C_c/C_h$.

2.9. Obtain the expression for exchanger heat transfer effectiveness, ε, for parallel flow given by Equation 2.47.

2.10. 5,000 kg/h of water will be heated from 20°C to 35°C by hot water at 140°C. A 15°C hot water temperature drop is allowed. A number of double-pipe heat exchangers, each connected in series, will be used. Hot water flows through the inner tube. The thermal conductivity of the material is 50 W/m·K.

Fouling factors: $R_i = 0.000176$ m²·K/W
$R_o = 0.000352$ m²·K/W
Inner tube diameters: ID = 0.0525 m, OD = 0.0603 m
Annulus diameters: ID = 0.0779 m, OD = 0.0889 m

The heat transfer coefficients in the inner tube and in the annulus are 4,620 W/m²·K and 1,600 W/m²·K, respectively. Calculate the overall heat transfer coefficient and the surface area of the heat exchanger for both parallel and counterflow arrangements.

2.11. Water at a rate of 45,500 kg/h is heated from 80°C to 150°C in a shell-and-tube heat exchanger having two shell passes and eight tube passes with a total outside heat transfer surface area of 925 m². Hot exhaust gases having approximately the same thermophysical properties as air enter at 350°C and exit at 175°C. Determine the overall heat transfer coefficient based on the outside surface area.

2.12. A shell-and-tube heat exchanger given in Problem 2.11 is used to heat 62,000 kg/h of water from 20°C to about 50°C. Hot water at 100°C is available. Determine how the heat transfer rate and the water outlet temperature vary with the hot water mass flow rate. Calculate the heat transfer rates and the outlet temperatures for hot water flow rates:

a. 80,000 kg/h
b. 40,000 kg/h

2.13. Water (c_p = 4182 J/kg · K) at a flow rate of 5,000 kg/h is heated from 10°C to 35°C in an oil cooler by engine oil (c_p = 2072 J/kg · K) with an inlet temperature of 65°C and a flow rate of 6,000 kg/h. Take the overall heat transfer coefficient to be 3,500 W/m² · K. What are the areas required for the following:

a. Parallel flow
b. Counterflow

2.14. In order to cool a mass flow rate of 9.4 kg/h of air from 616°C to 232°C, it is passed through the inner tube of double-pipe heat exchanger with counterflow, which is 1.5 m long with an outer diameter of the inner tube of 2 cm.

a. Calculate the heat transfer rate. For air, c_{ph} = 1060 J/kg · K.
b. The cooling water enters the annular side at 16°C with a volume flow rate of 0.3 L/min. Calculate the exit temperature of the water. For water, c_{pc} = 4,180 J/kg · K.
c. Determine the effectiveness of this heat exchanger, and then determine NTU. The overall heat transfer coefficient based on the outside heat transfer surface area is 38.5 W/m² · K. Calculate the surface area of the heat exchanger and the number of hairpins.

2.15. A shell-and-tube heat exchanger is designed to heat water from 40°C to 60°C with a mass flow rate of 20,000 kg/h. Water at 180°C flows through tubes with a mass flow rate of 10,000 kg/h. The tubes have an inner diameter of d_i = 20 mm; the Reynolds number is Re = 10,000. The overall heat transfer coefficient based outside heat transfer surface area is estimated to be U = 450 W/m² · K.

a. Calculate the heat transfer rate Q of the heat exchanger and the exit temperature of the hot fluid.
b. If the heat exchanger is counterflow with one tube and one shell pass, determine (by use of the LMTD and ε-NTU methods):
  i. The outer heat transfer area
  ii. The velocity of the fluid through the tubes
  iii. The total cross-sectional area of the tubes
  iv. The number of the tubes and the length of the heat exchanger

2.16. An oil cooler is used to cool lubricating oil from 70°C to 30°C. The cooling water enters the exchanger at 15°C and leaves at 25°C. The specific heat capacities of the oil and water are 2 and 4.2 kJ/kg·K, respectively, and the oil flow rate is 4 kg/s.
   a. Calculate the water flow rate required.
   b. Calculate the true mean temperature difference for two-shell-pass-and-four-tube passes and one-shell-pass-and-two-tube passes shell-and-tube heat exchangers and an unmixed–unmixed crossflow configuration, respectively.
   c. Find the effectiveness of the heat exchangers.

2.17. For the oil cooler described in Problem 2.16, calculate the surface area required for the shell-and-tube and unmixed–unmixed crossflow exchangers, assuming the overall heat transfer coefficient $U = 90$ W/m²·K. For the shell-and-tube exchanger, calculate the oil outlet temperature and compare it with the given values.

2.18. In an oil cooler, oil flows through the heat exchanger with a mass flow rate of 8 kg/s and inlet temperature of 70°C. The specific heat of oil is 2 kJ/kg·K. The cooling stream is treated cooling water that has a specific heat capacity of 4.2 kJ/kg·K, a flow rate of 20 kg/s, and an inlet temperature of 15°C. Assuming a total heat exchanger surface area of 150 m² and an overall heat transfer coefficient of 150 W/m²·K, calculate the outlet temperature for a heat exchanger with two-shell passes and four-tube passes and a heat exchanger with unmixed–unmixed crossflow, respectively. Estimate the respective F-correction factors.

2.19. An air blast cooler with a surface area for a heat transfer of 600 m² and an overall heat transfer coefficient of W/(m²·K) is fed with the following streams: air: $\dot{m}_c = 15$ kg/s, $c_{p,c} = 1050$ J/(kg·K), $T_{c,in} = 25$°C; oil: $\dot{m}_h = 5$ kg/s, $c_{p,h} = 2000$ J/(kg·K), $T_{h,in} = 90$°C. $U = 60$ W/m²·K. Calculate the stream exit temperatures, the F-factor, and the effectiveness for a two-shell-pass and four-tube-pass exchanger.

References

1. Shah, R. K., Classification of heat exchangers, in *Heat Exchangers: Thermal-Hydraulic Fundamentals and Design*, Kakaç, S., Bergles, A. E., and Mayinger, F., Eds., Hemisphere, Washington, DC, 1981, 9.
2. Chenoweth, J. M. and Impagliazzo, M., Eds., *Fouling in Hot Exchange Equipment*, ASME Symposium, Volume HTD-17, ASME, New York, 1981.

3. Kakaç, S., Shah, R. K., and Bergles, A. E., Eds., *Low Reynolds Number Flow Heat Exchangers*, Hemisphere, Washington, DC, 1981, 21.
4. Kakaç, S., Shah, R. K., and Aung, W., Eds., *Handbook of Single Phase Convective Heat Transfer*, John Wiley & Sons, New York, 1987, Ch. 4, 18.
5. Kern, D. Q. and Kraus, A. D., *Extended Surface Heat Transfer*, McGraw-Hill, New York, 1972.
6. Padet, J., *Echangeurs Thermiques*, Masson, Paris, France, 1994.
7. Tubular Exchanger Manufacturers Association (TEMA), *Standards of the Tubular Exchanger Manufacturers Association*, 7th ed., New York, 1988.
8. Bowman, R. A., Mueller, A. C., and Nagle, W. M., Mean temperature difference in design, *Trans. ASME*, 62, 283, 1940.
9. Kays, W. M. and London, A. L., *Compact Heat Exchangers*, 3rd ed., McGraw-Hill, New York, 1984.
10. Shah, R. K. and Mueller, A. C., Heat exchangers, in *Handbook of Heat Transfer Applications*, Rohsenow, W. M., Hartnett, J. P., and Gani, E. N., Eds., McGraw-Hill, New York, 1985, Ch. 4.
11. Kays, W. M., London, A. L., and Johnson, K. W., *Gas Turbine Plant Heat Exchangers*, ASME, New York, 1951.
12. Walker, G., *Industrial Heat Exchanger. A Basic Guide*, Hemisphere, Washington, DC, 1990.
13. Butterworth, D., Condensers: thermohydraulic design, in *Heat Exchangers: Thermal-Hydraulic Fundamentals and Design*, Kakaç, S., Bergles, A. E., and Mayinger, F., Eds., Hemisphere, Washington, DC, 1981, 547.
14. Colburn, A. P., Mean temperature difference and heat transfer coefficient in liquid heat exchangers, *Ind. Eng. Chem.*, 25, 873, 1933.
15. Shah, R. K., Heat exchanger design methodology — an overview, in *Heat Exchangers: Thermal-Hydraulic Fundamentals and Design*, Kakaç, S., Bergles, A. E., and Mayinger, F., Eds., Hemisphere, Washington, DC, 1981, 455.
16. Bell, K. J., Overall design methodology for shell-and-tube exchangers, in *Heat Transfer Equipment Design*, Shah, R. K., Subbarao, E. C., and Mashelkar, R. A., Eds., Hemisphere, Washington, DC, 1988, 131.
17. Shah, R. K. and Sekulić, D. P., *Fundamentals of Heat Exchanger Design*, John Wiley & Sons, New York, 2003.

3

Forced Convection Correlations for the Single-Phase Side of Heat Exchangers

3.1 Introduction

In this chapter, recommended correlations for the single-phase side (or sides) of heat exchangers are given. In many two-phase-flow heat exchangers, such as boilers, nuclear steam generators, power condensers, and air conditioning and refrigeration evaporators and condensers, one side has single-phase flow while the other side has two-phase flow. In general, the single-phase side represents higher thermal resistance, particularly with gas or oil flow. This chapter provides a comprehensive review of the available correlations for laminar and turbulent flows of a single-phase Newtonian fluid through circular and noncircular ducts with and without the effect of property variations. A large number of experimental and analytical correlations are available for heat transfer coefficients and flow friction factors for laminar and turbulent flows through ducts and across tube banks.

Laminar and turbulent forced convection correlations for single-phase fluids represent an important class of heat transfer solutions for heat exchanger applications. When a viscous fluid enters a duct, a boundary layer will form along the wall. The boundary layer gradually fills the entire duct cross section and the flow is then said to be fully developed. The distance at which the velocity becomes fully developed is called the hydrodynamic or velocity entrance length (L_{he}). Theoretically, the approach to the fully developed velocity profile is asymptotic, and it is therefore impossible to describe a definite location where the boundary layer completely fills the duct. But for all practical purpose, the hydrodynamic entrance length is finite.

If the walls of the duct are heated or cooled, then a thermal boundary layer will also develop along the duct wall. At a certain point downstream, one can talk about the fully developed temperature profile where the thickness of the thermal boundary layer is approximately equal to half the distance across the cross section. The distance at which the temperature profile becomes fully developed is called the thermal entrance length (L_{te}).

If heating or cooling starts from the inlet of the duct, then both the velocity and the temperature profiles develop simultaneously. The associated heat transfer problem is referred to as the combined hydrodynamic and thermal entry length problem or the simultaneously developing region problem. Therefore, there are four regimes in duct flows with heating/cooling, namely, hydrodynamically and thermally fully developed, hydrodynamically fully developed but thermally developing, thermally developed but hydrodynamically developing; simultaneously developing; and the design correlations should be selected accordingly.

The relative rates of development of the velocity and temperature profiles in the combined entrance region depend on the fluid Prandtl number ($Pr = v/\alpha$). For high Prandtl number fluids such as oils, even though the velocity and temperature profiles are uniform at the tube entrance, the velocity profile is established much more rapidly than the temperature profile. In contrast, for very low Prandtl number fluids such as liquid metals, the temperature profile is established much more rapidly than the velocity profile. However, for Prandtl numbers around 1, such as gases, the temperature and velocity profiles develop simultaneously at a similar rate along the duct, starting from uniform temperature and uniform velocity at the duct entrance.

For the limiting case of $Pr \to \infty$, the velocity profile is developed before the temperature profile starts developing. For the other limiting case of $Pr \to 0$, the velocity profile never develops and remains uniform while the temperature profile is developing. The idealized $Pr \to \infty$ and $Pr \to 0$ cases are good approximations for highly viscous fluids and liquid metals (high thermal conductivity), respectively.

When fluids flow at very low velocities, the fluid particles move in definite paths called streamlines. This type of flow is called laminar flow. There is no component of fluid velocity normal to the duct axis in the fully developed region. Depending on the roughness of the circular duct inlet and the inside surface, fully developed laminar flow will be obtained up to $Re_d \leq 2,300$ if the duct length L is longer than the hydrodynamic entry length L_{he}; however, if $L < L_{he}$, developing laminar flow would exist over the entire duct length. The hydrodynamic and thermal entrance lengths for laminar flow inside conduits are provided by Shah and co-workers.[1,2]

If the velocity of the fluid is gradually increased, there will be a point where the laminar flow becomes unstable in the presence of small disturbances and the fluid no longer flows along smooth lines (streamlines) but along a series of eddies, which results in a complete mixing of the entire flow field. This type of flow is called turbulent flow. The Reynolds number at which the flow ceases to be purely laminar is referred to as the critical (value of) Reynolds number. The critical Reynolds number in circular ducts is between 2,100 and 2,300. Although the value of the critical Reynolds number depends on the duct cross-sectional geometry and surface roughness, for particular applications, it can be assumed that the transition from laminar to turbulent flow

Forced Convection Correlations for the Single-Phase Side of Heat Exchangers

in noncircular ducts will also take place between $Re_{cr} = 2{,}100\text{–}2{,}300$ when the hydraulic diameter of the duct, which is defined as four times the cross-sectional (flow) area A_c divided by the wetted perimeter P of the duct, is used in calculation of the Reynolds number.

At a Reynolds number $Re > 10^4$, the flow is completely turbulent. Between the lower and upper limits lies the transition zone from laminar to turbulent flow.

The heat flux between the duct wall and a fluid flowing inside the duct can be calculated at any position along the duct by

$$\frac{\delta Q}{dA} = hx(T_x - T_b)_x \tag{3.1}$$

where h_x is called the local heat transfer coefficient or film coefficient and is defined based on the inner surface of the duct wall by using the convective boundary condition

$$h_x = \frac{-k(\partial T/\partial y)_w}{(T_w - T_b)_x} \tag{3.2}$$

where k is the thermal conductivity of the fluid, T is the temperature distribution in the fluid, and T_w and T_b are the local wall and the fluid bulk temperatures, respectively. The local Nusselt number is calculated from

$$Nu_x = \frac{h_x d}{k} = \frac{-d(\partial T/\partial y)_w}{(T_w - T_b)_x} \tag{3.3}$$

The local fluid bulk temperature T_{bx}, also referred to as the "mixing cup" or average fluid temperature, is defined for incompressible flow as

$$T_b = \frac{1}{A_c u_m} \int_{A_c} uT dA_c \tag{3.4}$$

where u_m is the mean velocity of the fluid, A_c is the flow cross section, and u and T are, respectively, the velocity and temperature profiles of the flow at position x along the duct.

In design problems, it is necessary to calculate the total heat transfer rate over the total (entire) length of a duct using a mean value of the heat transfer coefficient based on the mean value of the Nusselt number defined as

$$Nu = \frac{hL}{k} = \frac{1}{L}\int_0^L Nu_x \, dx \tag{3.5}$$

3.2 Laminar Forced Convection

Laminar duct flow is generally encountered in compact heat exchangers, cryogenic cooling systems, heating or cooling of heavy (highly viscous) fluids such as oils, and in many other applications. Different investigators performed extensive experimental and theoretical studies with various fluids for numerous duct geometries and under different wall and entrance conditions. As a result, they formulated relations for the Nusselt number versus the Reynolds and Prandtl numbers for a wide range of these dimensionless groups. Shah and London[1] and Shah and Bhatti[2] have compiled the laminar flow results.

Laminar flow can be obtained for a specified mass velocity $G = \rho u_m$ for (1) small hydraulic diameter D_h of the flow passage, or (2) high fluid viscosity μ. Flow passages with small hydraulic diameter are encountered in compact heat exchangers since they result in large surface area per unit volume of the exchanger. The internal flow of oils and other liquids with high viscosity in noncompact heat exchangers is generally of a laminar nature.

3.2.1 Hydrodynamically Developed and Thermally Developing Laminar Flow in Smooth Circular Ducts

The well-known Nusselt–Graetz problem for heat transfer to an incompressible fluid with constant properties flowing through a circular duct with a constant wall temperature boundary condition (subscript T) and fully developed laminar velocity profile was solved numerically by several investigators.[1,2] The asymptotes of the mean Nusselt number for a circular duct of length L are

$$Nu_T = 1.61\left(\frac{Pe_b d}{L}\right)^{1/3} \quad \text{for} \quad \frac{Pe_b d}{L} > 10^3 \quad (3.6)$$

and

$$Nu_T = 3.66 \quad \text{for} \quad \frac{Pe_b d}{L} < 10^2 \quad (3.7)$$

Properties are evaluated at mean bulk temperature.

The superposition of two asymptotes for the mean Nusselt number derived by Gnielinski[3] gives sufficiently good results for most of the practical cases:

$$Nu_T = \left[3.66^3 + 1.61^3\left(\frac{Pe_b d}{L}\right)\right]^{1/3} \quad (3.8)$$

An empirical correlation has also been developed by Hausen[4] for laminar flow in the thermal entrance region of a circular duct at constant wall temperature and is given as

$$Nu_T = 3.66 + \frac{0.19(Pe_b d/L)^{0.8}}{1 + 0.117(Pe_b d/L)^{0.467}} \qquad (3.9)$$

The results of Equations 3.8 and 3.9 are comparable to each other. These equations may be used for the laminar flow of gases and liquids in the range of $0.1 < Pe_b d/L < 10^4$. Axial conduction effects must be considered if $Pe_b d/L < 0.1$. All physical properties are evaluated at the fluid mean bulk temperature of T_b, defined as

$$T_b = \frac{T_i + T_o}{2} \qquad (3.10)$$

where T_i and T_o are the bulk temperatures of the fluid at the inlet and outlet of the duct, respectively.

The asymptotic mean Nusselt numbers in circular ducts with a constant wall heat flux boundary condition (subscript H) are[1,2]

$$Nu_H = 1.953\left(\frac{Pe_b d}{L}\right)^{1/3} \quad \text{for} \quad \frac{Pe_b d}{L} > 10^2 \qquad (3.11)$$

and

$$Nu_H = 4.36 \quad \text{for} \quad \frac{Pe_b d}{L} < 10 \qquad (3.12)$$

The fluid properties are evaluated at the mean bulk temperature T_b as defined by Equation 3.10.

The results given by Equations 3.7 and 3.12 represent the dimensionless heat transfer coefficients for laminar forced convection inside a circular duct in the hydrodynamically and thermally developed region (fully developed conditions) under constant wall temperature and constant wall heat flux boundary conditions, respectively.

3.2.2 Simultaneously Developing Laminar Flow in Smooth Ducts

When heat transfer starts as soon as the fluid enters a duct, the velocity and temperature profiles start developing simultaneously. The analysis of the temperature distribution in the flow and, hence, of the heat transfer between the fluid and the duct wall, for such situations is more complex because the velocity distribution varies in the axial direction as well as normal to the axial direction. Heat transfer problems involving simultaneously developing flow have been mostly solved by numerical methods for various duct cross sections. A comprehensive review of such solutions are given by Shah and Bhatti[2] and Kakaç.[5]

Shah and London[1] and Shah and Bhatti[2] presented the numerical values of the mean Nusselt number for this region. In the case of a short duct length, Nu values are represented by the asymptotic equation of Pohlhausen[6] for simultaneously developing flow over a flat plate; for a circular duct, this equation becomes

$$Nu_T = 0.664 \left(\frac{Pe_b d}{L} \right)^{1/2} Pr_b^{-1/6} \qquad (3.13)$$

The range of validity is $0.5 < Pr_b < 500$ and $Pe_b d/L > 10^3$.

For most engineering applications with short circular ducts ($d/L > 0.1$), it is recommended that whichever of Equations 3.8, 3.9, and 3.13 gives the highest Nusselt number be used.

3.2.3 Laminar Flow through Concentric Annular Smooth Ducts

Correlations for concentric annular ducts are very important in heat exchanger applications. The simplest form of a two-fluid heat exchanger is a double-pipe heat exchanger made up of two concentric circular tubes (Figure 3.1). One fluid flows inside the inner tube, while the other flows through the annular passage. Heat is usually transferred through the wall of the inner tube while the outer wall of the annular duct is insulated. The heat

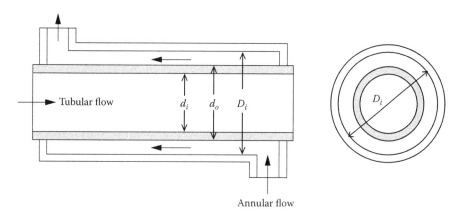

FIGURE 3.1
Concentric tube annulus.

transfer coefficient in the annular duct depends on the ratio of the diameters D_i/d_o because of the shape of the velocity profile.

The hydraulic (equivalent) diameter approach is the simplest method to calculate the heat transfer and the pressure drop in the annulus. In this approach, the hydraulic diameter of annulus D_h is substituted instead of the tube diameter in internal flow correlations:

$$D_h = 4\,\frac{\text{net free-flow area}}{\text{wetted (or heat transfer) perimeter}} \tag{3.14}$$

This approximation is acceptable for heat transfer and pressure drop calculations. The validity of the hydraulic diameter approach has been substantiated by the results of experiments performed with finned annuli.[7] The wetted perimeter for pressure drop calculation in the annulus is defined as

$$P_w = \pi(D_i + d_o) \tag{3.15}$$

and the heat transfer perimeter of the annulus can be calculated by

$$P_h = \pi d_o \tag{3.16}$$

The only difference between P_w and P_h is D_i, which is the inner diameter of the shell (outer tube) of the annulus. This difference is due to the fluid friction on the inner surface of the shell; however, this is not the case for the heat transfer perimeter since the heat transfer takes place only through the walls of the inner tube. The net free-flow area of the annulus is given by

$$A_c = \frac{\pi(D_i^2 - d_o^2)}{4} \tag{3.17}$$

The hydraulic diameter based on the total wetted perimeter for pressure drop calculation is

$$D_h = \frac{4A_c}{P_w} \tag{3.18}$$

and the hydraulic diameter based on the heat transfer perimeter is given by Equation 3.19, which is hereafter called the equivalent diameter:

$$D_e = \frac{4A_c}{P_h} \tag{3.19}$$

The Reynolds number, Graetz number, and the ratio d/L are to be calculated with D_h. D_e is used to calculate the heat transfer coefficient from the Nusselt number and to evaluate the Grashof number in natural convection. Slightly higher heat transfer coefficients arise when D_h is used instead of D_e for heat transfer calculations.

For the constant wall temperature boundary condition, Stephan[8] has developed a heat transfer correlation based on Equation 3.9. The Nusselt number for hydrodynamically developed laminar flow in the thermal entrance region of an isothermal annulus, the outer wall of which is insulated, may be calculated by the following correlation:

$$Nu_T = Nu_\infty + \left[1 + 0.14\left(\frac{d_o}{D_i}\right)^{-1/2}\right] \frac{0.19(Pe_b D_h/L)^{0.8}}{1 + 0.117(Pe_b D_h/L)^{0.467}} \quad (3.20a)$$

where Nu_∞ is the Nusselt number for fully developed flow, which is given by

$$Nu_\infty = 3.66 + 1.2\left(\frac{d_o}{D_i}\right)^{-1/2} \quad (3.20b)$$

where the outer wall of the annulus is insulated.[3]

A detailed review of laminar convective heat transfer in ducts for various hydrodynamic and thermal boundary conditions is given by Shah and Bhatti.[2]

3.3 Effect of Variable Physical Properties

When the previously mentioned correlations are applied to practical heat transfer problems with large temperature differences between the wall and the fluid mean bulk temperatures, the constant-property assumption could cause significant errors, since the transport properties of most fluids vary with temperature, which influences the variation of velocity and temperature through the boundary layer or over the flow cross section of a duct.

For practical applications, a reliable and appropriate correlation based on the constant-property assumption can be modified and/or corrected so that it may be used when the variable-property effect becomes important.

Two correction methods for constant-property correlations for the variable-property effect have been employed: namely, the reference temperature method and the property ratio method. In the former, a characteristic temperature is chosen at which the properties appearing in nondimensional groups are evaluated so that the constant-property results at that temperature may be used to account for the variable-property behavior; in the latter, all properties are taken at the bulk temperature and then all variable-property effects are lumped into a function of the ratio of one property evaluated at the wall (surface) temperature to that property evaluated at the average bulk temperature. Some correlations may involve a modification or combination of these two methods.

For liquids, the variation of viscosity with temperature is responsible for most of the property effects. Therefore, the variable-property Nusselt numbers and friction factors in the property ratio method for liquids are correlated by

$$\frac{Nu}{Nu_{cp}} = \left(\frac{\mu_b}{\mu_w}\right)^n \quad (3.21a)$$

$$\frac{f}{f_{cp}} = \left(\frac{\mu_b}{\mu_w}\right)^m \quad (3.21b)$$

where μ_b and k are the viscosity and conductivity evaluated at the bulk mean temperature, μ_w is the viscosity evaluated at the wall temperature, and c_p refers to the constant-property solution. The friction coefficient usually employed is the so-called Fanning friction factor based on the wall shear rather than the pressure drop.

For gases, the viscosity, thermal conductivity, and density vary with the absolute temperature. Therefore, in the property ratio method, temperature corrections of the following forms are found to be adequate in practical applications for the temperature-dependent property effects in gases:

$$\frac{Nu}{Nu_{cp}} = \left(\frac{T_w}{T_b}\right)^n \quad (3.22a)$$

$$\frac{f}{f_{cp}} = \left(\frac{T_w}{T_b}\right)^m \quad (3.22b)$$

where T_b and T_w are the absolute bulk mean and wall temperatures, respectively.

It must be noted that the constant-property portion of the specific correlation is evaluated in terms of the parameters and conditions defined by its author(s).

Extensive theoretical and experimental investigations on convective heat transfer of fluids with variable properties have been reported in the literature to obtain the values of the exponents n and m and will be cited in the following sections of this chapter.

3.3.1 Laminar Flow of Liquids

Desissler[9] carried out a numerical analysis as described previously for laminar flow through a circular duct at a constant heat flux boundary condition for liquid viscosity variation with temperature given by

$$\frac{\mu}{\mu_w} = \left(\frac{T}{T_w}\right)^{-1.6} \quad (3.23)$$

and obtained $n = 0.14$ to be used with Equation 3.21a. This has been widely used to correlate experimental data for laminar flow for $Pr > 0.6$.

Deissler[9] also obtained $m = -0.58$ for heating and $m = -0.50$ for cooling of liquids, to be used with Equation 3.21b.

A simple empirical correlation has been proposed by Seider and Tate[11] to predict the mean Nusselt number for laminar flow in a circular duct for the combined entry length with constant wall temperature as

$$Nu_T = 1.86 \left(\frac{Pe_b d_i}{L}\right)^{1/3} \left(\frac{\mu_b}{\mu_w}\right)^{0.14} \quad (3.24)$$

which is valid for smooth tubes for $0.48 < Pr_b < 16{,}700$ and $0.0044 < (\mu_b/\mu_w) < 9.75$. This correlation has been recommended by Whitaker[12] for values of

$$\left(\frac{Pe_b d_i}{L}\right)^{1/3} \left(\frac{\mu_b}{\mu_w}\right)^{0.14} \geq 2 \quad (3.25)$$

Below this limit, fully developed conditions will be established and Equation 3.7 may be used for a good approximation. All physical properties are evaluated at the fluid bulk mean temperature except μ_w, which is evaluated at the wall temperature.

Example 3.1

Determine the total heat transfer coefficient at 30 cm from the inlet of a heat exchanger where engine oil flows through tubes with an inner diameter of 0.5 in. Oil flows with a velocity of 0.5 m/s and at a local bulk temperature of 30°C, while the local tube wall temperature is 60°C.

Solution
From Appendix B, the properties of engine oil at $T_b = 30°C$ are

$\rho = 882.3$ kg/m³ $\quad c_p = 1{,}922$ J/kg · K
$\mu = 0.416$ N · s/m² $\quad k = 0.144$ W/m · K
$Pr = 5{,}550$ $\quad \mu_w = 0.074$ N · s/m² (at $T_w = 60°C$)

To determine the heat transfer coefficient, we first need to find the Reynolds number:

$$Re_b = \frac{\rho u_m d_i}{\mu} = \frac{882.3 \times 0.5 \times 0.0127}{0.416} = 13.47$$

Since $Re < 2{,}300$, the flow inside the tube is laminar. We can calculate the heat transfer coefficient from the Sieder and Tate correlation, Equation 3.24, as long as the following conditions are satisfied:

$$\left(Re_b Pr_b \frac{d}{L}\right)^{1/3} \left(\frac{\mu_b}{\mu_w}\right)^{0.14} = \left[\frac{13.47 \times 5550 \times 0.0127}{0.3}\right]^{1/3} \left(\frac{0.416}{0.074}\right)^{0.14}$$

$= 18.7 > 2 \quad \left(\frac{\mu_b}{\mu_w}\right) = \left(\frac{0.416}{0.074}\right) = 5.62 < 9.75$

Therefore,

$$Nu_T = 1.86 (Re\, Pr)_b^{1/3} \left(\frac{d_i}{L}\right)^{1/3} \left(\frac{\mu_b}{\mu_w}\right)^{0.14}$$

and

$$Nu_T = 1.86 \times 18.7 = 34.8$$
$$h = \frac{Nu_T k}{d_i} = \frac{34.8 \times 0.144}{0.0127} = 394.6 \text{ W/m}^2 \cdot \text{K}$$

The Nusselt–Graetz correlation given by Equation 3.11, which is applicable with constant heat flux boundary conditions, can also be used since

$$\frac{Pe_b d_i}{L} = \frac{Re_b Pr_b d_i}{L} = \frac{13.47 \times 5550 \times 0.0127}{0.3} = 3164.8 > 100$$

$$Nu_H = 1.953 \left(Pe_b \frac{d}{L}\right)^{1/3} = 1.953 \times \left(5550 \times 13.47 \times \frac{0.0127}{0.3}\right)^{1/3} = 28.67$$

$$h = \frac{Nu_H k}{d_i} = \frac{28.67 \times 0.144}{0.0127} = 325 \text{ W/m}^2 \cdot \text{K}$$

Equation 3.24 takes into account the fact that higher fluid temperature due to heating increases the heat transfer coefficient. Although a constant heat flux boundary condition gives a higher heat transfer coefficient than a constant wall temperature boundary condition, here, because of the heating of the fluid, the Nusselt–Graetz correlation gives a more conservative answer.

It is not surprising that alternative correlations have been proposed for specific fluids. Oskay and Kakaç[13] performed experimental studies with mineral oil in laminar flow through a circular duct under a constant wall heat flux boundary condition in the range of $0.8 \times 10^3 < Re_b < 1.8 \times 10^3$ and $1 < (T_w/T_b) < 3$ and suggested that the viscosity ratio exponent for Nu in Equation 3.24 should be increased to 0.152 for mineral oil.

Kuznestova[14] conducted experiments with transformer oil and fuel oil in the range of $400 < Re_b < 1,900$ and $170 < Pr_b < 640$ and recommended

$$Nu_b = 1.23 \left(\frac{Pe_b d}{L} \right)^{0.4} \left(\frac{\mu_b}{\mu_w} \right)^{1/6} \quad (3.26)$$

Test[15] conducted an analytical and experimental study on the heat transfer and fluid friction of laminar flow in a circular duct for liquids with temperature-dependent viscosity. The analytical approach is a numerical solution of the continuity, momentum, and energy equations. The experimental approach involves the use of a hot-wire technique for determination of the velocity profiles. Test obtained the following correlation for the local Nusselt number:

$$Nu_b = 1.4 \left(\frac{Pe_b d}{L} \right)^{1/3} \left(\frac{\mu_b}{\mu_w} \right)^{n} \quad (3.27)$$

where

$$n = \begin{cases} 0.05 & \text{for heating liquids} \\ \frac{1}{3} & \text{for cooling liquids} \end{cases}$$

He also obtained the friction factor as

$$f = \frac{16}{Re} \frac{1}{0.89} \left(\frac{\mu_b}{\mu_w} \right)^{0.2} \quad (3.28)$$

Equations 3.24 and 3.27 should not be applied to extremely long ducts.

3.3.2 Laminar Flow of Gases

The first reasonably complete solution for laminar heat transfer of a gas flowing in a tube with temperature-dependent properties was developed by Worsõe-Schmidt.[16] He solved the governing equations with a finite-difference technique for fully developed and developing gas flow through a circular duct. Heating and cooling with a constant surface temperature and heating with a constant heat flux were considered. In his entrance region solution, the radial velocity was included. He concluded that near the entrance, and also well downstream (thermally developed), the results could be satisfactorily correlated for heating $1 < (T_w/T_b) < 3$ by $n = 0$, $m = 1.00$, and for cooling $0.5 < (T_w/T_b) < 1$ by $n = 0$, $m = 0.81$ in Equations 3.22a and 3.22b.

Laminar forced convection and fluid flow in ducts have been studied extensively, and numerous results are available for circular and noncircular ducts under various boundary conditions. These results have been compiled by Shah and London[1] and Shah and Bhatti.[2] The laminar forced convection correlations discussed in previous sections are summarized in Table 3.1. The constant-property correlations can be corrected for the variable physical properties by the use of Table 3.2, in which the exponents m and n are summarized. For fully developed laminar flow, $n = 0.14$ is generally recommended for heating liquids.

3.4 Turbulent Forced Convection

Extensive experimental and theoretical efforts have been made to obtain the solutions for turbulent forced convection heat transfer and flow friction problems in ducts because of their frequent occurrence and application in heat transfer engineering. A compilation of such solutions and correlations for circular and noncircular ducts has been put together by Bhatti and Shah.[17] There are a large number of correlations available in the literature for the fully developed (hydrodynamically and thermally) turbulent flow of single-phase Newtonian fluids in smooth, straight, circular ducts with constant and temperature-dependent physical properties. The objective of this section is to highlight some of the existing correlations to be used in the design of heat exchange equipment and to emphasize the conditions or limitations imposed on the applicability of these correlations.

Extensive efforts have been made to obtain empirical correlations that represent a best-fit curve to experimental data or to adjust coefficients in the theoretical equations to best fit the experimental data. An example of the latter is the correlation given by Petukhov and Popov.[18] Their theoretical calculations for the case of fully developed turbulent flow with constant properties in a circular tube with constant heat flux boundary conditions yielded a correlation, which was based on the three-layer turbulent boundary layer model with constants adjusted to match the experimental data. Petukhov[19] also gave a simplified form of this correlation as

$$Nu_b = \frac{(f/2) Re_b Pr_b}{1.07 + 12.7(f/2)^{1/2} (Pr_b^{2/3} - 1)} \quad (3.29)$$

where $f = (1.58 \ln Re_b - 3.28)^{-2}$.

TABLE 3.1

Laminar Forced Convection Correlations in Smooth Straight Circular Ducts

Number	Correlation	Limitations and Remarks	Ref.
1	$Nu_T = 1.61(Pe_bd/L)^{1/3}$ $Nu_T = 3.66$	$Pe_bd/L > 10^3$, constant wall temperature $Pe_bd/L < 10^2$, fully developed flow in a circular duct, constant wall temperature	1, 2
2	$Nu_T = [(3.66)^3 + (1.61)^3 Pe_bd/L]^{1/3}$	Superposition of two asymptotics given in case 1 for the mean Nusselt number, $0.1 < Pe_bd/L < 10^4$	3
3	$Nu_T = 3.66 + \dfrac{0.19(Pe_bd/L)^{0.8}}{1+0.177(Pe_bd/L)^{0.467}}$	Thermal entrance region, constant wall temperature, $0.1 < Pe_bd/L < 10^4$	4
4	$Nu_H = 1.953(Pe_bd/L)^{1/3}$ $Nu_H = 4.36$	$Pe_bd/L > 10^2$, constant heat flux $Pe_bd/L > 10$, fully developed flow in a circular duct, constant heat flux	1, 2 1, 2
5	$Nu_T = 0.664 \dfrac{1}{(Pr)^{1/6}} \left(Pe_b \dfrac{d}{L}\right)^{1/2}$	$Pe_bd/L > 10^4$, $0.5 < Pr < 500$, simultaneously developing flow	6
6	$Nu_T = Nu + \phi\left(\dfrac{d_o}{D_i}\right) \dfrac{0.19(PeD_h/L)^{0.8}}{1+0.117(PeD_h/L)^{0.467}}$	Circular annular duct, constant wall temperature, thermal entrance region	8
	$\phi(d_o/D_i) = 1 + 0.14(d_o/D_i)^{-1/2}$ $\phi(d_o/D_i) = 1 + 0.14(d_o/D_i)^{0.1}$	Outer wall is insulated, heat transfer through the inner wall Heat transfer through outer and inner wall	
7	$Nu_T = 1.86(Re_b Pr_b d/L)^{1/3} (\mu_b/\mu_w)^{0.14}$	Thermal entrance region, constant wall temperature, $0.48 < Pr_b < 16{,}700$, $4.4 \times 10^{-3} < (\mu_b/\mu_w) < 9.75$, $(Re_b Pr_b d/L)^{1/3}(\mu_b/\mu_w)^{0.14} > 2$	11
8	$Nu_H = 1.86(Re_b Pr_b d/L)^{1/3} (\mu_b/\mu_w)^{0.152}$	Thermal entrance region, constant wall heat flux, for oils $0.8 \times 10^3 < Re_b < 1.8 \times 10^3$, $1 < (T_w/T_b) < 3$	13
9	$Nu_H = 1.23(Re_b Pr_b d/L)^{0.4} (\mu_b/\mu_w)^{1/6}$	Thermal entrance region, constant heat flux, $400 < Re_b < 1{,}900$, $170 < Pr_b < 640$, for oils	14
10	$Nu_b = 1.4(Re_b Pr_b d/L)^{1/3} (\mu_b/\mu_w)^n$	Thermal entrance region, $n = 0.05$ for heating liquids, $n = 1/3$ for cooling liquids	15

Note: Unless otherwise stated, fluid properties are evaluated at the bulk mean fluid temperature, $T_b = (T_i + T_o)/2$.

TABLE 3.2
Exponents n and m Associated with Equations 3.21 and 3.22 for Laminar Forced Convection Through Circular Ducts, $Pr > 0.5$

Number	Fluid	Condition	n	m^a	Limitations	Ref.
1	Liquid	Laminar, heating	0.14	−0.58	Fully developed flow, q''_w = constant $Pr > 0.6$, $\mu/\mu_w = (T/T_w)^{-1.6}$	9
	Liquid	Laminar, cooling	0.14	−0.50		
2	Liquid	Laminar, heating	0.11	—	Developing and fully developed regions of a circular duct, T_w = constant q''_w = constant	10
3	Gas	Laminar, heating	0	1.00	Developing and fully developed regions, q''_w = constant, T_w = constant, $1 < (T_w/T_b) < 3$	16
	Gas	Laminar, cooling	0	0.81	T_w = constant, $0.5 < (T_w/T_b) < 1$	

[a] Fanning friction factor f is defined as $f = 2\tau_w/(\rho u_m^2)$ and for hydrodynamically developed isothermal laminar flow as $f = 16/Re$.

Equation 3.29 predicts the results in the range $10^4 < Re_b < 5 \times 10^6$ and $0.5 < Pr_b < 200$ with 5 to 6% error, and in the range $0.5 < Pr_b < 2{,}000$ with 10% error.

Webb[20] examined a range of data for turbulent flow under fully developed conditions in smooth tubes; he concluded that the relation developed by Petukhov and Popov[18] provided the best agreements with the measurements (Table 3.3). Sleicher and Rouse[21] correlated analytical and experimental results for the range $0.1 < Pr_b < 10^4$ and $10^4 < Re_b < 10^6$ and obtained.

$$Nu_b = 5 + 0.015\, Re_b^m\, Pr_b^n \tag{3.30}$$

with

$$m = 0.88 - \frac{0.24}{4 + Pr_b}$$

$$n = \frac{1}{3} + 0.5\exp(-0.6\, Pr_b)$$

Equations 3.29 and 3.30 give the average Nusselt numbers and are not applicable in the transition region. Gnielinski[22] further modified the

TABLE 3.3

Correlations for Fully Developed Turbulent Forced Convection Through a Circular Duct with Constant Properties

Number	Correlation[a]	Remarks and Limitations	Ref.
1	$Nu_b = \dfrac{(f/2)Re_b Pr_b}{1+8.7(f/2)^{1/2}(Pr_b-1)}$	Based on three-layer turbulent boundary layer model, $Pr > 0.5$	23, 24
2	$Nu_b = 0.021\, Re_b^{0.8} Pr_b^{0.4}$	Based on data for common gases; recommended for Prandtl numbers ≈ 0.7	25
3	$Nu_b = \dfrac{(f/2)Re_b Pr_b}{1.07+12.7(f/2)^{1/2}(Pr_b^{2/3}-1)}$	Based on three-layer model with constants adjusted to match experimental data $0.5 < Pr_b < 2{,}000$, $10^4 < Re_b < 5\times 10^6$	19
4	$Nu_b = \dfrac{(f/2)Re_b Pr_b}{1.07+9(f/2)^{1/2}(Pr_b-1)Pr_b^{-1/4}}$	Theoretically based; Webb found case 3 better at high Pr and this one the same at other Pr	20
5	$Nu_b = 5 + 0.015\, Re_b^m Pr_b^n$ $m = 0.88 - 0.24/(4+Pr_b)$ $n = 1/3 + 0.5\exp(-0.6\,Pr_b)$ $Nu_b = 5 + 0.012\, Re_b^{0.87}(Pr_b + 0.29)$	Based on numerical results obtained for $0.1 < Pr_b < 10^4$, $10^4 < Re_b < 10^6$ Within 10% of case 6 for $Re_b > 10^4$ Simplified correlation for gases, $0.6 < Pr_b < 0.9$	21
6	$Nu_b = \dfrac{(f/2)(Re_b - 1000)Pr_b}{1+12.7(f/2)^{1/2}(Pr_b^{2/3}-1)}$ $f = (1.58 \ln Re_b - 3.28)^{-2}$ $Nu_b = 0.0214\,(Re_b^{0.8}-100)Pr_b^{0.4}$ $Nu_b = 0.012\,(Re_b^{0.87}-280)Pr_b^{0.4}$	Modification of case 3 to fit experimental data at low Re ($2{,}300 < Re_b < 10^4$) Valid for $2{,}300 < Re_b < 5\times 10^6$ and $0.5 < Pr_b < 2{,}000$ Simplified correlation for $0.5 < Pr < 1.5$; agrees with case 4 within −6% and +4% Simplified correlation for $1.5 < Pr < 500$; agrees with case 4 within −10% and +0% for $3 \times 10^3 < Re_b < 10^6$	22
7	$Nu_b = 0.022\, Re_b^{0.8} Pr_b^{0.5}$	Modified Dittus–Boelter correlation for gases ($Pr \approx 0.5 - 1.0$); agrees with case 6 within 0 to 4% for $Re_b \geq 5{,}000$	23

[a] Properties are evaluated at bulk temperatures.

Petukhov–Kirillov correlation by comparing it with the experimental data so that the correlation covers a lower Reynolds number range. Gnielinski recommended the following correlation for the average Nusselt number, which is also applicable in the transition region where the Reynolds numbers are between 2,300 and 10^4:

Forced Convection Correlations for the Single-Phase Side of Heat Exchangers

$$Nu_b = \frac{(f/2)(Re_b - 1000) Pr_b}{1 + 12.7(f/2)^{1/2} (Pr_b^{2/3} - 1)} \tag{3.31}$$

where

$$f = (1.58 \ln Re_b - 3.28)^{-2} \tag{3.32}$$

The effect of thermal boundary conditions is almost negligible in turbulent forced convection[23]; therefore, the empirical correlations given in Table 3.3 can be used for both constant wall temperature and constant wall heat flux boundary conditions.

Example 3.2

A two-tube pass, baffled single-pass shell-and-tube heat exchanger is used as an oil cooler. Cooling water enters the tubes at 25°C at a total flow rate of 8.154 kg/s and leaves at 32°C. The inlet and the outlet temperatures of the engine oil are 65°C and 55°C, respectively. The heat exchanger has 12.25 in. ID shell, and the tubes have 0.652 in. ID and 0.75 in. OD. A total of 160 tubes are laid out on 15/16 in. equilateral triangular pitch. $R_{fo} = 1.76 \times 10^{-4}$ m² K/W, $A_o R_w = 1.084 \times 10^{-5}$ m² · K/W, $h_o = 686$ W/m² · K, $A_o/A_i = 1.1476$, and $R_{fi} = 0.00008$ m² · K/W; find:

1. The heat transfer coefficient inside the tube
2. The total surface area of the heat exchanger using the LMTD method

Solution

From Appendix B, for water at 28.5°C, $c_p = 4.18$ kJ/kg · K, $k = 0.611$ W/m · K, $Pr = 5.64$, $\mu = 8.24 \times 10^{-4}$ Pa · s. The proper heat transfer correlations can be selected from knowledge of the Reynolds number:

$$Re_b = \frac{\rho u_m d_i}{\mu}$$

or

$$Re_b = \frac{4 \dot{m}_c}{\left(\dfrac{N_T}{2}\right) \pi \mu d_i}$$

where N_T is the total number of tubes and ID = 0.652 in. = 0.01656 m.

$$Re_b = \frac{4 \times 8.154}{80 \times \pi \times 8.24 \times 10^{-4} \times 0.0165} = 9510$$

Since $Re > 2,300$, the flow inside the tube is turbulent.
By working with Gnielinski's correlation, Equation 3.31,

$$Nu_b = \frac{(f/2)(Re-1000)\,Pr}{1+12.7(f/2)^{1/2}\left(Pr^{2/3}-1\right)}$$

where

$$f = (1.58 \ln Re_b - 3.28)^{-2} = (1.58 \ln 9510 - 3.28)^{-2} = 7.982 \times 10^{-3}$$

Thus,

$$Nu_b = \frac{(7.982 \times 10^{-3})/2(9510-1000)5.64}{1+12.7\left(\frac{7.982 \times 10^{-3}}{2}\right)^{1/2}\left(5.64^{2/3}-1\right)} = 70$$

Hence,

$$h_i = \frac{Nu_b\,k}{d_i} = 70 \times \frac{0.611}{0.01656} = 2583\ \text{W/m}^2\cdot\text{K}$$

The required heat transfer surface area can be obtained from Equation 2.36.

$$Q = U_o A_o F \Delta T_{lm,cf}$$

By using Equation 2.11, U_o can be calculated:

$$U_o = \left[\frac{A_o}{A_i}\frac{1}{h_i} + A_o R_w + R_{fo} + \frac{A_o}{A_i}R_{fi} + \frac{1}{h_o}\right]^{-1}$$

$$U_o = \left[1.1476 \times \frac{1}{2583} + 1.084 \times 10^{-5} + 1.76 \times 10^{-4} + (1.1476)(0.00008) + \frac{1}{686}\right]^{-1}$$

$$= 458.6\ \text{W/m}^2\cdot\text{K}$$

The correction factor F may be obtained from Figure 2.7, where

$$P = \frac{T_{c2}-T_{c1}}{T_{h1}-T_{c1}} = \frac{32-25}{65-25} = 0.175$$

$$R = \frac{T_{h1}-T_{h2}}{T_{c2}-T_{c1}} = \frac{65-55}{32-25} = 1.43$$

Hence, $F \approx 0.98$.
From Equation 2.33, it follows that

$$\Delta T_{lm,cf} = \frac{\Delta T_1 - \Delta T_2}{\ln \dfrac{\Delta T_1}{\Delta T_2}} = \frac{(65-32)-(55-25)}{\ln \dfrac{65-32}{55-25}} = 31.48°\ C$$

From the heat balance,

$$Q = \dot{m}_c c_{pc}(T_{c2} - T_{c1}) = 8.154 \times 4.18 \times (32-25) = 238586\ W$$

and

$$A_o = \pi d_o L N_T$$

$$A_o = \frac{Q}{U_o F \Delta T_{lm,cf}} = \frac{238586}{(458.6)(0.98)(31.5)} = 16.85\ m^2$$

The length of the heat exchanger is calculated by

$$L = \frac{A_o}{\pi d_o N_T} = \frac{16.85}{\pi \times 0.75 \times 2.54 \times 10^{-2} \times 160} = 1.76\ m$$

3.5 Turbulent Flow in Smooth Straight Noncircular Ducts

The heat transfer and friction coefficients for turbulent flow in noncircular ducts were compiled by Bhatti and Shah.[17] A common practice is to employ the hydraulic diameter in the circular duct correlations to predict Nu and f for turbulent flow in noncircular ducts. For most of the noncircular smooth ducts, the accurate constant-property experimental friction factors are within ±10% of those predicted using the smooth circular duct correlation with hydraulic (equivalent) diameter D_h in the place of circular duct diameter d. The constant-property experimental Nusselt numbers are also within ±10 to ±15% except for some sharp-cornered and narrow channels. This order of accuracy is adequate for the overall heat transfer coefficient and the pressure drop calculations in most of the practical design problems.

The literature describes many attempts to arrive at a universal characteristic dimension for internal turbulent flows that would correlate the constant-property friction factors and Nusselt numbers for all noncircular ducts.[28–30] It must be emphasized that any improvement made by these attempts is only a small percentage, and, therefore, the circular duct correlations may be adequate for many engineering applications.

The correlations given in Table 3.3 do not account for entrance effects occurring in short ducts. Gnielinski[3] recommends the entrance correction factor derived by Hausen[27] for obtaining the Nusselt number for short ducts from the following correlation:

$$Nu_L = Nu_\infty \left[1 + \left(\frac{d}{L}\right)^{2/3} \right] \tag{3.33}$$

where Nu_∞ represents the fully developed Nusselt number calculated from the correlations given in Table 3.3. It should be noted that the entrance length depends on the Reynolds and Prandtl numbers and the thermal boundary condition. Thus, Equation 3.33 should be used cautiously.

Example 3.3

Water flowing at 5,000 kg/h will be heated from 20°C to 35°C by hot water at 140°C. A 15°C hot water temperature drop is allowed. A number of 15-ft (4.5-m) hairpins (Figure 6.4) of 3 in. (ID = 3.068 in., OD = 3.5 in.) by 2 in. (ID = 2.067 in., OD = 2.357 in.) double-pipe heat exchanger with annuli and pipes, each connected in series, will be used. Hot water flows through the inner tube and the outside of the annulus is insulated against heat loss. Calculate: (1) the heat transfer coefficient in the inner tube and (2) the heat transfer coefficient inside the annulus.

Solution

1. First calculate the Reynolds number to determine if the flow is laminar or turbulent, and then select the proper correlation to calculate the heat transfer coefficient. From Appendix B (Table B.2), the properties of hot water at $T_b = 132.5°C$ are

$$\rho = 932.4 \text{ kg/m}^3 \quad c_p = 4271 \text{ J/kg} \cdot \text{K}$$
$$k = 0.688 \text{ W/m} \cdot \text{K} \quad \mu = 0.213 \times 10^{-3} \text{ N} \cdot \text{s/m}^2$$
$$Pr_b = 1.325$$

We now make an energy balance to calculate the hot water mass flow rate:

$$(\dot{m}c_p)_h \Delta T_h = (\dot{m}c_p)_c \Delta T_c$$

$$\dot{m}_h = \frac{\dot{m}_c c_{pc}}{c_{ph}} = \frac{(5000/3600) \times 4179}{4271} = 1.36 \text{ kg/s}$$

where

$$c_{pc} = 4179 \text{ J/kg·K at } T_b = 27.5° \text{ C}$$

$$Re_b = \frac{\rho u_m d_i}{\mu} = \frac{4\dot{m}}{\pi \mu d_i} = \frac{4 \times 1.36}{\pi \times 0.0525 \times 0.213 \times 10^{-3}} = 154850$$

Therefore, the flow is turbulent and a correlation can be selected from Table 3.3. The Petukhov–Kirillov correlation is used here:

$$Nu_b = \frac{(f/2) Re_b Pr_b}{1.07 + 12.7(f/2)^{0.5} (Pr_b^{2/3} - 1)}$$

where f is the Filonenko's friction factor (Table 3.4), calculated as

$$f = (1.58 \ln Re_b - 3.28)^{-2} = [1.58 \ln(154850) - 3.28]^{-2} = 0.0041$$

$$Nu_b = \frac{(4.1 \times 10^{-3}/2)(154850 \times 1.325)}{1.07 + 12.7(4.1 \times 10^{-3}/2)^{0.5}(1.325^{2/3} - 1)} = 353.9$$

$$h_i = \frac{Nu_b k}{d_i} = \frac{353.9 \times 0.688}{2.067 \times 2.54 \times 10^{-2}} = 4637.6 \text{ W/m}^2 \cdot \text{K}$$

The effect of property variations can be found from Equation 3.21a with $n = 0.25$ for cooling of a liquid in turbulent flow (see also Table 3.5), and the heat transfer coefficient can be corrected.

2. Calculate the heat transfer coefficient in the annulus. From Appendix B (Table B.2), the properties of cold water at $T_b = 27.5°C$ are

$\rho = 996.3 \text{ kg/m}^3$ $c_p = 4{,}179 \text{ J/kg · K}$
$k = 0.61 \text{ W/m · K}$ $\mu = 0.848 \times 10^{-3} \text{ N · s/m}^2$
$Pr = 5.825$

The hydraulic diameter of the annulus from Equation 3.17 is

$$D_i - d_o = (3.068 - 2.375) \times 2.54 \times 10^{-2} = 0.0176 \text{ m}$$

$$Re_b = \frac{4 D_h \dot{m}_c}{\pi \mu (D_i^2 - d_o^2)} = \frac{4(0.0176)(5000/3600)}{\pi \times (848 \times 10^{-6})(0.002434)} = 15{,}079$$

Therefore, the flow inside the annulus is turbulent. One of the correlations can be selected from Table 3.3. The correlation

TABLE 3.4

Turbulent Flow Isothermal Fanning Friction Factor Correlations for Smooth Circular Ducts

Number	Correlation[a]	Remarks and Limitations	Ref.[b]
1	$f = \tau_w/1/2\,\rho u_m^2 = 0.0791\,Re^{-1/4}$	This approximate explicit equation agrees with case 3 within ±2.5%, $4 \times 10^3 < Re < 10^5$	Blasius
2	$f = 0.00140 + 0.125\,Re^{-0.32}$	This correlation agrees with case 3 within −0.5% and + 3%, $4 \times 10^3 < Re < 5 \times 10^6$	Drew, Koo, and McAdams
3	$1/\sqrt{f} = 1.737\ln(Re\sqrt{f}) - 0.4$	von Kármán's theoretical equation with the constants adjusted to best fit Nikuradse's experimental data, also referred to as the Prandtl correlation, should be valid for very high values of Re. $4 \times 10^3 < Re < 3 \times 10^6$	von Kármán and Nikuradse
	or $1/\sqrt{f} = 4\log(Re\sqrt{f}) - 0.4$ approximated as $f = (3.64\log Re - 3.28)^{-2}$ $f = 0.046\,Re^{-0.25}$	This approximate explicit equation agrees with the preceding within −0.4 and + 2.2% for $3 \times 10^4 < Re < 10^6$	
4	$f = 1/(1.58\ln Re - 3.28)^2$	Agrees with case 3 within ±0.5% for $3 \times 10^4 < Re < 10^7$ and within ±1.8% at $Re = 10^4$. $10^4 < Re < 5 \times 10^5$	Filonenko
5	$1/f = \left(1.7372\ln\dfrac{Re}{1.964\ln Re - 3.8215}\right)^2$	An explicit form of case 3; agrees with it within ±0.1%, $10^4 < Re < 2.5 \times 10^8$	Techo, Tickner, and James

[a] Cited in References 17, 23, 24, 26.
[b] Properties are evaluated at bulk temperatures.

given by Petukhov and Kirillov (No. 3) is used here. It should be noted that, for the annulus, the Nusselt number should be based on the equivalent diameter calculated from Equation 3.19:

$$D_e = \frac{4A_c}{P_h} = \frac{4[\pi/4(D_i^2 - d_o^2)]}{\pi d_o}$$

$$= \frac{[0.0779^2 - 0.0603^2]}{0.0603} = 0.0403 \text{ m}$$

$$Nu_b = \frac{(f/2) Re_b Pr_b}{1.07 + 12.7(f/2)^{1/2}(Pr_b^{2/3} - 1)}$$

$$f = (1.58 \ln Re_b - 3.28)^{-2} = [1.58 \ln(15079) - 3.28]^{-2} = 0.007037$$

$$Nu_b = \frac{(0.007037/2) \times 15079 \times 5.825}{1.07 + 12.7(0.007037/2)^{1/2}(5.825^{2/3} - 1)} = 112.2$$

$$h_o = \frac{Nu_b k}{D_e} = \frac{112.2 \times 0.614}{0.0403} = 1709.4 \text{ W/m}^2 \cdot \text{K}.$$

3.6 Effect of Variable Physical Properties in Turbulent Forced Convection

When there is a large difference between the duct wall and fluid bulk temperatures, heating and cooling influence heat transfer and the fluid friction in turbulent duct flow because of the distortion of turbulent transport mechanisms, in addition to the variation of fluid properties with temperature as for laminar flow.

3.6.1 Turbulent Liquid Flow in Ducts

Petukhov[19] reviewed the status of heat transfer and friction coefficient in fully developed turbulent pipe flow with both constant and variable physical properties.

To choose the correct value of n in Equation 3.21a, the heat transfer experimental data corresponding to heating and cooling for several liquids over a wide range of values (μ_w/μ_b) were collected by Petukhov,[19] who found that the data were well correlated by

$$\frac{\mu_w}{\mu_b} < 1, \quad n = 0.11 \text{ for heating liquids} \qquad (3.34)$$

$$\frac{\mu_w}{\mu_b} > 1, \quad n = 0.25 \text{ for cooling liquids} \qquad (3.35)$$

which are applicable for fully developed turbulent flow in the range $10^4 < Re_b < 5 \times 10^6$, $2 < Pr_b < 140$, and $0.08 < (\mu_w/\mu_b) < 40$. The value of Nu_{cp} in Equation 3.21a is calculated from Equations 3.30 or 3.31. The value of Nu_{cp} can also be calculated from the correlations listed in Table 3.3.

Petukhov[19] collected data from various investigators for the variable viscosity influence on friction in water for both heating and cooling, and he suggested the following correlations for the friction factor:

$$\frac{\mu_w}{\mu_b} < 1, \quad \frac{f}{f_{cp}} = \frac{1}{6}\left(7 - \frac{\mu_b}{\mu_w}\right) \quad \text{for heating liquids} \qquad (3.36)$$

$$\frac{\mu_w}{\mu_b} > 1, \quad \frac{f}{f_{cp}} = \left(\frac{\mu_w}{\mu_b}\right)^{0.24} \quad \text{for cooling liquids} \qquad (3.37)$$

The friction factor for an isothermal (constant-property) flow f_{cp} can be calculated by the use of Table 3.4 or directly from Equation 3.32 for the range $0.35 < (\mu_w/\mu_b) < 2$, $10^4 < Re_b < 23 \times 10^4$, and $1.3 < Pr_b < 10$.

3.6.2 Turbulent Gas Flow in Ducts

The heat transfer and friction coefficients for turbulent fully developed gas flow in a circular duct were obtained theoretically by Petukhov and Popov[18] by assuming physical properties ρ, c_p, k, and μ as given functions of temperature. This analysis is valid only for small subsonic velocities, since the variations of density with pressure and heat dissipation in the flow were neglected. The eddy diffusivity of momentum was extended to the case of variable properties. The turbulent Prandtl number was taken to be 1 (i.e., $\varepsilon_H = \varepsilon_M$). The analyses were carried out for hydrogen and air for the following range of parameters: $0.37 < (T_w/T_b) < 3.1$ and $10^4 < Re_b < 4.3 \times 10^6$ for air and $0.37 < (T_w/T_b) < 3.7$ and $10^4 < Re_b < 5.8 \times 10^6$ for hydrogen. The analytical results are correlated by Equation 3.22a, where Nu_{cp} is given by Equation 3.30 or 3.31, and the following values for n are obtained:

$$\frac{T_w}{T_b} < 1 \quad n = -0.36 \quad \text{for cooling gases} \qquad (3.38)$$

$$\frac{T_w}{T_b} > 1 \quad n = -\left[0.3 \log\left(\frac{T_w}{T_b}\right) + 0.36\right] \quad \text{for heating gases} \qquad (3.39)$$

With these values for n, Equation 3.22a describes the solution for air and hydrogen within an accuracy of ±4%. For simplicity, one can take n to be

Forced Convection Correlations for the Single-Phase Side of Heat Exchangers

constant for heating as $n = -0.47$ (Table 3.5); then Equation 3.22a describes the solution for air and hydrogen within ±6%. These results have also been confirmed experimentally and can be used for practical calculations when $1 < (T_w/T_b) < 4$. The values of m can be chosen from Table 3.5.

Example 3.4

Air at a mean bulk temperature of 40°C flows through a heated pipe section with a velocity of 6 m/s. The length and diameter of the pipe are 300 cm and 2.54 cm, respectively. The average pipe wall temperature is 300°C. Determine the average heat transfer coefficient.

Solution

Because the wall temperature is so much greater than the initial air temperature, variable-property flow must be considered. From Appendix B (Table B.1), the properties of air at $T_b = 40°C$ are

$$\rho = 1.128 \text{ kg/m}^3 \qquad c_p = 1005.3 \text{ J/kg} \cdot \text{K}$$
$$k = 0.0267 \text{ W/m} \cdot \text{K} \qquad \mu = 1.912 \times 10^{-5} \text{ N} \cdot \text{s/m}^2$$
$$Pr = 0.719$$

TABLE 3.5

Exponents n and m Associated with Equations 3.21 and 3.22 for Turbulent Forced Convection Through Circular Ducts

Number	Fluid	Condition	n	m	Limitations	Ref.
1	Liquid	Turbulent heating	0.11	—	$10^4 < Re_b < 1.25 \times 10^5$, $2 < Pr_b < 140, 0.08 < \mu_w/\mu_b < 1$	19
	Liquid	Turbulent cooling	0.25	—	$1 < \mu_w/\mu_b < 40$	
	Liquid	Turbulent heating	—	Equation 3.36 or −0.25	$10^4 Re_b < 23 \times 10^4$, $1.3 < Pr_b < 10^4, 0.35 < \mu_w/\mu_b < 1$	
	Liquid	Turbulent cooling	—	−0.24	$1 < \mu_w/\mu_b < 2$	
2	Gas	Turbulent heating	−0.47	—	$10^4 < Re_b < 4.3 \times 10^6$, $1 < T_w/T_b < 3.1$	18
	Gas	Turbulent cooling	−0.36	—	$0.37 < T_w/T_b < 1$	
	Gas	Turbulent heating	—	−0.52	$14 \times 10^4 < Re_w \leq 10^6$, $1 < T_w/T_b < 3.7$	
	Gas	Turbulent cooling	—	−0.38	$0.37 < T_w/T_b < 1$	
3	Gas	Turbulent heating	—	−0.264	$1 \leq T_w/T_b \leq 4$	31
4	Gas	Turbulent heating	—	−0.1	$1 < T_w/T_b < 2.4$	32

The inside heat transfer coefficient can be obtained from knowledge of the flow regime, that is, the Reynolds number

$$Re_b = \frac{\rho U_m d_i}{\mu} = \frac{1.128 \times 6 \times 0.0254}{1.912 \times 10^{-5}} = 8991$$

Hence, the flow in the tube can be assumed to be turbulent. On the other hand, $L/d = 3/0.0254 = 118 > 60$, so fully developed conditions can be assumed. Since $Pr > 0.6$, we can use one of the correlations given in Table 3.3. Gnielinsky's correlation, Equation 3.31, with constant properties,

$$Nu_b = \frac{(f/2)(Re_b - 1000)Pr_b}{1 + 12.7(f/2)^{1/2}(Pr_b^{2/3} - 1)}$$

is used here to determine the Nusselt number:

$$f = (1.58 \ln Re - 3.28)^{-2} = [1.58 \ln(8991) - 3.28]^{-2} = 0.00811$$

$$Nu_b = \frac{hd}{k} = \frac{(0.00811/2)(8991 - 1000)(0.719)}{1 + 12.7(0.00811/2)^{0.5}(0.719^{2/3} - 1)} = 27.72$$

$$h = Nu_b \frac{k}{d} = \frac{27.72 \times 0.0267}{0.0254} = 29.14 \text{ W/m}^2 \cdot \text{K}$$

The heat transfer coefficient with variable properties can be calculated from Equation 3.22a, where n is given in Table 3.5 as $n = -0.47$:

$$Nu_b = Nu_{cp}\left(\frac{T_w}{T_b}\right)^{-0.47} = 27.72\left(\frac{573}{313}\right)^{-0.47} = 20.856$$

Then,

$$h = \frac{Nu_b k}{d} = \frac{(20.856) \times (0.0267)}{0.0254} = 21.92 \text{ W/m}^2 \cdot \text{K}$$

As can be seen in the case of a gas with temperature-dependent properties, heating a gas decreases the heat transfer coefficient.

A large number of experimental studies are available in the literature for heat transfer between tube walls and gas flows with large temperature differences and temperature-dependent physical properties. The majority of the work deals with gas heating at constant wall temperature in a circular duct; experimental studies on gas cooling are limited.

The results of heat transfer measurements at large temperature differences between the wall and the gas flow are usually presented as

$$Nu = CRe_b^{0.8} Pr_b^{0.4}\left(\frac{T_w}{T_b}\right)^n \tag{3.40}$$

For fully developed temperature and velocity profiles (i.e., $L/d > 60$), C becomes constant and independent of L/d (Table 3.7).

A number of heat transfer correlations have been developed for variable-property fully developed turbulent liquid and gas flows in a circular duct, some of which are also summarized in Tables 3.6 and 3.7.

Comprehensive information and correlations for convective heat transfer and friction factor in noncircular ducts, coils, and various fittings in various flow arrangements are given by Kakaç et al.[40] Comparisons of the important correlations for forced convection in ducts are also given by Kakaç et al.[41]

3.7 Summary of Forced Convection in Straight Ducts

Important and reliable correlations for Newtonian fluids in single-phase laminar and turbulent flows through ducts have been summarized and can be used in the design of heat transfer equipment.

The tables presented in this chapter cover the recommended specific correlations for laminar and turbulent forced convection heat transfer through ducts with constant and variable fluid properties. Tables 3.2 and 3.5 provide exponents m and n associated with Equations 3.21 and 3.22 for laminar and turbulent forced convection in circular ducts, respectively. Using these tables, the effects of variable properties are incorporated by the property ratio method. The correlations for turbulent flow of liquids and gases with temperature dependent properties are also summarized in Tables 3.6 and 3.7.

Turbulent forced convection heat transfer correlations for fully developed flow (hydrodynamically and thermally) through a circular duct with constant properties are summarized in Table 3.3. The correlations by Gnielinski, Petukhov and Kirillov, Webb, Sleicher, and Rouse are recommended for constant-property Nusselt number evaluation for gases and liquids, and the entrance correction factor is given by Equation 3.33. Table 3.4 lists recommended Fanning friction factor correlations for isothermal turbulent flow in smooth circular ducts. The correlations given in Tables 3.3, 3.4, 3.6, and 3.7 can also be utilized for turbulent flow in smooth, straight, noncircular ducts for engineering applications by use of the hydraulic diameter concept for heat transfer and pressure drop calculations, as discussed in Section 3.2.3. Except for sharp-cornered and/or very irregular duct cross sections, the fully developed turbulent Nusselt number and friction factor vary from their actual values within ±15 and ±10%, respectively, when the hydraulic diameter is used in circular duct correlations; for laminar flows, however, the use of circular tube correlations are less accurate with the hydraulic diameter concept, particularly with cross sections characterized by sharp corners.

TABLE 3.6

Turbulent Forced Convection Correlations in Circular Ducts for Liquids with Variable Properties

Number	Correlation	Comments and Limitations	Ref.
1	$St_b Pr_f^{2/3} = 0.023\, Re_f^{-0.2}$	$L/d > 60$, $Pr_b > 0.6$, $T_f = (T_b + T_w)/2$; inadequate for large $(T_w - T_b)$	33
2	$Nu_b = 0.023\, Re_b^{0.8} Pr_b^{1/3} \left(\dfrac{\mu_b}{\mu_w}\right)^{0.14}$	$L/d > 60$, $Pr_b > 0.6$, for moderate $(T_w - T_b)$	11
3	$Nu_b = \dfrac{(f/8) Re_b Pr_b}{1.07 + 12.7\sqrt{f/8}\,(Pr_b^{2/3} - 1)} \left(\dfrac{\mu_b}{\mu_w}\right)^n$	$L/d > 60$, $0.08 < \mu_w/\mu_b < 40$, $10^4 < Re_b < 5 \times 10^6$, $2 < Pr_b < 140$, $f = (1.82\log Re_b - 1.64)^{-2}$, $n = 0.11$ (heating), $n = 0.25$ (cooling)	19
4	$Nu_b = \dfrac{(f/8) Re_b Pr_b}{1.07 + 12.7\sqrt{f/8}\,(Pr_b^{2/3} - 1)} \left(\dfrac{Pr_b}{Pr_w}\right)^{0.11}$	Water, $2 \times 10^4 < Re_b < 6.4 \times 10^5$, $2 < Pr_b < 5.5$, $f = (1.82 \log Re_b - 1.64)^{-2}$, $0.1 < Pr_b/Pr_w < 10$	34
5	$Nu_b = 0.0277\, Re_b^{0.8} Pr_b^{0.36} \left(\dfrac{Pr_b}{Pr_w}\right)^{0.11}$	Fully developed conditions, the use of the Prandtl group was first suggested by the author in 1960	35
6	$Nu_b = 0.023\, Re_b^{0.8} Pr_b^{0.4} \left(\dfrac{\mu_b}{\mu_w}\right)^{0.262}$	Water, $L/d > 10$, $1.2 \times 10^4 < Re_b < 4 \times 10^4$	13
	$Nu_b = 0.023\, Re_b^{0.8} Pr_b^{0.4} \left(\dfrac{\mu_b}{\mu_w}\right)^{0.487}$	30% glycerine–water mixture $L/d > 10$, $0.89 \times 10^4 < Re_b < 2.0 \times 10^4$	
7	$Nu_b = 0.0235\,(Re_b^{0.8} - 230 \times 1.8 Pr_b^{0.3} - 0.8)$ $\times \left[1 + \left(\dfrac{d}{L}\right)^{2/3}\right]\left(\dfrac{\mu}{\mu_w}\right)^{0.14}$	Altered form of equation presented in 1959[4]	36
8	$Nu_b = 5 + 0.015\, Re_f^m Pr_w^n$		21
	$m = 0.88 - 0.24/(4 + Pr_w)$ $n = 1/3 + 0.5 e^{-0.6 Pr_w}$	$L/d > 60$, $0.1 < Pr_b < 10^4$, $10^4 < Re_b < 10^6$	
	$Nu_b = 0.015\, Re_f^{0.88} Pr_w^{1/3}$	$Pr_b > 50$	
	$Nu_b = 4.8 + 0.015\, Re_f^{0.85} Pr_w^{0.93}$	$Pr_b < 0.1$, uniform wall temperature	
	$Nu_b = 6.3 + 0.0167\, Re_f^{0.85} Pr_w^{0.93}$	$Pr_b < 0.1$, uniform wall heat flux	

Forced Convection Correlations for the Single-Phase Side of Heat Exchangers

TABLE 3.7

Turbulent Forced Convection Correlations in Circular Ducts for Gases with Variable Properties

Number	Correlation	Gas	Comments and Limitations	Ref.
1	$Nu_b = 0.023\, Re_b^{0.8}\, Pr_b^{0.4} \left(\dfrac{T_w}{T_b}\right)^n$ $T_w/T_b < 1, n = 0$ (cooling) $T_w/T_b > 1, n = -0.55$ (heating)	Air	$30 < L/d < 120$, $7 \times 10^3 < Re_b < 3 \times 10^5$, $0.46 < T_w/T_b < 3.5$	37
2	$Nu_b = 0.022\, Re_b^{0.8}\, Pr_b^{0.4} \left(\dfrac{T_w}{T_b}\right)^{-0.5}$	Air	$29 < L/d < 72$, $1.24 \times 10^5 < Re_b < 4.35 \times 10^5$, $1.1 < T_w/T_b < 1.73$	19
3	$Nu_b = 0.023\, Re_b^{0.8}\, Pr_b^{0.4} \left(\dfrac{T_w}{T_b}\right)^n$ $n = -0.4$ for air, $n = -0.185$ for helium, $n = -0.27$ for carbon dioxide	Air, helium, carbon dioxide	$1.2 < T_w/T_b < 2.2$, $4 \times 10^3 < Re_b < 6 \times 10^4$, $L/d > 60$	38
4	$Nu_b = 0.021\, Re_b^{0.8}\, Pr_b^{0.4} \left(\dfrac{T_w}{T_b}\right)^{-0.5}$ $Nu_b = 0.021\, Re_b^{0.8}\, Pr_b^{0.4} \left(\dfrac{T_w}{T_b}\right)^{-0.5}$ $\times \left[1 + \left(\dfrac{L}{d}\right)^{-0.7}\right]$	Air, helium, nitrogen	$L/d > 30$, $1 < T_w/T_b < 2.5$, $1.5 \times 10^4 < Re_b < 2.33 \times 10^5$ $L/d > 5$, local values	32
5	$Nu_b = 0.024\, Re_b^{0.8}\, Pr_b^{0.4} \left(\dfrac{T_w}{T_b}\right)^{-0.7}$ $Nu_b = 0.023\, Re_b^{0.8}\, Pr_w^{0.4}$ $Nu_b = 0.024\, Re_b^{0.8}\, Pr_b^{0.4} \left(\dfrac{T_w}{T_b}\right)^{-0.7}$ $\times \left[1 + \left(\dfrac{L}{d}\right)^{-0.7}\left(\dfrac{T_w}{T_b}\right)^{0.7}\right]$	Nitrogen	$L/d > 40$, $1.24 < T_w/T_b < 7.54$, $18.3 \times 10^3 < Re_b < 2.8 \times 10^5$ Properties evaluated at wall temperature, $L/d > 24$ $1.2 \leq L/d \leq 144$	31
6	$Nu_b = 0.021\, Re_b^{0.8}\, Pr_b^{0.4} \left(\dfrac{T_w}{T_b}\right)^n$ $n = -\left(0.9 \log \dfrac{T_w}{T_b} + 0.205\right)$	Nitrogen	$80 < L/d < 100$, $13 \times 10^3 < Re_b < 3 \times 10^5$, $1 < T_w/T_b < 6$	19
7	$Nu_b = 5 + 0.012\, Re_f^{0.83}\, (Pr_w + 0.29)$		For gases $0.6 < Pr_b < 0.9$	21

(Continued)

TABLE 3.7 (CONTINUED)

Turbulent Forced Convection Correlations in Circular Ducts for Gases with Variable Properties

Number	Correlation	Gas	Comments and Limitations	Ref.
8	$Nu_b = 0.0214(Re_b^{0.8} - 100)Pr_b^{0.4}\left(\dfrac{T_b}{T_w}\right)^{0.45}$ $\times \left[1+\left(\dfrac{d}{L}\right)^{2/3}\right]$ $Nu_b = 0.012(Re_b^{0.87} - 280)Pr^{0.4}\left(\dfrac{T_b}{T_w}\right)^{0.4}$ $\times \left[1+\left(\dfrac{d}{L}\right)^{2/3}\right]$	Air, helium, carbon dioxide	$0.5 < Pr_b < 1.5$, for heating of gases; the author collected the data from the literature; second for $1.5 < Pr_b < 500$	3
9	$Nu_b = 0.022 Re_b^{0.8} Pr_b^{0.4}\left(\dfrac{T_w}{T_b}\right)^{-10.29+0.0019 L/d}$	Air, helium	$10^4 < Re_b < 10^5$, $18 < L/d < 316$	39

TABLE 3.8

Nusselt Number and Friction Factor for Hydrodynamically and Thermally Developed Laminar Flow in Ducts of Various Cross Sections

Geometry ($L/D_h > 100$)	Nu_T	Nu_{H1}	Nu_{H2}	$4f\,Re$

Source: From Shah, R. K. and Bhatti, M. S., *Handbook of Single-Phase Convective Heat Transfer*, John Wiley & Sons, New York, pp. 3.1–3.137, 1987. With permission.

For such cases, the Nusselt number for fully developed conditions may be obtained from Table 3.8 using the following nomenclatures:

Nu_T = average Nusselt number for uniform wall temperature

Nu_{H1} = average Nusselt number for uniform heat flux in flow direction and uniform wall temperature at any cross section

Nu_{H2} = average Nusselt number for uniform heat flux both axially and circumferentially

The calculation of heat transfer coefficient of plain and finned tubes for various arrangements in external flow is important in heat exchanger design. Various correlations for these are given by Gnielinski.[3]

3.8 Heat Transfer from Smooth-Tube Bundles

A circular tube array is one of the most complicated and common heat transfer surfaces, particularly in shell-and-tube heat exchangers. The most common tube arrays are staggered and inline, as shown in Figure 3.2. Other arrangements are also possible (see Chapter 8). The bundle is characterized by the cylinder diameter, d_o, the longitudinal spacing of the consecutive rows, X_l, and the transversal spacing of two consecutive cylinders, X_t.

A considerable amount of work has been published on heat transfer in tube bundles. The most up-to-date review of this work was presented by Zukauskas.[42] The average heat transfer from bundles of smooth tubes is generally determined from the following equation:

$$Nu_b = c\, Re_b^m\, Pr_b^n \left(\frac{Pr_b}{Pr_w}\right)^p \qquad (3.41)$$

where Pr_w stands for the Prandtl number estimated at the wall temperature. Fluid properties are evaluated at the bulk mean temperature.

The variation of c in Equation 3.41 for staggered arrangements may be represented by a geometrical parameter X_t^*/X_l^* with an exponent of 0.2 for $X_t^*/X_l^* < 2$. For $X_t^*/X_l^* > 2$, a constant value of $c = 0.40$ may be assumed. For inline bundles, the effect of change in either longitudinal or transverse pitch is not so evident, and $c = 0.27$ may be assumed for the whole subcritical regime ($10^3 < Re < 2 \times 10^5$).[42]

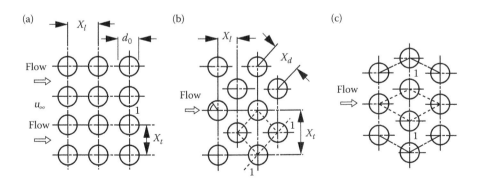

FIGURE 3.2
Tube bundle arrangements: (a) inline array, (b)–(c) staggered array. Minimum intertube spacing at Section I-I between two tubes. (From Zukauskas, A. A., *Handbook of Single-Phase Convective Heat Transfer*, John Wiley & Sons, New York, pp. 6.1–6.45, 1986. With permission.)

For inline tube bundles in crossflow, the array averaged Nusselt numbers are given by[42]

$$\overline{Nu}_b = 0.9\, c_n Re_b^{0.4} Pr_b^{0.36} \left(\frac{Pr_b}{Pr_w}\right)^{0.25} \quad \text{for} \quad Re_b = 1-10^2 \quad (3.42a)$$

$$\overline{Nu}_b = 0.52\, c_n Re_b^{0.5} Pr_b^{0.36} \left(\frac{Pr_b}{Pr_w}\right)^{0.25} \quad \text{for} \quad Re_b = 10^2-10^3 \quad (3.42b)$$

$$\overline{Nu}_b = 0.27\, c_n Re_b^{0.63} Pr_b^{0.36} \left(\frac{Pr_b}{Pr_w}\right)^{0.25} \quad \text{for} \quad Re_b = 10^3-2\times10^5 \quad (3.42c)$$

$$\overline{Nu}_b = 0.033\, c_n Re_b^{0.8} Pr_b^{0.4} \left(\frac{Pr_b}{Pr_w}\right)^{0.25} \quad \text{for} \quad Re_b = 2\times10^5-2\times10^6 \quad (3.42d)$$

where c_n is a correction factor for the number of tube rows because of the shorter bundle; the effect of the number of tubes becomes negligible only when $n > 16$, as shown in Figure 3.3.

The average heat transfer from staggered bundles in crossflow is presented by Zukauskas[42] as

$$\overline{Nu}_b = 1.04\, c_n Re_b^{0.4} Pr_b^{0.36} \left(\frac{Pr_b}{Pr_w}\right)^{0.25} \quad \text{for} \quad Re_b = 1-500 \quad (3.43a)$$

$$\overline{Nu}_b = 0.71\, c_n Re_b^{0.5} Pr_b^{0.36} \left(\frac{Pr_b}{Pr_w}\right)^{0.25} \quad \text{for} \quad Re_b = 500-10^3 \quad (3.43b)$$

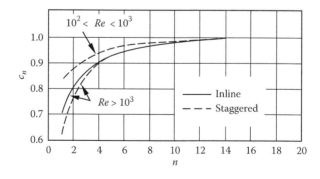

FIGURE 3.3
Correlation factor for the number of rows for the average heat transfer from tube bundles. (From Zukauskas, A. A., *Handbook of Single-Phase Convective Heat Transfer*, John Wiley & Sons, New York, pp. 6.1–6.45, 1987. With permission.)

Forced Convection Correlations for the Single-Phase Side of Heat Exchangers

$$\overline{Nu}_b = 0.35 \, c_n Re_b^{0.6} \, Pr_b^{0.36} \left(\frac{Pr_b}{Pr_w}\right)^{0.25} \left(\frac{X_t}{X_l}\right)^{0.2} \quad \text{for} \quad Re_b = 10^3 - 2\times 10^5 \quad (3.43c)$$

$$\overline{Nu}_b = 0.031 \, c_n Re_b^{0.8} \, Pr_b^{0.4} \left(\frac{Pr_b}{Pr_w}\right)^{0.25} \left(\frac{X_t}{X_l}\right)^{0.2} \quad \text{for} \quad Re_b = 2\times 10^5 - 2\times 10^6 \quad (3.43d)$$

All physical properties except Pr_w are evaluated at the bulk mean temperature of the fluid that flows around the cylinders in the bundle. The uncertainty of the results given by Equations 3.42 and 3.43 is within ±15%.

The Reynolds number, Re, is based on the average velocity through the narrowest cross section formed by the array, that is, the maximum average velocity, u_o.

$$Re_b = \frac{u_o d_o \rho}{\mu} \quad (3.44)$$

For inline arrays, the narrowest cross section is $(X_t - d)$; however, for staggered arrays, it is either $(X_t - d)$ or $2(X_d - d)$ due to the conservation of mass. Heat transfer over a tube in a bundle and heat transfer for rough-tube bundles and finned-tube bundles are also given by Zukauskas[42]; pressure drop analysis of bundles is given in Chapter 4.

Example 3.5

Air at atmospheric pressure flows across a staggered tube bank consisting of eight rows of tubes in the flow direction. The tube diameter $d = 1.0$ cm, the longitudinal spacing and the transversal spacing of two consecutive cylinders are $X_l = 1.50$ cm and $X_t = 2.54$ cm. The upstream velocity $U\infty$ is 6 m/s, and the upstream temperature T_∞ is 20°C. If the surface temperature of the tube is 180°C, find the average heat transfer coefficient for the tube bank.

Solution

The air properties can be obtained from Table B.1 in Appendix B based on the inlet temperature of 20°C.

$\rho = 1.2045 \text{ kg/m}^3$ $k = 0.0257 \text{ W/(m·K)}$

$c_p = 1005 \text{ J/(kg·K)}$ $\mu = 1.82 \times 10^{-5} \text{ kg/(m·s)}$

$\nu = 15.11 \times 10^{-6} \text{ m}^2/\text{s}$ $Pr = 0.713$

The Prandtl number evaluated at a tube surface temperature of 180°C is $Pr_w = 0.685$. First the maximum air velocity in the tube bank must be determined. The tube spacing in Figure 3.2b is

$$X_d - d = \sqrt{X_L^2 + \left(\frac{X_t}{2}\right)^2} - d$$

$$= \sqrt{1.5^2 + \left(\frac{2.54}{2}\right)^2} - 1.0$$

$$= 0.95 \text{ cm}$$

The tube spacing in the transverse direction

$$\frac{X_t - d}{2} = \frac{2.54 - 1.0}{2} = 0.77 \text{ cm} < X_d - d$$

so the minimum section is in the front area and the maximum velocity is

$$u_{max} = \frac{X_t}{X_t - d} u_\infty = \frac{2.54}{2.54 - 1} \times 6 = 9.90 \text{ m/s}$$

The Reynolds number based on the maximum velocity is

$$Re_{D,max} = \frac{u_{max} d_o}{v} = \frac{9.90 \times 0.01}{15.11 \times 10^{-6}} = 6549$$

From Figure 3.3, for $n = 8$ rows, $c_n = 0.98$. Thus, the average Nusselt number, given by Equation 3.43c, is

$$\overline{Nu}_b = 0.35\, c_n Re_b^{0.6} Pr_b^{0.36} \left(\frac{Pr_b}{Pr_w}\right)^{0.25} \left(\frac{X_t}{X_l}\right)^{0.2}$$

$$= 0.35 \times 0.98 \times 6549.4^{0.6} \times 0.713^{0.36} \times \left(\frac{0.713}{0.69}\right)^{0.25} \times \left(\frac{2.54}{1.50}\right)^{0.2}$$

$$= 66.3$$

The average heat transfer coefficient

$$\overline{h} = \frac{\overline{Nu}_b k}{d} = \frac{66.3 \times 0.0257}{0.01} = 170.4 \text{ W/m}^2 \cdot \text{K}$$

3.9 Heat Transfer in Helical Coils and Spirals

Helical coils and spirals (curved tubes) are used in a variety of heat exchangers such as chemical reactors, agitated vessels, storage tanks, and heat recovery systems. Curved-tube heat exchangers are used extensively in food processing, dairy, refrigeration and air-conditioning, and hydrocarbon processing. An enormous amount of work has been published on heat transfer and pressure drop analysis as it relates to this subject; the most up-to-date review of the work in this area was presented by Shah and Joshi.[43]

Nomenclature for helical and spiral coils is illustrated in Figure 3.4. A horizontal coil with tube diameter 2a, coiled curvature radius R, and coil pitch b is shown in Figure 3.4a. It is called a horizontal coil since the tube in each turn is approximately horizontal. If the coil in Figure 3.4a were turned 90°, it would be referred to as a vertical coil. A spiral coil is shown in Figure 3.4b. A simple spiral coil of circular cross section is characterized by the tube diameter 2a, the constant pitch b, and the minimum and maximum radii of curvature (R_{min} and R_{max}) of the beginning and the end of the spiral.

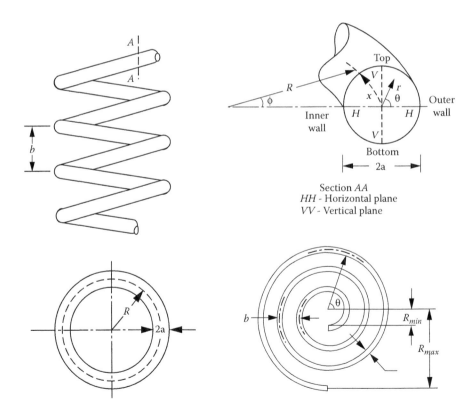

FIGURE 3.4
(a) A schematic of a helical coil; (b) a schematic of a spiral. (From Shah, R. K. and Joshi, *Handbook of Single-Phase Convective Heat Transfer*, John Wiley & Sons, New York, pp. 5.1–5.46, 1987. With permission.)

3.9.1 Nusselt Numbers of Helical Coils—Laminar Flow

Several theoretical and experimental studies have reported Nusselt numbers for Newtonian fluids through a helical coil subjected to constant temperature boundary conditions.[44,45]

The experimental and theoretical results are compared with the following Manlapaz–Churchill correlation[46] based on a regression analysis of the available data, and acceptable agreement is obtained.

$$Nu_T = \left[\left(3.657 + \frac{4.343}{x_1} \right)^3 + 1.158 \left(\frac{De}{x_2} \right)^{3/2} \right]^{1/3} \quad (3.45)$$

where

$$x_1 = \left(1.0 + \frac{957}{De^2 Pr} \right)^2, \quad x_2 = 1.0 + \frac{0.477}{Pr} \quad (3.46)$$

and De is the Dean number, defined as

$$De = Re \left(\frac{a}{R} \right)^{1/2} \quad (3.47)$$

The characteristic length in the Nusselt and Reynolds numbers is the inside diameter of the coil.

Manlapaz and Churchill[46] derived the following correlation under constant heat flux boundary conditions by comparing theoretical predictions with available experimental Nusselt number data:

$$Nu_H = \left[\left(4.364 + \frac{4.636}{x_3} \right)^3 + 1.816 \left(\frac{De}{x_4} \right)^{3/2} \right]^{1/3} \quad (3.48)$$

where

$$x_3 = \left(1.0 + \frac{1342}{De^2 Pr} \right)^2, \quad x_4 = 1.0 + \frac{1.15}{Pr} \quad (3.49)$$

In the above correlations, the properties are evaluated at the average bulk temperature or at an arithmetic-mean temperature of the inlet and outlet temperatures.

3.9.2 Nusselt Numbers for Spiral Coils—Laminar Flow

Kubair and Kuloor[47,48] obtained the Nusselt number for two spirals that were enclosed in a steam chamber using glycerol solutions. They recommended the following correlation:

$$Nu_T = \left(1.98 + \frac{1.8}{R_{ave}}\right) Gz^{0.7} \qquad (3.50)$$

for $9 \leq Gz < 1{,}000$, $80 < Re < 6{,}000$, and $20 < Pr < 100$. The Graetz number, Gz, is given by $Gz = (P/4x) \times RePr$. The properties in Equation 3.50 should be evaluated at an arithmetic mean of the inlet and outlet temperatures.

3.9.3 Nusselt Numbers for Helical Coils—Turbulent Flow

A large number of experimental and theoretical correlations are available to predict Nusselt numbers for a turbulent flow through a helical coil, but experimental data for spiral coils are very limited.

Most of the turbulent fluid flow and heat transfer analyses are limited to fully developed flow. Limited data on turbulent developing flow indicate the flow becoming fully developed within the first half turn of the coil.[43] The entrance length for developing turbulent flow is usually shorter than that for laminar flow. Various correlations are available.

Schmidt's correlation[43] has the largest application range:

$$\frac{Nu_c}{Nu_s} = 1.0 + 3.6\left[1 - \left(\frac{a}{R}\right)\right]\left(\frac{a}{R}\right)^{0.8} \qquad (3.51)$$

where Nu_c is the Nusselt number for the curved coil and Nu_s is the straight tube Nusselt number. It is applicable for $2 \times 10^4 < Re < 1.5 \times 10^5$ and $5 < R/a < 84$. This correlation was developed using air and water in coils under constant wall temperature boundary conditions. The properties were evaluated at fluid mean bulk temperature. For a low Reynolds number, Pratt's correlation[49] is recommended:

$$\frac{Nu_c}{Nu_s} = 1.0 + 3.4\left(\frac{a}{R}\right) \quad \text{for} \quad 1.5 \times 10^3 < Re < 2 \times 10^4 \qquad (3.52)$$

For the influence of temperature-dependent properties, especially viscosity for liquid, Orlov and Tselishchev[50] recommend the following correlation by Mikheev:

$$\frac{Nu_c}{Nu_s} = \left(1.0 + 3.4\frac{a}{R}\right)\left(\frac{Pr_b}{Pr_w}\right)^{0.25} \quad \text{for} \quad 1.5 \times 10^3 < Re < 2 \times 10^4 \qquad (3.53)$$

The only experimental data for spiral coils are reported by Orlov and Tselishchev.[50] They indicate that Mikheev's correlation, Equation 3.53, represents their data within ±15% when an average radius of curvature of a spiral was used in the correlation. This indicates that most helical coil correlations can be used for spiral coils if the average radius of curvature of the spiral, R_{ave}, is used in the correlations.

3.10 Heat Transfer in Bends

Bends are used in pipelines, in circuit lines of heat exchangers, and in tubular heat exchangers. In some cases, bends are heated; in other applications, for example, inner tube returns in hairpin type (double pipe) exchangers, the bend is not heated, but tubes leading to and from a bend are heated. There are various investigations on heat transfer and pressure drop analysis available in the open literature; an excellent review of heat transfer in bends and fittings is given by Shah and Joshi.[51]

Bends are characterized by a bend angle ϕ in degrees and a bend curvature ratio R/a. Bends with circular and rectangular cross sections are shown in Figure 3.5. In this section, heat transfer in 90° and 180° smooth bends for Newtonian fluids will be briefly outlined. As an engineering approximation, the heat transfer results for bends may be used for geometrically similar smooth fittings.

3.10.1 Heat Transfer in 90° Bends

Presently, no experimental data are available to calculate the heat transfer coefficient in laminar flow through a 90° bend. As an engineering application, it is suggested that helical coil thermal entrance length Nusselt number correlations which are given by Shah and Joshi[43] can be used to calculate bend heat transfer coefficients. Experimental studies indicate that a bend has a negligible effect on the heat transfer in a tube upstream of a bend, but it does have a significant influence downstream.[52,53]

Ede[52] and Tailby and Staddon[53] measured the Nusselt number for turbulent flow in a 90° bend at a constant wall temperature; they reported a 20 to 30% increase above the straight-tube Nusselt numbers.

Tailby and Stadon[53] proposed the following correlation for the local heat transfer coefficient in turbulent air cooling in a 90° bend:

$$Nu_x = 0.0336 \, Re^{0.81} \, Pr^{0.4} \left(\frac{R}{a}\right)^{-0.06} \left(\frac{x}{d_i}\right)^{-0.06} \quad (3.54)$$

$$2.3 \leq R/a \leq 14.7, \; 7 \leq x/D \leq 30, \; 10^4 \leq Re \leq 5 \times 10^4$$

where x is measured along the axis of a bend starting from the bend inlet.

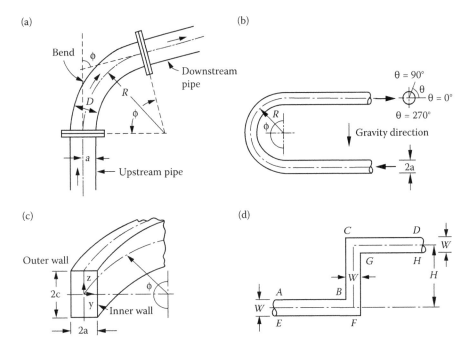

FIGURE 3.5
Schematic diagrams of bends: (a) a bend with an angle $\phi < 90°$; (b) 180° bend; (c) rectangular cross-section bend; and (d) two 90° bends in series. (With permission from Shah, R. K. and Joshi, S. D., Convective heat transfer in curved pipes, in *Handbook of Single-Phase Convective Heat Transfer*, Kakaç, S., Shah, R. K., and Aung, W., Eds., John Wiley & Sons, New York, 1987, ch. 5, pp. 5.1–5.46.)

For a turbulent flow fluid cooling, the average heat transfer in a bend is only about 20 to 30% above the straight-tube values. It is important to note that the above correlation is valid only for fluid cooling. For practical applications, the fully developed turbulent flow fluid heating heat transfer correlation given by Equation 3.51 may be used.[51]

For turbulent flow through a 90° bend, the overall effect of incremental heat transfer in a bend and its downstream pipe is equivalent to increasing the length of a heated pipe by about eight diameters for an elbow and three diameters for bends with $R/a = 8$ and 4.[52,53]

Experimental results show that a bend affects heat transfer in a downstream straight pipe to a distance of about 10 diameters.[46] In turbulent flow, heat transfer in the downstream pipe was found to be independent of upstream heating.[52,53]

3.10.2 Heat Transfer in 180° Bends

For laminar flow through 180° bends, the most extensive results are reported by Mashfeghian and Bell[54] for $R/a = 4.84, 7.66, 12.32,$ and 25.62, but they could not reach a satisfying correlation for laminar flow. Because of the lack of information, helical coil entrance region correlations given by Kubair and Kuloor[48] may be used to estimate laminar flow Nusselt numbers in a 180° bend.

Mashfeghian and Bell[54] measured turbulent flow heat transfer coefficients in a heated bend using different fluids, and they presented the following correlation for turbulent flow local Nusselt numbers:

$$Nu_x = 0.0285\, Re^{0.81}\, Pr^{0.4} \left(\frac{x}{d_i}\right)^{0.046} \left(\frac{R}{a}\right)^{-0.133} \left(\frac{\mu_b}{\mu_w}\right)^{0.14} \quad (3.55)$$

for $4.8 \leq R/a \leq 26, 10^4 \leq Re \leq 3\times 10^4, 0 < x/d_i \leq \pi R/(2a)$

The properties are evaluated at local mean temperature. The results show less than 8% deviation from the experimental data.

Pressure drops in helical and spiral coils and bends are discussed in Chapter 4.

Nomenclature

a	coil tube inside radius, m
A	constant
A_c	net free-flow cross-sectional area, m²
b	coil pitch, m
c_p	specific heat at constant pressure, J/kg · K
c_v	specific heat at constant volume, J/kg · K
De	Dean number, $Re(a/R)^{1/2}$
D_e	equivalent diameter for heat transfer, $4A_c/P_h$, m
D_i	inner diameter of a circular annulus, m
D_h	hydraulic diameter for pressure drop, $4A_c/P_w$, m
d	circular duct diameter, or distance between parallel plates, m
d_i	inside diameter of circular or curved duct, m
d_o	outside diameter of the inner tube of an annular duct, m
Eu	n-row average Euler number $= 2\Delta p/\rho U_o^2 n$
f	Fanning friction factor, $\tau_w/\tfrac{1}{2}\rho u_m^2$
G	fluid mass velocity, ρu_m, kg/m² · s
Gz	Graetz number, $(P/4x) \cdot RePr$
h	average heat transfer coefficient, W/m² · K
h_x	local heat transfer coefficient, W/m² · K

k	thermal conductivity of fluid, W/m · K
L	distance along the duct, or tube length, m
L_{he}	hydrodynamic entrance length, m
L_{te}	thermal entrance length, m
m	exponent, Equations 3.21b and 3.22b
$\dot{m}$	mass flow rate, kg/s
n	number of tube rows in the flow direction
n	exponent, Equations 3.21a and 3.22a
Nu	average Nusselt number, hd/k
P	duct wetted perimeter, m
Pe	Péclet number, $RePr$
Pr	Prandtl number, $c_p\mu/k = \nu/\alpha$
R	radius of curvature (see Figures 3.4 and 3.5), m
R_{ave}	mean radius of curvature for a spiral, m
Re	Reynolds number, $\rho u_m d/\mu$, $\rho u_m D_h/\mu$
T	temperature, °C, K
T_f	film temperature, $(T_w + T_b)/2$, °C, K
u	velocity component in axial direction, m/s
u_m	mean axial velocity, m/s
u_o	mean velocity in the minimum free-flow area or intertube spacing, m/s
x	Cartesian coordinate, axial distance, or axial distance along the axis of the coil, m
X_l	tube longitudinal pitch, m
X_t	tube transfer pitch, m
X^*_l	ratio of longitudinal pitch to tube diameter for crossflow to a tube bundle, X_l/d_o
X^*_t	ratio of transverse pitch to tube diameter for crossflow to a tube bundle, X_t/d_o
y	Cartesian coordinate, distance normal to the surface, m

Greek Symbols

α	thermal diffusivity of fluid, m²/s
μ	dynamic viscosity of fluid, Pa · s
υ	kinematic viscosity of fluid, m²/s
ρ	density of fluid, kg/m³
τ_w	shear stress at the wall, Pa
φ	bend angle, or angle measured along the tube perimeter from the front stagnation point, rad or deg

Subscripts

a	arithmetic mean

b	bulk fluid condition or properties evaluated at bulk mean temperature
c	curved coil or duct
cp	constant property
e	equivalent
f	film fluid condition or properties evaluated at film temperature
H	constant heat flux boundary condition
l	laminar
i	inlet condition or inner
o	outlet condition or outer
r	reference condition
s	straight duct
T	constant temperature boundary condition
t	turbulent
w	wall condition or wetted
x	local value at distance x
∞	fully developed condition

PROBLEMS

3.1. A fluid flows steadily with a velocity of 6 m/s through a commercial iron rectangular duct whose sides are 1 in. by 2 in., and the length of the duct is 6 m. The average temperature of the fluid is 60°C. The fluid completely fills the duct. Calculate the surface heat transfer coefficient if the fluid is
   a. Water
   b. Air at atmospheric pressure
   c. Engine oil ($\rho = 864$ kg/m³, $c_p = 2{,}047$ J/(kg · K), $\nu = 0.0839 \times 10^{-3}$ m²/s, $Pr = 1{,}050$, $k = 0.140$ W/m · K).

3.2. Air at 1.5 atm and 40°C flows through a 10 m rectangular duct of 40 cm by 25 cm with a velocity of 5 m/s. The surface temperature of the duct is maintained at 120°C, and the average air temperature at exit is 80°C. Calculate the total heat transfer using Gnielinski's correlation and check your result by energy balance.

3.3. Calculate the heat transfer coefficient for water flowing through a 2 cm diameter tube with a velocity of 1 m/s. The average temperature of water is 60°C and the surface temperature is
   a. Slightly over 60°C
   b. 120°C

3.4. An oil with $k = 0.120$ W/m · K, $c_p = 2{,}000$ J/kg · K, $\rho = 895$ kg/m³, and $\mu = 0.0041$ kg/m · s flows through a 2 cm diameter tube which is 2 m long. The oil is cooled from 60°C to 30°C. The mean flow velocity is 0.4 m/s, and the tube wall temperature is maintained at 24°C ($\mu_w = 0.021$ kg/m · s). Calculate the heat transfer rate.

3.5. A shell-and-tube type condenser is to be made with 3/4 in. outer diameter (0.654 in. inner diameter) brass tubes, and the length of the tubes between tube plates is 3 m. Under the worst conditions, cooling water is available at 21°C and the outlet temperature is 31°C. Water velocity inside the tubes is approximately 2 m/s. The vapor side film coefficient can be taken as 10,000 W/m² · K. Calculate the overall heat transfer coefficient for this heat exchanger.

3.6. Water at 1.15 bar and 30°C is heated as it flows through a 1 in. inside diameter tube at a velocity of 3 m/s. The pipe surface temperature is kept constant by condensing steam outside the tube. If the water outlet temperature is 80°C, calculate the surface temperature of the tube.

3.7. A counterflow double pipe heat exchanger is used to cool the lubricating oil for a large industrial gas turbine engine. The flow rate of cooling water through the inner tube is $m_c = 0.2$ kg/s, while the flow rate of oil through the outer annulus is $m_h = 0.4$ kg/s. The oil and water enter at temperatures of 60°C and 30°C, respectively. The heat transfer coefficient in the annulus is calculated to be 8 W/m · K. The inner diameter of the tube and the annulus are 25 and 45 mm, respectively. The outlet temperature of the oil is 40°C. Take $c_p = 4{,}178$ J/kg · K for water and $c_p = 2{,}006$ J/kg · K for oil.

   a. Calculate the heat transfer coefficients in the inner tube.

   b. Neglect the wall resistance and the curvature of the tube wall (assume a flat plate surface), and calculate the overall heat transfer coefficient assuming the water used is the city water with a fouling factor of 0.000176 m² · K/W inside the tube. The oil side is clean.

3.8. City water flowing at 0.5 kg/s will be heated from 20°C to 35°C by hot treated boiler water at 140°C. A 15°C hot water temperature drop is allowed. A number of 4.50 m hairpins of 3 in. by 2 in. (schedule 40, see Table 8.2) double-pipe heat exchangers with annuli and tubes, each connected in series, will be used. Hot water flows through the inner tube. Calculate the heat transfer coefficient in the tube and in the annulus.

3.9. It is proposed that the stationary diesel exhaust gases (CO_2) be used for heating air by the use of a doublepipe heat exchanger. From the measurements of the velocity, it is calculated that the mass flow rate of gases from the engine is 100 kg/h. The exhaust gas temperature is 600 K. The air is available at 20°C and will be heated to 80°C at a mass flow rate of 90 kg/h. A standard tube size of 4 × 3 in. for a doublepipe heat exchanger (hairpins) in 2 m length (copper tubes) will be used. The air side heat transfer coefficient in the tube is 25 W/m² · K. Neglecting the tube wall resistance, and assuming clean and smooth surfaces, calculate the overall heat transfer coefficient.

3.10. Consider the laminar flow of an oil inside a duct with a Reynolds number of 1,000. The length of the duct is 2.5 m and the diameter is 2 cm. The duct is heated electrically by the use of its walls as an electrical resistance. Properties of the oil at the average oil temperature are $\rho = 870$ kg/m³, $\mu = 0.004$ N · s/m², and $c_p = 1,959$ kJ/kg · K, and $k = 0.128$ W/m · K. Obtain the local Nusselt number at the end of the duct.

3.11. In a crossflow heat exchanger, hot air at atmosphere pressure with an average velocity of 3 m/s flows across a bank of tubes in an in-line arrangement with $X_l = X_t = 5$ cm (see Figure 3.2). The tube diameter is 2.5 cm. The array has 20 rows in the direction of flow. The tube wall temperature is 30°C and the average air temperature in the bundle is assumed to be 300°C. Calculate the average heat transfer coefficient and repeat the calculation if the array has six rows in the flow direction.

3.12. Repeat Problem 3.11 for a heat exchanger that employs a bank of staggered bare tubes with a longitudinal pitch of 4 cm and transverse pitch of 5 cm.

3.13. A shell-and-tube heat exchanger is used to cool 20 kg/s of water from 40°C to 20°C. The exchanger has one shell-side pass and two tube-side passes. The hot water flows through the tubes and the cooling water flows through the shell. The cooling water enters at 15°C and leaves at 25°C. The maximum permissible pressure drop is 10 kPa. The tube wall thickness is 1.25 mm and is selected as 18 BWG copper (ID = 0.652 in, cross-section area $A_c = 0.3339$ in²). The length of the heat exchanger is 5 m. Assume that the pressure losses at the inlet and outlet are equal to twice the velocity head ($\rho u_m^2/2$). Find the number of tubes and the proper tube diameter to expand the available pressure drop. (Hint: assume a tube diameter and average velocity inside the tubes.)

3.14. Repeat Problem 3.13, assuming that an overall heat transfer coefficient is given or estimated as 2,000 W/m² · K.

3.15. Calculate the average heat transfer coefficient for water flowing at 1.5 m/s with an average temperature of 20°C in a long, 2.5 cm inside diameter tube by four different correlations from

Table 3.6 (Nos. 2, 3, 4, and 6), considering the effect of temperature-dependent properties. The inside wall temperature of the tube is 70°C.

3.16. A double-pipe heat exchanger is used to condense steam at 40°C saturation temperature. Water at an average bulk temperature of 20°C flows at 2 m/s through the inner tube (copper, 2.54 cm ID, 3.05 cm OD). Steam at its saturation temperature flows in the annulus formed between the outer surface of the inner tube and outer tube of 6 cm ID. The average heat transfer coefficient of the condensing steam is 6,000 W/m² · K, and the thermal resistance of a surface scale on the outer surface of the copper pipe is 0.000176 m² · K/W.

  a. Determine the overall heat transfer coefficient between the steam and the water based on the outer area of the copper tube.
  b. Evaluate the temperature at the inner surface of the tube.
  c. Estimate the length required to condense 0.5 kg/s of steam.

3.17. Carbon dioxide gas at 1 atm pressure is to be heated from 30°C to 75°C by pumping it through a tube bank at a velocity of 4 m/s. The tubes are heated by steam condensing within them at 200°C. The tubes have a 2.5 cm OD, are in an in-line arrangement, and have a longitudinal spacing of 4 cm and a transverse spacing of 4.5 cm. If 15 tube rows are required, what is the average heat transfer coefficient?

References

1. Shah, R. K. and London, A. L., *Laminar Forced Convection in Ducts*, Academic Press, New York, 1978.
2. Shah, R. K. and Bhatti, M. S., Laminar convective heat transfer in ducts, in *Handbook of Single-Phase Convective Heat Transfer*, Kakaç, S., Shah, R. K., and Aung, W., Eds., John Wiley & Sons, New York, 1987, ch. 3.
3. Gnielinski, V., Forced convection ducts, in *Heat Exchanger Design Handbook*, Schlünder, E. U., Ed., Hemisphere, Washington, DC, 1983, 2.5.1–2.5.3.
4. Hausen, H., Neue Gleichungen für die Warmeübertragung bei freier oder erzwungener Strömung, *Allg. Wärmetech.*, 9, 75, 1959.
5. Kakaç, S., Laminar forced convection in the combined entrance region of ducts, in *Natural Convection: Fundamentals and Applications*, Kakaç, S., Aung, W., and Viskanta, R., Eds., Hemisphere, Washington, DC, 1985, 165.
6. Pohlhausen, E., Der Wärmeaustausch Zwischen festen Körpern und Flüssigkeiten mit Kleiner Reibung und Kleiner Wärmeleitung, *Z. Angew. Math. Mech.*, 1, 115, 1921.

7. Delorenzo, B. and Anderson, E. D., Heat transfer and pressure drop of liquids in double pipe fintube exchangers, *Trans. ASME*, 67, 697, 1945.
8. Stephan, K., Warmeübergang und Druckabfall bei nichtausgebildeter Laminar Stömung in Rohren und ebenen Spalten, *Chem. Ing. Tech.*, 31, 773, 1959.
9. Deissler, R. G., Analytical investigation of fully developed laminar flow in tubes with heat transfer with fluid properties variable along the radius, NACA TN 2410, 1951.
10. Yang, K. T., Laminar forced convection of liquids in tubes with variable viscosity, *J. Heat Transfer*, 84, 353, 1962.
11. Sieder, E. N. and Tate, G. E., Heat transfer and pressure drop of liquids in tubes, *Ind. Eng. Chem.*, 28, 1429, 1936.
12. Whitaker, S., Forced convection heat-transfer correlations for flow in pipes, past flat plates, single cylinders, single spheres, and flow in packed beds and tube bundles, *AIChE J.*, 18, 361, 1972.
13. Oskay, R. and Kakaç, S., Effect of viscosity variations on turbulent and laminar forced convection in pipes, *METU J. Pure Appl. Sci.*, 6, 211, 1973.
14. Kuzneltsova, V. V., Convective heat transfer with flow of a viscous liquid in a horizontal tube (in Russian), *Teploenergetika*, 19(5), 84, 1972.
15. Test, F. L., Laminar flow heat transfer and fluid flow for liquids with a temperature dependent viscosity, *J. Heat Transfer*, 90, 385, 1968.
16. Worsøe-Schmidt, P. M., Heat transfer and friction for laminar flow of helium and carbon dioxide in a circular tube at high heating rate, *Int. J. Heat Mass Transfer*, 9, 1291, 1966.
17. Bhatti, M. S. and Shah, R. K., Turbulent forced convection in ducts, in *Handbook of Single-Phase Convective Heat Transfer*, Kakaç, S., Shah, R. K., and Aung, W., Eds., John Wiley & Sons, New York, 1987, ch. 4.
18. Petukhov, B. S. and Popov, V. N., Theoretical calculation of heat exchanger and frictional resistance in turbulent flow in tubes of incompressible fluid with variable physical properties, *High Temp.*, 1(1), 69, 1963.
19. Petukhov, B. S., Heat transfer and friction in turbulent pipe flow with variable physical properties, in *Advances in Heat Transfer*, Vol. 6, Hartnett, J. P. and Irvine, T. V., Eds., Academic Press, New York, 1970, 504.
20. Webb, R. I., A critical evaluation of analytical solutions and Reynolds analogy equations for heat and mass transfer in smooth tubes, *Wärme und Stoffübertragung*, 4, 197, 1971.
21. Sleicher, C. A. and Rouse, M. W., A convenient correlation for heat transfer to constant and variable property fluids in turbulent pipe flow, *Int. J. Heat Mass Transfer*, 18, 677, 1975.
22. Gnielinski, V., New equations for heat and mass transfer in turbulent pipe and channel flow, *Int. Chem. Eng.*, 16, 359, 1976.
23. Kays, W. M. and Crawford, M. E., *Convective Heat and Mass Transfer*, 2nd ed., McGraw-Hill, New York, 1981.
24. Kakaç, S. and Yener, Y., *Convective Heat Transfer*, 2nd ed., CRC Press, Boca Raton, FL, 1994.
25. McAdams, W. H., *Heat Transmission*, 3rd ed., McGraw-Hill, New York, 1954.
26. Kakaç, S., The effects of temperature-dependent fluid properties on convective heat transfer, in *Handbook of Single-Phase Convective Heat Transfer*, Kakaç, S., Shah, R. K., and Aung, W., Eds., John Wiley & Sons, New York, 1987, ch. 18.

27. Hausen, H., Darstellung des Warmeüberganges in Rohren durch verallgemeinerte Potenzbeziehungen, *Z. Ver. Dtsch. Ing. Beigeft Verfahrenstech.*, 4, 91, 1943.
28. Rehme, K., A simple method of predicting friction factors of turbulent flow in noncircular channels, *Int. J. Heat Mass Transfer*, 16, 933, 1973.
29. Malak, J., Hejna, J., and Schmid, J., Pressure losses and heat transfer in noncircular channels with hydraulically smooth walls, *Int. J. Heat Mass Transfer*, 18, 139, 1975.
30. Brundrett, E., Modified hydraulic diameter, in *Turbulent Forced Convection in Channels and Bundles*, Vol. 1, Kakaç, S. and Spalding, D. B., Eds., Hemisphere, Washington, DC, 1979, 361.
31. Perkins, H. C. and Worsøe-Schmidt, P., Turbulent heat and momentum transfer for gases in a circular tube at wall to bulk temperature ratios to seven, *Int. J. Heat Mass Transfer*, 8, 1011, 1965.
32. McElligot, D. M., Magee, P. M., and Leppert, G., Effect of large temperature gradients on convective heat transfer: the downstream region, *J. Heat Transfer*, 87, 67, 1965.
33. Colburn, A. P., A method of correlating forced convection heat transfer data and comparison with fluid friction, *Trans. AIChE*, 29, 174, 1933.
34. Hufschmidt, W., Burck, E., and Riebold, W., Die Bestimmung örtlicher und Warmeübergangs-Zahlen in Rohren bei Hohen Warmestromdichten, *Int. J. Heat Mass Transfer*, 9, 539, 1966.
35. Rogers, D. G., Forced convection heat transfer in single phase flow of a Newtonian fluid in a circular pipe, CSIR Report CENG 322, Pretoria, South Africa, 1980.
36. Hausen, H., Extended equation for heat transfer in tubes at turbulent flow, *Wärme und Stoübertragung*, 7, 222, 1974.
37. Humble, L. V., Lowdermilk, W. H., and Desmon, L. G., Measurement of average heat transfer and friction coefficients for subsonic flow of air in smooth tubes at high surface and fluid temperature, NACA Report 1020, 1951.
38. Barnes, J. F. and Jackson, J. D., Heat transfer to air, carbon dioxide and helium flowing through smooth circular tubes under conditions of large surface/gas temperature ratio, *J. Mech. Eng. Sci.*, 3(4), 303, 1961.
39. Dalle-Donne, M. and Bowditch, P. W., Experimental local heat transfer and friction coefficients for subsonic laminar transitional and turbulent flow of air or helium in a tube at high temperatures, Dragon Project Report 184, Winfirth, Dorchester, Dorset, U.K., 1963.
40. Kakaç, S., Shah, R. K., and Aung, W., Eds., *Handbook of Single-Phase Convective Heat Transfer*, John Wiley & Sons, New York, 1987.
41. Kakaç, S., Bergles, A. E., and Fernandes, E. O., Eds., *Two-Phase Flow Heat Exchangers*, Kluwer Publishers, The Netherlands, 1988.
42. Zukauskas, A. A., Convective heat transfer in cross flow, in *Handbook of Single-Phase Convective Heat Transfer*, Kakaç, S., Shah, R. K., and Aung, W., Eds., John Wiley & Sons, New York, 1987, ch. 6.
43. Shah, R. K. and Joshi, S. D., Convective heat transfer in curved pipes, in *Handbook of Single-Phase Convective Heat Transfer*, Kakaç, S., Shah, R. K., and Aung, W., Eds., John Wiley & Sons, New York, 1987, ch. 5.
44. Mori, Y. and Nakayoma, W., Study on forced convective heat transfer in curved pipes (3rd Report, Theoretical Analysis under the Condition of Uniform Wall Temperature and Practical Formulae), *Int. J. Heat Mass Transfer*, 10, 1426, 1967.

45. Akiyama, M. and Cheng, K. C., Laminar forced convection heat transfer in curved pipes with uniform wall temperature, *Int. J. Heat Mass Transfer*, 15, 1426, 1972.
46. Manlapaz, R. L. and Churchill, S. W., Fully developed laminar convection from a helical coil, *Chem. Eng. Commun.*, 9, 185, 1981.
47. Kubair, V. and Kuloor, N. R., Heat transfer to Newtonian fluids in coiled pipes in laminar flow, *Int. J. Heat Mass Transfer*, 9, 63, 1966.
48. Kubair, V. and Kuloor, N. R., Heat transfer to Newtonian fluids in spiral coils at constant tube wall temperature in laminar flow, *Indian J. Technol.*, 3, 144, 1965.
49. Pratt, N. H., The heat transfer in a reaction tank cooled by means of a coil, *Trans. Inst. Chem. Eng.*, 25, 163, 1947.
50. Orlov, V. K. and Tselishchev, P. A., Heat exchange in a spiral coil with turbulent flow of water, *Thermal Eng.*, 11(12), 97, 1964.
51. Shah, R. K. and Joshi, S. D., Convective heat transfer in bends and fittings, in *Handbook of Single-Phase Convective Heat Transfer*, Kakaç, S., Shah, R. K., and Aung, W., Eds., John Wiley & Sons, New York, 1987, ch. 10.
52. Ede, A. J., The effect of a right-angled bend on heat transfer in a pipe, *Int. Dev. Heat Transfer*, 634, 1962.
53. Tailby, S. R. and Staddon, P. W., The influence of 90° and 180° pipe bends on heat transfer from an internally flowing gas stream, *Heat Transfer*, Vol. 2, Paper No. FC4.5, 1970.
54. Moshfeghian, M. and Bell, K. J., Local heat transfer measurements in and downstream from a U-bend, ASME Paper No. 79-HT-82, 1979.

4

Heat Exchanger Pressure Drop and Pumping Power

4.1 Introduction

The thermal design of a heat exchanger is directed to calculating an adequate surface area to handle the thermal duty for a given specification. Fluid friction effects in the heat exchanger are equally important since they determine the pressure drop of the fluids flowing in the system and, consequently, the pumping power or fan work input necessary to maintain the flow. Provision of pumps or fans adds to the capital cost and is a major part of the operating cost of the exchanger. Savings in exchanger capital cost achieved by designing a compact unit with high fluid velocities may soon be lost by increased operating costs. The final design and selection of a unit will, therefore, be influenced just as much by effective use of the permissible pressure drop and the cost of pump or fan power as they are influenced by the temperature distribution and provision of adequate area for heat transfer.

4.2 Tube-Side Pressure Drop

4.2.1 Circular Cross-Sectional Tubes

In fully developed flow in a tube, the following functional relationship can be written for the frictional pressure drop for either laminar or turbulent flow:

$$\frac{\Delta p}{L} = \phi(u_m, d_i, \rho, \mu, e) \tag{4.1}$$

where the quantity e is a statistical measure of the surface roughness of the tube and has the dimension of length. It is assumed that Δp is proportional to the length L of the tube. With mass M, length L, and time θ as the fundamental dimensions and u_m, d_i, and ρ as the set of maximum number of quantities which, in and of themselves, cannot form a dimensionless group, the π theorem leads to

$$\frac{\Delta p}{4(L/d_i)(\rho u_m^2/2)} = \phi\left(\frac{u_m d_i \rho}{\mu}, \frac{e}{d_i}\right) \quad (4.2)$$

where the dimensionless numerical constants 4 and 2 are added for convenience.

This dimensionless group involving Δp has been defined as the Fanning friction factor f:

$$f = \frac{\Delta p}{4(L/d_i)(\rho u_m^2/2)} \quad (4.3)$$

Equation 4.2 becomes

$$f = \phi(Re, e/d_i) \quad (4.4)$$

Figure 4.1 shows this relationship as deduced by Moody[1] from experimental data for fully developed flow. In the laminar region, both theoretical analysis

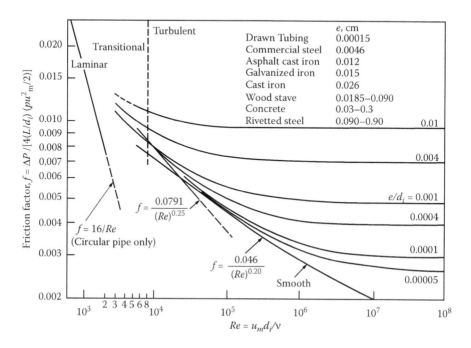

FIGURE 4.1
Friction factor for fully developed flow inside a circular duct. (From Moody, L. F., *Trans. ASME*, 66, 671–684, 1994. With permission.)

and experimental data result in a simple relationship between f and Re, independent of the surface roughness:

$$f = \frac{16}{Re} \tag{4.5}$$

The transition from laminar to turbulent flow is somewhere in the neighborhood of a Reynolds number from 2,300 to 4,000.

The f versus Re relation for smooth tubes in turbulent flow has a slight curvature on a log–log plot. A few recommended correlations for turbulent flow in smooth pipes are given in Table 4.1.[1] The two approximations shown in Figure 4.1 for turbulent flow are

$$f = 0.046\, Re^{-0.2} \quad \text{for} \quad 3 \times 10^4 < Re < 10^6 \tag{4.6}$$

and

$$f = 0.079\, Re^{-0.25} \quad \text{for} \quad 4 \times 10^3 < Re < 10^5 \tag{4.7}$$

f can be read from the graph, but the correlations for f are useful for computer analysis of a heat exchanger, and they also show the functional relationships of various quantities.

For fully developed flow in a tube, a simple force balance yields (Figure 4.2)

$$\Delta p\, \frac{\pi}{4}\, d_i^2 = \tau_w \left(\pi d_i L\right) \tag{4.8}$$

TABLE 4.1

Turbulent Flow Isothermal Fanning Friction Factor Correlations for Smooth Circular Ducts[3]

Source	Correlation[a]	Limitations
Blasius	$f = \dfrac{\tau_w}{\rho u_m^2/2} = 0.0791\, Re^{-0.25}$	$4 \times 10^3 < Re < 10^5$
Drew, Koo, and McAdams	$f = 0.00140 + 0.125\, Re^{-0.32}$	$4 \times 10^3 < Re < 5 \times 10^6$
Karman–Nikuradse	$\dfrac{1}{\sqrt{f}} = 1.737\, \ln\left(Re\sqrt{f}\right) - 0.4$	$4 \times 10^3 < Re < 3 \times 10^6$
	or	
	$\dfrac{1}{\sqrt{f}} = 4\, \log_{10}\left(Re\sqrt{f}\right) - 0.4$	
	approximated as $f = 0.046\, Re^{-0.2}$	$3 \times 10^4 < Re < 10^6$
Filonenko	$f = (3.64\, \log_{10} Re - 3.28)^{-2}$	

[a] Properties are evaluated at the bulk temperature.

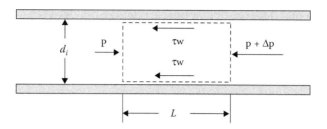

FIGURE 4.2
Force balance of a cylindrical fluid element within a pipe. (From Rohsenow, W. M., *Heat Exchangers: Thermal-Hydraulic Fundamentals and Design*, Hemisphere, New York, 1981. With permission.)

which may be combined with Equation 4.3 to get an equivalent form for the friction factor

$$f = \frac{\tau_w}{\rho u_m^2/2} \qquad (4.9)$$

Equations 4.9 and 4.3 are valid for either laminar or turbulent flows, provided that the flow is fully developed within the length L.

4.2.2 Noncircular Cross-Sectional Ducts

A duct of noncircular cross section is not geometrically similar to a circular duct; hence, dimensional analysis does not relate the performance of these two geometrical shapes. However, in turbulent flow, f for noncircular cross sections (annular spaces, rectangular and triangular ducts, etc.) may be evaluated from the data for circular ducts if d_i is replaced by the hydraulic diameter, D_h, defined by Equation 3.14:

$$D_h = 4\frac{A_c}{P_w} = \frac{4(\text{net free flow area})}{\text{wetted perimeter}} \qquad (4.10)$$

Using the hydraulic diameter in turbulent flow gives f values within about ±8% of the measured values.[2]

The hydraulic diameter of an annulus of inner and outer diameters D_i and d_o, respectively, is

$$D_h = \frac{4(\pi/4)(D_i^2 - d_o^2)}{\pi(D_i + d_o)} = (D_i - d_o) \qquad (4.11)$$

For a circular duct, Equation 4.10 reduces to $D_h = d_i$.

In using the correlations in Table 4.1, the effect of property variations has been neglected. When these correlations are applied to the design of heat exchange equipment with large temperature differences between the heat transfer surface and fluid, the friction coefficient must be corrected according to Equations 3.21b and 3.22b.

The transition Reynolds number for noncircular ducts ($\rho u_m D_h/\mu$) is also found to be approximately 2,300, as it is for circular ducts.

For laminar flows, however, the results for noncircular cross sections are not universally correlated. In a thin annulus, the flow has a parabolic distribution perpendicular to the wall and has this same distribution at every circumferential position. If this flow is treated as a flow between two parallel flat plates separated by a distance b, one obtains

$$\frac{\Delta p}{\Delta x} = \frac{12\mu u_m}{b^2} \tag{4.12}$$

Here $D_h = 2b$ and Equation 4.12 can be written in the form

$$f = \frac{24}{Re} \tag{4.13}$$

with D_h replacing d_i in the definitions of f and Re. This equation is different from Equation 4.5, which applies to laminar flow in circular ducts.

Flow in a rectangular duct (dimensions $a \times b$) in which $b \ll a$ is similar to this annular flow. For rectangular ducts of other aspect ratios (b/a)

$$f = \frac{16}{\phi Re} \tag{4.14}$$

where

$$D_h = \frac{4ab}{2(a+b)} \tag{4.15}$$

and ϕ is given in Figure 4.3.[3,4]

For laminar flow in ducts of triangular and trapezoidal cross sections, Nikuradse[4] showed that f is approximated by $16/Re$ with D_h given by Equation 4.10, and transition occurs approximately at $Re = 2,300$.

The frictional pressure drop for flow through a duct of length L is generally expressed as

$$\Delta p = 4f \frac{L}{D_h} \frac{\rho u_m^2}{2} \tag{4.16a}$$

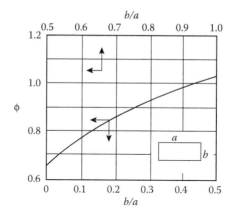

FIGURE 4.3
Values of ϕ for rectangular ducts.

or

$$\Delta p = 4f \frac{L}{D_h} \frac{G^2}{2\rho} \qquad (4.16b)$$

where $G = u_m \rho$ is referred to as the mass velocity.

The exact form of Equation 4.16 depends on the flow situation. The pressure drop experienced by a single-phase fluid in the tubes of a shell-and-tube heat exchanger must include the effects of all the tubes. The pressure drop for all the tubes can be calculated by

$$\Delta p_t = 4f \frac{LN_p}{D_h} \frac{G^2}{2\rho} \qquad (4.17)$$

where N_p is the number of tube passes and $D_h = d_i$. The fluid will experience additional pressure losses due returns at the end of the exchangers. Experiments show that the return pressure loss is given by[5]

$$\Delta p_r = 4N_p \frac{\rho u_m^2}{2} \qquad (4.18)$$

where N_p is the number of tube passes, $4N_p$ is the return pressure loss coefficient, and $\rho u_m^2/2$ is called the dynamic head.

The pumping power expenditure that is required to circulate each stream to flow through its heat exchange passage is as important as the heat transfer characteristics of a heat exchanger. The pumping power expenditure is proportional to the total pressure drop, Δp, that is experienced by the stream across the passage. For an incompressible stream with a mass flow rate $\dot{m}$, the power required by an adiabatic pump is[6]

$$W_p = \frac{1}{\eta_p} \frac{\dot{m}}{\rho} \Delta p \qquad (4.19)$$

In this expression, Δp is the pressure rise through the pump, ρ is the liquid density, and η_p is the isentropic efficiency of the pump. The group $\dot{m}\Delta p/\rho$ is the isentropic (minimum) power requirement.

4.3 Pressure Drop in Tube Bundles in Crossflow

A circular tube bundle is one of the most common heat transfer surfaces, especially in shell-and-tube heat exchangers. The two most common tube arrangements are staggered and inline, both of which are shown in Figure 3.2. Inline tubes form rectangles with the centers of their cross sections, where staggered tubes form isosceles triangles.

Bundles are characterized by the cylinder diameter d_o and the ratios of pitches to the tube diameter in the transverse $X_t^* = X_t/d_o$, longitudinal $X_l^* = X_l/d_o$, or diagonal $X_d^* = X_d/d_o$ directions. An enormous amount of work has been published on pressure drop across the tube bundles and on heat transfer performance. The experimental results obtained by Zukauskas[7,8] with air, water, and several oils will be presented here.

The pressure drop of a multirow tube bundle is given by

$$\Delta p = \left(\frac{Eu}{\chi}\right) \chi \frac{1}{2} \rho u_0^2 \cdot n \qquad (4.20)$$

where the Euler number, $Eu = 2\Delta p/(\rho u_0^2 \cdot n)$, is defined per tube row; the velocity, u_o, is the mean velocity in the minimum intertube spacing; and n is the number of tube rows counted in the flow direction.

The charts of average Euler number per tube row for multirow inline bundles are presented in Figure 4.4 as a function of $Re = u_o d_o/\mu$ and X_t^* (= X_l^*). For other tube pitches $(X_l^* \neq X_t^*)$, a correction factor x is determined from the inset of Figure 4.4, and then Eu/χ is determined from the main part of Figure 4.4 to find Eu for a specified inline tube bundle.

The charts of average Euler number per tube row of multirow staggered tube bundles with 30° tube layout $(X_t = X_d)$ and $X_l = (\sqrt{3}/2)X_t$ are presented in Figure 4.5 as a function of Reynolds number and X_t^*. For other tube

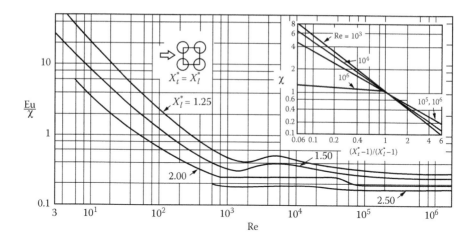

FIGURE 4.4
The hydraulic drag coefficient for inline bundles for $n > 9$. (From Zukauskas, A., *Handbook of Convective Heat Transfer*, John Wiley & Sons, New York, 1987. With permission.)

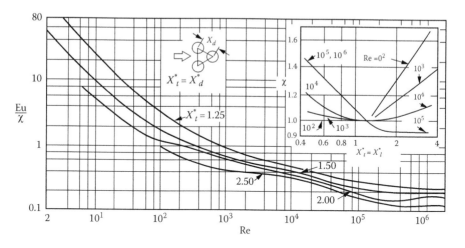

FIGURE 4.5
The hydraulic drag coefficient for staggered bundles for $n > 9$. (From Zukauskas, A., *Handbook of Convective Heat Transfer*, John Wiley & Sons, New York, 1987. With permission.)

pitches $[X_l \neq (\sqrt{3}/2)X_t]$, the correction factor χ is obtained from the upper right inset of Figure 4.5. The Reynolds number, Re, is based on the average velocity through the narrow cross section formed by the array, that is, the maximum average velocity u_o:

$$Re = \frac{u_o d_o \rho}{\mu} \quad (4.21)$$

In the case of inline tubes, the conservation of mass gives

$$u_\infty X_t = u_o (X_t - d_o) \qquad (4.22)$$

For a staggered arrangement, the minimum cross-sectional area may be either between two tubes in a row or between one tube and a neighboring tube in the succeeding row (see Example 3.5 or Problem 4.9).

Equation 4.20 and Figures 4.4 and 4.5 are valid for $n > 9$, and these two charts refer to isothermal conditions. They also apply to nonisothermal flows if the fluid physical properties are evaluated at the bulk mean temperature and a correction is applied to account for variable fluid properties for liquids as follows:

$$Eu_b = Eu \left(\frac{\mu_w}{\mu_b} \right)^p \qquad (4.23)$$

Here, Eu_b is the Euler number for both heating and cooling, Eu is the Euler number for isothermal conditions, and μ_w and μ_b are the liquid dynamic viscosities at the wall temperature and bulk mean temperature, respectively. For $Re > 10^3$, $p \approx 0$. For $Re \leq 10^3$,

$$\Delta p = -0.0018\, Re + 0.28 \quad \text{for liquid heating} \qquad (4.24)$$

$$\Delta p = -0.0026\, Re + 0.43 \quad \text{for liquid cooling} \qquad (4.25)$$

Gaddis and Gnielinski[9] also suggested curves and formulas for the pressure drop for inline and staggered tube bundles.

4.4 Pressure Drop in Helical and Spiral Coils

Helical and spiral coils are curved pipes which are used as curved tube heat exchangers in various applications such as dairy and food processing, refrigeration and air conditioning industries, and in medical equipment such as kidney dialysis machines. Nomenclature for a helical coil and a spiral are given in Figure 3.4.

Experimental and theoretical studies show that coiled tube friction factors are higher than those in a straight tube for a given Reynolds number. In this section, friction factors for laminar and turbulent flows in heat exchanger design application will be given. Various studies have been conducted, and an up-to-date review of this subject was presented by Shah and Joshi.[10]

4.4.1 Helical Coils—Laminar Flow

Srinivasan et al.[11] proposed the following correlation for laminar flow in terms of the Dean number for their experimental data for several coils ($7 < R/a < 104$):

$$\frac{f_c}{f_s} = \begin{cases} 1 & \text{for } De < 30 \\ 0.419 De^{0.275} & \text{for } 30 < De < 300 \\ 0.1125 De^{0.5} & \text{for } De > 300 \end{cases} \quad (4.26)$$

where $De = Re\sqrt{(a/R)}$, and c and s stand for curved duct and straight duct, respectively.

Some correlations also include R/a as a separate term. Manlapaz and Churchill[12] reviewed available theoretical and experimental studies on laminar flow and recommended the following correlation:

$$\frac{f_c}{f_s} = \left[\left(1.0 + \frac{0.18}{\left[(1+35/De)^2\right]^{0.5}}\right)^m + \left(1.0 + \frac{a/R}{3}\right)^2 \left(\frac{De}{88.33}\right)\right]^{0.5} \quad (4.27)$$

where $m = 2$ for $De < 20$, $m = 1$ for $20 < De < 40$, and $m = 0$ for $De > 40$.

The above two correlations are in excellent agreement for $R/a > 7$. If $R/a < 7$, the correlation expressed in Equation 4.27 is recommended.[10]

Kubair and Kuloor[13] proposed the following correlation to account for the effect of temperature-dependent fluid properties (for liquid viscosity):

$$\frac{f_c}{f_{cp}} = 0.91 \left(\frac{\mu_w}{\mu_b}\right)^{0.25} \quad (4.28)$$

where f_{cp} is the friction coefficient for constant properties.

4.4.2 Spiral Coils—Laminar Flow

Kubair and Kuloor[14–16] and Srinivasan et al.[11] have measured friction factors for different spirals. For design purposes, the following correlation[10,11] is recommended:

$$f_c = \frac{0.63\left(n_2^{0.7} - n_1^{0.7}\right)^2}{Re^{0.6}(b/a)^{0.3}} \quad (4.29)$$

for $500 < Re(b/a) < 20{,}000$ and $7.3 < b/a < 15.5$

where n_1 and n_2 are the number of turns from the origin to the start and the end of a spiral, respectively.

4.4.3 Helical Coils—Turbulent Flow

Several experimental studies have proposed correlations to calculate the friction factor for turbulent flow in a helical coil. Ito[17] obtained the following correlation:

$$f_c \left(\frac{R}{a}\right)^{0.5} = 0.00725 + 0.076 \left[Re \left(\frac{R}{a}\right)^{-2} \right]^{-0.25} \quad (4.30)$$

$$\text{for } 0.034 < Re \left(\frac{R}{a}\right)^{-2} < 300$$

As a result of extensive friction factor data, Srinivasan et al.[11] proposed the following correlation:

$$f_c \left(\frac{R}{a}\right)^{0.5} = 0.0084 \left[Re \left(\frac{R}{a}\right)^{-2} \right]^{-0.2} \quad (4.31)$$

$$\text{for } Re \left(\frac{R}{a}\right)^{-2} < 700 \text{ and } 7 < \frac{R}{a} < 104$$

These two correlations for smooth coils are in good agreement with each other and also show good agreement with experimental data for air and water (within ±10%).[18,19]

For temperature-dependent fluid properties, Rogers and Mayhew[19] proposed the following relation:

$$\frac{f_c}{f_{cp}} = \left(\frac{Pr_b}{Pr_w}\right)^{-0.33} \quad (4.32)$$

where f_c is calculated using Ito's correlation, Equation 4.30.

4.4.4 Spiral Coils—Turbulent Flow

Kubair and Kuloor[14] and Srinivasan et al.[11] measured friction factors in five spirals for water and fuel-oil, and their experimental correlation is

$$f_c = \frac{0.0074 \left(n_2^{0.9} - n_1^{0.9}\right)^{1.5}}{\left[Re(b/a)^{0.5} \right]^{0.2}} \quad (4.33)$$

4.5 Pressure Drop in Bends and Fittings

4.5.1 Pressure Drop in Bends

Bends are used in heat exchanger piping circuits and in turbulent heat exchangers. The schematical diagram of bend geometry is shown in Figure 3.5. Friction factor correlations in smooth bends with circular cross section for laminar flow of Newtonian fluids are given in this section.

The total pressure drop in a bend is the sum of the frictional head loss due to the length of the bend, head loss due to curvature, and head loss due to excess pressure drop in the downstream pipe because of the velocity profile distortion.

Ito[20] defined a total loss coefficient K as

$$\Delta p = K \frac{\rho u_m^2}{2} \tag{4.34}$$

where

$$K = \frac{4 f_c L}{D_h} = \frac{4 f L}{D_h} + K^* \tag{4.35}$$

and f_c is the bend friction factor, f is the friction factor for a straight pipe at the Reynolds number in the bend, and K^* represents a combined loss coefficient other than friction loss.

Here, f is given by

$$f = 0.0791 \, Re^{-0.25} \quad \text{for} \quad 10^4 < Re < 10^5 \tag{4.36}$$

and

$$f = 0.0008 + 0.005525 \, Re^{-0.237} \quad \text{for} \quad 10^5 < Re < 10^7 \tag{4.37}$$

Idelchik[21] reported the following correlation to calculate laminar friction factors in smooth bends of any degree, $\phi < 360°$:

$$f_c = 5 \, Re^{-6.5} (R/a)^{-0.175} \quad \text{for} \quad 50 < De \leq 600 \tag{4.38}$$

$$f_c = 2.6 \, Re^{-0.55} (R/a)^{-0.225} \quad \text{for} \quad 600 < De \leq 1400 \tag{4.39}$$

$$f_c = 1.25 \, Re^{-0.45} (R/a)^{-0.275} \quad \text{for} \quad 1400 < De \leq 5000 \tag{4.40}$$

For turbulent flow, Ito[20] obtained extensive experimental data for 45°, 90°, and 180° bends for different values of R/a, and he proposed the following correlation for $2 \times 10^4 < Re < 4 \times 10^5$:

$$K = \begin{cases} 0.00873 \, B\phi \, f_c (R/a) & \text{for } Re(R/a)^{-2} < 91 \\ 0.00241 \, B\phi \, Re^{-0.17} (R/a)^{0.84} & \text{for } Re(R/a)^{-2} > 91 \end{cases} \quad (4.41)$$

where ϕ is a bend angle in degrees and f_c is the curved tube friction factor obtained from

$$f_c (R/a)^{0.5} = 0.00725 + 0.076 \left[Re(a/R)^2 \right]^{-0.25} \quad (4.42)$$

for $0.034 < Re \, (a/R)^2 < 300$, and B is a numerical constant which is determined from the following relations:
For $\phi = 45°$,

$$B = 1 + 14.2(R/a)^{-1.47} \quad (4.43)$$

For $\phi = 90°$,

$$B = \begin{cases} 0.95 + 17.2(R/a)^{-1.96} & \text{for } R/a < 19.7 \\ 1 & \text{for } R/a > 19.7 \end{cases} \quad (4.44)$$

For $\phi = 180°$,

$$B = 1 + 116(R/a)^{-4.52} \quad (4.45)$$

For turbulent flow, Powle[22] also recommended the use of the following loss coefficient of K^* in Equation 4.35:

$$K^* = B(\phi) \left[0.051 + 0.38(R/a)^{-1} \right] \quad (4.46)$$

where

$$B(\phi) = \begin{cases} 1 & \text{for } \phi = 90° \\ 0.9 \sin \phi & \text{for } \phi \leq 70° \\ 0.7 + 0.35 \sin(\phi/90) & \text{for } \phi \geq 70° \end{cases} \quad (4.47)$$

It is recommended that Powle's correlation be used for other values of ϕ rather than $\phi = 45°, 90°$, and $180°$.

Friction factors for bends with rectangular cross sections and non-Newtonian fluids are also given by Shah and Joshi.[10]

4.5.2 Pressure Drop in Fittings

In the absence of irreversibility, the energy balance per unit mass for flow requires that $p/\rho + u^2/2 + gz$ must be conserved. In real fluid flows, some of the mechanical energy is converted into heat due to friction in the fluid itself. This loss of mechanical energy can be expressed as a loss in the total pressure as

$$\left(p_1 + \frac{\rho}{2}u_1^2 + \rho g z_1\right) - \left(p_2 + \frac{\rho}{2}u_2^2 + \rho g z_2\right) = K\frac{1}{2}\rho u_m^2 \qquad (4.48a)$$

The pressure drop in fittings is usually given as the equivalent length in pipe diameters of straight pipe, L_e/d_i, which gives the same pressure drop as the piece of fittings in question (Table 4.2), and the pressure drop is calculated using Equation 4.3. The pressure drop can also be calculated as a total loss coefficient K, by which the dynamic head

$$\frac{\rho u_m^2}{2} \qquad (4.48b)$$

has to be multiplied to give the pressure drop (Table 4.3).[23]

Hooper[24] also compiled available laminar and turbulent flow pressure drop data for standard and long radius elbows (45°, 90°, 180°), miter bends, tees, and valves. The compiled data were presented as the following single correlation:

$$K^* = K_1 Re^{-1} + K_\infty(1 + 0.5a^*) \qquad (4.49)$$

where a^* is the fitting radius in inches. The values of K_1 and K_∞ for various fittings are given by Shah and Joshi.[10]

Example 4.1

Consider a piping circuit of a heat exchanger. In the circuit there are four 90° standard elbows, three close pattern return bends, 2 check valves (clearway swing type), 2 angle valves (with no obstruction in flat type seat), and three gate valves (conventional wedge type); the valves are fully open. The straight part of the circuit pipe is 150 m, and water at

TABLE 4.2
Equivalent Length in Pipe Diameters (L_e/d_i) of Various Valves and Fittings[a]

Product	Description of Product		Equivalent Length in Pipe Diameters, L_e/d_i
Angle Valves	Conventional with no obstruction in flat-, bevel-, or plug-type seat	Fully open	145
	Conventional with wing or pin-guided disc	Fully open	200
Gate Valves	Conventional wedge disc, double disc, or plug disc	Fully open	13
		Three-quarters open	35
		One-half open	160
Check Valves	Conventional swing	Fully open	135
	Clearway swing	Fully open	50
Fittings	90° standard elbow		30
	45° standard elbow		16
	90° long radius elbow		20
	90° street elbow (one male & one female)		50
	45° street elbow (one male & one female)		26
	Square corner elbow		57
	Standard tee with flow through run		20
	Standard tee with flow through branch		60
	Close pattern return bend		50

[a] Minimum calculated pressure drop (psi) across the valve to provide sufficient flow to lift the disc fully. (Courtesy of The Crane Co.)

50°C flows with a velocity of 4 m/s. The pressure drop through the heat exchanger is 12 kPa. Nominal pipe size is 2 in. (ID = 0.052 m).

1. Calculate the total pressure drop in the system.
2. Calculate the mass flow rate and the pumping power (kW), assuming the isentropic efficiency of the pump to be 0.8.

Solution

The pressure drop can be calculated using Equation 4.3:

$$\Delta p = 4f \frac{L}{d_i} \rho \frac{u_m^2}{2}$$

TABLE 4.3
Pressure Drop Δp due to Friction in Fittings, Valves, Bends, Contraction, and Enlargement

Sudden Contraction			Sudden Enlargement
$D/D_0 = 0.0$	0.5	0.75	$\Delta p_c = \rho \dfrac{V^2 - V_0^2}{2}$
$L_e/D = 25$	20	14	
$k = 0.5$	0.4	0.3	

Standard Tee			Standard Elbow	Sharp Elbow
			45°	
60	70	46	$L_e/D = 15$	60
1.3	1.5	1.0	$k = 0.3$	1.3

90° Bend				180° Bend	
				Small radius	Large radius
$R/D = 0.5$	1.0	2.0	4.0	8.0	
$L_e/D = 36$	16.5	10	10	14.5	
$R/D = 0.5$	1.0	2.0	4.0	8.0	
$L_e/D = 50$	23	20	26	35	$L_e/D = 75$	$L_e/D = 50$
				$k = 1.7$	$k = 1.2$

Gate Valve

Opening	L_e/D	K
Full	7	0.13
3/4	40	0.8
1/2	200	3.8
1/4	800	15

Diaphragm Valve

Opening	L_e/D	K
Full	125	2.3
3/4	140	2.6
1/2	235	4.3
1/4	1,140	21

Check valve fully open

	L_e/D	K
Hinged	110	2
Disk	500	10
Ball	3,500	65

Globe Valve

Opening	L_e/D	K
Full	330	6
Half	470	8.5

Fully $L_e/D = 170$
Open $k = 3$

Plug Cock

α	L_e/D	K
5	2.7	0.05
10	16	0.29
20	85	1.56
40	950	17.3
60	11,200	206

Water meter

	L_e/D	K
Wheel	300	6
Disk	400	8
Piston	600	12

Note: Given as length of straight pipe, L_e, with the same pressure drop, or as factor K in the equation $\Delta p = K\rho V^2/2$. For further details, the reader is referred to the literature.[23]

From Table 4.2, the following equivalent length in pipe diameters can be obtained:

Pipe	L_e/d_i
90° standard elbows	30
Close pattern return bends	50
Check valve clearway swing	50
Angles valves	145
Gate valves	13

Properties of water at 50°C, from Appendix B (Table B.2), are

$$c_p = 4181.2 \text{ J}/(\text{kg} \cdot \text{K}) \qquad k = 0.640 \text{ W}/(\text{m} \cdot \text{K})$$
$$\rho = 987.9 \text{ kg/m}^3 \qquad Pr = 3.62$$
$$\mu = 5.531 \times 10^{-4} \text{ Pa} \cdot \text{s}$$

with the following Reynolds number:

$$Re = \frac{u_m d_i \rho}{\mu} = \frac{4 \times 0.052 \times 987.9}{5.531 \times 10^{-4}} = 371512$$

Therefore, the flow is turbulent. The friction coefficient can be calculated using one of the correlations given in Table 4.1. Equation 4.6 is used here:

$$f = 0.046(Re)^{-0.2}$$
$$f = 0.046(371512)^{-0.2} = 0.00354$$

$$\Delta p = 4f \frac{L}{d_i} \rho \frac{u_m^2}{2} = 4 \times 0.00354 \times \frac{L}{d_i} \times 987.9 \times \frac{16}{2} = \Delta p_1 = 111.91 \left(\frac{L}{d_i} \right)$$

For the minor losses,

$$\Delta p_1 = 111.91 \sum \left(\frac{L_e}{d_i} \right)$$
$$= 111.91 [4 \times 30 + 3 \times 50 + 2 \times 50 + 2 \times 145 + 3 \times 13] = 78225 \text{ Pa}$$

For the straight part of the pipe,

$$\Delta p_2 = 4f \left(\frac{L}{d_i} \right) \rho \frac{u_m^2}{2} = 111.91 \left(\frac{L}{d_i} \right) = 111.91(150/0.052) = 322817 Pa$$

Then, the total pressure drop is

$$\Delta p_t = \Delta p_1 + \Delta p_2 + \Delta p_{HE} = 78.225 + 322.817 + 12 = 413.042 \text{ kPa}$$

The pumping power can be calculated from Equation 4.19:

$$\dot{m} = \frac{\pi}{4} d_i^2 u_m \rho = \frac{\pi}{4} \times (0.052)^2 \times 4 \times 987.9 = 8.4 \text{ kg/s}$$

$$W_P = \frac{\dot{m} \Delta p_t}{\rho \eta_p} = \frac{8.4 \times 413.042}{987.9 \times 0.80} = 4.36 \text{ kW}$$

4.6 Pressure Drop for Abrupt Contraction, Expansion, and Momentum Change

In heat exchangers, streams can experience a sudden contraction followed by a sudden enlargement when flowing in and out of a heat exchanger core, such as is the case in the tube side of shell-and-tube heat exchangers and compact tube- and plate-fin heat exchangers. Pressure drop is expressed in terms of the contracting loss coefficient, K_c, the enlargement loss coefficient, K_e, and the dynamic head; the values of K_c and K_e are given by Kays and London.[26] Therefore, the total pressure drop of the stream is

$$\Delta p_i = K_c \frac{1}{2} \rho u^2 + \Delta p_s + K_e \frac{1}{2} \rho u^2 \tag{4.50}$$

where Δp_s is the straight-duct pressure drop. The velocity represents the mean velocity in the narrowest portion of the flow channel. It is assumed that the density is constant and the contraction ratio equals the enlargement ratio. It should be noted that the pressure rises during enlargement.

For an incompressible flow, Δp_s is given by Equation 4.16. The incompressibility assumption is a good approximation for liquids such as water and oil, but it is not adequate for gases. As a result of change in density in heating, the momentum increases from $\dot{m}u_i$ to $\dot{m}u_o$. Therefore, the change in longitudinal momentum must be balanced by the pressure difference applied due to the acceleration of the stream between the inlet and the outlet of the heat exchanger passage:

$$\dot{m}(u_o - u_i) = \Delta p_a A_c \tag{4.51}$$

or

$$\Delta p_a = G^2 \left(\frac{1}{\rho_a} - \frac{1}{\rho_i} \right) \tag{4.52}$$

The total pressure drop in the straight section, Δp_s, is then

$$\Delta p_s = 4f \frac{L}{D_h} \rho \frac{u_m^2}{2} + G^2 \left(\frac{1}{\rho_o} - \frac{1}{\rho_i} \right) \tag{4.53}$$

The density in the wall frictional effect is the average density between the inlet and outlet.

It should be noted that when the gas is cooled, Δp_a is negative, which is the case of the deceleration of the stream, which decreases the overall pressure drop, Δp, across the passage.

4.7 Heat Transfer and Pumping Power Relationship

The fluid pumping power, p, is proportional to the pressure drop in the fluid across a heat exchanger. It is given by

$$W_p = \frac{\dot{m}\Delta p}{\rho \eta_p} \quad (4.54)$$

where η_p is the pump or fan efficiency ($\eta_p = 0.80$–0.85).

Frequently, the cost in terms of increased fluid friction requires an input of pumping work greater than the realized benefit of increased heat transfer.

In the design of heat exchangers involving high-density fluids, the fluid pumping power requirement is usually quite small relative to the heat transfer rate, and, thus, the friction power expenditure (i.e., pressure drop) has hardly any influence on the design. However, for gases and low-density fluids, and also for very high viscosity fluids, pressure drop is always of equal importance to the heat transfer rate and it has a strong influence on the design of heat exchangers.

Let us consider single-phase side passage (a duct) of a heat exchanger where the flow is turbulent and the surface is smooth. Forced convection correlation and the friction coefficient can be expressed as[24,25]

$$St \cdot Pr^{2/3} = \phi_h(Re) \quad (4.55)$$

and

$$f = \phi_f(Re) \quad (4.56)$$

Equation 4.55 gives the heat transfer coefficient

$$h = (\mu c_p) Pr^{-2/3} \frac{Re}{D_h} \phi_h \quad (4.57)$$

h can be expressed in W/m² · K and it can be interpreted as the heat transfer power per unit surface area.

If the pressure drop through the passage is Δp and the associated heat transfer surface area is A, the pumping power per unit heat transfer area (W/m²) is given by

$$\frac{W_p}{A} = \frac{\Delta p\, \dot{m}}{\rho \eta_p}\frac{1}{A} \tag{4.58}$$

By substituting Δp from Equation 4.16 into Equation 4.58 and noting that $D_h = 4A_c/P_w$,

$$\frac{W_p}{A} = 8\left(\frac{\mu^3}{\eta_p \rho^2}\right)\left(\frac{1}{D_h}\right)^3 Re^3 \phi_f \tag{4.59}$$

If it is assumed for simplicity that the friction coefficient is given by Equation 4.5 and the Reynolds analogy is applicable, then

$$\phi_h = 0.023\, Re^{-0.2} \tag{4.60}$$

$$\phi_f = 0.046\, Re^{-0.2} \tag{4.61}$$

Equations 4.60 and 4.61 approximate the typical characteristics of fully developed turbulent flow in smooth tubes. Substituting these relations into Equations 4.57 and 4.59 and combining them to eliminate the Reynolds number, the pumping power per unit heat transfer area (W/m²) is obtained as

$$\frac{W_p}{A} = \frac{C h^{3.5} \mu^{1.83} (D_h)^{0.5}}{k^{2.33} c_p^{1.17} \rho^2 \eta_p} \tag{4.62}$$

where $C = 1.2465 \times 10^4$.

As can be seen from Equation 4.62, the pumping power depends strongly on fluid properties as well as on the hydraulic diameter of the flow passage. Some important conclusions can be drawn from Equation 4.62 (see Table 4.4):

1. With a high-density fluid such as a liquid, the heat exchanger surface can be operated at large values of h without excessive pumping power requirements.
2. A gas with its very low density results in high values of pumping power for even very moderate values of heat transfer coefficient.
3. A large value of viscosity causes the friction power to be large even though density may be high. Thus, heat exchangers using oils must

TABLE 4.4

Pumping Power Expenditure for Various Fluid Conditions

Fluid Conditions	Power Expenditure, W/m²
Water at 300 K	
$h = 3{,}850$ W/m² K	3.85
Ammonia at 500 K, atmospheric pressure	
$h = 100$ W/m² K	29.1
$h = 248$ W/m² K	697
Engine oil at 300 K	
$h = 250$ W/m² K	0.270×10^4
$h = 500$ W/m² K	3.06×10^4
$h = 1{,}200$ W/m² K	65.5×10^4

$\eta_p = 80\%$, $D_e = 0.0241$ m

be designed for relatively low values of h to hold the pumping power within acceptable limits.

4. The thermal conductivity, k, also has a very strong influence and, therefore, for liquid metals with very large values of thermal conductivity, the pumping power is seldom of significance.

5. Small values of hydraulic diameter, D_h, tend to minimize the pumping power.

Nomenclature

A	area, m²
A_c	heat exchanger passage area, m²
a	dimension of rectangular duct, or tube inside radius, m
b	dimension of rectangular duct, m
c_p	specific heat at constant pressure, J/(kg · K)
De	Dean number, Re $(0.5\, D_h/R)^{1/2}$
D_e	equivalent diameter, m
D_h	hydraulic diameter of the duct, m
d	diameter, m
e	tube surface roughness, m
f	Fanning friction factor, $\tau_w/(\rho u_w^2/2)$
G	mass velocity, kg/m² · s
h	heat transfer coefficient, W/(m² · K)
k	thermal conductivity, W/(m · K)

K	modified pressure loss coefficient, $4f_c L/D_h$
K^*	pressure loss coefficient, defined in Equation 4.35
L	tube length or bend length, m
$\dot{m}$	mass flow rate, kg/s
N_p	number of tube passes
p	pressure, Pa
Pr	Prandtl number, $c_p\mu/k = \nu/a$
P	wetted perimeter, m
P_T	pitch, m
R	radius of curvature, m
Re	Reynolds number, $\rho u_m d/\mu$
r	radius, m
St	Stanton number, $h/\rho c_p u$
T	temperature, °C, K
ΔT_m	mean temperature difference, °C, K
t	wall thickness, m
u_m	mean velocity, m/s
W_p	pumping power, W
x	axial distance along the axis of a bend measured from the bend inlet, m

Greek Symbols

η_p	pump efficiency
μ	dynamic viscosity, kg/ms
ν	kinematic viscosity, m²/s
ρ	density, kg/m³
τ	shear stress, Pa
ϕ	parameter, function of or bend angle, deg

Subscripts

a	acceleration
c	curved coil or duct
i	inner, inside
m	mean
o	outer, outside, overall
p	pump
r	return
s	straight tube
t	tube, thermal, total
w	wall, wetted
x	axial value
1	inlet
2	outlet

PROBLEMS

4.1. Consider the flow of 20°C water through a circular duct with an inner diameter of 2.54 cm. The average velocity of the water is 4 m/s. Calculate the pressure drop per unit length ($\Delta p/L$).

4.2. Water at 5°C flows through a parallel-plate channel in a flat-plate heat exchanger. The spacing between the plates is 2 cm, and the mean velocity of the water is 3.5 m/s. Calculate the pressure drop per unit length in the hydrodynamically fully developed region.

4.3. It is proposed that the stationary diesel exhaust gases (CO_2) be used for heating air by the use of a double-pipe heat exchanger. CO_2 flows through the annulus. From the measurements of the velocity, it is calculated that the mass flow rate of gases from the engine is 100 kg/h. The exhaust gas temperature is 600 K. The air is available at 20°C and it will be heated to 80°C at a mass flow rate of 90 kg/h. One standard tube size of 4 × 3 in. (The outer pipe has a nominal diameter of 4 in. and the inner pipe has a nominal diameter of 3 in.) in a double-pipe exchanger (hairpin) of 2 m length (copper tubes) will be used.

   a. Calculate the frictional pressure drop in the inner tube.
   b. Calculate the pressure drop (Pa) in the annulus.
   c. Calculate the pumping power for both streams.

   (Assuming the average temperature of CO_2 at 500 K and the following properties: $\rho = 1.0732$ kg/m³, $c_p = 1{,}013$ J/kg · K, $k = 0.03352$ W/m · K, $Pr = 0.702$, $\upsilon = 21.67 \times 10^{-6}$ m²/s. For air: $c_p = 1{,}007$ J/kg · K.)

4.4. Calculate the pumping power for both streams given in Problem 3.8.

4.5. Consider a shell-and-tube heat exchanger. Air with a flow rate of 1.5 kg/s at 500°C and at atmospheric pressure flows through 200 parallel tubes. Each tube has an internal diameter of 2 cm and length of 4 m. Assume that the $D/D_0 = 0.5$ data can be used for the contraction and enlargement. The outlet temperature of the cooled air is 100°C. Calculate the pressure drop for

   a. The abrupt contraction
   b. Friction
   c. Acceleration
   d. Enlargement

   and compare the total pressure drop with the frictional pressure drop.

4.6. The core of a shell-and-tube heat exchanger contains 60 tubes (single-tube pass) with an inside diameter of 2.5 cm and length of 2 m. The shell inside diameter is 35 cm. The air, which flows through the tubes at a flow rate of 1.5 kg/s, is to be heated from 100°C to 300°C.

a. Calculate the total pressure drop as the sum of the frictional pressure drop in the tube and the pressure drop due to acceleration, abrupt contraction ($K_c = 0.25$), and abrupt enlargement ($K_e = 0.27$). Is the total pressure drop smaller than the frictional pressure drop in the tube? Why?

b. Calculate the fan power needed.

4.7. Consider a tube bundle arrangement in a crossflow heat exchanger. Tubes are staggered bare tubes with a longitudinal pitch of 20 mm, transverse pitch of 24.5 mm and are 1/2 in. in diameter. The length of the heat exchanger is 0.80 m. The frontal area seen by the air stream is a 0.6 m × 0.6 m square. The air flows at 2 atm pressure with a mass flow rate of 1,500 kg/h. Assume that the mean air temperature is 200°C. Calculate the air frictional pressure drop across the core of the heat exchanger.

4.8. Assume that the tube bundle arrangements given in Problems 3.11 and 3.12 are the core of a heat exchanger. Calculate the pressure drop caused by the tubes for inline and staggered arrangements with 20 rows of tubes in the flow direction, and compare the two pressure drops and the two pumping powers.

4.9. In a crossflow heat exchanger with inline tubes, air flows across a bundle of tubes at 5°C and is heated to 32°C. The inlet velocity of air is 15 m/s. Dimensions of tubes are: $d_o = 25$ mm and $X_t = X_l = 50$ mm. There are 20 rows in the flow direction and 20 columns counted in the heat exchanger. Air properties may be evaluated at 20°C and 1 atm. Calculate the frictional pressure drop and the pumping power.

4.10. In a heat exchanger for two different fluids, power expenditure per heating surface area is to be calculated. Fluids flow through the channels of the heat exchanger. The hydraulic diameter of the channel is 0.0241 m. For

a. Air at an average temperature of 30°C and 2 atm pressure, the heat transfer coefficient is 55 W/m² · K;

b. Water at an average temperature of 30°C, the heat transfer coefficient is 3,850 W/m² · K.

Comparing the results of these two fluids, outline your conclusions.

4.11. City water will be cooled in a heat exchanger by sea water entering at 15°C. The outlet temperature of the sea water is 20°C. City water will be recirculated to reduce water consumption. The suction line of the pump has an inner diameter of 154 mm, is 22 m long, and has two 90° bends and a hinged check valve. The pipe

from the pump to the heat exchanger has an inner diameter of 127 mm, is 140 m long, and has six 90° bends. The 90° bends are all made of steel with a radius equal to the inner diameter of the pipe, $R/d = 1.0$. The heat exchanger has 62 tubes in parallel, each tube 6 m long. The inner diameter of the tubes is 18 mm. All pipes are made of drawn mild steel ($\varepsilon = 0.0445$ mm). The sea water flow rate is 120 m³/h. Assume that there is one velocity head loss at the inlet and 0.5 velocity head loss at the outlet of the heat exchanger. The elevation difference is 10.5 m.
Calculate

a. The total pressure drop in the system (kPa and m liquid head = H m)

b. The power of the sea water pump (pump efficiency $\eta = 60\%$).

Plot the pumping power as a function of the sea water flow rate (for three flow rate values: 2,000 L/min, 2,500 L/min, and 3,000 L/min).

4.12. For the situation given in Problem 4.11,

a. Plot the total pressure head vs. flow rate (m liquid head vs. L/min);

b. Change the pipe diameter before and after the pump and repeat (a), (b), and (c) in Problem 4.11;

c. Obtain the optimum pipe diameter considering the cost of pumping power (operational cost) plus the fixed charges based on capital investment for the pipe installed.

4.13. A shell-and-tube heat exchanger is used to cool 25.3 kg/s of water from 38°C to 32°C. The exchanger has one shell-side pass and two tube-side passes. The hot water flows through the tubes, and the cooling water flows through the shell. The cooling water enters at 24°C and leaves at 30°C. The shell-side (outside) heat transfer coefficient is estimated to be 5,678 W/m² · K. Design specifications require that the pressure drop through the tubes be as close to 13.8 kPa as possible, that the tubes be 18 BWG copper tubing (1.24 mm wall thickness), and that each pass be 4.9 m long. Assume that the pressure losses at the inlet and outlet are equal to one and one half of a velocity head ($\rho u_m^2/2$), respectively. For these specifications, what tube diameter and how many tubes are needed?

4.14. Water is to be heated from 10°C to 30°C at a rate of 300 kg/s by atmospheric pressure steam in a single-pass shell-and-tube heat exchanger consisting of 1-in. schedule 40 steel pipe. The surface coefficient on the steam side is estimated to be 11,350 W/m² · K. An available pump can deliver the desired quantity of water provided that the pressure drop through the pipes does not exceed

15 psi. Calculate the number of tubes in parallel and the length of each tube necessary to operate the heat exchanger with the available pump.

References

1. Moody, L. F., Friction factor for pipe flow, *Trans. ASME*, 66, 671, 1994.
2. Brundrett, E., Modified hydraulic diameter for turbulent flow, in *Turbulent Forced Convection in Channels and Bundles*, Vol. 1, Kakaç, S. and Spalding, D. B., Eds., Hemisphere, Washington, DC, 1979, 361.
3. Rohsenow, W. M., Heat exchangers — basic methods, in *Heat Exchangers: Thermal-Hydraulic Fundamentals and Design*, Kakaç, S., Bergles, A. E., and Mayinger, F., Eds., Hemisphere, Washington, DC, 1981, 429–459.
4. Kakaç, S., Shah, R. K., and Aung, W., Eds., *Handbook of SinglePhase Convective Heat Transfer*, John Wiley & Sons, New York, 1987, ch. 4 and 8.
5. Kern, D. Q., *Process Heat Transfer*, McGraw-Hill, New York, 1950.
6. Moran, J. M. and Shapiro, N. H., *Fundamentals of Engineering Thermodynamics*, 3rd ed., John Wiley & Sons, New York, 1995.
7. Zukauskas, A., Convective heat transfer in cross flow, in *Handbook of Convective Heat Transfer*, Kakaç, S., Shah, R. K., and Aung, W., Eds., John Wiley & Sons, New York, 1987, ch. 6.
8. Zukauskas, A. and Uliniskas, R., Banks of plane and finned tubes, in *Heat Exchanger Design Handbook*, Vol. 2, Schlünder, E. U., Ed., Hemisphere, Washington, DC, 1983, 2.5.3.1.
9. Gaddis, E. S. and Gnielinski, V., Pressure drop in cross flow across tube bundles, *Int. J. Eng.*, 25(1), 1, 1985.
10. Shah, R. K. and Joshi, S. D., Convective heat transfer in curved ducts, in *Handbook of Convective Heat Transfer*, Kakaç, S., Shah, R. K., and Aung, W., Eds., John Wiley & Sons, New York, 1987, ch. 5 and 10.
11. Srinivasan, P. S., Nandapurkar, S. S., and Holland, F. A., Friction for coils. *Trans. Inst. Chem. Eng.*, 48, T156, 1970.
12. Manlapaz, R. L. and Churchill, S. W., Fully developed laminar flow in a helically coiled tube of finite pitch, *Chem. Eng. Commun.*, 7, 557, 1980.
13. Kubair, V. and Kuloor, N. R., Non-isothermal pressure drop data for liquid flow in helical coils, *Indian J. Technol.*, 3, 5, 1965.
14. Kubair, V. and Kuloor, N. R., Flow of Newtonian fluids in Archimedean spiral tube coils, *Indian J. Technol.*, 4, 3, 1966.
15. Kubair, V. and Kuloor, N. R., Non-isothermal pressure drop data for spiral tube coils, *Indian J. Technol.*, 3, 382, 1965.
16. Kubair, V. and Kuloor, N.R., Heat transfer for Newtonian fluids in coiled pipes in laminar flow, *Int. J. Heat Transfer*, 9, 63, 1966.
17. Ito, H., Friction factors for turbulent flow in curved pipes, *J. Basic Eng.*, 81, 123, 1959.
18. Boyce, B. E., Collier, J. G., and Levy, J., Eds., Hole-up and pressure drop measurements in the two-phase flow of air-water mixtures in helical coils, in *Co-current Gas Liquid Flow*, Plenum Press, UK, 1969, 203.

19. Rogers, G. F. C. and Mayhew, Y. R., Heat transfer and pressure loss in helically coiled tubes with turbulent flow, *Int. J. Heat Mass Transfer*, 7, 1207, 1964.
20. Ito, H., Pressure losses in smooth bends, *J. Basic Eng.*, 82, 131, 1960.
21. Idelchik, I. E., *Handbook of Hydraulic Resistance*, 2nd ed., Hemisphere, Washington, DC, 1986.
22. Powle, U. S., Energy losses in smooth bends, *Mech. Eng. Bull. (India)*, 12(4), 104, 1981.
23. Perry, R. N. and Chilton, C. H., *Chemical Engineers' Handbook*, 5th ed., McGraw-Hill, New York, 1973.
24. Hooper, W. B., The two-K method predicts head losses in pipe fittings, *Chem. Eng.*, 24, 96–100, 1981.
25. Kakaç, S., Ed., *Boilers, Evaporators and Condensers*, John Wiley & Sons, New York, 1991.
26. Kays, W. M. and London, A. L., *Compact Heat Exchangers*, McGraw-Hill, New York, 1984.

5

Micro/Nano Heat Transfer

5.1 PART A—Heat Transfer for Gaseous and Liquid Flow in Microchannels

5.1.1 Introduction of Heat Transfer in Microchannels

With the development of miniaturized manufacturing, microdevices are successfully employed in a variety of fields such as aerospace, biomedical, and electronic industries. Since 1980s, a large number of microdevices such as microbiochips, micropumps, micromotors, microfuel cells, microheat exchangers are used widely. Generally, all devices with dimensions between 1 μm and 1 mm are called microdevices, and these microdevices can be divided into three main categories[1]:

- MEMS: Microelectromechanical systems (air bag acceleration sensors, HD reader, etc.)
- MOEMS: Microoptoelectromechanical systems (microendoscope, etc.)
- MFD: Microflow devices (microheat exchangers, micropumps, etc.)

Because of the miniaturization trend, especially in electronics, problems associated with overheating of the electronic circuits have emerged. These problems increased the demand for a space efficient and high performance heat dissipation devices. To provide a solution for high flux cooling problem, the use of convective heat transfer in microchannels is believed to be the most suitable way. However, if the flow is considered to be in microchannels, it may not be valid to use conventional techniques for heat transfer and flow characteristics. Thus, many experimental and analytical researches have been performed to understand the heat transfer and fluid flow at microscale for the development and the design of microdevices. However, there are no generalized solutions for the determination of heat transfer and flow characteristics in designing microdevices. On the other hand, the predictions of the friction coefficient and convective heat transfer in microchannels are important in the application of electronics where such channel form microheat exchangers.

In numerous fields of medicine, microheat exchangers are used particularly in minimally invasive therapy. Furthermore, based on new fabrication technologies, new micromedical instruments are feasible in a variety of usage.[2,3]

The convective heat transfer in liquid and gaseous flows should be considered separately. Experimental and theoretical results for liquid flows cannot be considered or modified for gaseous flows in microchannels because of the reason that the flow regime boundaries are significantly different as well as flow and heat transfer characteristics. The results of the gaseous flows in macrochannels are not applicable for flow in microchannels.[4]

This section aims to provide a better understanding of heat transfer and fluid flow for gases in microchannels based on several experimental, analytical, and numerical results of many researches.

5.1.2 Fundamentals of Gaseous Flow in Microchannels

Recent studies proved that heat transfer and fluid flow in microchannels cannot be explained by the conventional transport theory. The conventional transport theory is based on continuum assumption. Under the continuum assumption, matter is continuous and indefinitely divisible. The continuum assumption is often not valid for microstructures since the interaction between molecules and solid surface becomes important.[5]

Although, classical studies may help to understand the fluid flow in microstructures, there are several properties that make microfluidics a unique phenomenon. Surface to volume ratio in microstructures are greater than it is for macrostructures; so surface effects become dominant. This effects an important pressure drop and greater mass flow rate than that is predicted with the continuum theory. The velocity in microchannels is not very high due to large pressure drops. Because of small dimensions, Reynolds number becomes smaller. Moreover, the transition from laminar to turbulent flow occurs earlier; viscous effects and friction factors may also be higher than that is predicted by conventional theory. All these discrepancies affect both the fluid flow and the convective heat transfer in microchannels.[5]

For gas flow in microchannels, surface effects become more important. There is a fluid slip along the channel wall with a finite tangential velocity in rarefied gas flow. Therefore, fluid velocity cannot be considered equal to the relevant wall values as it is in macrochannels. Furthermore, there is also a temperature jump between the wall and fluid so the temperature of the fluid cannot be taken to be equal to the wall values.[6]

5.1.2.1 Knudsen Number

When the gas is at very low pressure, the interaction between gas molecules and the wall becomes as frequent as intermolecular collisions so the boundaries and the molecular structure become more effective on flow. This type of flow is known as rarefied flow, and these effects are called rarefaction effects. The Knudsen number, which is the ratio of the mean free path to the characteristic length, is used to represent rarefaction effects, and the Knudsen number is defined as[4]

$$Kn = \frac{\lambda}{L} \qquad (5.1)$$

where L is a characteristic flow dimension such as hydraulic diameter and λ is mean free molecular path, which corresponds to the distance traveled by the molecules between collision. Knudsen number is a well-established criterion to indicate whether fluid flow problem can be solved by the continuum approach.

The local value of Knudsen determines the degree of rarefaction and the degree of deviation from continuum model. Therefore, the physics of flow depends on the magnitude of Knudsen number. Flow regimes are separated according to Knudsen number as shown in Table 5.1.[5]

In continuum flow regime ($Kn \leq 0.001$) for gaseous flow in microchannels, continuum assumption which is widely used for macroscopic heat transfer problems becomes valid. In the slip flow regime ($0.001 < Kn \leq 0.1$), continuum model is applicable except in the layer next to the wall, which can be identified as Knudsen layer. For the Knudsen layer, slip boundary conditions should be considered. If the flow is in the transition regime ($0.1 < Kn \leq 10$) and continues into free molecular flow regime, molecular approach should be used. In other words, Boltzmann equation should be considered for atomic-level studies of gaseous flow in transition regime.[8]

The definition of the Reynolds number and the Mach number, which are $Re = \rho U\infty L/\mu$ and $Ma = U\infty/U_a$, where $U\infty$ is free stream velocity and $U_a = \sqrt{kRT}$ is the speed of sound in the gas. Here, k_r is the ratio of specific heats of the gas and R is the gas constant. According to the kinetic theory,

TABLE 5.1

Flow Regimes Based on the Knudsen Number[7]

Regime	Method of Calculation	Kn Range
Continuum	Navier-Stokes and energy equations with no-slip/no-jump boundary conditions	$Kn \leq 0.001$
Slip flow	Navier-Stokes and energy equations with slip/jump boundary conditions, DSMC	$0.001 < Kn \leq 0.1$
Transition	BTE, DSMC	$0.1 < Kn \leq 10$
Free molecule	BTE, DSMC	$Kn > 10$

Source: From Gad-El-Hak, M., *J. Fluids Eng.*, 121, 5–33, 1999. With permission.
DSMC = direct simulation Monte Carlo, BTE = Boltzmann transport equations.

viscosity can be expressed as a function of mean free path $\left(\mu = \rho\lambda\sqrt{2RT/\pi}\right)$. Knudsen number can also be expressed as[9]

$$Kn = \frac{Ma}{Re}\sqrt{\frac{\pi k_r}{2}} \qquad (5.2)$$

Mach number is a measure of fluid compressibility. Conventionally, the flow can be assumed incompressible if $Ma < 0.3$. However, in microdevices, density changes can be significant; therefore, the compressibility must be taken into account.

5.1.2.2 Velocity Slip

As the characteristic length of microchannels approaches to the mean free path, the gas flow becomes rarefied. Therefore, the interactions between fluid and the wall become more significant than the intermolecular collisions. When gas molecules impinge on the wall, gas molecules will be reflected after collision and the molecules leave some of their momentum and create shear stress on the wall. If specular reflection becomes, the tangential momentum remains same whereas the normal momentum will be reserved. If diffuse reflection becomes, the tangential momentum vanishes. By neglecting thermal creep, the tangential momentum balance at the wall gives the first-order slip velocity as[9,10]

$$u_s = \frac{2 - F_m}{F_m} \lambda \left(\frac{du}{dy}\right)_w \qquad (5.3)$$

where F_m is the tangential momentum accommodation coefficient and λ is mean free molecular path, which corresponds to the distance traveled by the molecules between collision. Molecules will be reflected specularly if they collide to an ideal smooth surface. Therefore, the tangential momentum of the molecules will be conserved and no shear stress will be created in the wall and thus $F_m = 0$. If diffusive reflection occurs which means that tangential momentum will be lost at the wall, F_m will equal to 1. For real surfaces, molecules reflect from the wall both diffusively and specularly. According to experiments, there is a ratio between diffusive and specular reflection (diffusive/specular) between the range of 0.2–0.8.[6]

5.1.2.3 Temperature Jump

As mentioned before, there is a finite difference between the fluid temperature at the wall and the wall temperature, and it is called temperature jump. A temperature jump coefficient is proposed to be[6]

$$T_s - T_w = c_{jump}\left(\frac{\partial T}{\partial y}\right)_w \qquad (5.4)$$

The thermal accommodation coefficient is defined as $F_T = Q_i - Q_r / Q_i - Q_w$, where Q_i is the impinging stream energy, Q_r is the energy carried by the reflected molecules, and Q_w is the energy of the molecules leaving the surface at the wall temperature.[6]

For a perfect gas, the temperature jump coefficient can be obtained as[6]

$$c_{jump} = \frac{2-F_T}{F_T}\frac{2k}{(k+1)}\frac{\lambda}{Pr} \qquad (5.5)$$

Then the temperature jump can be expressed as in the form of[6]

$$T_s - T_w = \frac{2-F_T}{F_T}\frac{2k}{(k+1)}\frac{\lambda}{Pr}\left(\frac{\partial T}{\partial y}\right)_w \qquad (5.6)$$

For slip flow, the fully developed velocity profile can be obtained from the momentum equations for laminar flow with constant thermophysical properties in a parallel-plate channel and a circular duct, respectively, as

$$u = \frac{3}{2}u_m \frac{\left[1-\left(\frac{y}{d}\right)^2 + 4Kn\right]}{1+8Kn} \qquad (5.7)$$

and

$$u = 2u_m \frac{\left[1-\left(\frac{r}{R}\right)^2 + 4Kn\right]}{1+8Kn} \qquad (5.8)$$

In the slip flow model, governing equations for fluid flow and heat transfer used are still continuum equations, but the boundary conditions are modified as given by Equations 5.3 and 5.6.

5.1.2.4 Brinkman Number

The friction between the layers of the fluid results in viscous heat generation. Although viscous heat generation can be neglected for continuum flow, it is an important factor for fluid flow in microchannels because of the large surface area to volume ratio. Brinkman number is a dimensionless parameter, which represents the importance of heat generated by viscous dissipation. The definition of Brinkman number varies with the boundary condition at the wall. When constant wall temperature boundary condition is considered, Brinkman number is given by Equation 5.9a. In addition, if constant wall heat flux boundary condition is considered, Brinkman number is defined as in the form of Equation 5.9b.

$$Br = \frac{\mu u_m^2}{k(T_w - T_i)} \tag{5.9a}$$

$$Br = \frac{\mu u_m^2}{q_w d} \tag{5.9b}$$

where $T_w - T_i$ is the wall-fluid temperature difference at a particular axial location. It measures the relative importance at viscous heating to heat conduction in the fluid along the microchannel. For heating the fluid, the value at Brinkman number is positive, and for cooling, the value is negative. For cooling the gas, the heat transfer coefficient increases.

The effects of Brinkman and Knudsen number on heat transfer are also illustrated in Table 5.2. It can be clearly seen that Nusselt numbers decreases with increasing Knudsen number for both Brinkman numbers. This situation can be explained with the increasing temperature jump.

Example 5.1

Air flows at a mean temperature of 40°C through a heated micropipe with diameter of 0.24 mm Knudsen and Brinkman numbers are 0.06 and 0.01, respectively. Calculate the fully developed heat transfer coefficient

TABLE 5.2

Nusselt Numbers for Developed Laminar Flow

	T_w = Constant		q_w = Constant		
Kn	Br = 0.00	Br = 0.01	Br = 0.00	Br = 0.01	Br = −0.01
0.00	3.6566	9.5985	4.3649	4.1825	4.564
0.02	3.4163	7.4270	4.1088	4.0022	4.2212
0.04	3.1706	6.0313	3.8036	3.7398	3.8695
0.06	2.9377	5.0651	3.4992	3.4498	3.5395
0.08	2.7244	4.3594	3.2163	3.1912	3.2419
0.10	2.5323	3.8227	2.9616	2.944	2.9784

Source: From Çetin, B., Yuncu, H., and Kakaç, S., *International Journal of Transport Phenomena*, 8, 297–315, 2007. With permission.
q_w = constant, T_w = constant, Pr = 0.7.

Micro/Nano Heat Transfer

of air for constant wall temperature and constant wall heat flux boundary conditions. Reynolds number is calculated to be 600. At 40°C, k equals to 0.067 W/m × K, μ equals to 1.912 × 10⁻⁵ N × s/m², and Pr equals to 0.719.

Solution

Properties of air, from Appendix B (Table B.1), are

$$\rho = 1.1267 \text{ kg/m}^3$$
$$\mu = 1.91 \times 10^{-5} \text{ kg}/(\text{m} \cdot \text{s})$$
$$c_p = 1.009 \text{ kJ}/(\text{kg} \cdot \text{K})$$
$$c_v = 0.7185 \text{ kJ}/(\text{kg} \cdot \text{K})$$
$$k = 0.0271 \text{ W}/(\text{m} \cdot \text{K})$$
$$\text{Pr} = 0.711$$

From Table 5.2, for given Kn and Br values, Nusselt number for the constant heat flux condition can be obtained as

$$Nu = 3.4598$$

So the heat transfer coefficient for the constant heat flux condition is

$$h = \frac{Nu \times k}{d} = \frac{3.4598 \times 0.0271}{0.24 \times 10^{-3}} = 390.66 \text{ W/m}^2\text{K}$$

From Table 5.2, for given Kn and Br values, Nusselt number for the constant wall temperature condition can be obtained as

$$Nu = 5.0651$$

So the heat transfer coefficient for the constant wall temperature condition is

$$h = \frac{Nu \times k}{d} = \frac{5.0651 \times 0.0271}{0.24 \times 10^{-3}} = 571.93 \text{ W/m}^2\text{K}$$

5.1.3 Engineering Applications for Gas Flow

As the use of microchannels becomes a suitable solution for high flux cooling problem, the fabrication and design technology of microchannels have improved. Microchannels are designed and manufactured in a variety of geometries according to requirements of the systems. The geometry of the channels is highly dependent on the fabrication technology. In literature, there are four main geometries for microchannels which are rectangular, trapezoidal, hexagonal, and circular channels (see Figure 5.1).

The flow in microchannels can be induced in two different ways. One way is to apply an external pressure gradient to microchannel, and the other way is to apply an external electric field. When pressure gradient is applied, generally the velocity profile varies across the entire cross-section area of the channel. On the other hand, if an electric field is used to induce the flow, the velocity profile is generally uniform across the entire cross section.[1]

First, brief fundamentals of single phase gaseous flow and heat transfer in microchannels is introduced, and then experimental and analytical results are analyzed to predict Nusselt numbers and friction factors, which are essential for design of microheat exchangers.

Microchannel heat exchangers have several advantages over conventional heat exchangers such as[8]

- Compact size
- High surface area density
- Enhanced heat transfer coefficients

Microchannel heat exchangers are made using materials such as silicon, aluminum, stainless steel, glass, and ceramics. Fabrication material is selected dependently with the temperature and pressure range of operation. Microchannel heat exchangers are mostly used as recuperators in microreactors, microchannel fuel cells, and microminiature refrigerators. Recently, they are used in the cooling of cells. Microchannel heat exchangers are also used in medicine, electronics, avionics, metrology, robotics, telecommunication, and automotive industries to enhance heat transfer for heating and cooling.

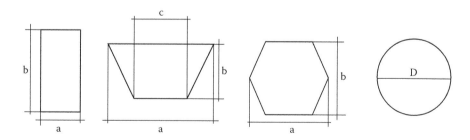

FIGURE 5.1
Rectangular, trapezoidal, hexagonal, and circular microchannel geometries.

5.1.3.1 Heat Transfer in Gas Flow

For the engineering applications and to design microheat exchangers, the prediction of the heat transfer coefficient and the friction factor are important. Beginning from early 1980s, several researches have been conducted. Wu and Little[12] tested the heat transfer characteristics of gases flowing in the trapezoidal silicon/glass microchannels of widths 130 to 300 µm and depths of 30–60 µm. The tests involved both laminar and turbulent flow regimes. The authors found that average Nusselt numbers are higher than those predicted by the conventional correlations for fully developed laminar flows and for fully developed turbulent flows. Their also indicated that transition from laminar to turbulent flow occurs at Reynolds numbers of 400–900 depending on the various test configurations. Therefore, it is reported that the heat transfer is improved because of early transition in flow regime. As a result of their research, a Nusselt number correlation was proposed for trapezoidal microchannels as in the form of

$$Nu = 0.00222 Pr^{0.4} Re^{1.08} \quad \text{for } Re > 3000 \quad (5.10)$$

Choi et al.[13] presented experimental data on the Nusselt numbers for both turbulent and laminar regimes. They studied for Reynolds numbers between the range of 50 and 20,000. Nitrogen gas is used as the working fluid in the experiments. In this study, microtubes with diameters ranging from 3 to 81 µm and all experiments were fully developed for both hydraulically and thermally. They found that their experimental results differ from the conventional theory. Although, in conventional theory, heat transfer coefficients are independent from Reynolds number for fully established laminar flow, in this study, it is reported that heat transfer coefficients are dependent on Reynolds number in laminar flow. Moreover, for turbulent flow in microtubes, heat transfer coefficients were measured to be larger than those predicted by conventional correlations. The authors proposed two new correlations for the average Nusselt number in laminar and turbulent regimes.

$$Nu = 0.000972 Re^{1.17} Pr^{1/3} \quad \text{for } Re < 2000 \quad (5.11)$$

$$Nu = 3.82 \times 10^{-6} Re^{1.96} Pr^{1/3} \quad \text{for } 2500 < Re < 20,000 \quad (5.12)$$

In this study, it is also reported that the ratio of micro- to macroturbulent Nusselt numbers were obtained as a function of Reynolds number; $Nu_{micro}/Nu_{macro} = 0.000166 Re^{1.16}$. However, those correlations proposed by Choi et al.[13] are not in agreement with the correlation proposed by Wu and Little.[12]

Yu et al.[14] investigated the heat transfer in microtubes with diameters 19, 52, and 102 µm. They used dry nitrogen gas as the operating fluid in the experiments, and they also used water as a working fluid. They studied gaseous flow in microtubes for Reynolds numbers between the range of 6,000 and 20,000, and for Prandtl numbers between the range of 0.7 and 5. It is reported that the heat transfer coefficient was enhanced in turbulent flow regime. They also investigated the turbulent flow interacting with the wall. As a result, they proposed a correlation for Nusselt number in microtubes as in following form:

$$Nu = 0.007 Re^{1.2} Pr^{0.2} \quad \text{for } 6000 < Re < 20{,}000 \quad (5.13)$$

Peng and Peterson[15] studied the effect of dimensions of the microchannel geometry on heat transfer for laminar and turbulent regimes, and it is reported that Nusselt number depends on Reynolds number, Prandtl number, and the microchannel aspect ratio. The authors determined that geometric configuration of the microchannel is a critical parameter for heat transfer phenomena. They also proposed two correlations for laminar and turbulent regime.

$$Nu = 0.1165 (D_h/W_c)^{0.81} (H/W)^{0.79} Re^{0.62} Pr^{1/3} \quad \text{for } Re < 2200 \quad (5.14)$$

$$Nu = 0.072 (D_h/W_c)^{1.15} \left(1 - 2.421(z - 0.5)^2\right) Re^{0.8} Pr^{1/3} \quad \text{for } Re > 2200 \quad (5.15)$$

where H and W are the height and width of the microchannels, respectively, and W_c is the center-to-center distance of microchannels. $z = \min(H,W)/\max(H,W)$, for $z = 0.5$ value, turbulent convective heat transfer is maximum regardless of H/W or W/H.

Ameel et al.[16] studied hydrodynamically fully developed laminar gaseous flow in microtubes analytically. They solved the problem under constant heat flux boundary condition. They used two approximations in order to simplify the problem. First, temperature jump boundary condition was not taken into account directly. Second, the effect of momentum and thermal accommodation coefficients were assumed to be unity. The flow was considered to be incompressible with constant physical properties, two dimensional and steady state. In addition, viscous effects were neglected. They developed a relation for Nusselt number by considering previous work of Baron et al.[17]

$$Nu = \frac{48(2\beta - 1)^2}{(24\beta^2 - 16\beta + 3)\left[1 + \dfrac{24\gamma(\beta - 1)(2\beta - 1)^2}{(24\beta^2 - 16\beta + 3)(\gamma + 1)Pr}\right]} \quad (5.16)$$

In Equation 5.16, β is defined as $\beta = 1 + 4Kn$. They determined that Nusselt number decreases with increasing Knudsen number. It can be said that rarefaction effects decrease the Nusselt number. Moreover, they developed a relation between Knudsen number and the entrance length as

$$x_e^* = 0.0828 + 0.14 Kn^{0.69} \tag{5.17}$$

It is also reported that with the increasing rarefaction effects thermal length increases too. For $Kn = 0$, Equation 5.16 reduces to $Nu = 4.364$, which is the Nusselt number under fully developed conditions.

Example 5.2

Using the Nu number expression given by Equation 5.16, repeat the example 5.1 for air in engineering applications with Pr equals to 0.7, the ratio between c_p and c_v equals to 1.4 and constant heat flux boundary conditions when Kn equals to 0.02 and 0.6.

Solution

The Nusselt correlation for constant heat flux condition is

$$Nu = \frac{48(2\beta-1)^2}{(24\beta^2 - 16\beta + 3)\left[1 + \frac{24\gamma(\beta-1)(2\beta-1)^2}{(24\beta^2 - 16\beta + 3)(\gamma+1)Pr}\right]}$$

For $Kn = 0.02$,

$$\beta = 1 + (4 \times Kn) = 1 + (4 \times 0.02) = 1.08$$

$$Nu = \frac{48 \times (2 \times 1.08 - 1)^2}{(24 \times 1.08^2 - 16 \times 1.08 + 3) \times \left[1 + \frac{24 \times 1.4 \times (1.08-1) \times (2 \times 1.08 - 1)^2}{(24 \times 1.08^2 - 16 \times 1.08 + 3) \times (1.4+1) \times 0.7}\right]}$$

$$= 4.0707$$

So, the heat transfer coefficient for the constant heat flux condition is

$$h = \frac{Nu \times k}{d} = \frac{3.4598 \times 0.0271}{0.24 \times 10^{-3}} = 390.66 \text{ W/m}^2\text{K}$$

For $Kn = 0.06$,

$$\beta = 1 + (4 \times Kn) = 1 + (4 \times 0.06) = 1.24$$

$$Nu = \frac{48 \times (2 \times 1.24 - 1)^2}{(24 \times 1.24^2 - 16 \times 1.24 + 3) \times \left[1 + \frac{24 \times 1.4 \times (1.24 - 1) \times (2 \times 1.24 - 1)^2}{(24 \times 1.24^2 - 16 \times 1.24 + 3) \times (1.4 + 1) \times 0.7}\right]}$$

$$= 3.4386$$

So, the heat transfer coefficient for the constant heat flux condition is

$$h = \frac{Nu \times k}{d} = \frac{3.4386 \times 0.0271}{0.24 \times 10^{-3}} = 388.28 \text{ W/m}^2\text{K}$$

Yu and Ameel[18] used the same integral transform technique with Bayazitoglu et al.[4] and obtained the Nusselt number for rectangular microchannels. In their solution, they used uniform temperature boundary condition. They also included slip flow effects but viscous effects were neglected. It is reported that Knudsen number, Prandtl number, aspect ratio, and temperature jump may cause the Nusselt number to deviate from conventional value. By integrating the studies of Bayazitoglu et al.[4] and Yu and Ameel,[18] the results for fully developed Nusselt number under uniform heat flux or uniform temperature boundary conditions were given for different geometries in Table 5.3. In the calculations, to constitute the Table 5.3, viscous effects were neglected ($Br = 0$).

Renksizbulut et al.[19] studied rarefied gas flow and heat transfer in the entrance region of rectangular microchannels. They made investigations in the slip flow regime numerically. Both velocity slip and temperature jump boundary conditions are considered in the study. It is reported that Nusselt number decreases in the entrance region due to rarefaction effects. For different channel aspect ratios and $Kn \leq 0.1$, they worked in the range of $0.1 \leq Re \leq 1,000$ with $Pr = 1$. Under these conditions,

TABLE 5.3

Nusselt Number for Different Geometries Subject to Slip-Flow[6]

		$Kn = 0.00$		$Kn = 0.04$		$Kn = 0.08$		$Kn = 0.12$	
$Br = 0.0$		Nu_T	Nu_q	Nu_T	Nu_q	Nu_T	Nu_q	Nu_T	Nu_q
Cylindrical		3.67	4.36	3.18	3.75	2.73	3.16	2.37	2.68
Rectangular	$\gamma = 1$	2.98	3.10	2.71	2.85	2.44	2.53	2.17	2.24
	$\gamma = 0.84$	3.00	3.09	2.73	2.82	2.46	2.48	2.19	2.17
	$\gamma = 0.75$	3.05	3.08	2.77	2.81	2.49	2.44	2.22	2.12
	$\gamma = 0.5$	3.39	3.03	2.92	2.71	2.55	2.26	2.24	2.18
	$\gamma = 0.25$	4.44	2.93	3.55	2.42	2.89	1.81	2.44	1.68
Aspect Ratio $\gamma = a/b$	$\gamma = 0.125$	5.59	2.85	4.30	1.92	3.47	1.25	2.80	1.12
Cylindrical		7.54	8.23	6.26	6.82	5.29	5.72	4.56	4.89

Source: From Kakac S. et al., *Microscale Heat Transfer- Fundamentals and Applications,* Kluwer Academic Publishers, The Netherlands, 125–148, 2005.

they also modified a correlation of earlier study of Renksizbulut et al.[20] for trapezoidal and rectangular microchannels under fully developed conditions.

$$(Nu)_{fd} = \left[2.87\left(\frac{90°}{\theta}\right)^{-0.26} + 4.8\exp\left(-3.9\gamma\left(\frac{90°}{\theta}\right)^{0.21}\right)\right]G_2G_3 \quad (5.18a)$$

$$G_2 = 1 + 0.075(1+\gamma)\exp(-0.45Re) \quad (5.18b)$$

$$G_3 = 1 - 1.75Kn^{0.64}(1 - 0.72\tanh(2\gamma)) \quad (5.18c)$$

where θ is the side angle and γ is defined as the aspect ratio of the short side to the long side of the channel. For fully developed flows, an expression for the friction coefficient is also obtained:

$$f Re = 13.9\left(\frac{90°}{\theta}\right) + 10.4\exp\left[-3.25\gamma\left(\frac{90°}{\theta}\right)^{0.23}\right] \quad (5.19)$$

where f is the Fanning friction factor defined as in Equation 4.3.

Table 5.3 shows the fully developed Nusselt number values for rectangular and circular channels for various values of Knudson number under two boundary conditions.

5.1.3.2 Friction Factor

As it was discussed in Chapter 4, the product of the Reynolds number and the friction factor should be constant for laminar flow and is depended on Poiseuille number as

$$Po = 4f Re \quad (5.20a)$$

and for laminar flow in circular tubes

$$f Re = 16 \quad (5.20b)$$

where f is the Fanning friction factor defined as Equation 4.3.

Poiseuille number can be alternately defined with Darcy coefficient

$$Po = \lambda_D Re \quad (5.21)$$

where Re is Reynolds number defined as

$$Re = \frac{u_m D_h}{\nu} \quad (5.22)$$

ν is the kinematic viscosity of the fluid.

Friction factors and Nusselt numbers for fully developed laminar flow for different duct geometries are given in Tables 3.8 and 5.4. Where the frictional pressure drop is obtained as

$$\Delta p = \frac{4 f L \rho u_m^2}{2 D_h} \qquad (5.23)$$

Various experimental investigations reported in microchannels for the variation of $f\,Re$ for different Reynolds numbers exhibit increasing and decreasing trends, both in laminar and turbulent flow regimes.

Since the early 1980s, calculation of the friction factor for fluid flow in microchannels has captivated great interest. The first work on this subject was presented by Wu and Little[12] who investigated frictional losses for gas flow in glass and silicon trapezoidal microchannels with hydraulic diameters equal to 55.81, 55.92, and 72.38 μm. Nitrogen, hydrogen, and argon gases were used as working fluids. They found the friction factor to be 10–30% larger than those predicted by Moody's chart. This situation is a result of great relative roughness and asymmetric distribution of the surface roughness on the microchannel walls. Flow transition occurred at $Re = 400$. They proposed correlations for various Reynolds numbers and flow regimes to calculate friction factors.

$$f = (110 \pm 8) Re^{-1} \quad \text{for } Re \le 900 \qquad (5.24a)$$

$$f = 0.165(3.48 - \log Re)^{2.4} + (0.081 \pm 0.007) \quad \text{for } 900 < Re < 3000 \qquad (5.24b)$$

$$f = (0.195 \pm 0.017) Re^{-0.11} \quad \text{for } 3000 < Re < 15{,}000 \qquad (5.24c)$$

TABLE 5.4

Nusselt Number for Ellipse Subject to Slip-Flow[6]

Duct Shape	Ellipse Major/Minor Axis a/b	Nu_H	Nu_T	Po
	1	4.36	3.66	64
	2	4.56	3.74	67.28
	4	4.88	3.79	72.96
	8	5.09	3.72	76.60
	16	5.18	3.65	78.16

Source: From Kakac S. et al., *Microscale Heat Transfer- Fundamentals and Applications,* Kluwer Academic Publishers, The Netherlands, 125–148, 2005.

$Nu = hD_h/k$; $Re = \rho u_m D_h/\mu$; Nu_H = Nusselt number under a constant heat flux boundary condition, constant axial heat flux, and circumferential temperature; Nu_T = Nusselt number under a constant wall temperature boundary condition.

The pressure drop can be calculated from Equation 5.23. Acosta et al.[21] studied friction factors for rectangular microchannels with a hydraulic diameter ranging between 368.9 and 990.4 µm. The rectangular channels used in the investigations have very small aspect ratios ($0.019 < \gamma < 0.05$). All channels with different aspect ratios showed similar friction factor values. Therefore, it is reported that those behavior is consistent with the conventional asymptotic behavior of Poiseuille number ($f\,Re$) for channels with small aspect ratios.

Pfhaler et al.[22–24] investigated the friction factor in microchannels for liquids and gases. They conducted their researches on very small microchannels (1.6–3.4 µm hydraulic diameters) and on rectangular channels, 100 µm wide and 0.5–50 µm deep with Reynolds numbers varying between 50 and 300. They found that the friction factor is lower than the conventional theory would predict. Their friction factor results were opposite to the results of Wu and Little.[12] In this study, it is also reported that the viscosity becomes size dependent at microscales. Therefore, it can be concluded that small friction value for fluids is due to the reduction of viscosity with decreasing size and rarefaction effects.

Choi et al.[13] investigated the friction factor in silica microtubes with diameters of 3, 7, 10, 53, and 81 µm. Nitrogen gas was used as the operating fluid. The experiments were conducted for Reynolds numbers between the range of 30 and 20,000. It is reported that no variation was observed in friction factor with the wall roughness. They found that the Poiseuille number is lower than the conventional value. They proposed the following correlations for different Reynolds numbers:

$$f = \frac{64}{Re}\left[1 + 30\left(\frac{v}{dC_a}\right)\right]^{-1} \quad \text{for } Re < 2300,\ C_a = 30 \pm 7 \quad (5.25a)$$

$$f = 0.14 Re^{-0.192} \quad \text{for } 4000 < Re < 18{,}000 \quad (5.25b)$$

The pressure drop can be calculated from Equation 5.23.

Arklic et al.[25] investigated the flow in rectangular silicon microchannel with a hydraulic diameter of 2.6 µm. Helium and nitrogen gases were used as working fluids. They reported that lower friction factors were obtained in microchannels compared to conventional results for macrochannels. They modeled the relationship between mass flow rate and pressure by including a slip flow boundary condition at the wall.

Pong et al.[26] measured the pressure along the 4.5 mm length of rectangular microchannels 5–40 µm wide and 1.2 µm deep for flows of helium and nitrogen. They found that pressure distribution in the channel was nonlinear. They concluded that this fact is a result of rarefaction and compressibility effects. Liu et al.[27] studied on pressure drop along the rectangular microchannels, and helium was used as the testing fluid. They used same procedure of the Pong et al.[26] It was reported that the pressure drop was smaller than the conventional values. These findings were completed by Shih et al.[28] They used nitrogen and helium as the working fluids. It is concluded that when inlet pressures lower than 0.25 MPa, the pressure drop can be predicted by slip flow model.

Yu et al.[14] experimentally studied on silica microtubes with diameters of 19.52 and 102 µm. They investigated both gas and liquid flow in their research. Nitrogen was used as the working fluid for gaseous flow. They made observations in the range of Reynolds numbers 250 and 20,000. They concluded that friction factors were lower than conventional macroscale predictions. They also investigated how the roughness of tubes affects flow. With this study, they proposed the following correlations to calculate friction factor defined as Equation 5.23 for laminar and transition flow regimes.

$$f = 50.13/Re \quad \text{for } Re < 2000 \tag{5.26a}$$

$$f = 0.302/Re^{0.25} \quad \text{for } 2000 < Re < 6000 \tag{5.26b}$$

Harley et al.[29] investigated the flow experimentally in trapezoidal silicon microchannels with 100 µm in width and 0.5–20 µm in depth. Nitrogen, helium, and argon gases were used as testing fluids. The hydraulic diameters of the microchannels were in the range between 1.01 and 35.91 µm and the aspect ratios between the range of 0.0053 and 0.161. Compressibility and rarefaction effects on Poiseuille number were investigated. It is concluded that friction factors were lower than predicted according to conventional theory. They also explained all these observation by isothermal, fully developed, first-order slip model.

Araki et al.[30] investigated the flow characteristics of nitrogen and helium along three different trapezoidal microchannel geometries with hydraulic diameter differing from 3 and 10 µm. It is reported that the investigated friction factors were lower than that predicted by the conventional theory. It is explained that this situation was a result of rarefaction effects.

Li et al.[31] studied on friction factor for the flow in five different microtubes with diameters from 80 to 166.6 µm. They used nitrogen gas as the testing fluid. They found a nonlinear distribution of the pressure along microtubes when Mach number exceeded 0.3. However, their results showed that the friction factors were higher than those predicted by the conventional theory.

Yang et al.[32] investigated the air flow in circular microtubes with diameters ranging between 173 and 4,010 µm and lengths ranging from 15 to 1 mm. Although the laminar regime friction factors were in a good agreement with conventional theory, the turbulent friction factors were lower than those predicted by conventional correlation for macrotubes.

Lalonde et al.[33] studied experimentally on the friction factor for flow in microtube with a diameter of 52.8 µm. Air was used as the testing fluid.

It is reported that the friction factors obtained from the experiments indicated a good agreement with the predictions of conventional theory. Turner et al.[34] investigated compressible gas flow in rough and smooth rectangular microchannels experimentally. Nitrogen, helium, and air were used as testing fluids for microchannels with hydraulic diameter ranging between 4 and 100 μm. They investigated just the laminar flow regime for the Reynolds numbers between the ranges of 0.02 to 1,000. Similarly to Lalonde et al., they reported that the friction factors obtained in the experiments were in agreement with the conventional theory.

Hsieh et al.[35] studied the characteristics of nitrogen flow in a rectangular microchannel with the length of 24 mm, width of 200 μm, and depth of 50 μm. They investigated the behavior of the flow between the range of Reynolds number 2.6 and 89.4 and the range of Knudsen number from 0.001 to 0.02. It is reported that friction factors were lower than that was predicted by the conventional theory due to compressibility and rarefaction effects. Moreover, they proposed the following correlation for friction factor defined as in the form of:

$$f = 48.44/Re^{1.02} \quad \text{for } 2.6 < Re < 89.4 \tag{5.27}$$

Tang and He[36] studied on friction factors in silica microtubes and square microchannels with hydraulic diameters ranging from 50 to 210 μm. They found that the measured friction factors were in a good agreement with conventional predictions, but it is also concluded that in smaller microtubes compressibility, roughness, and rarefaction effects cannot be neglected.

Celata et al.[37] investigated the helium gas flow along the microtubes with diameters between 30 and 254 μm and high L/D ratios to observe how compressibility affects the friction factor. They found that the quantitative behavior can be predicted by the incompressible theory and the frictional losses at the inlet and outlet only give a minor contribution to the total pressure drop.

All the observed literature results of friction factors are summarized in Table 5.5. It is clearly seen that there is not a general solution for the friction factor behavior in microchannels. It can also be concluded that, at the lower values of hydraulic diameters, there is no agreement with the conventional theory due to rarefaction effects.

5.1.3.3 Laminar to Turbulent Transition Regime

Continuum flow regime description fails when flow enters the transition flow regime. The molecular approach is not enough to express the transition flow regime, and the effect of Knudsen number becomes more important. Because of the failure of continuum description, Boltzmann equation should be used to investigate the atomic-level flows in transition regime. Boltzmann equation can be solved by several methods such as molecular dynamics (MD)

TABLE 5.5

Experimental Results on Friction Factor for Gas Flows in Microchannels

Literatures	fRe	D_h (μm)	Cross-Section	Test Fluids
Wu and Little[12]	↑↑	55.8–72.4	Trap.	N_2, H_2, Ar
Acosta et al.[21]	≈	368.9–990.4	Rect.	He
Phfaler et al.[22–24]	↓↓	1.6–65	Rect.-Trap.	N_2, He
Choi et al.[13]	↓↓	3–81	Circ.	N_2
Arkilic et al.[25]	↓↓	2.6	Rect.	He
Pong et al.[26]	↓↓	1.94–2.33	Rect.	N_2, He
Liu et al.[27]	↓↓	2.33	Rect.	N_2, He
Shih et al.[28]	↓↓	2.33	Rect.	N_2, He
Yu et al.[18]	↓↓	19–102	Circ.	N_2
Harley et al.[29]	↓↓	1.01–35.91	Rect.-Trap.	N_2, He, Ar
Araki et al.[30]	↓↓	3–10	Trap.	N_2, He
Li et al.[31]	↑↑	128.8–179.8	Circ.	N_2
Yang et al.[32]	↓↓	173–4010	Circ.	Air
Lalonde et al.[33]	≈	58.2	Circ.	Air
Turner et al.[34]	≈	4–100	Rect.	Air, N_2, He
Hsieh et al.[35]	↓↓	80	Rect.	N_2
Tang and He[36]	≈	50–201	Rect.-Circ.	N_2
Celata et al.[37]	≈	30–254	Circ.	He
Kohl et al.[38]	≈	24.9–99.8	Rect.	Air

Source: From Morini, G.L., *Int. J. Therm. Sci.*, 631-651, 2004; Çetin, B., Yuncu, H., and Kakaç, S., *Int. J. Transp. Phenom.*, 8, 297–315, 2007.

↑↑ = fRe higher than the conventional theory; ↓↓ = fRe lower than the conventional theory; ≈ = fRe agrees with the conventional theory.

and direct simulation Monte Carlo (DSMC). In addition, the simplified Boltzmann equation can be solved by using Lattice Boltzmann Method (LBM). However, the current computational methods (MD and DSMC) cannot provide an effective solution for transition flow regime.[9] Thermal creep is an important phenomenon for transition flow regime. Therefore, it should be considered while studying laminar to turbulent transition regime analytically.[10]

Hadjiconstantinou et al.[39] investigated the convective heat transfer characteristics of a model gaseous flow of two-dimensional microchannels. They focused on monoatomic gases and used the simplest monoatomic gas model, the hard sphere gas. They performed simulations by using DSMC. Simulations were for $0.02 < Kn < 2$. They concluded a weak dependence of Nusselt number on Peclet number. The dependence of Nusselt number on Peclet number was expressed as in the form of

$$Nu_T^s(Pe, Kn=0) = 8.11742(1 - 0.0154295 Pe + 0.0017359 Pe^2) \quad (5.28)$$

In Equation 5.28, Nu_T^s represents slip-flow constant wall temperature Nusselt number.

Besides the analytical solutions, several experimental researches have been conducted to understand the transition flow regime. Wu and Little[40] experimentally studied the gaseous transition flow regime. They tested silicon and glass microchannels. Silicon microchannels were considered to be smooth and compared with glass microchannels which have a high value of relative roughness. Therefore, it is concluded that the roughness of the microchannel have an impact on the transition. It is also stated that the transition occurs at Reynolds number ranging from 1,000 to 3,000.

Acosta et al.[21] studied experimentally isothermal gas flows in rectangular microchannels with very small aspect ratios ($0.019 < \gamma < 0.05$). Test results indicated very similar behaviors with respect to conventional transition. It is concluded that the transition Reynolds number is about 2,770.

Li et al.[31] conducted experiments with gases to calculate the friction factors for circular microtubes. Their results are in a good agreement with the conventional theory. They found that the transition from laminar to turbulence in microtubes occurs at Reynolds number equal to 2,300. In a further work of these authors,[41] they concluded that the transition flow regime occurs at Reynolds numbers between 1,700 and 2,000.

According to literature survey that is conducted about the transition flow regime for gaseous flow in microchannels, experimental results are summarized in Table 5.6.

Microscale heat transfer has been an important issue for researchers in the last decades. In this particular study, the gaseous flow is investigated with a comprehensive literature survey of experimental and analytical works. The basics of microscale heat transfer and fluid flow such as Knudsen number, Brinkman number, temperature jump, velocity slip, and boundary conditions related to them are presented. The differences and similarities between micro- and macrotransport phenomena are also mentioned. Variety of available experimental data on convective heat transfer in microchannels are stated and compared with each other.

TABLE 5.6

The Experimental Data on the Laminar to Turbulent Flow Transition in Microchannels

Literatures	Cross-Section	ε/D_h	γ	Critical Reynolds
Wu and Little[40]	Trap.	–	0.444	1,000–3,000
Acosta et al.[21]	Rect.	–	0.019, 0.033, 0.05	2,770
Choi et al.[13]	Circ.	–	–	500–2,000
Li et al.[31]	Circ.	–	–	2,300
Li et al.[41]	Circ.	0.04	–	1,700–2,000
Yang et al.[32]	Circ.	–	–	1,200–3,800

The conclusion drawn for microscale gaseous flow can be summarized as follows:

- The behavior of heat transfer in microchannels is very different when compared with that of conventional sized situation.
- Convective heat transfer in microchannels is highly dependent on Knudsen number, Prandtl number, Brinkman number, and size and geometry of microchannels.
- Velocity slip and temperature jump affects heat transfer in microchannels while velocity slip increases the heat transfer, temperature jump reduces the heat transfer by reducing the temperature gradient at the wall. The magnitude of temperature jump is directly dependent on Prandtl number.
- Rarefaction effects have an important role especially in gaseous flows in microchannels and the degree of rarefaction is defined by Knudsen number. Knudsen number also determines the degree of deviation from conventional model. As the Knudsen number increases, a reduction in Nusselt number and heat transfer occurs.
- Viscous heat dissipation is a significant factor for microchannel heat transfer. The viscous heat dissipation effects are defined by Brinkman number.
- In microchannels, the Nusselt number can be correlated as a function of Br, Re, Pr, and geometric parameters of microchannels.
- According to experimental review for correlations of Nusselt number, in the laminar regime, the Nusselt number increases with the Reynolds number.[12,13] However, in a different study, it is stated that the Nusselt number decreases with increasing Reynolds number in laminar regime.[15]
- In microchannels, the friction factor can be correlated as a function of Re and geometric parameters of microchannels.
- Experimental review for friction number can be summarized that, in several studies, the friction factor for fully developed flow is found to be lower than conventional value[14,22–24,29,30]; however, in other studies, the friction factor is found to be lower than conventional value.[12,31]
- For gaseous fully developed laminar flow, friction factor decreases with the increasing Knudsen number.[25,30]
- In several experimental works, it is concluded that the friction factor depends on the material and relative roughness of microchannel.[12,14,22–24]
- The critical Reynolds number for transition regime is affected by the fluid temperature, velocity, and geometric parameters of microchannel.

- Axial conduction, in other words thermal creep, should be considered for transition flow regime.
- In transition flow regime, critical Reynolds numbers depend on wall roughness in microchannels.[12]
- The critical Reynolds numbers decrease with the microchannel hydraulic diameter.[13,32]

5.1.4 Engineering Applications of Single-Phase Liquid Flow in Microchannels

The application of the liquid flow in microchannels for cooling of electronics and microdevices has increased steadily during the last 25 years. Because of the theoretical and experimental investigations in recent years, interesting observations have been obtained in liquid flow in microchannels, which is essential for thermal design of microheat exchangers.

Convective heat transfer in liquids flowing through microchannels has been extensively experimented to obtain the characteristics in the laminar, transitional, and turbulent regimes. The observations, however, indicate significant departures from the classical correlations for the conventionally sized tubes, which have not been explained.

Experimental investigations on convective heat transfer in liquid flows in microchannels have been in the continuum regime. Hence, the conventional Navier-Stokes equations are applicable.

The geometric parameters of individual rectangular microchannels, namely the hydraulic diameter and the aspect ratio, and the geometry of the microchannel plate have significant influence on the single-phase convective heat transfer characteristics.

In the laminar and transition regimes in microchannels, the behavior of convective heat transfer coefficient is very different compared with the conventionally sized situation. In the laminar regime, Nu decreases with increasing Re, which has not been explained.

In microchannels, the flow transition point and range are functions of the heating rate or the wall temperature conditions. The transitions are also a direct result of the large liquid temperature rise in the microchannels, which causes significant liquid thermophysical property variations and, hence, significant increases in the relevant flow parameters, such as the Reynolds number. Hence, the transition point and range are affected by the liquid temperature, velocity, and geometric parameters of the microchannel.

Some of the correlations for the prediction of the friction factor and Nusselt number for liquid flow in microchannels are summarized in the following sections.

The mechanism by which heat is transferred in an energy conversion system is complex; however, three basic and distinct modes of heat transfer have been classified: conduction, convection, and radiation.[2] Convection is the heat transfer mechanism, which occurs in a fluid by the mixing of one portion of the fluid with another portion due to gross movements of the mass of fluid. Although the actual process of energy transfer from one fluid particle or molecule to another is still heat conduction, the energy may be transported from one point in space to another by the displacement of the fluid itself. Hence, heat transfer in microchannel heat sinks as seen in Figure 5.2 belongs to forced convection heat transfer with or without phase change.[42] However, the flow without phase change or in other words the single phase liquid flow will be investigated in this study.

The Nusselt number in the fully developed laminar flow is expected to be constant as predicted by the classical theory. For basic Nusselt correlations, the Nusselt number is considered as constant. Table 5.3 presents the Nusselt numbers for commonly used geometries under constant heat flux and constant wall temperature boundary conditions.

The thermal and hydrodynamic entry lengths are expressed by the following form for flow in ducts:

$$\frac{L_t}{D_h} = 0.056 Re\, Pr \qquad (5.29)$$

$$\frac{L_h}{D_h} = 0.056 Re \qquad (5.30)$$

The local heat transfer in the developing region of a laminar flow in a circular tube is given by the following equations[18]:

$$Nu_x = 4.363 + 8.68\left(10^3 x^*\right)^{-0.506} e^{-41 x^*} \qquad (5.31)$$

$$x^* = \frac{x/D_h}{Re\, Pr} \qquad (5.32)$$

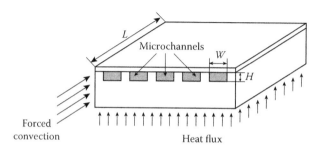

FIGURE 5.2
A schematic of a typical microchannel heat sink. (From Stephan, K., *Heat Transfer in Condensation and Boiling*, Springer-Verlag, Berlin, 1992. With permission.)

where Re is Reynolds number and Pr is Prandtl number.

$$Re = \frac{u_m D_h}{\nu}, \quad Pr = \frac{\mu C_p}{k}$$

C_p is the specific heat and k is the thermal conductivity. The correlations that are found in open literature are given in the following section.

5.1.4.1 Nusselt Number and Friction Factor Correlations for Single-Phase Liquid Flow

Most of the studies reported in the open literature indicate that conventional correlations for liquid flow and heat transfer in ducts with various cross sections can predict the behavior of heat transfer and friction coefficients with acceptable accuracy. The experimental studies in microchannels are quite complex because of channel dimensions, various measurement techniques used, and the fabrication techniques which make the experimental results are so different from each other. The improvement in the experimental techniques indicates that for the majority of the macrochannel correlations are applicable to the studies in microchannels.

A deep literature survey has been done and generalized Nusselt number and friction factor correlations from the literature are obtained for different cases.

Gnielinski[43] developed Nusselt number and friction factor correlation for the transition regime from laminar to turbulent:

$$Nu = \frac{(f/8)(Re-1000)Pr}{1+12.7(f/8)^{1/2}(Pr^{2/3}-1)} \quad \begin{array}{l} 3000 < Re < 5 \times 10^6 \\ 0.5 < Pr < 2000 \end{array} \quad (5.33)$$

$$f = [1.82 \log(Re) - 1.64]^{-2}$$

Adams et al.[2,44] modified the Gnielinski correlation, which is a generalized correlation for the Nusselt number for turbulent flow in small diameter tubes (0.76–1.09 mm)

$$Nu_{Adams} = Nu_{Gnielinski}(1+F) \quad \begin{cases} F = 7.6 \times 10^{-6} Re\left(1 - \left(\frac{d}{D_0}\right)^2\right) \\ 2600 < Re < 23000 \\ 1.53 < Pr < 6.43 \\ D_0 = 1.164 \text{ mm is the reference diameter} \end{cases} \quad (5.34)$$

where F is an enhancement factor. D_0 represents the diameter for which Equation 5.34 predicts no enhancement of the Nusselt number over that predicted by Equation 5.33.

Tsuzuki et al.[45] found Nusselt number correlation for water and CO_2 in their experimental study. The correlation for water is in Equation 5.35 and for CO_2 is in Equation 5.36.

$$Nu = 0.253 Re^{0.597} Pr^{0.349} \quad \begin{cases} 100 < Re < 1500 \\ 2 < Pr < 11 \end{cases} \tag{5.35}$$

$$Nu = 0.207 Re^{0.627} Pr^{0.34} \quad \begin{cases} 1500 < Re < 15,000 \\ 1 < Pr < 3 \end{cases} \tag{5.36}$$

Friction factor correlation given by Blasius is presented as in Equation 5.21 from White[46]

$$f = 0.0791 Re^{-1/4} \quad \{10^4 < Re < 10^5\} \tag{5.37}$$

Jiang et al.[47] found correlations for friction factor for large relative roughness with respect to their experiments.

$$f = \frac{1639}{Re^{1.48}} \quad \{Re < 600\} \tag{5.38}$$

$$f = \frac{5.45}{Re^{0.55}} \quad \{600 < Re < 2800\} \tag{5.39}$$

Note that f is defined as in Equation 5.23.

Example 5.3

Water flows at a temperature of 30°C through a microchannel with a hydraulic diameter of 0.04 mm and the length of the channel is 20 mm. The Reynolds number for flow is 500. By the use of correlation in Equation 5.23, (a) calculate the friction coefficient, Poiseuille number, and Darcy coefficient, and (b) by the use of Equation 5.23, calculate the frictional pressure drop in bar. Kinematic viscosity of water at 30°C can be taken as v and equals to 0.8315×10^{-6} m²/s.

Solution
From properties of water at $T = 30°C$

$$\rho = 995.65 \, kg/m^3$$

$$\mu = 0.0007973 \, Pa \cdot s$$

$$k = 0.6155 \, W/(m \cdot K)$$

$$Pr = 5.43$$

$$v = 0.8315 \, m^2/s$$

The water velocity can be found from the given Reynolds number

$$Re = 500 = \frac{D_h \times u_m}{\nu}$$

$$500 = \frac{0.04 \times 10^{-3} \times u_m}{0.8315 \times 10^{-6}}$$

$$u_m = 10.4 \text{ m/s}$$

Since $Re = 500 < 600$, Jiang et al.[47] correlations, Equation 5.38, can be used to calculate the friction factor

$$f = \frac{1639}{Re^{1.48}} \quad \text{for } Re < 600$$

$$= \frac{1639}{500^{1.48}} = 0.166$$

$$\lambda_{Dr} = 4f = 0.664$$

For laminar flow, the Poiseuille number is introduced, Equation 5.20a, as

$$Po = f\,Re$$

$$= 0.166 \times 500$$

$$= 83$$

Poiseuille number can be defined with Darcy coefficient, Equation 5.21, as

$$Po = \lambda_D\,Re$$

$$83 = \lambda_D \times 500$$

$$\lambda_D = 83/500$$

The frictional pressure drop, Equation 5.23, can be obtained as

$$\Delta p = \frac{\lambda_{Dr} L \rho u_m^2}{2 D_h}$$

$$= \frac{0.664 \times 0.02 \times 995.65 \times 10.4^2}{2 \times 0.04 \times 10^{-3}}$$

$$= 17876.46 \text{ kPa}$$

$$= 178.73 \text{ bar}$$

The following correlations are also given for turbulent flow regime[48]:

$$f = (1.0875 - 0.1125\alpha^*) f_c \quad \begin{cases} 4 \times 10^3 < Re < 10 \\ f_c = 0.00128 + 0.1143 Re^{-0.311} \\ \alpha^* = \dfrac{a}{b} \end{cases} \quad (5.40)$$

Choi et al.[49] found Nusselt number correlations (Equation 5.25) for laminar and turbulent flow in microchannels. Yu et al.[50] also found a Nusselt number correlation is given in Equation 5.27 for turbulent flow in microchannels.

$$Nu = 0.000972 Re^{1.17} Pr^{1/3} \quad \{Re < 2000 \quad (5.41)$$

$$Nu = 3.82 \times 10^{-6} Re^{1.96} Pr^{1/3} \quad \{2500 < Re < 20,000 \quad (5.42)$$

$$Nu = 0.007 Re^{1.2} Pr^{0.2} \quad \{6000 < Re < 20,000 \quad (5.43)$$

Bejan[51] recommends the equation given by Gnielinski, Equation 5.27 over the traditional Dittus–Boelter equation, because the errors are usually limited to about ±10%.

$$Nu = 0.0.12 (Re^{0.87} - 280) Pr^{0.4} \quad \begin{cases} 3 \times 10^3 \le Re \le 10^6 \\ 1.5 \le Pr < 500 \end{cases} \quad (5.44)$$

Park and Punch[52] found a new correlation for Nusselt number with respect to their experimental data.

$$Nu = 0.015 Br^{-0.22} Re^{0.62} Pr^{0.33} \quad \begin{cases} 69 < Re < 800 \\ 4.44 < Pr < 5.69 \\ 106\,\mu m < D_h < 307\,\mu m \\ Br < 0.0006 \end{cases} \quad (5.44)$$

Br represents Brinkman number, which is given by Equations 5.9a and 5.9b.

Lee et al.[53] investigated heat transfer in rectangular microchannels in their study and found a new interval for Sieder–Tate correlation that is given by[48]

$$Nu = 1.86 \left(\dfrac{Re\, Pr\, d}{L} \right)^{1/3} \left(\dfrac{\mu_f}{\mu_w} \right)^{0.14} \quad \{Re < 2200 \quad (5.46)$$

Wu and Little[54] found a Nusselt number correlation for turbulent liquid "Nitrogen" flow in microchannels:

$$Nu = 0.0022 Re^{1.09} Pr^{0.4} \quad \{Re > 3000 \quad (5.47)$$

Wang and Peng[55] found a Nusselt number correlation with respect to their experimental study:

$$Nu = 0.00805 Re^{0.8} Pr^{1/3} \quad \{Re > 1500 \quad (5.48)$$

Additional correlations are given in Refs. 56 and 57, which can be further used to obtain heat transfer coefficients in the design of microchannel heat exchangers which are

$$Nu = 0.00805 Re^{4/5} Pr^{1/3} \quad (5.49)$$

for fully developed turbulent flow $1{,}000 < Re < 1{,}500$,

$$Nu = 0.038 Re^{0.62} Pr^{1/3} \quad (5.50)$$

for laminar flow $Re < 700$,
and

$$Nu = 0.072 Re^{0.8} Pr^{1/3} \quad (5.51)$$

for turbulent $Re > 700$.

Example 5.4

Water flows at a temperature of 30°C through a rectangular microchannel, 0.04 mm by 320 µm, used to cool a chip with 15 mm length of the channel. The water flow rate is 21.6×10^{-6} kg/s.

1. Determine the heat transfer coefficient for the constant heat flux and constant wall temperature boundary conditions.
2. Determine the friction factor and the pressure drop in the channel.

Solution
From properties of water at $T = 30°C$

$$\rho = 995.65 \text{ kg/m}^3$$
$$\mu = 0.0007973 \text{ Pa} \cdot \text{s}$$
$$c_p = 4.1798 \text{ kJ/kg} \cdot \text{K}$$
$$k = 0.6155 \text{ W/(m} \cdot \text{K)}$$
$$Pr = 5.43$$

The hydraulic diameter can be calculated as

$$D_h = \frac{2 \times a \times b}{a+b} = \frac{2 \times 40 \times 320}{40+320} = 71.11\ \mu m$$

and the Reynolds number can be obtained as

$$Re = \frac{4 \times \dot{m}}{\pi \times \mu \times D_h} = \frac{4 \times 21.6 \times 10^{-6}}{\pi \times 0.0007973 \times 71.11 \times 10^{-6}} = 485$$

To check the hydrodynamically fully developed flow condition, L_h has to be found:

$$L_h = 0.056 \times Re \times D_h = 0.056 \times 485 \times 71.11 \times 10^{-6}$$
$$= 1.92 \times 10^{-3}\ m$$
$$= 1.92\ mm < 15\ mm$$

so, the flow is hydrodynamic fully developed.

To check the thermal fully developed flow condition, L_t has to be found:

$$L_t = 0.05 \times Re \times Pr \times D_h = 0.05 \times 485 \times 5.43 \times 71.11 \times 10^{-6}$$
$$= 9.36 \times 10^{-3}\ m$$
$$= 9.36\ mm < 15\ mm$$

so, the flow is thermal fully developed.

Choi et al.[13] correlation, Equation 5.11, can be used to calculate the average Nusselt number:

$$Nu = 0.000972 Re^{1.17} Pr^{1/3} \quad \text{for } Re < 2000$$
$$= 0.0384 \times 485^{1.17} \times 5.43^{1/3}$$
$$= 2.37$$

$$h = \frac{Nu_\infty \times k}{d} = \frac{2.37 \times 0.6155}{71.11 \times 10^{-6}} = 20521.77\ W/m^2 \cdot K$$

Jiang et al.[47] correlations, Equation 5.38, can be used to calculate the friction factor for rough surfaces:

$$f = \frac{1639}{Re^{1.48}} \quad \text{for } Re < 600$$
$$= \frac{1639}{485^{1.48}} = 0.173$$

The pressure drop can be calculated from Equation 5.23.

$$\Delta p = f \times \frac{L}{D_h} \times \frac{\rho \times u_m^2}{2}$$

$$= 0.173 \times \frac{0,015}{71.11 \times 10^{-6}} \times \frac{995.65 \times 2.87}{2}$$

$$= 52.34 \text{ kPa}$$

$$= 0.52 \text{ bar}$$

5.1.4.2 Roughness Effect on Friction Factor

For microchannels, the relative roughness values are expected to be higher than the limit of 0.05 used in the Moody diagram. Celeta et al.[58] measured the friction factor for R114 flowing in a channel 130 μm in diameter. The Reynolds number varied from 100 to 8,000 and the relative channel surface roughness was 2.65%. Their experimental results show that for laminar flow, the friction factor was in good agreement with Poiseuille theory for $Re < 583$; while for higher values of Re, the experimental data were higher than Poiseuille theory; the transition from laminar to turbulent occurred for Re in the range of 1,881–2,479. Kandlikar et al.[59,60] considered the effect of cross-sectional area reduction due to protrusive roughness elements and suggested using the constricted flow area in calculating the friction factor. Using a constricted diameter, $D_{cf} = D - 2\varepsilon$, a modified Moody diagram.[60] In the turbulent region, it was calculated that a constant value of a friction factor was above $\varepsilon/D_{cf} > 0.03$.

In the turbulent fully rough region, $0.03 \leq \varepsilon/D_{cf} \leq 0.05$, the friction factor can be calculated as

$$f_{\text{Darcy},cf} = \lambda_D = 0.042 \tag{5.52}$$

Fanning friction factor can be calculated from Darcy coefficient as

$$\lambda_D = 4f = 0.042 \tag{5.53a}$$

$$f = \frac{0.042}{4} \tag{5.53b}$$

$$f = 0.0105 \tag{5.53c}$$

For relative roughness values that are higher than 0.05 is not compatible with Equation 5.36 because of the reason that experimental data is not available beyond $\varepsilon/D_{cf} > 0.03$. Constricted friction factors can be seen in Table 5.7 for different cases.

TABLE 5.7

Constricted friction factors

Flow Region	f_{cf}	Interval
Laminar region	$f_{cf} = \dfrac{Po}{Re_{cf}}$	$0 \leq \varepsilon/D_{h,cf} \leq 0.15$
Fully developed turbulent region	$f_{cf} = \dfrac{0.25\left[\log_{10}\left(\dfrac{\varepsilon/D_{h,cf}}{3.7} + \dfrac{5.74}{Re_{cf}^{0.9}}\right)\right]^{-2}}{4}$	$0 \leq \varepsilon/D_{h,cf} \leq 0.03$
Fully developed turbulent region	$f_{cf} = 0.0105$	$0.03 \leq \varepsilon/D_{h,cf} \leq 0.05$

5.2 PART B—Single-Phase Convective Heat Transfer with Nanofluids

5.2.1 Introduction of Convective Heat Transfer with Nanofluids

There are several methods to improve the heat transfer efficiency. Some methods are utilization of extended surfaces, application of vibration to the heat transfer surfaces, and usage of microchannels. Heat transfer efficiency can also be improved by increasing the thermal conductivity of the working fluid. Commonly used heat transfer fluids such as water, ethylene glycol, and engine oil have relatively low thermal conductivities, when compared to the thermal conductivity of solids. High thermal conductivity of solids can be used to increase the thermal conductivity of a fluid by adding small solid particles to that fluid. These solid particles can also cause sedimentation, clogging of channels, and erosion in channel walls. These drawbacks prevented the practical application of suspensions of solid particles in base fluids as advanced working fluids in heat transfer applications.[61,62]

With the improvements in nanotechnology, the production of particles with sizes on the order of nanometers, also called nanoparticles, can be achieved with relative ease. Because of the suspension mechanism of nanoparticles in a base fluid, this combination is called a nanofluid. It is even possible to use nanofluids in microchannels.[63,64] When it comes to the stability of the suspension, it was shown that sedimentation of particles can be prevented by utilizing proper dispersants. Therefore the effective thermal conductivity of nanofluids is expected the enhanced heat transfer compared with conventional heat transfer liquids.

Nanofluid is envisioned to describe a fluid in which nanometer-sized particles are suspended in conventional heat transfer basic fluids. Conventional heat transfer fluids, including oil, water, and ethylene glycol mixture, are poor heat transfer fluids, since the thermal conductivity of these fluids play important role on the heat transfer coefficient between the heat transfer medium and the heat transfer surface.

5.2.1.1 Particle Materials and Base Fluids

Many different particle materials are used for nanofluid preparation. Since the thermal conductivity of the particle materials, metallic or nonmetallic such as Al_2O_3, CuO, TiO_2, SiC, TiC, Ag, Au, Cu, and Fe, are typically order-of-magnitude higher than the base fluids (as shown in Table 5.8) even at low concentrations, result in significant increases in the heat transfer coefficient. Carbon nanotubes, frequently used in nanofluid research, are also utilized due to their extremely high thermal conductivity in the longitudinal (axial) direction.

5.2.1.2 Particle Size and Shape

The solid nanoparticles with typical length scales of 1–100 nm with high thermal conductivity are suspended in the base fluid (low thermal conductivity), have been shown to enhance effective thermal conductivity and the convective heat transfer coefficient of the base fluid. Spherical particles are mostly used in nanofluids. However, rod-, tube-, and disk-shaped nanoparticles are also used. On the other hand, the clusters formed by nanoparticles may have fractal-like shapes. When particles are not spherical but rod or tube-shaped,

TABLE 5.8

Thermal Conductivities of Various Solids and Liquids[65]

Solids/Liquids	Materials	Thermal Conductivity (W/m. K)
Metallic solids	Silver	429
	Copper	401
	Aluminum	237
	Copper Oxide (CuO)	20
Nonmetallic solids	Diamond	3,300
	Carbon nanotubes	3,000
	Silicon	148
	Alumina (Al_2O_3)	40
Metallic liquids	Sodium @ 644 K	72.3
Nonmetallic liquids	Water	0.613
	Ethylene glycol (EG)	0.253
	Engine oil (EO)	0.144

Source: From Eastman, J. et al., *Proceedings of the Symposium on Nanophase and Nanocomposite Materials II*, Materials Research Society, Boston, 447, 3–11, 1997. With permission.

the diameter is still below 100 nm, but the length of the particles may be on the order of micrometers. It should also be noted that due to the clustering phenomenon, particles may form clusters with sizes on the order of micrometers.

5.2.1.3 Nanofluid Preparation Methods

Production of nanoparticles can be divided into two main categories, namely, physical and chemical syntheses. The common production techniques of nanofluids have been listed as follows[66]:

- Physical Synthesis: Mechanical grinding, inert-gas-condensation technique.
- Chemical Synthesis: Chemical precipitation, chemical vapor deposition, microemulsions, spray pyrolysis, and thermal spraying.

There are mainly two methods of nanofluid preparation, namely, two-step technique and one-step technique. In the two-step technique, the first step is the production of nanoparticles and the second step is the dispersion of the nanoparticles in a base fluid. Two-step technique is advantageous when mass production of nanofluids is considered; nanoparticles can be produced in large quantities by utilizing the technique of inert gas condensation.[67] The main disadvantage of the two-step technique is that the nanoparticles form clusters during the preparation of the nanofluid which prevents the proper dispersion of nanoparticles inside the base fluid.[66]

One-step technique combines the production of nanoparticles and dispersion of nanoparticles in the base fluid into a single step. There are some variations of this technique. In one of the common methods, named direct evaporation one-step method, the nanofluid is produced by the solidification of the nanoparticles, which are initially gas phase, inside the base fluid.[68] The dispersion characteristics of nanofluids produced with one-step techniques are better than those produced with two-step technique.[66] The main drawback of one-step techniques is that they are not proper for mass production, which limits their commercialization.[66]

5.2.2 Thermal Conductivity of Nanofluids

The high thermal conductivity nanoparticles suspended in the base fluid, which has a low thermal conductivity, remarkably increase thermal conductivity of nanofluids. Researchers developed many models to tell how much that increase would be and many experiments have been conducted to compare experimental data with those analytical models as shown in Table 5.8. In thermal conductivity measurements of nanofluids, the transient hot-wire technique is the most commonly used method.[69–72] A modified transient hot-wire method is required in the measurements, since nanofluids conduct electricity. The modification is made by insulating the wire. Some other methods such as steady-state parallel-plate technique, temperature oscillation technique, microhot strip method, and optical beam deflection technique have also been utilized by some researchers.[73–76]

5.2.2.1 Classical Models

In 1873, Maxwell proposed an equation for calculating the effective thermal conductivity of solid–liquid mixtures consisting of spherical particles[77]:

$$k_{nf} = \frac{k_p + 2k_f + 2(k_p - k_f)\phi}{k_p + 2k_f - (k_p - k_f)\phi} k_f \quad (5.54)$$

where k_{nf}, k_p, and k_f are the thermal conductivity of the nanofluid, nanoparticles, and base fluid, respectively. ϕ is the volume fraction of particles in the mixture. Maxwell derived his model based on the assumption that the discontinuous phase is spherical in shape and the thermal conductivity of nanofluids depend on the thermal conductivity of spherical particles, the base fluid, and the particle volume fraction. As seen from the expression, the effect of the size and shape of the particles was not included in the analysis. It should be noted that the interaction between the particles was also neglected in the derivation.

Maxwell work was extended by Hamilton and Crosser[78] to cover none spherical particles and introduced the shape factor (n) which can be determined experimentally for different type of materials. The goal of their research was to develop a model as a function of particle shape, composition, and the conductivity of both continuous and discontinuous phases. Hamilton and Crosser model for a discontinuous phase (particles) dispersed in a continuous phase is

$$k_{nf} = \frac{k_p + (n-1)k_f - (n-1)\phi(k_f - k_p)}{k_p + (n-1)k_f + \phi(k_f - k_p)} k_f \quad (5.55)$$

where n is the empirical shape factor and it is defined as

$$n = \frac{3}{\psi} \quad (5.56)$$

where ψ is the sphericity. Sphericity is the ratio of the surface area of a sphere with a volume equal to that of the particle to the surface area of the particle.

Therefore, $n = 3$ for a sphere and in that case the Hamilton and Crosser model becomes identical to the Maxwell model.[77]

Example 5.5

Determine the thermal conductivities of Nanofluid, Alumina(Al_2O_3)/Water, by using information from Table 5.8 and two classical models, the Maxwell and the Hamilton and Crosser models, as well as compare these results with the use of the base fluid.

1. At 5% of particle volume fraction and sphericity value equals to 1
2. At 5% of particle volume fraction and sphericity value equals to 3.

Solution

Properties of alumina (Al_2O_3) and water, from Table 5.8, and water are

$$k_p = 40 \ W/(m \cdot K)$$
$$k_f = 0.613 \ W/(m \cdot K)$$
$$\phi = 0.05$$

The thermal conductivities of nanofluid, alumina (Al_2O_3)/water, can be calculated by using classical models; Maxwell model as

$$k_{nf} = \frac{k_p + 2k_f + 2(k_p - k_f)\phi}{k_p + 2k_f - (k_p - k_f)\phi} k_f$$

$$= \frac{(40) + (2)(0.613) + (2)(40 - 0.613)(0.05)}{(40) + (2)(0.613) - (40 - 0.613)(0.05)} (0.613)$$

$$= 0.705 \ W/(m \cdot K)$$

and Hamilton and Crosser model where

$$n = \frac{3}{\psi} \quad \text{and} \quad \psi = 1$$

as

$$k_{nf} = \frac{k_p + (n-1)k_f - (n-1)\phi(k_f - k_p)}{k_p + (n-1)k_f + \phi(k_f - k_p)} k_f$$

$$= \frac{(40) + (3-1)(0.613) - (3-1)(0.05)(0.613 - 40)}{(40) + (3-1)(0.613) + (0.05)(0.613 - 40)} (0.613)$$

$$= 0.705 \ W/(m \cdot K)$$

and where

$$n = \frac{3}{\psi} \quad \text{and} \quad \psi = 3$$

as

$$\begin{aligned}
k_{nf} &= \frac{k_p + (n-1)k_f - (n-1)\phi(k_f - k_p)}{k_p + (n-1)k_f + \phi(k_f - k_p)} k_f \\
&= \frac{(40) + (1-1)(0.613) - (1-1)(0.05)(0.613 - 40)}{(40) + (1-1)(0.613) + (0.05)(0.613 - 40)} (0.613) \\
&= 0.645 \text{ W}/(\text{m} \cdot \text{K})
\end{aligned}$$

As can be seen the enhancement is not anomalous.

5.2.2.2 Brownian Motion of Nanoparticles

Brownian motion is the random motion of particles suspended in a fluid. When nanofluids are considered, this random motion transports energy directly by nanoparticles. In addition, a microconvection effect, which is due to the fluid mixing around nanoparticles, is also proposed to be important. Brownian dynamics simulation is used to determine the effective thermal conductivity of nanofluids[79] by considering the Brownian motion of the nanoparticles.

The thermal conductivity of nanofluids can be assumed as the combination of two parts[80]:

$$k_{nf} = k_{static} + k_{Brownian} \tag{5.57}$$

For the static part, the classical Maxwell model[77] was proposed as

$$\frac{k_{nf}}{k_f} = \frac{k_p + 2k_f + 2(k_p - k_f)\phi}{k_p + 2k_f - (k_p - k_f)\phi} \tag{5.58}$$

For $k_{Brownian}$, Brownian motion of particles was considered together with the effect of fluid particles moving with nanoparticles around them. As a result, the following expression was proposed:

$$k_{Brownian} = 5 \times 10^4 \beta_\beta \phi \rho_f c_{p,f} \sqrt{\frac{\kappa_B T}{\rho_p d_p}} f \tag{5.59}$$

where κ_B is the Boltzmann constant ($1.3806503 \times 10^{-23}$ J/K):

TABLE 5.9
β_β Values for Different Nanoparticles to be Used in Equation 5.59

Type of Particles	β_β	Remarks
Au-citrate, Ag-citrate and CuO	$0.0137(100\phi)^{-0.8229}$	$\phi < 1\%$
CuO	$0.0011(100\phi)^{-0.7272}$	$\phi > 1\%$
Al_2O_3	$0.0017(100\phi)^{-0.0841}$	$\phi > 1\%$

Source: From Koo, J. and Kleinstreuer, C., *J. Nanopart. Res.*, 6(6), 577–588, 2004. With permission.

$$f = (-134.63 + 1722.3\phi) + (0.4705 - 6.04\phi)T \qquad (5.60)$$

which is obtained by using the results of the study of Das et al.[81] for CuO nanofluids. For other nanofluids, f can be taken as 1 due to lack of experimental data. Associated β_β values are listed in Table 5.9. It is difficult to determine theoretical expressions for f and β_β due to the complexities involved and this can be considered as a drawback of the model.[82] In the analysis, the interactions between nanoparticles and fluid volumes moving around them were not considered and an additional term, β_β, was introduced in order to take that effect into account.

Example 5.6

By considering the combination of static and Brownian motion parts, determine the thermal conductivity of nanofluid, alumina (Al_2O_3) nanoparticles with a mean diameter of 36 nm dispersed in water at 273.15 K and 2% particle volume fraction by using Maxwell model to calculate the static part. Density of alumina is considered to be 4 g/cm³. Properties of water can be obtained from the water properties at pressure equals to 1 bar.

Solution

Properties of alumina (Al_2O_3), from Table 5.8, are

$$k_p = 40 \text{ W}/(\text{m} \cdot \text{K})$$

$$\phi = 0.02$$

$$\rho_p = 4000 \text{ kg}/\text{m}^3$$

$$d_p = 36 \times 10^{-9} \text{ m}$$

$$\kappa_B = 1.3807 \times 10^{-23} \text{ m}^2\text{kg}/\text{s}^2\text{K}$$

From properties of water at $P = 1$ bar (100 kPa) and $T = 273.15$ K (0°C):

$$k_f = 0.552 \text{ W}/(\text{m}\cdot\text{K})$$

$$\rho_f = 999.8 \text{ kg}/\text{m}^3$$

$$c_{p,f} = 4217.9 \text{ J}/\text{kg}\cdot\text{K}$$

β_β value of alumina (Al_2O_3), from Table 5.9, are

$$\beta_\beta = 0.0017(100\phi)^{-0.0841}$$

$$= 0.0017[(100)(0.02)]^{-0.0841}$$

$$= 0.0016$$

$$k_{static} = k_{nf} = \frac{k_p + 2k_f + 2(k_p - k_f)\phi}{k_p + 2k_f - (k_p - k_f)\phi} k_f$$

$$= \frac{(40) + (2)(0.552) + (2)(40 - 0.552)(0.02)}{(40) + (2)(0.552) - (40 - 0.552)(0.02)}(0.552)$$

$$= 0.584 \text{ W}/(\text{m}\cdot\text{K})$$

f can be taken as 1 due to lack of experimental data, so

$$k_{Brownian} = 5 \times 10^4 \beta_\beta \phi \rho_f c_{p,f} \sqrt{\frac{\kappa_B T}{\rho_p d_p}} f$$

$$= (5 \times 10^4)(0.0016)(0.02)(998.8)(4217.9)\sqrt{\frac{(1.3806503 \times 10^{-23})(273.15)}{(4000)(36 \times 10^{-9})}} \quad (1)$$

$$= 0.035 \text{ W}/(\text{m}\cdot\text{K})$$

$$k_{nf} = k_{static} + k_{Brownian}$$

$$= 0.584 + 0.035$$

$$= 0.619 \text{ W}/(\text{m}\cdot\text{K})$$

5.2.2.3 Clustering of Nanoparticles

Nanoparticles are known to form clusters.[83,84] These clusters can be handled by using fractal theory.[61] Clustering can result in fast transport of heat along relatively large distances since heat can be conducted much faster by solid particles when compared to liquid matrix.[85]

Effect of clusters was analyzed in three steps by using Bruggeman model[83,86] and Maxwell–Garnett (M–G) model.[61,83] The resulting thermal conductivity ratio expression is

$$\frac{k_{nf}}{k_f} = \frac{(k_{cl} + 2k_f) + 2\phi_{cl}(k_{cl} - k_f)}{(k_{cl} + 2k_f) - \phi_{cl}(k_{cl} - k_f)} \tag{5.61}$$

It was shown that the effective thermal conductivity increased with increasing cluster size. However, as particle volume fraction increased, the nanofluid with clusters showed relatively smaller thermal conductivity enhancement. When it comes to interfacial resistance, it was found that interfacial resistance decreases the enhancement in thermal conductivity, but this decrease diminishes for nanofluids with large clusters. Fiber-shaped nanoparticles are more effective in thermal conductivity enhancement when compared to spherical particles. However, it was also noted that such fiber shaped particles or clusters increase the viscosity of the nanofluids significantly. It should be noted that excessive clustering of nanoparticles may result in sedimentation, which adversely affects the thermal conductivity.[83] Therefore, there should be an optimum level of clustering for maximum thermal conductivity enhancement.[82]

Enhancement mechanisms such as microconvection created by Brownian motion of nanoparticles, nanolayer formation around particles, and near field radiation were concluded not to be the major cause of the enhancement. It was noted that effective medium theories can predict the experimental data well when the effect of clustering is taken into account. The effect of clustering by taking the effect of particle size into account is modeled.[87] It was found that clustering improves thermal conductivity enhancement and formation of clusters is more pronounced in nanofluids with smaller nanoparticles since distances between nanoparticles are smaller in those nanofluids, which increases the importance of van der Waals forces attracting particles to each other. The thermal conductivity study of nanofluids by considering Brownian motion and clustering of nanoparticles propose an equation to predict the thermal conductivity of nanofluids[88]:

$$\frac{k_{nf}}{k_f} = \frac{k_p + 2k_f - 2\phi(k_f - k_p)}{k_p + 2k_f + \phi(k_f - k_p)} + \frac{\rho_p \phi c_{p,p}}{2k_f} \sqrt{\frac{\kappa_B T}{3\pi r_{cl} \mu_f}} \tag{5.62}$$

It can be seen that the first term on the right-hand side of Equation 5.62 is the Maxwell model[77] for thermal conductivity of suspensions of solid particles in fluids and the second term on the right-hand side of Equation 5.62 adds the effect of the random motion of the nanoparticles into account.

Example 5.7

Determine the thermal conductivity of nanofluid in Example 5.6 when considering the apparent radius of the nanoparticle clusters equal to $r_{cl1} = 10$ nm and $r_{cl2} = 100$ nm respectively. Compare results from both cluster radius values.

Solution

Properties of alumina (Al_2O_3), from Table 5.8, are

$$k_p = 40 \ W/(m \cdot K)$$
$$\phi = 0.02$$
$$\rho_p = 4000 \ kg/m^3$$
$$\kappa_B = 1.3807 \times 10^{-23} \ m^2 kg/s^2 K$$
$$c_{p,p} = 718 J/kg \cdot K$$

From properties of water at $P = 1$ bar (100 kPa) and $T = 273.15$ K (0°C):

$$k_f = 0.552 \ W/(m \cdot K)$$
$$\rho_f = 999.8 \ kg/m^3$$
$$c_{p,f} = 4217.9 J/kg \cdot K$$
$$\mu_f = 0.0017909 Pa \cdot s$$

when

$$r_{cl1} = 10 \ nm = 10 \times 10^{-9} m$$

so

$$\frac{k_{nf}}{k_f} = \frac{k_p + 2k_f - 2\phi(k_f - k_p)}{k_p + 2k_f + \phi(k_f - k_p)} + \frac{\rho_p \phi c_{p,p}}{2k_f} \sqrt{\frac{\kappa_B T}{3\pi r_{cl} \mu_f}}$$

$$= \frac{(40) + (2)(0.552) - (2)(0.02)(0.552 - 40)}{(40) + (2)(0.552) + (0.02)(0.552 - 40)}$$

$$+ \frac{(4000)(0.02)(718)}{(2)(0.552)} \sqrt{\frac{(1.3806503 \times 10^{-23})(273.15)}{(3)(22/7)(10 \times 10^{-9})(0.0017909)}}$$

$$= 1.3046 \quad \text{or} \quad 30.46\% \text{ enhancement}$$

when

$$r_{cl1} = 100 \ nm = 100 \times 10^{-9} m$$

so

$$\frac{k_{nf}}{k_f} = \frac{k_p + 2k_f - 2\phi(k_f - k_p)}{k_p + 2k_f + \phi(k_f - k_p)} + \frac{\rho_p \phi c_{p,p}}{2k_f} \sqrt{\frac{\kappa_B T}{3\pi r_{cl} \mu_f}}$$

$$= \frac{(40) + (2)(0.552) - (2)(0.02)(0.552 - 40)}{(40) + (2)(0.552) + (0.02)(0.552 - 40)}$$

$$+ \frac{(4000)(0.02)(718)}{(2)(0.552)} \sqrt{\frac{(1.3806503 \times 10^{-23})(273.15)}{(3)(22/7)(100 \times 10^{-9})(0.0017909)}}$$

$$= 1.1365 \quad \text{or} \quad 13.65\% \text{ enhancement}$$

5.2.2.4 Liquid Layering around Nanoparticles

Liquid molecules form layered structures around solid surfaces,[89] and it is expected that those nanolayers have a larger effective thermal conductivity than the liquid matrix.[90] The layered structures that form around nanoparticles are proposed to be responsible for the thermal conductivity enhancement of nanofluids.[90]

Yu and Choi[90] presented a model for the determination of the effective thermal conductivity of nanofluids by modifying the Maxwell model.[7] In the modification, the effect of the liquid nanolayers formed around nanoparticles was taken into account. The nanoparticle and the layer around it were considered as a single particle and the thermal conductivity of this particle was determined by using effective medium theory.[91] The result was substituted into the Maxwell model and the following expression was obtained:

$$k_{nf} = \frac{k_{pe} + 2k_f + 2(k_{pe} - k_f)(1+\beta_L)^3 \phi}{k_{pe} + 2k_f - (k_{pe} - k_f)(1+\beta_L)^3 \phi} k_f \quad (5.63)$$

where

$$k_{pe} = \frac{[2(1-\gamma_k) + (1+\beta_L)^3(1+2\gamma_k)]\gamma_k}{-(1-\gamma_k) + (1+\beta_L)^3(1+2\gamma_k)} k_p \quad (5.64)$$

$$\gamma_k = \frac{k_l}{k_p} \quad (5.65)$$

and k_l is thermal conductivity of the nanolayer. β is defined as

$$\beta_L = \frac{t}{r_p} \quad (5.66)$$

Surface adsorption can also be considered and it was proposed that the thickness of the adsorption layer can be found by using the formula[92]

$$t = \frac{1}{\sqrt{3}}\left(\frac{4M_f}{\rho_f N_A}\right)^{1/3} \qquad (5.67)$$

where N_A the Avogadro constant (6.023×10^{23}/mol). Since k_l is difficult to predict, this model can be used as an upper bound for the determination of effective thermal conductivity by letting $k_l = k_p$.

Example 5.8

Determine the effective thermal conductivity of nanofluid, CuO nanoparticles with a particle radius of 100 nm dispersed in water at 273.15 K and 2% particle volume fraction by using Yu and Choi model. Properties of water can be obtained from the water properties at pressure equals to 1 bar.

Solution

Properties of alumina (Al_2O_3), from Table 5.8, are

$$k_p = 40 \ \text{W}/(\text{m} \cdot \text{K})$$

$$\phi = 0.02$$

$$r_p = 100 \times 10^{-9} \text{m}$$

$$N_A = 6.023 \times 10^{23} \text{mol}^{-1}$$

From properties of water at P = 1 bar (100 kPa) and T = 273.15 K (0°C):

$$k_f = 0.552 \ \text{W}/(\text{m} \cdot \text{K})$$

$$\rho_f = 999.8 \text{kg}/\text{m}^3$$

$$M_f = 18 \times 10^{-3} \text{kg}/\text{mol}$$

where

$$\varepsilon_p = k_p/k_f = 20/0.552 = 36.232$$

$$t = \frac{1}{\sqrt{3}}\left(\frac{4M_f}{\rho_f N_A}\right)^{1/3}$$

$$= \frac{1}{\sqrt{3}}\left(\frac{(4)(18 \times 10^{-3})}{(999.8)(6.023 \times 10^{23})}\right)^{1/3}$$

$$= 2.8443 \times 10^{-10} \text{m}$$

$$\beta_L = t/r_p = 2.8443 \times 10^{-10}/100 \times 10^{-9} = 0.0028443$$

$$Z = \varepsilon_p(1+\beta_L) - 1 = 36.232(1+0.0028443) - 1 = 35.335$$

$$k_l = \frac{k_f Z^2}{(Z-\beta_L)\ln(1+Z)+\beta_L Z}$$

$$= \frac{(0.552)(35.335)^2}{(35.335-0.0028443)\ln(1+35.335)+(0.0028443)(35.335)}$$

$$= 5.425 \text{ W}/(\text{m}\cdot\text{K})$$

$$\gamma_k = k_l/k_p = 5.425/20 = 0.271$$

$$k_{pe} = \frac{[2(1-\gamma_k)+(1+\beta_L)^3(1+2\gamma_k)]\gamma_k}{-(1-\gamma_k)+(1+\beta_L)^3(1+2\gamma_k)}k_p$$

$$= \frac{[(2)(1-0.271)+(1+0.0028443)^3(1+(2)(0.271))](0.271)}{-(1-0.271)+(1+0.0028443)^3(1+(2)(0.271))}(20)$$

$$= 19.767 \text{ W}/(\text{m}\cdot\text{K})$$

$$k_{nf} = \frac{k_{pe}+2k_f+2(k_{pe}-k_f)(1+\beta_L)^3\phi}{k_{pe}+2k_f-(k_{pe}-k_f)(1+\beta_L)^3\phi}k_f$$

$$= \frac{(19.767)+(2)(0.552)+(2)(19.767-0.552)(1+0.0028443)^3(0.02)}{(19.767)+(2)(0.552)-(19.767-0.552)(1+0.0028443)^3(0.02)}(0.552)$$

$$= 0.583 \text{ W}/(\text{m}\cdot\text{K})$$

A nanolayer was modeled as a spherical shell with thickness t around the nanoparticle, by assuming that the thermal conductivity changes linearly across the radial direction, so that it is equal to thermal conductivity of base liquid at the nanolayer–liquid interface and equal to thermal conductivity of the nanoparticle at the nanolayer–nanoparticle interface. The associated expression for the determination of the thermal conductivity of nanofluid was given as

$$\frac{k_{nf}-k_f}{k_f} = 3\Theta\phi_T + \frac{3\Theta^2\phi_T^2}{1-\Theta\phi_T} \tag{5.68}$$

where

$$\Theta = \frac{\beta_{lf}\left[(1+\beta_L)^3 - \frac{\beta_{pl}}{\beta_{fl}}\right]}{(1+\beta_L)^3 + 2\beta_{lf}\beta_{pl}} \tag{5.69}$$

and

$$\beta_{lf} = \frac{k_l - k_f}{k_l + 2k_f} \tag{5.70}$$

$$\beta_{pl} = \frac{k_p - k_l}{k_p + 2k_l} \tag{5.71}$$

$$\beta_{fl} = \frac{k_f - k_l}{k_f + 2k_l} \tag{5.72}$$

$$\phi_T = \phi(1+\beta_L)^3 \tag{5.73}$$

$$k_l = \frac{k_f Z^2}{(Z-\beta_L)\ln(1+Z) + \beta_L Z} \tag{5.74}$$

$$Z = \varepsilon_p(1+\gamma) - 1 \tag{5.75}$$

$$\varepsilon_p = k_p/k_f \tag{5.76}$$

Example 5.9

Calculate the thermal conductivity of nanofluid, CuO nanoparticles with a particle radius of 100 nm dispersed in water at 273.15 K and 2% particle volume fraction, by considering a nanolayer as a spherical shell from Equation 5.68, t from Equation 5.67 and letting $k_l = k_p$. Properties of water can be obtained from the water properties at pressure equals to 1 bar.

Solution

Properties of copper oxide (CuO), from Table 5.8, are

$$k_p = 20 \text{ W}/(\text{m}\cdot\text{K})$$

$$\phi = 0.02$$

$$r_p = 100 \times 10^{-9} \text{ m}$$

$$N_A = 6.023 \times 10^{23} \text{ mol}^{-1}$$

From properties of water at $P = 1$ bar (100 kPa) and $T = 273.15$ K (0°C):

$$k_f = 0.552 \ \text{W}/(\text{m} \cdot \text{K})$$
$$\rho_f = 999.8 \ \text{kg}/\text{m}^3$$
$$M_f = 18 \times 10^{-3} \ \text{kg}/\text{mol}$$

A nanolayer was modeled as a spherical shell by considering these following parameters and letting $k_l = k_p$:
From

$$\beta_{lf} = \frac{k_l - k_f}{k_l + 2k_f} = \frac{20 - 0.552}{20 + (2)(0.552)} = 0.922$$

$$\beta_{pl} = \frac{k_p - k_l}{k_p + 2k_l} = \frac{20 - 20}{20 + (2)(20)} = 0$$

$$\beta_{fl} = \frac{k_f - k_l}{k_f + 2k_l} = \frac{0.552 - 20}{0.552 + (2)(20)} = -0.480$$

$$t = \frac{1}{\sqrt{3}} \left(\frac{4M_f}{\rho_f N_A} \right)^{1/3} = \frac{1}{\sqrt{3}} \left(\frac{(4)(18 \times 10^{-3})}{(999.8)(6.023 \times 10^{23})} \right)^{1/3}$$
$$= 2.8443 \times 10^{-10} \ \text{m}$$

$$\beta_L = t/r_p = \frac{(2.8443 \times 10^{-10})}{(100 \times 10^{-9})} = 0.0028443$$

$$\phi_T = \phi(1 + \beta_L)^3 = (0.02)(1 + 0.0028443)^3 = 0.02017$$

$$\Theta = \frac{\beta_{lf} \left[(1+\beta_L)^3 - \dfrac{\beta_{pl}}{\beta_{fl}} \right]}{(1+\beta_L)^3 + 2\beta_{lf}\beta_{pl}} = \frac{(0.922) \left[(1+0.0028443)^3 - \dfrac{(0)}{(-0.480)} \right]}{(1+0.0028443)^3 + (2)(0.922)(0)}$$
$$= 0.922$$

$$\frac{k_{nf} - k_f}{k_f} = 3\Theta\phi_T + \frac{3\Theta^2 \phi_T^2}{1 - \Theta\phi_T}$$

$$\frac{k_{nf} - 0.552}{0.552} = (3)(0.922)(0.02017) + \frac{(3)(0.922)^2(0.02017)^2}{1 - (0.922)(0.02017)}$$

$$k_{nf} = (0.552) \left[(3)(0.922)(0.02017) + \frac{(3)(0.922)^2(0.02017)^2}{1 - (0.922)(0.02017)} \right] + (0.552)$$

$$k_{nf} = 0.583 \ \text{W}/(\text{m} \cdot \text{K})$$

The nanoparticles can be assumed to have a liquid layer around them with a specific thermal conductivity.[93] First, an expression for the effective thermal conductivity of the "complex particle," which was defined as the combination of the nanoparticle and nanolayer, was determined. Then, by using Bruggeman's effective media theory,[92] the effective thermal conductivity of the nanofluid was determined. The resulting implicit expression for thermal conductivity of nanofluids is

$$\left(1-\frac{\phi}{\alpha_r}\right)\frac{k_{nf}-k_f}{2k_{nf}+k_f}+\frac{\phi}{\alpha_r}\frac{(k_{nf}-k_l)(2k_l+k_p)-\alpha_r(k_p-k_l)(2k_l+k_{nf})}{\alpha_r(2k_{nf}+k_l)(2k_l+k_p)+2\alpha_r(k_p-k_l)(k_l-k_{nf})}=0 \quad (5.77)$$

where

$$\alpha_r = \left(\frac{r_p}{r_p+t}\right)^3 \quad (5.78)$$

The effect of temperature on clustering according to the fact that decreasing temperature results in a decrease in particle surface energy, which decreases the severity of clustering. As a result, the average cluster size was proposed to be calculated by[88]

$$r_{cl} = r_{cl0}(1+z\Delta T) \quad (5.79)$$

where ΔT is defined as the difference between the nanofluid temperature and reference temperature. z is a negative constant.

Example 5.10

Calculate the thermal conductivity of nanofluid in Example 5.9 by considering the combination of the nanoparticle and nanolayer.

Solution

Properties of copper oxide (CuO), from Table 5.8, are

$$k_p = 20 \text{ W}/(\text{m}\cdot\text{K})$$
$$\phi = 0.02$$
$$r_p = 100\times 10^{-9}\text{m}$$
$$N_A = 6.023\times 10^{23}\text{mol}^{-1}$$

From properties of water at $P = 1$ bar (100 kPa) and $T = 273.15$ K (0°C):

$$k_f = 0.552 \text{ W}/(\text{m} \cdot \text{K})$$

$$\rho_f = 999.8 \text{ kg/m}^3$$

$$M_f = 18 \times 10^{-3} \text{ kg/mol}$$

From

$$t = \frac{1}{\sqrt{3}} \left(\frac{4M_f}{\rho_f N_A} \right)^{1/3} = \frac{1}{\sqrt{3}} \left(\frac{(4)(18 \times 10^{-3})}{(999.8)(6.023 \times 10^{23})} \right)^{1/3}$$

$$= 2.8443 \times 10^{-10} \text{ m}$$

$$\alpha_r = \left(\frac{r_p}{r_p + t} \right)^3 = \left(\frac{100 \times 10^{-9}}{(100 \times 10^{-9}) + (2.8443 \times 10^{-10})} \right)^3 = 0.992$$

$$\left(1 - \frac{\phi}{\alpha_r}\right) \frac{k_{nf} - k_f}{2k_{nf} + k_f} + \frac{\phi}{\alpha_r} \frac{(k_{nf} - k_l)(2k_l + k_p) - \alpha_r(k_p - k_l)(2k_l + k_{nf})}{(2k_{nf} + k_l)(2k_l + k_p) + 2\alpha_r(k_p - k_l)(k_l - k_{nf})} = 0$$

$$\left(1 - \frac{0.02}{0.992}\right) \frac{k_{nf} - 0.552}{2k_{nf} + 0.552}$$

$$+ \left(\frac{0.02}{0.992}\right) \frac{(k_{nf} - 20)((2)(20) + 20) - (0.992)(20 - 20)((2)(20) + k_{nf})}{(2k_{nf} + 20)((2)(20) + 20) + (2)(0.992)(20 - 20)(20 - k_{nf})} = 0$$

$$(0.980) \frac{k_{nf} - 0.552}{2k_{nf} + 0.552} + (0.0202) \frac{(k_{nf} - 20)(60)}{(2k_{nf} + 20)(60)} = 0$$

$$\frac{0.980 k_{nf} - 0.54096}{2k_{nf} + 0.552} + \frac{1.212 k_{nf} - 24.24}{120 k_{nf} + 1200} = 0$$

$$\frac{(0.980 k_{nf} - 0.54096)(120 k_{nf} + 1200) + (1.212 k_{nf} - 24.24)(2k_{nf} + 0.552)}{(2k_{nf} + 0.552)(120 k_{nf} + 1200)} = 0$$

$$(117.6 k_{nf}^2 + 1176 k_{nf} - 64.9152 k_{nf} - 649.152)$$

$$+ (2.424 k_{nf}^2 + 0.669024 k_{nf} - 48.48 k_{nf} - 13.38048) = 0$$

$$120.024 k_{nf}^2 + 1063.273824 k_{nf} - 662.53248 = 0$$

$$k_{nf}^2 + 8.859 k_{nf} - 5.52 = 0$$

$$k_{nf} = 0.5845 \text{ W}/(\text{m} \cdot \text{K})$$

5.2.3 Thermal Conductivity Experimental Studies of Nanofluids

Since the high thermal conductivity nanoparticles suspended in the base Experimental studies also show that thermal conductivity of nanofluids depends on many factors such as particle volume fraction, particle material, particle size, particle shape, base fluid material, and temperature. The effects of these factors can be concluded as following:

Particle volume fraction. Particle volume fraction is a parameter that is investigated in almost all of the experimental studies and the results are usually in agreement qualitatively. Most of the researchers report increasing thermal conductivity with increasing particle volume fraction and the relation found is usually linear.[82] However, there are also some studies which indicate nonlinear behavior.[94,95]

Particle materials. The experimental studies from literatures show that particle material is an important parameter that affects the thermal conductivity of nanofluids.[95,96] The heat is transported ballistically inside the nanotubes improves the conduction of heat in the tubes; it should also be noted that the shape of nanotubes might also be effective in the anomalous enhancement values. It should also be noted that the mean-free path of phonons in nanoparticles may be smaller than the size of the nanoparticles. In such a condition, heat transfer mechanism inside the particles is not diffusion but heat is transported ballistically. When this fact is considered, relating the superior enhancement characteristics of a specific nanoparticle material to its high bulk thermal conductivity is not reasonable.[82]

Base fluid. According to the conventional thermal conductivity models such as the Maxwell model,[77] as the base fluid thermal conductivity of a mixture decreases, the thermal conductivity ratio (thermal conductivity of nanofluid (k_{nf}) divided by the thermal conductivity of base fluid (k_f)) increases. It should also be noted that the viscosity of the base fluid affects the Brownian motion of nanoparticles and that in turn affects the thermal conductivity of the nanofluid[88] and the effect of electric double layer forming around nanoparticles on the thermal conductivity of nanofluids shows that the thermal conductivity and thickness of the layer depends on the base fluid.[97]

Particle size. The general trend in the experimental data is that the thermal conductivity of nanofluids increases with decreasing particle size.[78,98–100] It should be noted that clustering may increase or decrease the thermal conductivity enhancement. If a network of nanoparticles is formed as a result of clustering, this may enable fast heat transport along nanoparticles. On the other hand, excessive clustering may result in sedimentation, which decreases the effective particle volume fraction of the nanofluid.[82] Although clustering at a certain level may improve thermal conductivity enhancement, excessive clustering creates an opposite effect due to associated sedimentation.[83]

Particle shape. There are mainly two particle shapes used in nanofluid research: spherical particles and cylindrical particles. It should be noted that nanofluids with cylindrical particles usually have much larger viscosities than those with spherical nanoparticles.[101] As a result, the associated increase in pumping power is large and this reduces the feasibility of usage of nanofluids with cylindrical particles.

Temperature. In case of nanofluids, change of temperature affects the Brownian motion of nanoparticles and clustering of nanoparticles,[102] which results in dramatic changes of thermal conductivity of nanofluids with temperature.[103–106] The experimental results generally suggest that thermal conductivity ratio increases with temperature.

Clustering. It is the formation of larger particles through aggregation of nanoparticles. Clustering effect is always present in nanofluids and it is an effective parameter in thermal conductivity. It was seen that thermal conductivity ratio increased with increasing vibration time and the rate of this increase became smaller for longer vibration time.[107] The size of the clusters formed by the nanoparticles had a major influence on the thermal conductivity.

The experimental studies on thermal conductivity of nanofluids are summarized in Table 5.10.[66] When Table 5.10 is observed, it is seen that there exists significant discrepancies in experimental data. An important issue regarding this discrepancy in experimental data is the debate about the measurement techniques. Natural convection effect in the transient hot-wire method starts to deviate the results in a sense that higher values are measured by the method. Another important reason of discrepancy in experimental data is clustering of nanoparticles. Although there are no universally accepted quantitative values, it is known that the level of clustering affects the thermal conductivity of nanofluids.[107] The level of clustering depends on many parameters. It was shown that adding some surfactants and adjusting the pH value of the nanofluid provide better dispersion and prevent clustering to some extent.[108] As a consequence, two nanofluid samples with all of the parameters being the same can lead to completely different experimental results if their surfactant parameters and pH values are not the same.[82] Therefore, when performing experiments, researchers should also consider the type and amount of additives used in the samples and pH value of the samples.

A commonly utilized way of obtaining good dispersion and breaking the clusters is to apply ultrasonic vibration to the samples. The duration and the intensity of the vibration affect the dispersion characteristics. Moreover, immediately after the application of vibration, clusters start to form again and size of the clusters increases as time progresses.[107] Therefore, the time between the application of vibration and measurement of thermal conductivity, duration of vibration, and intensity of vibration also affect the thermal conductivity of nanofluids, which creates discrepancy in experimental data in the literature.[82]

TABLE 5.10

Summary of Experimental Studies of Thermal Conductivity Enhancement

	Particle Type	Base Fluid[a]	Particle Volume Fraction (%)	Particle Size (nm)	Max. Enhance-ment (%)[b]	Notes
Masuda et al.[109]	Al_2O_3	Water	1.30–4.30	13	32.4	31.85°C
	SiO_2	Water	1.10–2.40	12	1.1	–
	TiO_2	Water	3.10–4.30	27	10.8	86.85°C
Lee et al.[96]	Al_2O_3	water / EG	1.00–4.30 / 1.00–5.00	38.4	10 / 18	Room temperature
	CuO	water / EG	1.00–3.41 / 1.00–4.00	23.6	12 / 23	
Wang et al.[73]	Al_2O_3	water / EG	3.00–5.50 / 5.00–8.00	28	16 / 41	Room temperature
	Al_2O_3	EO/PO	2.25–7.40 / 5.00–7.10	28	30 / 20	
	CuO	water / EG	4.50–9.70 / 6.20–14.80	23	34 / 54	
Eastman et al.[68]	Cu	EG	0.01–0.56	<10	41	Room temperature
Xie et al.[110]	SiC	water / EG	0.78–4.18 / 0.89–3.50	26 sphere	17 / 13	Effect of particle shape and size is examined.
	SiC	water / EG	1.00–4.00	600 cylinder	24 / 23	
Xie et al.[111]	Al_2O_3	water / EG	5.00	60.4	23 / 29	Room temperature
	Al_2O_3	PO/glycerol	5.00	60.4	38 / 27	
Das et al.[81]	Al_2O_3	Water	1.00–4.00	38.4	24	21°C – 51°C
	CuO	water	1.00–4.00	28.6	36	
Murshed et al.[94]	TiO_2	water	0.50–5.00	15 sphere	30	Room temperature
	TiO_2	water	0.50–5.00	10 × 40 rod	33	
Hong et al.[107]	Fe	EG	0.10–0.55	10	18	Effect of clustering was investigated.
Li and Peterson[112]	Al_2O_3	water	2.00–10.00	36	29	27.5°C – 34.7°C
	CuO	water	2.00–6.00	29	51	28.9°C – 33.4°C

(Continued)

TABLE 5.10 (CONTINUED)

Summary of Experimental Studies of Thermal Conductivity Enhancement

	Particle Type	Base Fluid[a]	Particle Volume Fraction (%)	Particle Size (nm)	Max. Enhance-ment (%)[b]	Notes
Chopkar et al.[98]	Al$_2$Cu	water/EG	1.00–2.00	31/68/101	96/76/61	Effect of particle size was examined.
	Ag$_2$Al	water/EG	1.00–2.00	33/80/120	106/93/75	
Beck et al.[72]	Al$_2$O$_3$	water	1.86–4.00	8–282	20	Effect of particle shape and size is examined.
	Al$_2$O$_3$	EG	2.00–3.01	12–282	19	
Mintsa et al.[113]	Al$_2$O$_3$	water	0–18	36/47	31/31	20°C–48°C
	CuO	water	0–16	29	24	
Turgut et al.[114]	TiO$_2$	water	0.2–3.0	21	7.4	13°C–55°C
Choi et al.[95]	MWCNT	PAO	0.04–1.02	25 × 50,000	57	Room temperature
Assael et al.[115]	DWCNT	water	0.75–1.00	5 (diameter)	8	Effect of sonication time was examined.
	MWCNT	water	0.60	130 × 10,000	34	
Liu et al.[116]	MWCNT	EG/EO	0.20–1.00/1.00–2.00	20–50 (diameter)	12/30	Room temperature
Ding et al.[69]	MWCNT	water	0.05–0.49	40 nm (diameter)	79	20°C–30°C

[a] EG: ethylene glycol, EO: engine oil, PO: pump oil, TO: transformer oil, PAO: polyalphaolefin
[b] The percentage values indicated are according to the expression $100(k_{nf} - k_f)/k_f$

5.2.4 Convective Heat Trasfer of Nanofluids

When forced convection in tubes is considered, it is expected that heat transfer coefficient enhancement obtained by using a nanofluid is equal to the enhancement in thermal conductivity of the nanofluid, due to the definition of Nusselt number which is not the case.

In the analysis of convective heat transfer of nanofluids, accurate determination of the thermophysical properties is a key issue. The thermal conductivity values, calculated as described above, can be used by considering different perspectives and assumptions.

Density. Density of nanofluids can be determined by using the following expression[117]:

$$\rho_{nf} = \phi \rho_p + (1-\phi)\rho_f \qquad (5.90)$$

Example 5.11

Determine the density of nanofluid, CuO/Water at 2% and 3% particle volume fractions. Properties of water can be obtained from the water properties at pressure equals to 1 bar. Density of CuO is considered to be 6.31 g/cm³.

Solution

Given

$$\rho_p = 6310 \text{ kg/m}^3$$

From a property of water at $P = 1$ bar (100 kPa) and $T = 273.15$ K (0°C):

$$\rho_f = 999.8 \text{ kg/m}^3$$

when

$$\phi = 0.02$$

$$\rho_{nf} = \phi \rho_p + (1-\phi)\rho_f$$
$$= (0.02)(6310) + (1-0.02)(999.8) = 1106.004 \text{ kg/m}^3$$

when

$$\phi = 0.03$$

$$\rho_{nf} = \phi\rho_p + (1-\phi)\rho_f$$
$$= (0.03)(6310) + (1-0.03)(999.8) = 1159.106 \text{ kg/m}^3$$

Specific heat. There are two expressions for determining the specific heat of nanofluids[117,118]:

$$c_{p,nf} = \phi c_{p,p} + (1-\phi)c_{p,f} \tag{5.81}$$

and

$$(\rho c_p)_{nf} = \phi(\rho c_p)_p + (1-\phi)(\rho c_p)_f \tag{5.82}$$

It is thought that Equation 5.82 is theoretically more consistent since specific heat is a mass specific quantity whose effect depends on the density of the components of a mixture.

Example 5.12

Compare specific heat values of nanofluid, CuO/Water, which are calculated by using Equations 5.81 and 5.82 at 273.15 K and 2% particle volume fractions. Properties of water can be obtained from the water properties at pressure equals to 1 bar where $c_{p,CuO} = 460 \text{ J/kg} \cdot \text{K}$.[119]

Solution

Given

$$\rho_p = 6310 \text{ kg/m}^3$$
$$\phi = 0.02$$
$$c_{p,CuO} = 460 \text{ J/kg} \cdot \text{K}$$

From a property of water at $P = 1$ bar (100 kPa) and $T = 273.15$ K (0°C):

$$\rho_f = 999.8 \text{ kg/m}^3$$
$$c_{p,f} = 4217.9 \text{ J/kg} \cdot \text{K}$$

$$\rho_{nf} = \phi\rho_p + (1-\phi)\rho_f$$
$$= (0.02)(6310) + (1-0.02)(999.8)$$
$$= 1106.004 \text{ kg/m}^3$$

From Equation 5.81, the specific heat value of the nanofluid is obtained as

$$c_{p,nf} = \phi c_{p,p} + (1-\phi)c_{p,f}$$
$$= (0.02)(460) + (1-0.02)(4217.9)$$
$$= 3383.52 \, \text{J/kg} \cdot \text{K}$$

and, from Equation 5.82, the specific heat value of the nanofluid is obtained as

$$(\rho c_p)_{nf} = \phi(\rho c_p)_p + (1-\phi)(\rho c_p)_f$$
$$((1106.004)c_p)_{nf} = (0.02)(6310)(460) + (1-0.02)(999.8)(4217.9)$$
$$((1106.004)c_p)_{nf} = 4190767.29$$
$$c_{p,nf} = \frac{4190767.29}{1106.004} = 3789.11 \, \text{J/kg} \cdot \text{K}$$

The value obtained from Equation 5.81 is lower than that obtained from Equation 5.82 because the former does not take the density values into account.

Viscosity. Nanofluid viscosity is an important parameter for practical applications since it directly affects the pressure drop in forced convection. Therefore, for enabling the usage of nanofluids in practical applications, the increasing viscosity of nanofluids with respect to pure fluids should be thoroughly investigated.

The general trend is that the increase in the viscosity by the addition of nanoparticles to the base fluid is significant. It is known that nanofluid viscosity depends on many parameters such as; particle volume fraction, particle size, temperature, and extent of clustering. The general trend from experimental data shows that the increase in viscosity with the addition of nanoparticles into the base fluid is larger than the theoretical predictions obtained by using conventional theoretical models.

An expression for determining the dynamic viscosity of dilute suspensions that contain spherical particles has been proposed.[120] In the model, the interactions between the particles are neglected as

$$\mu_{nf} = (1 + 2.5\phi)\mu_f \tag{5.83}$$

When the viscosity is considered to increase unboundedly as particle volume fraction reaches its maximum, the following equation has been suggested[121]

$$\mu_{nf} = \frac{1}{(1-\phi)^{2.5}} \mu_f \tag{5.84}$$

TABLE 5.11

Viscosity Models for Nanofluids

Tseng and Lin[122], TiO_2/water nanofluids

$\mu_{nf} = 13.47 e^{35.98\phi} \mu_f$

Maïga et al.[123], Al_2O_3/water nanofluids

$\mu_{nf} = (1 + 7.3\phi + 123\phi^2)\mu_f$

Maïga et al.[123], Al_2O_3/ethylene glycol nanofluids

$\mu_{nf} = (1 - 0.19\phi + 306\phi^2)\mu_f$

Koo and Kleinstreuer (2005), CuO-based nanofluids

$\mu_{eff} = \mu_f + 5 \times 10^4 \beta \rho_f \phi \sqrt{\dfrac{\kappa_B T}{2\rho_p r_p}} \left[(-134.63 + 1722.3\phi) + (0.4705 - 6.04\phi)T\right]$

where

$\beta = \begin{cases} 0.0137(100\phi)^{-0.8229} \text{ for } \phi < 0.01 \\ 0.0011(100\phi)^{-0.7272} \text{ for } \phi > 0.01 \end{cases}$

κ_B is Boltzmann constant, T is temperature in K, r_p is nanoparticle radius in m.

Kulkarni et al.[124], CuO/water nanofluids

$\mu_{nf} = \exp\left[-(2.8751 + 53.548\phi - 107.12\phi^2) + (1078.3 + 15857\phi + 20587\phi^2)(1/T)\right]$

Source: From Koo, J. and Kleinstreuer, C., *J. Nanopart. Res.*, 6(6), 577–588, 2004. With permission.

The equations are provided in Table 5.11 together with their applicable nanofluid type.

Example 5.13

Compare viscosity values of nanofluid, CuO/Water, which are calculated by using Equations 5.83 and 5.84 at 273.15 K and 2% particle volume fractions. Properties of water can be obtained from the water properties at pressure equals to 1 bar.

Solution

Given

$$\phi = 0.02$$

From a property of water at $P = 1$ bar (100 kPa) and $T = 273.15$ K (0°C):

$$c_{p,f} = 4217.9 \text{ J/kg} \cdot \text{K}$$

$$\mu_f = 0.0017909 \text{ Pa} \cdot \text{s}$$

Micro/Nano Heat Transfer

by using Equation 5.83, we obtain

$$\mu_{nf} = (1+2.5\phi)\mu_f$$
$$= (1+(2.5)(0.02))(0.0017909)$$
$$= 0.001880445 \text{ Pa}\cdot\text{s}$$

and by using Equation 5.84, we obtain

$$\mu_{nf} = \frac{1}{(1-\phi)^{2.5}}\mu_f$$
$$= \frac{1}{(1-0.02)^{2.5}}(0.0017909)$$
$$= 0.001883676 \text{ Pa}\cdot\text{s}$$

Example 5.14

Using information from Example 5.13 with following data to calculate the viscosity of the nanofluid at 323.15 K. Density of CuO is considered to be 6.31 g/cm³ and CuO nanoparticles have particle radius of 100 nm.

Solution

Given

$$\phi = 0.02$$
$$\rho_p = 6310 \text{ kg/m}^3$$
$$\kappa_B = 1.3807 \times 10^{-23} \text{ m}^2\text{kg/s}^2\text{K}$$
$$r_p = 100 \times 10^{-9} \text{ m}$$
$$T = 323.15 \text{ K}$$

From a property of water at $P = 1$ bar (100 kPa) and $T = 273.15$ K (0°C):

$$c_{p,f} = 4217.9 \text{ J/kg}\cdot\text{K}$$
$$\mu_f = 0.0017909 \text{ Pa}\cdot\text{s}$$
$$\rho_f = 988.1 \text{ kg/m}^3$$
$$\beta = 0.0011(100\phi)^{-0.7272} \text{ for } \phi > 0.01$$
$$= (0.0011)((100)(0.02))^{-0.7272}$$
$$= 0.000664483$$

From Koo and Kleinstreuer (2005), the dynamic viscosity of CuO based nanofluids can be obtained from

$$\mu_{eff} = \mu_f + 5 \times 10^4 \beta \rho_f \phi \sqrt{\frac{\kappa_B T}{2\rho_p r_p}} \left[(-134.63 + 1722.3\phi) + (0.4705 - 6.04\phi)T\right]$$

$$= (0.0005468) + (5 \times 10^4)(0.000664483)(988.1)(0.02)$$

$$\times \sqrt{\frac{(1.3806503 \times 10^{-23})(323.15)}{(2)(6310)(100 \times 10^{-9})}} \left[((-134.63) + (1722.3)(0.02))\right.$$

$$+ ((0.4705) - (6.04)(0.02))(323.15)]$$

$$= 0.000562628 \text{ Pa} \cdot \text{s}$$

5.2.5 Analysis of Convective Heat Transfer of Nanofluids

Accurate theoretical analysis of convective heat transfer of nanofluids is very important for the practical application of nanofluids in thermal devices. One may assume that the heat transfer performance of nanofluids can be analyzed by using classical correlations developed for the determination of Nusselt number for the flow of pure fluids. Then, in the calculations, one should substitute the thermophysical properties of the nanofluid to the associated correlations. This approach is a very practical way of analyzing convective heat transfer of nanofluids. By comparing the results of such an analysis with the experimental data available for nanofluids in the literature, to examine the validity of the approach. It should be noted that this approach underpredict the heat transfer coefficient.

5.2.5.1 Constant Wall Heat Flux Boundary Condition

One of the most commonly used empirical correlations for the determination of convective heat transfer in laminar flow regime inside circular tubes is the Shah correlation.[125] The associated expression for the calculation of local Nusselt number is as follows:

$$\begin{aligned}
Nu_x &= 1.302 X^{-1/3} - 1 & \text{for } X \leq 5 \times 10^{-5} \\
&\; 1.302 X^{-1/3} - 0.5 & \text{for } 5 \times 10^{-5} < X \leq 1.5 \times 10^{-3} \\
&\; 4.364 + 0.263 X^{-0.506} e^{-41X} & \text{for } X > 1.5 \times 10^{-3}
\end{aligned} \quad (5.85)$$

where

$$X = \frac{x/d}{Re\ Pr} = \frac{x/d}{Pe} = \frac{1}{Gz_x} \qquad (5.86)$$

At this point, it should be noted that for the same tube diameter and same flow velocity, Pe_f and Pe_{nf} are different.

$$\frac{Pe_{nf}}{Pe_f} = \frac{\alpha_f}{\alpha_{nf}} = \frac{k_f}{k_{nf}} \frac{\rho_{nf}}{\rho_f} \frac{c_{p,nf}}{c_{p,f}} \qquad (5.87)$$

Enhancement in density and specific heat increases Pe_{nf} whereas thermal conductivity increase results in a decrease in Pe_{nf}. Therefore, when the pure working fluid in a system is replaced with a nanofluid, the flow velocity should be adjusted in order to operate the system at the same Peclet number. In this case, heat transfer enhancements obtained with nanofluids are calculated by comparing the associated results with the pure fluid case for the same flow velocity and tube diameter, so that the effect of the change in Peclet number is also examined.

In order to examine the heat transfer enhancement obtained with nanofluids for the same flow velocity, axial position, and tube diameter, one can use the correlation given by Equation 5.85 and obtain the local Nusselt number enhancement ratio:

$$\frac{Nu_{x,nf}}{Nu_{x,f}} = \frac{1.302 X_{nf}^{-1/3} - 1}{1.302 X_f^{-1/3} - 1} \quad \text{for } X \leq 5 \times 10^{-5}$$

$$\frac{1.302 X_{nf}^{-1/3} - 0.5}{1.302 X_f^{-1/3} - 0.5} \quad \text{for } 5 \times 10^{-5} < X \leq 1.5 \times 10^{-3} \qquad (5.88)$$

$$\frac{4.364 + 0.263 X_{nf}^{-0.506} e^{-41 X_{nf}}}{4.364 + 0.263 X_f^{-0.506} e^{-41 X_f}} \quad \text{for } X > 1.5 \times 10^{-3}$$

It should be noted that the increases in the density and specific heat of the nanofluid increase values of the terms with X_{nf} in Equation 5.88, which increases local Nusselt number enhancement ratio. On the other hand, increase in the thermal conductivity of the nanofluid decreases values of the terms with X_{nf}, and the local Nusselt number enhancement ratio decreases as a consequence. One can integrate the numerator and denominator of Equation 5.88 along the tube and obtain the average Nusselt number enhancement ratio for a specific case. The effects of thermophysical properties on average Nusselt number enhancement ratio are qualitatively the same as the local Nusselt number case. For a pure fluid, average heat transfer coefficient can be defined as follows:

$$h_f = \frac{Nu_f k_f}{d} \tag{5.89}$$

For a nanofluid, average heat transfer coefficient becomes the following expression:

$$h_{nf} = \frac{Nu_{nf} k_{nf}}{d} \tag{5.90}$$

By using Equations 5.89 and 5.90, one can obtain the average heat transfer coefficient enhancement ratio as follows:

$$\frac{h_{nf}}{h_f} = \frac{k_{nf}}{k_f} \frac{Nu_{nf}}{Nu_f} \tag{5.91}$$

One should use Equation 5.88 and perform the associated integrations for calculating average Nusselt number enhancement ratio in Equation 5.91. Although the increase in thermal conductivity of the nanofluid decreases Nusselt number enhancement ratio, due to the multiplicative effect of thermal conductivity in the definition of heat transfer coefficient, it is expected that the increase in the thermal conductivity of the nanofluid improves heat transfer coefficient enhancement ratio.

5.2.5.2 Constant Wall Temperature Boundary Condition

For constant wall temperature boundary condition, the determination of the average Nusselt number using the classical Sieder-Tate correlation is introduced as[126]

$$Nu = 1.86 \left(Pe \frac{d}{L} \right)^{1/3} \left(\frac{\mu_b}{\mu_w} \right) \tag{5.92}$$

The bulk mean temperature is defined as

$$T_b = \frac{T_i + T_o}{2} \tag{5.93}$$

Neglecting the variation of viscosity enhancement ratio of the nanofluid (μ_{nf}/μ_f) with temperature, and applying Sieder–Tate correlation

Micro/Nano Heat Transfer

(Equation 5.92) for a pure fluid and nanofluid, under the same geometrical parameter following expression can be obtained:

$$\frac{Nu_{nf}}{Nu_f} = \left(\frac{Pe_{nf}}{Pe_f}\right)^{1/3} = \left(\frac{k_f\, \rho_{nf}\, c_{p,nf}}{k_{nf}\, \rho_f\, c_{p,f}}\right)^{1/3} \quad (5.94)$$

Using Equation 5.92, heat transfer coefficient enhancement ratio becomes as

$$\frac{h_{nf}}{h_f} = \left(\frac{\rho_{nf}\, c_{p,nf}}{\rho_f\, c_{p,f}}\right)^{1/3}\left(\frac{k_{nf}}{k_f}\right)^{2/3} \quad (5.95)$$

By examining this expression, it can be observed that the enhancements in the thermophysical properties of the nanofluid; density, specific heat, and thermal conductivity, improve the heat transfer coefficient. It should be noted that the effect of thermal conductivity enhancement is more pronounced when compared to density and specific heat.

Example 5.15

Calculate the heat transfer coefficient enhancement ratio of nanofluid, CuO/water, flowing in a tube by using Equation 5.95 and information from Examples 5.8, 5.11, and 5.12.

Solution

From Example 5.8,

$$k_{nf} = 0.583\,\text{W}/(\text{m}\cdot\text{K})$$

From Example 5.11,

$$\rho_{nf} = 1159.106\,\text{kg}/\text{m}^3$$

From Example 5.12 using Equation 5.82,

$$c_{p,nf} = 3789.11\,\text{J}/\text{kg}\cdot\text{K}$$

From properties of water at $P = 1$ bar (100 kPa) and $T = 273.15$ K (0°C):

$$k_f = 0.552\,\text{W}/(\text{m}\cdot\text{K})$$
$$\rho_f = 999.8\,\text{kg}/\text{m}^3$$
$$c_{p,f} = 4217.9\,\text{J}/\text{kg}\cdot\text{K}$$

$$\frac{h_{nf}}{h_f} = \left(\frac{\rho_{nf}\, c_{p,nf}}{\rho_f\, c_{p,f}}\right)^{1/3} \left(\frac{k_{nf}}{k_f}\right)^{2/3}$$

$$= \left(\frac{(1106.004)}{(999.8)} \frac{(3789.11)}{(4217.9)}\right)^{1/3} \left(\frac{0.583}{0.552}\right)^{2/3}$$

$$= 1.035$$

A more detailed convective heat transfer performance analysis of nanofluids, together with two additional cases (slug flow case and linear wall temperature boundary condition case) is provided by Kakaç and Pramuanjaroenkij.[127,128]

5.2.6 Experimental Correlations of Convective Heat Transfer of Nanofluids

There are many other experimental studies in the literature about the convective heat transfer of nanofluids[127–129]; the general trend is that the heat transfer enhancement obtained with nanofluids exceeds the enhancement due to the thermal conductivity enhancement. This fact indicates that there should be additional mechanisms of heat transfer enhancement with nanofluids.

Although the thermal dispersion approach provides higher accuracy, the lack of empirical constants for specific nanofluid types and flow configurations may limit its applicability for the time being. Therefore, the correlations summarized in this section provide another practical alternative for the prediction of forced convection heat transfer of nanofluids.

Pak and Cho[117] provided a Nusselt number correlation that correctly predicts the experimental data for the turbulent flow of Al_2O_3(13 nm)/water and TiO_2(27 nm)/water nanofluids inside a straight circular tube, values in parentheses indicate the nanoparticle size. The associated expression is as follows:

$$Nu_{nf} = 0.021 Re_{nf}^{0.8} Pr_{nf}^{0.5} \tag{5.96}$$

The ranges of experimental data are $10{,}000 < Re < 100{,}000$, $6.54 < Pr < 12.33$, $0.99\% < \phi < 3.16\%$. These ranges can be considered as the applicability intervals of the provided correlation.

Another correlation for the heat transfer performance of nanofluids in the turbulent flow regime is given by Xuan and Li.[130] They investigated turbulent convective heat transfer of the nanofluid in a tube, and a new correlation for heat transfer is proposed to correlate the experimental data for nanofluids as

$$Nu_{nf} = 0.0059\left(1.0 + 7.6286 \phi^{0.6886} Pe_d^{0.001}\right) Re_{nf}^{0.9238} Pr_{nf}^{0.4} \tag{5.97}$$

where $Pe_{nf} = u_m d_p/\alpha_{nf}$; d_p is the diameter of the nanoparticles and $\alpha_{nf} = k_{nf}/\rho_{nf} c_{p,nf}$.

Yang et al.[131] examined the laminar flow heat transfer of nanofluids by using a horizontal tube heat exchanger. The graphite particles were used with two base fluids, namely, a commercial automatic transmission fluid, and a mixture of two synthetic baseoils. By using the experimental data obtained in the study, a modified Sieder–Tate correlation was proposed for the determination of nanofluid Nusselt number given by Equation 3.24.

This correlation gives the average Nusselt number for the laminar flow inside a straight circular tube under constant wall temperature boundary condition. In the application of Equation 3.24 to nanofluids, Pe number should be replaced by Pe_{nf} by the use of the properties of nanofluids.

Maïga et al.[132] investigated the problem by considering the nanofluids, the laminar flow of Al_2O_3/water and Al_2O_3/ethylene glycol nanofluids inside a straight circular tube, as single phase fluids with enhanced thermophysical properties.[132] As a result of the numerical analysis, the following expressions are proposed for the determination of Nusselt number, depending on the boundary condition:

$$Nu_{nf} = 0.086 Re_{nf}^{0.55} Pr_{nf}^{0.5}, \quad \text{for constant wall heat flux} \quad (5.97)$$

$$Nu_{nf} = 0.28 Re_{nf}^{0.35} Pr_{nf}^{0.36}, \quad \text{for constant wall temperature} \quad (5.98)$$

Validity ranges of the correlations are $Re \leq 1{,}000$, $6 \leq Pr \leq 753$, and $\phi \leq 10\%$.

Numerical results were utilized for obtaining an empirical correlation that provides the fully developed Nusselt number for the turbulent flow regime.[132] The associated expression is as follows:

$$Nu_{fd,nf} = 0.085 Re_{nf}^{0.71} Pr_{nf}^{0.35} \quad (5.99)$$

Validity ranges of Equation 5.99 are $10{,}000 < Re < 500{,}000$, $6.6 < Pr < 13.9$, and $\phi < 10\%$.

Example 5.16

Consider the solved Example 3.3 where the heat transfer coefficient in the inner tube is calculated; if a hot nanofluid of Al_2O_3/water flows through the inner tube with 5% particle volume fraction and sphericity value equals to 1.

1. Calculate heat transfer coefficient and compare the value with that of Example 3.3.
2. Find the heat transfer enhancement of the nanofluid.

Solution
From Example 3.3,

$$\Delta T_1 = \Delta T_2 = \Delta T_m = 105°C = 378 \text{ K}$$

$$T_{\text{bulk}} = \frac{140+125}{2} = 132°C = 405 \text{ K} \quad \text{and} \quad T_{\text{wall}} = 105°C = 378 \text{ K}$$

From properties of water at the wall temperature, $T_{\text{wall}} = 378$ K (105°C):

$$k_{f,w} = 0.682 \text{ W}/(\text{m} \cdot \text{K})$$
$$\rho_{f,w} = 955.11 \text{ kg}/\text{m}^3$$
$$\mu_{f,w} = 268 \times 10^{-6} \text{ Pa} \cdot \text{s}$$

and properties of water at the bulk temperature, $T_{\text{bulk}} = 405$ K (132°C):

$$k_{f,b} = 0.688 \text{ W}/(\text{m} \cdot \text{K})$$
$$\rho_{f,b} = 932.836 \text{ kg}/\text{m}^3$$
$$\mu_{f,b} = 209 \times 10^{-6} \text{ Pa} \cdot \text{s}$$
$$Pr_b = 1.325$$
$$c_{p,fb} = 4271 \text{ J}/\text{kg} \cdot \text{K}$$

$$Re_{f,b} = \frac{\rho u_m d_i}{\mu} = \frac{4\dot{m}_f}{\pi \mu_f d_i} = \frac{4 \times 1.36}{\pi \times 0.0525 \times 0.209 \times 10^{-3}} = 15,485 > 2300$$

Turbulent flow

For the Turbulent flow, from Table 3.6,

$$Nu_{f,b} = 0.023 \times Re_{f,b}^{0.8} \times Pr_{f,b}^{1/3} \times \left(\frac{\mu_{f,b}}{\mu_{f,w}}\right)^{0.14} = 347$$

$$h_f = \frac{Nu_{f,b} k_{f,b}}{d_i} = \frac{347 \times 0.688}{0.0525} = 4547.35 \text{ W}/\text{m}^2 \cdot \text{K}$$

The nanofluid properties can be obtained from

$$\mu_{nf,b} = (1 + 7.3\phi + 123\phi^2)\mu_{f,b} = 3.496 \times 10^{-4} \text{ Pa} \cdot \text{s}$$
$$\mu_{nf,w} = (1 + 7.3\phi + 123\phi^2)\mu_{f,w} = 4.480 \times 10^{-4} \text{ Pa} \cdot \text{s}$$

$$c_{p,h} = c_{p,nf} = \phi c_{p,p} + (1-\phi)c_{p,f} = 4087.45 \text{ J/kg}\cdot\text{K}; \quad c_{p,c} = 4179 \text{ J/kg}\cdot\text{K}$$

$$(\dot{m}\cdot c_p)_h \cdot \Delta T_h = (\dot{m}\cdot c_p)_c \cdot \Delta T_c$$

$$\dot{m}_{h,nf} = \frac{\dot{m}_c \cdot c_{p,c}}{c_{p,h}} = 1.42 \text{ kg/s}$$

$$Re_{h,b} = \frac{\rho u_m d_i}{\mu} = \frac{4\dot{m}_{nf,h}}{\pi \mu_{nf,h} d_i} = \frac{4 \times 1.42}{\pi \times 0.0525 \times 0.3496 \times 10^{-3}} = 98{,}557.11 > 2300$$

Turbulent flow

$$k_{nf} = \frac{k_p + (n-1)k_f - (n-1)\phi(k_f - k_p)}{k_p + (n-1)k_f + \phi(k_f - k_p)} k_f$$

$$= \frac{(30) + (3-1)(0.688) - (3-1)(0.05)(0.688 - 30)}{(30) + (3-1)(0.688) + (0.05)(0.688 - 30)}(0.688)$$

$$= 0.7891 \text{ W/(m}\cdot\text{K)}$$

$$Pr_{nf} = \frac{c_{p,nf}\mu_{nf}}{k_{nf}} = 1.81$$

$$\rho_{nf} = \phi\rho_p + (1-\phi)\rho_f = 1086.1942 \text{ kg/m}^3$$

From the Seider–Tate correlation with the same geometrical parameters,

$$Nu_{nf} = \left(\frac{Pe_{nf}}{Pe_f}\right)^{1/3} = \left(\frac{k_f \rho_{nf} c_{p,nf}}{k_{nf} \rho_f c_{p,f}}\right)^{1/3} = 342.68$$

$$\frac{h_{nf}}{h_f} = \left(\frac{\rho_{nf}\ c_{p,nf}}{\rho_f\ c_{p,f}}\right)^{1/3} \left(\frac{k_{nf}}{k_f}\right)^{2/3} = 1.136$$

Although Nusselt Number decreases in nanofluid, but the heat transfer coefficient is increased:

$$\frac{h_{nf}}{h_f} = 1.136$$

$$h_{nf} = 1.136 \times 4547.35 = 5165.79 \text{ W/m}^2\cdot\text{K}$$

The heat transfer enhancement of the nanofluid is 13.6%.

Example 5.17

Using the experimental correlation with nanofluids given in Equation 5.99 for turbulent flow conditions, calculate the heat transfer coefficient under the same nanofluid conditions as in Example 5.16 and compare the heat transfer coefficients from both cases.

Solution

From Example 5.16,

$$k_{f,b} = 0.688 \text{ W}/(\text{m}\cdot\text{K})$$
$$\rho_{f,b} = 932.836 \text{ kg/m}^3$$
$$\mu_{f,b} = 209 \times 10^{-6} \text{ Pa}\cdot\text{s}$$
$$Pr_b = 1.325$$
$$c_{p,fb} = 4271 \text{ J/kg}\cdot\text{K}$$
$$Re_{f,b} = 15485$$
$$Nu_{f,b} = 353.9$$
$$h_{fi} = 4637.6 \text{ W/m}^2\cdot\text{K}$$

The nanofluid properties can be obtained from Example 5.16 as.

$$\mu_{nf,b} = 3.496 \times 10^{-4} \text{ Pa}\cdot\text{s}$$
$$\mu_{nf,w} = 4.480 \times 10^{-4} \text{ Pa}\cdot\text{s}$$
$$c_{p,nf} = 4087.45 \text{ J/kg}\cdot\text{K}$$
$$Re_{h,b} = 98467.51$$
$$k_{nf} = 0.7891 \text{ W}/(\text{m}\cdot\text{K})$$
$$Pr_{nf} = 1.81$$
$$\rho_{nf} = 1086.1942 \text{ kg/m}^3$$

From Equation 5.99, the Maïga et al. correlation,

$$Nu_{nf} = 0.085 \times Re_{nf}^{0.71} \times Pr_{nf}^{0.35} = 540.28$$

$$h_{nf} = \frac{k_{nf} \times Nu_{nf}}{d_i} = \frac{0.7891 \times 540.28}{0.0525} = 8120.61 \text{ W/m}^2\cdot\text{K}$$

Therefore, the conventional correlations with the nanofluid properties underestimate the heat transfer coefficient.

Vasu et al.[133] proposed a correlation for the turbulent forced convection of nanofluids inside circular tubes. The correlation has the same form as the Dittus–Boelter equation:

$$Nu_{nf} = c_{DB} Re_{nf}^{m_{DB}} Pr_{nf}^{n_{DB}} \qquad (5.100)$$

The empirical constant in Equation 5.100, namely, c_{DB} was determined by applying a regression analysis to the experimental data[117,133,134]

$$Nu_{nf} = 0.0256 Re_{nf}^{0.8} Pr_{nf}^{0.4}, \quad \text{for } Al_2O_3/\text{water nanofluids} \qquad (5.101)$$

$$Nu_{nf} = 0.027 Re_{nf}^{0.8} Pr_{nf}^{0.4}, \quad \text{for } CuO/\text{water nanofluids} \qquad (5.102)$$

The validity ranges of the above correlations were not provided, but it can be said that they are applicable in the ranges of the associated experimental studies whose results were used in the determination of the correlations.

A similar analysis for laminar flow has been performed and the study proposed a Nusselt number correlation for the flow of nanofluids inside circular tubes. The correlation has the same form to that of the Sieder–Tate correlation, Equation 3.24 with nanofluid properties.

Chun et al.[135] proposed that the general form of heat transfer coefficient correlations for the flow of pure fluids can be applied to the case of nanofluids by modifying the associated constants. The laminar flow of Al_2O_3/transformer oil inside a double pipe heat exchanger was investigated:

$$h_{nf} = \alpha_{Chun} \frac{k_{nf}}{d} Re_{nf}^{\beta_{Chun}} Pr_{nf}^{\gamma_{Chun}} \qquad (5.103)$$

Constants α_{Chun}, β_{Chun}, and γ_{Chun} were determined by applying a regression analysis to the experimental data and they obtained the following expression for the heat transfer coefficient:

$$h_{nf} = 1.7 \frac{k_{nf}}{d} Re_{nf}^{0.4} \qquad (5.104)$$

The researchers indicated that Prandtl number is not an effective parameter for the proposed correlation. In the experiments, following ranges of parameters were considered, which can be considered as the applicability ranges of the proposed correlation, $100 < Re < 440$ and $0.25\% < \phi < 0.5\%$.

The convective heat transfer performance of laminar flow of Al_2O_3/water nanofluids inside a circular tube under constant wall heat flux boundary condition has been investigated by Anoop et al.[136]. A correlation for local Nusselt number was proposed based on the experimental data, as follows:

$$Nu_{x,nf} = 4.36 + \left[6.219 \cdot 10^{-3} x_+^{-1.1522}\left(1+\phi^{0.1533}\right)\exp(-2.5228x_+)\right]$$

$$\left[1+0.57825\left(\frac{d_p}{100\text{ nm}}\right)^{-0.2183}\right] \qquad (5.105)$$

where

$$x_+ = \frac{x}{d\, Re_{nf} Pr_{nf}} \qquad (5.106)$$

The validity range of the correlation is $50 < x/d < 200$, $500 < Re < 2{,}000$.

Asirvatham et al.[137] proposed a correlation which provides the local Nusselt number for the convective heat transfer of laminar CuO/water nanofluid inside a circular tube under constant wall heat flux boundary condition as follows:

$$Nu_{x,nf} = 0.155 Re_{nf}^{0.59} Pr_{nf}^{0.35}\left(\frac{d}{x}\right)^{0.38} \qquad (5.107)$$

Validity ranges are $Re < 2{,}300$, $\phi = 0.003\%$.

Jung et al.[138] modified the Dittus–Boelter correlation. They experimentally investigated the Al_2O_3(170 nm)/water nanofluid flow inside a rectangular microchannel under constant wall heat flux boundary condition:

$$Nu_{nf} = 0.014\phi^{0.095} Re_{nf}^{0.4} Pr_{nf}^{0.6} \qquad (5.108)$$

Ranges of Reynolds number and particle volume fraction for the experimental study were $5 < Re < 300$, and $0.6\% < \phi < 1.8\%$. It was found that the friction factor of the nanofluid flow is close to that of the pure fluid flow, which can be determined by utilizing the following expression:

$$f_{nf} = \frac{56.9}{Re_{nf}} \qquad (5.109)$$

where f is defined as in Equation 5.23.

The laminar flow of Al_2O_3/water nanofluid inside a straight circular tube with wire coil inserts under constant wall heat flux boundary condition is investigated by Chandrasekar et al.[139]. As a result of the experiments, following correlations were proposed for the determination of Nusselt number and friction factor:

$$Nu_{nf} = 0.279(Re\ Pr)_{nf}^{0.558} \left(\frac{p_i}{d}\right)^{-0.477} (1+\phi)^{134.65} \quad (5.110)$$

$$f_{nf} = 530.8 Re_{nf}^{-0.909} \left(\frac{p_i}{d}\right)^{-1.388} (1+\phi)^{-512.26} \quad (5.111)$$

where p_i is pitch ratio and validity ranges of the correlations are $Re < 2{,}300$, $\phi = 0.1\%$ and $2 \le p_i/d \le 3$ where f is defined as in Equation 5.23.

Duangthongsuk and Wongwises[140] analyzed the turbulent flow of TiO_2/water nanofluid inside a horizontal double tube counter-flow heat exchanger experimentally, and the following correlation Nusselt number was introduced:

$$Nu_{nf} = 0.074 Re_{nf}^{0.707} Pr_{nf}^{0.385} \phi^{0.074} \quad (5.112)$$

Validity ranges are $\phi \le 1\%$, and $3{,}000 < Re < 18{,}000$. In addition, a new correlation for the determination of friction factor was also proposed:

$$f_{nf} = 0.961 \phi^{0.052} Re_{nf}^{-0.375} \quad (5.113)$$

where f is defined as in Equation 5.23. This correlation can be used for particle volume fractions up to 2%.

Wu et al.[141] examined the heat transfer performance of Al_2O_3(56 nm)/water nanofluid flowing inside trapezoidal microchannels under constant wall heat flux boundary condition. The flow regime was laminar. As a result of the experimental analysis, following correlation was provided for the determination of Nusselt number:

$$Nu_{nf} = 0.566(1+100\phi)^{0.57} Re_{nf}^{0.20} Pr_{nf}^{0.18} \quad (5.114)$$

The above correlation is valid for the following ranges of parameters, $200 < Re < 1{,}300$, $4.6 \le Pr \le 5.8$ and $0 \le \phi \le 0.26\%$. In addition, a correlation for friction factor was also provided:

$$f_{nf} = 5.43(1+100\phi)^{0.18} Re_{nf}^{-0.81} \quad (5.115)$$

where f is defined as in Equation 4.3. Associated validity ranges are $190 < Re < 1{,}020$ and $0 \le \phi \le 0.26\%$.

Nomenclature

a	coefficient in Equation 5.40
b	coefficient in Equation 5.40
Br	Brinkman number, $\mu u^2/k(T_w - T_b)$
c_{jump}	temperature jump coefficient
c_p	specific heat at constant pressure, J/kg · K
d	tube diameter or distance, m
d_p	nanoparticle diameter, m
D_h	hydraulic diameter, m
f	friction factor
F	Enhancement factor (correction factor in Adam's correlation)
F_m	tangential momentum, kg m/s
F_T	thermal accommodation coefficient
Gz	Graetz number, $ud^2/\alpha L$
h	heat transfer coefficient, W/m² · K
H	Channel height, m
I	fractal index
k	thermal conductivity, W/m · K
k_r	ratio of specific heats
Kn	Knudsen number, λ/L
L	characteristic length, m
M	Molecular weight, kg/mol
Ma	Mach number, u/V_a
n	empirical shape factor
N_A	Avogadro constant, 6.023 × 10²³/mol
Nu	Nusselt number, hd/k
p	pressure, Pa
P_{as}	aspect ratio of nanotube
Pe	Peclet number, $Re\,Pr$ or ud/α
Po	Poiseuille number, $f\,Re$
Pr	Prandtl number, $\mu c_p/k$
ΔT	pressure drop, Pa
q_w	heat flux, W/m²
Q_i	impinging stream energy, W
Q_r	energy carried by reflected molecules, W
Q_w	energy of molecules leaving the surface at wall temperature, W
r	radius, m
R	Gas constant, 8.314472 J/mol K
R_d	duct radius, m
Re	Reynolds number, $\rho u_m d/\mu$
t	nanolayer thickness, m
T	temperature, K
ΔT	temperature difference, K

u	velocity, m/s
u_m	mean flow velocity, m/s
u_s	slip velocity, m/s
V_a	speed of sound, m/s
W	channel width, m
W_c	center to center distance between two adjacent microchannels, m
x	x-direction distance, m
x_e^*	entrance length, m
y	y-direction distance, m

Greek Symbols

α	thermal diffusivity, m²/s
β	coefficient representing velocity slip and temperature jump
γ	aspect ratio
δ_v^+	dimensionless thickness of the laminar sublayer
ε	average roughness, m
κ_B	Boltzmann constant, 1.3807×10^{-23} J/K
λ	mean free molecule path, m
λ_D	Darcy coefficient, $4f$
λ_{Dr}	drag coefficient
λ_n	Graetz eigenvalues
λ_0	Graetz eigenvalue at n equals to 0
μ	dynamics viscosity, Pa · s
ν	kinematic viscosity, m²/s
ρ	density, kg/m³
ϕ	particle volume fraction
ψ	sphericity
θ	the side angle of the trapezoidal microchannel

Subscripts

b	bulk mean
cf	constricted flow
cl	cluster
d	dispersed
e	entrance
eff	effective
f	base fluid
fd	fully developed
h	hydrodynamic
i	inlet
l	liquid nanolayer
nf	nanofluid
o	outlet or tube radius in m
p	nanoparticle
r	r-direction

s	surface
t	thermal
w	wall
x	local

Superscripts

s	slip-flow

PROBLEMS

5.1. Using the Table 3.8 to calculate the Nusselt number in a rectangular channel having aspect ratio of 0.5, if the width of the channel is 100 µm and Kn number is 0.12, calculate the heat transfer coefficient for air at 40°C for two different boundary conditions.

5.2. A microchannel heat exchanger has been designed for cooling an electronic chip of size 30 mm by 30 mm. The transition Reynolds number is 800. The average Nusselt numbers, can be calculated from the following correlations:

$$Nu = 0.0384 Re^{0.67} Pr_{nf}^{0.33} \quad \text{for Laminar Flow}$$

$$Nu = 0.00726 Re^{0.8} Pr_{nf}^{0.33} \quad \text{for Turbulent Flow}$$

To cool the chips, a square microchannel of 250 µm is chosen where the permissible water velocity is 1 m/s and fluid temperature is 40°C.

   a. Calculate the heat transfer coefficient
   b. If the length of the microchannel has been given as 20 mm, calculate the friction coefficient and the pressure drop

5.3. Repeat the problem 5.2, if the permissible water velocity is changed to be 2 m/s but the fluid temperature is still the same. By using a proper correlation, determine the average heat transfer coefficient.

5.4. If the microchannel from the problem 5.2 is changed from the square microchannel to be the trapezoidal microchannel. The width of the microchannel is 125 µm, the length of the microchannel is 62.5 µm, the side angle is 44 degree, the permissible water velocity is 1.5 m/s and the fluid temperature is 40°C. By selecting a proper correlation, calculate

   a. The average heat transfer coefficient,
   b. The pressure drop.

5.5. Consider the properties of the nanofluid, Al_2O_3/water, at 30°C, calculate the effective thermal conductivity of 6% volume fraction of nanofluid by using the Hamilton and Crosser model and assuming that the particles are spherical. Repeat the problem with 8% volume fraction nanofluid at the same water temperature.

5.6. Consider the Nusselt number for fully developed laminar flow under constant heat flux and constant wall temperature boundary conditions are constant as 4.36 and 3.66, respectively. If the nanofluid of Al_2O_3/water flows in a tube with a diameter of 0.5 cm at a bulk temperature of 40°C, calculate the heat transfer coefficients for both boundary conditions and compare these values with the heat transfer coefficients of pure water in both conditions to find the heat transfer enhancement, if the volume fraction is

   a. 0.05% and sphericity value equals to 1,
   b. 0.05% and sphericity value equals to 3.

5.7. Consider the solved Example 3.3 where the heat transfer coefficient in the inner tube is calculated; if a hot nanofluid of Al_2O_3/water flows through the inner tube with 5% of particle volume fraction and sphericity value equals to 1, calculate the heat transfer coefficient by the use of an experimentally obtained nanofluid correlation given by $Nu_{nf} = 0.0059(1.0 + 7.6286\phi^{0.6886} Pe_d^{0.001}) Re_{nf}^{0.9238} Pr_{nf}^{0.4}$, Equation 5.97.

5.8. The CuO/water nanofluid, $k_{CuO} = 401$ W/(m·K) and $k_{water} = 0.613$ W/(m·K), flows in a tube under the fully developed turbulent flow condition at 30°C. If an inside diameter of the tube is 0.5 cm:

   a. Using a proper correlation, determine the effective heat transfer coefficient of the nanofluid for the constant wall temperature and constant heat flux boundary conditions, assuming the volume fraction of the spherical nanoparticles is 8%,
   b. Compare the values with the heat transfer coefficients of water under the same conditions to find the enhancement.

5.9. Repeat the Problem 5.6 with the Al_2O_3/ethylene glycol nanofluid flows in the tube, calculate heat transfer coefficients with

   a. 0.05% of particle volume fraction and sphericity value equals to 1,
   b. 0.05% of particle volume fraction and sphericity value equals to 3.

5.10. Determine the local heat transfer coefficient at 30 cm from the entrance of a heat exchanger where engine oil flows through tubes with a diameter of 0.5 in. The oil flows with a velocity of 0.5 m/s and a local bulk temperature of 30°C, while the local tube wall temperature is at 60°C. If the oil is replaced with the nanofluid of Al_2O_3/oil with 5% particle volume fraction under the same condition, find the heat transfer enhancement.

5.11. Compare the Nusselt number values obtained from Equations 5.96 and 5.99, when the nanofluid, Al_2O_3/water, flows with the Reynolds and Prandtl numbers equal to 10,000 and 7, respectively, and 1% volume fraction.

5.12. Using the experimental correlation with nanofluids given in Equation 5.99 for turbulent flow conditions, calculate the heat transfer coefficient under the same nanofluid conditions as in Example 5.16 and compare the heat transfer coefficients from both cases.

References

1. Morini, G. L., Single-phase convective heat transfer in microchannels: a rewiev of experimental results, *International Journal of Thermal Sciences*, 631–651, 2004.
2. Kakaç, S. and Yener, Y., *Convective Heat Transfer*, 2nd ed., CRC Press, Florida, 1995.
3. Adams, T. M., Dowling, M. F., Abdel-Khalik, S. I., and Jeter, S. M., Applicability of traditional turbulent single-phase forced convection correlations to non-circular microchannels, *Int. J. Heat Mass Transfer*, 42, 4411–4415, 1999.
4. Bayazitoglu, Y., Tunc, G., Wilson, K., and Tjahjono, I., Convective heat transfer for single-phase gases in microchannel slip flow: Analytical solutions, in *Microscale Heat Transfer – Fundamentals and Applications*, Kakaç, S., Vasiliev, L., Bayazitoglu, Y., and Yener., Y., Eds., Kluwer Academic Publishers, The Netherlands, 2005, 125–148.
5. Zhang, M. Z., *Nano/Microscale Heat Transfer*, McGraw-Hill, New York, 2007.
6. Yener, Y., Kakaç, S., Avelino, M., and Okutucu, T., Single-phase forced convection in microchannels A State-of-the-Art Review, in *Microscale Heat Transfer- Fundamentals and Applications*, Kakaç, S., Vasiliev, L., Bayazitoglu, Y., and Yener., Y., Eds., Kluwer Academic Publishers, The Netherlands, 2005, 1–24.
7. Gad-El-Hak, M., The fluid mechanics of microdevices—the freshman scholar lecture, *J. Fluids Eng.*, 121, 5–33, 1999.
8. Bayazitoglu, Y. and Kakaç, S., Flow regimes in microchannel single-phase gaseous fluid flow, in *Microscale Heat Transfer- Fundamentals and Applications*, Kakaç, S., Vasiliev, L., Bayazitoglu, Y., and Yener., Y., Eds., Kluwer Academic Publishers, The Netherlands, 2005, 75–92.
9. Gad-El-Hak, M., Momentum and heat transfer in MEMS, *Congrs francais de Thermique*, SFT, Grenoble, France, 3–6 June, Elsevier, 2003.
10. Beskok, A., Trimmer, W., and Karniadakis, G., Rarefaction compressibility and thermal creep effects in gas microflows, *Proc. ASME DSC*, 57(2), 877–892, 1995.
11. Çetin, B., Yuncu, H., and Kakaç, S., Gaseous flow in microconduits with viscous dissipation, *Int. J. Transp. Phenom.*, 8, 297–315, 2007.

12. Wu, P. Y. and Little, W. A., Measurement of heat transfer characteristics of gas flow in fine channels heat exchangers used for microminiature refrigerators, *Cryogenics*, 24(5), 415–420, 1985.
13. Choi, S. B., Barron, R. F., and Warrington, R. O., Fluid flow and heat transfer in microtubes, in *Micromechanical Sensors, Actuators and Systems*, Vol. DSC–32, ASME, Georgia, 1991, 123–134.
14. Yu, D., Warrington, R., Barron, R., and Ameel, T., An experimental and theoretical investigation and heat Transfer in microtubes, *Proceedings of ASME/JSME Thermal Engineering Conference 1*, 523–530, 1995.
15. Peng, X. F. and Peterson, G. P., Convective heat transfer and flow friction for water flow in microchannel structures, *Int. J. Heat Mass Transfer*, 39, 2599–2608, 1996.
16. Ameel, T. A., Wang, X., Barron, R. F., and Warrington, R. O., Laminar forced convection in a circular tube with constant heat flux and slip flow, *Microscale Thermophys. Eng*, 1(4), 303–320, 1997.
17. Barron, R. F., Wang, X., Ameel, T. A., and Warrington, R. O., The Graetz problem extended to slip-flow, *Int. J. Heat Mass Transfer*, 40(8), 1817–1823, 1997.
18. Yu, S. and Ameel, T., Slip-flow heat transfer in rectangular microchannels, *Int. J. Heat Mass Transfer*, 44, 4225–4234, 2001.
19. Renksizbulut, M., Niazmand, H., and Tercan, G., Slip-flow and heat transfer in rectangular microchannels with constant wall temperature, *Int. J. Therm. Sci.*, 44, 870–881, 2006.
20. Renksizbulut, M. and Niazmand, H., Laminar flow and heat transfer in the entrance region of trapezoidal channels with constant wall temperature, *J. Heat Transfer*, 128, 63–74, 2006.
21. Acosta, R. E., Muller, R. H., and Tobias, W. C., Transport process in narrow (Capillary) channels, *AICHE J.*, 31, 473–482, 1985.
22. Pfalher, J., Harley, J., Bau, H. H., and Zemel, N., Liquid transport in micron and submicron channels, *Sensor. Actuat. A*, 21–23, 431–434, 1990.
23. Pfalher, J., Harley, J., Bau, H. H., and Zemel, N., Liquid and gas transport in small channels, Vol. DSC–31, ASME, Georgia, 1990, 149–157.
24. Pfalher, J., Harley, J., Bau, H. H., and Zemel, N., Gas and liquid flow in small channels, in *Micromechanical Sensors, Actuator and Systems*, Vol. DSC–32, ASME, Georgia, 1991, 49–60.
25. Arkilic, E. B., Breuer, K. S., and Schmidt, M. A., Gaseous flow in microchannels, in *Application of Microfabrication to Fluid Mechanics*, Vol. FED–197, ASME, New York, 1994, 57–66.
26. Pong, K., Ho, C., Liu, J., and Tai, Y., Non-linear pressure distribution in uniform microchannels, in *Application of Microfabrication to Fluid Mechanics*, Vol. FED–197, ASME, New York, 1994, 51–56.
27. Liu, J., Tai, Y. C., and Ho, C. M., MEMS for pressure distribution studies of gaseous flows through uniform microchannels, *Proceedings of Eighth Annual International Workshop MEMS*, IEEE, 209–215, 1995.
28. Shih, J. C., Ho, C. M., Liu, J., and Tai, Y. C., Monatomic and polyatomic gas flow through uniform microchannels, Vol. DSC–59, ASME, Georgia, 1996, 197–203.
29. Harley, J., Huang, Y., Bau, H. H., and Zemel, J. N., Gas flow in microchannels, *J. Fluid Mech.*. 284, 257–274, 1995.

30. Araki, T., Soo, K. M., Hiroshi, I., and Kenjiro, S., An experimental investigation of gaseous flow characteristics in microchannels, *Proceedings of the International Conference on Heat Transfer and Transport Phenomena in Microscale*, 155–161, 2000.
31. Li, Z. X., Du, D. X., and Guo, Z. Y., Characteristics of frictional resistance for gas flows in microtubes, *Proceedings of Symposium on Energy Engineering in the 21st Century*, 2, 658–664, 2000.
32. Yang, C. Y., Chien, H. T., Lu, S. R., and Shyu, R. J., Friction characteristics of water, R-134a and air in small tubes, *Proceedings of the International Conference on Heat Transfer and Transport Phenomena in Microscale*, 168–174, 2000.
33. Lalonde, P., Colin, S., and Caen, R., Mesure de Debit de Gaz dans les Microsystems, *Mech. Ind.*, 2, 355–362, 2001.
34. Turner, S. E., Sun, H., Faghri, M., and Gregory, O.J., Compressible gas flow through smooth and rough microchannels, *Proceedings of IMECE ASME 2001*, HTD–24144, 2001.
35. Hsieh, S. S., Tsai, H. H., Lin, C. Y., Huang, C. F., and Chien, C. M., Gas flow in long microchannel, *Int. J. Heat Mass Transfer*, 47, 3877–3887, 2004.
36. Tang, G. H. and He, Y. L., An experimental investigation of gaseous flow characteristics in microchannels, *Proceedings of Second International Conference on Microchannels and Minichannels*, Rochester, 359–366, 2004.
37. Celata, G. P., Cumo, M., McPhail, S. J., Tesfagabir, L., and Zummo, G., Experimental study on compressibility effects in microtubes, *Proceedings of the XXIII UIT Italian National Conference*, 53–60, 2005.
38. Morini, G. L., Lorenzini, M., and Salvigni, S., Friction characteristics of compressible gas flows in microtubes, *Exp. Therm. Fluid Sci.*, 30, 773–744, 2006.
39. Hadjiconstantinou, N. G. and Simsek, O., Constant-wall-temperature Nusselt number in micro and nano-channels, *Trans. ASME*, 57(2), 877–892, 2002.
40. Wu, P. and Little, W. A., Measurement of friction factors of the flow of gases in very fine channels used for mcrominiature Joule-Thompson refrigerators, *Cryogenics*, 23, 3–17, 2003.
41. Li, Z. X., Du, D. X., and Guo, Z. Y., Experimental study on flow characteristics of liquid in circular microtubes, *Microsc. Thermophys. Eng.*, 7, 253–265, 2003.
42. Stephan, K., *Heat Transfer in Condensation and Boiling*, Springer-Verlag, Berlin, 1992.
43. Gnielinski, V., New equations for heat transfer in turbulent pipe and channel flow, *Int. Chem. Eng.*, 16, 359–368, 1976.
44. Adams, T. M., Abdel-Khalik, S. I., Jeter, S. M., and Qureshi, Z. H., An experimental investigation of single-phase forced convection in microchannels, *Int. J. Heat Mass Transfer*, 41(6–7), 851–857, 1998.
45. Tsuzuki, N., Utamura, M., and Ngo, T. L., Nusselt number correlations for a microchannel heat exchanger hot water supplier with s-shaped fins, *Appl. Therm. Eng.*, 29, 3299–3308, 2009.
46. White, F. M., *Viscous Fluid Flow*, McGraw-Hill, New York, 1991.
47. Jiang, P. X., Fan, M. H., Si, G. S., and Ren, Z. P., Thermal-hydraulic performance of small scale microchannel and porous-media heat exchangers, *Int. J. Heat Mass Transfer*, 44, 1039–1051, 2001.
48. Kakaç, S., Shah, R. K., and Aung, W., *Handbook of Single Phase Convective Heat Transfer*, John Wiley and Sons, New York, 1987.

49. Choi, S. B., Barron, R. F., and Warrington, R. O., Fluid flow and heat transfer, in *Micromechanical Sensors, Actuator and Systems,* Vol. DSC–32, ASME, Georgia, 1991, 123–134.
50. Yu, D., Warrington, R. O., Barron, R., and Ameel, T., An experimental and theoretical investigation of fluid flow and heat transfer in microtubes, in *Proceedings of ASME/JSME Thermal Engineering Joint Conference,* Hawaii, 523–530, 1995.
51. Bejan, A., *Heat Transfer,* John Wiley & Sons, New Jersey, 1993.
52. Park, H. S. and Punch, J., Friction factor and heat transfer in multiple microchannels with uniform flow distribution, *Int. J. Heat Mass Transfer,* 51, 4435–4443, 2008.
53. Lee, P. S., Garimella, S. V., and Liu, D., Investigation of heat transfer in rectangular microchannels, *Int. J. Heat Mass Transfer,* 48, 1688–1704, 2005.
54. Wu, P. and Little, W. A., Measurement of the heat transfer characteristics of gas flow in fine channel heat exchangers used for microminiature refrigerators, *Cryogenics,* 24(8), 415–420, 1984.
55. Wang, B. X. and Peng, X. F., Experimental investigation of liquid forced convection heat transfer through microchannels, *Int. J. Heat Mass Transfer,* 37(1), 73–82, 1994.
56. Peng, X. F., Peterson, G. P., and Wang, B. X., Heat transfer characteristics of water flowing through microchannels, *Exp. Heat Transfer,* 7, 265–283, 1994.
57. Sobhan, C. B. and Peterson, G. P., *Microscale and Nanoscale Heat Transfer,* CRC Press, Florida, 2008.
58. Celata, G. P., Cumo, M., Gulielmi, M., and Zummo, G., Experimental investigation of hydraulic and single phase heat transfer in 0.130 mm capillary tube, *Nano. Micro. Thermophys. Eng.,* 6(2), 85–97, 2002.
59. Kandlikar, S. G. and Upadhye, H. R., Extending the heat flux limit with enhanced microchannels in direct single-phase cooling of computer chips, *21st Annual IEEE Semiconductor Thermal Measurement and Management Symposium,* 8–15, 2005.
60. Kandlikar, S. G., Schmitt, D., Carrano, A. L., and Taylor, J. B., Characterization of surface roughness effects on pressure drop in single-phase flow in minichannels, *Phys. Fluid.,* 17, 2005.
61. Wang, B., Zhou, L., and Peng, X., A fractal model for predicting the effective thermal conductivity of liquid with suspension of nanoparticles, *Int. J. Heat Mass Transfer,* 46(14), 2665–2672, 2003.
62. Keblinski, P., Phillpot, S. R., Choi, S. U. S., and Eastman, J. A., Mechanisms of heat flow in suspensions of nano-sized particles (Nanofluids), *Int. J. Heat Mass Transfer,* 44(4), 855–863, 2002.
63. Chein, R. and Chuang, J., Experimental microchannel heat sink performance studies using nanofluids, *Int. J. Therm. Sci.,* 46(1), 57–66, 2007.
64. Lee, J. and Mudawar, I., Assessment of the effectiveness of nanofluids for single-phase and two-phase heat transfer in micro-channels, *Int. J. Heat Mass Transfer,* 50(3–4), 442–463, 2007.
65. Eastman, J., Choi, S. U. S., Li, S., Thompson, L., and Lee, S., Enhanced thermal conductivity through the development of nanofluids, *Proceedings of the Symposium on Nanophase and Nanocomposite Materials II,* Boston, 447, 3–11, 1997.
66. Yu, W., France, D. M., Routbort, J. L., and Choi, S. U. S., Review and comparison of nanofluid thermal conductivity and heat transfer enhancements, *Heat Transfer Eng.,* 29(5), 432–460, 2008.

67. Romano, J. M., Parker, J. C., and Ford, Q. B., Application opportunities for nanoparticles made from the condensation of physical vapors, *Adv. Pm. Part.*, 2, 12–13, 1997.
68. Eastman, J. A., Choi, S. U. S., Li, S., Yu, W., and Thompson, L. J., Anomalously increased effective thermal conductivities of ethylene glycol-based nanofluids containing copper nanoparticles, *Appl. Phys. Lett.*, 78(6), 718–720, 2001.
69. Ding, Y., Alias, H., Wen, D., and Williams, R. A., Heat transfer of aqueous suspensions of carbon nanotubes (CNT nanofluids), *Int. J. Heat Mass Transfer*, 49(1–2), 240–250, 2006.
70. Zhu, H. T., Zhang, C. Y., Tang, Y. M., and Wang, J. X., Novel synthesis and thermal conductivity of CuO nanofluid, *J. Phys. Chem. C*, 111(4), 1646–1650, 2007.
71. Hong, T., Yang, H., and Choi, C. J., Study of the enhanced thermal conductivity of Fe nanofluids, *J. Appl. Phys.*, 97(6), 064311–4, 2005.
72. Beck, M., Yuan, Y., Warrier, P., and Teja, A., The effect of particle size on the thermal conductivity of alumina nanofluids, *J. Nanopart. Res.*, 11(5), 1129–1136, 2009.
73. Wang, X., Choi, S. U. S., and Xu, X., Thermal conductivity of nnoparticle - fluid mixture, *J. Thermophys. Heat Transfer*, 13(4), 474–480, 1999.
74. Czarnetzki, W. and Roetzel, W., Temperature oscillation techniques for simultaneous measurement of thermal diffusivity and conductivity, *Int. J. Thermophys.*, 16(2), 413–422, 1995.
75. Ju, Y. S., Kim, J., and Hung, M., Experimental study of heat conduction in aqueous suspensions of aluminum oxide nanoparticles, *J. Heat Transfer*, 130(9), 092403–6, 2008.
76. Putnam, S. A., Cahill, D. G., Braun, P. V., Ge, Z., and Shimmin, R. G., Thermal conductivity of nanoparticle suspensions, *J. Appl. Phys.*, 99(8), 084308–6, 2006.
77. Maxwell, J. C., *A Treatise on Electricity and Magnetism*, Clarendon Press, Oxford, 1873.
78. Hamilton, R. L. and Crosser, O. K., Thermal conductivity of heterogeneous two-component systems, *Ind. Eng. Chem. Fundam.*, 1(3), 187–191, 1962.
79. Bhattacharya, P., Saha, S. K., Yadav, A., Phelan, P. E., and Prasher, R. S., Brownian dynamics simulation to determine the effective thermal conductivity of nanofluids, *J. Appl. Phys.*, 95(11), 6492–6494, 2004.
80. Koo, J. and Kleinstreuer, C., A new thermal conductivity model for nanofluids, *J. Nanopart. Res.*, 6(6), 577–588, 2004.
81. Das, S. K., Putra, N., Thiesen, P., and Roetzel, W., Temperature dependence of thermal conductivity enhancement for nanofluids, *J. Heat Transfer*, 125(4), 567–574, 2003.
82. Özerinç, S., Kakaç, S., and Yazıcıoğlu, A. G., Enhanced thermal conductivity of nanofluids: A State-of-the-Art Review, *Microfluid. Nanofluid.*, 8(2), 144–170, 2010.
83. Prasher, R., Phelan, P. E., and Bhattacharya, P., Effect of aggregation kinetics on the thermal conductivity of nanoscale colloidal solutions (Nanofluid), *Nano Lett.*, 6(7), 1529–1534, 2006.
84. He, Y., Jin, Y., Chen, H., Ding, Y., Cang, D., and Lu, H., Heat transfer and flow behaviour of aqueous suspensions of TiO_2 nanoparticles (Nanofluids) flowing upward through a vertical pipe, *Int. J. Heat Mass Transfer*, 50(11–12), 2272–2281, 2007.

85. Evans, W., Prasher, R., Fish, J., Meakin, P., Phelan, P., and Keblinski, P., Effect of aggregation and interfacial thermal resistance on thermal conductivity of nanocomposites and colloidal nanofluids, *Int. J. Heat Mass Transfer*, 51(5–6), 1431–1438, 2008.
86. Nan, C., Birringer, R., Clarke, D. R., and Gleiter, H., Effective thermal conductivity of particulate composites with interfacial thermal resistance, *J. Appl. Phys.*, 81(10), 6692–6699, 1997.
87. Feng, Y., Yu, B., Xu, P., and Zou, M., The effective thermal conductivity of nanofluids based on the nanolayer and the aggregation of nanoparticles, *J. Phys. D: Appl. Phys.*, 40(10), 3164–3171, 2007.
88. Xuan, Y., Li, Q., and Hu, W., Aggregation structure and thermal conductivity of nanofluids, *AIChE J.*, 49(4), 1038–1043, 2003.
89. Yu, C., Richter, A. G., Datta, A., Durbin, M. K., and Dutta, P., Observation of molecular layering in thin liquid films using X-ray reflectivity, *Phys. Rev. Lett.*, 82(11), 2326–2329, 1999.
90. Yu, W. and Choi, S. U. S., The role of interfacial layers in the enhanced thermal conductivity of nanofluids: A renovated Maxwell model, *J. Nanopart. Res.*, 5(1), 167–171, 2003.
91. Schwartz, L. M., Garboczi, E. J., and Bentz, D. P., Interfacial transport in porous media: application to DC electrical conductivity of mortars, *J. Appl. Phys.*, 78(10), 5898–5908, 1995.
92. Bruggeman, D. A. G., The calculation of various physical constants of heterogeneous substances, 1. The dielectric constants and conductivities of mixtures Composed of Isotropic Substances, *Ann. Phys.*, 416(7), 636–664, 1935.
93. Xue, Q. and Xu, W., A model of thermal conductivity of nanofluids with interfacial shells, *Mater. Chem. Phys.*, 90(2–3), 298–301, 2005.
94. Murshed, S., Leong, K., and Yang, C., Enhanced thermal conductivity of TiO_2-water based nanofluids, *Int. J. Therm. Sci.*, 44(4), 367–373, 2005.
95. Choi, S. U. S., Zhang, Z. G., Yu, W., Lockwood, F. E., and Grulke, E. A., Anomalous thermal conductivity enhancement in nanotube suspensions, *Appl. Phys. Lett.*, 79(14), 2252–2254, 2001.
96. Lee, S., Choi, S. U. S., Li, S., and Eastman, J. A., Measuring thermal conductivity of fluids containing oxide nanoparticles, *J. Heat Transfer*, 121(2), 280–289, 1999.
97. Lee, D., Thermophysical properties of interfacial layer in nanofluids, *Langmuir*, 23(11), 6011–6018, 2007.
98. Chopkar, M., Sudarshan, S., Das, P., and Manna, I., Effect of particle size on thermal conductivity of nanofluid, *Metall. Mater. Trans. A*, 39(7), 1535–1542, 2008.
99. Eastman, J. A., Choi, S. U. S., Li, S., Yu, W., and Thompson, L. J., Anomalously increased effective thermal conductivities of ethylene glycol-based nanofluids containing copper nanoparticles, *Appl. Phys. Lett.*, 78(6), 718–720, 2001.
100. Chopkar, M., Das, P. K., and Manna, I., Synthesis and characterization of nanofluid for advanced heat transfer applications, *Scripta Mater.*, 55(6), 549–552, 2006.
101. Timofeeva, E. V., Routbort, J. L., and Singh, D., Particle shape effects on thermophysical properties of alumina nanofluids, *J. Appl. Phys.*, 106(1), 014304–10, 2009.
102. Li, C. H., Williams, W., Buongiorno, J., Hu, L., and Peterson, G. P., Transient and steady-state experimental comparison study of effective thermal conductivity of Al_2O_3/water nanofluids, *J. Heat Transfer*, 130(4), 042407–7, 2008.

103. Murshed, S., Leong, K., and Yang, C., Investigations of thermal conductivity and viscosity of nanofluids, *Int. J. Therm. Sci.*, 47(5), 560–568, 2008.
104. Zhang, X., Gu, H., and Fujii, M., Effective thermal conductivity and thermal diffusivity of nanofluids containing spherical and cylindrical nanoparticles, *J. Appl. Phys.*, 100(4), 044325-5, 2006.
105. Zhang, X., Gu, H., and Fujii, M., Experimental study on the effective thermal conductivity and thermal diffusivity of nanofluids, *Int. J. Thermophys.*, 27(2), 569–580, 2006.
106. Roy, G., Nguyen, C. T., Doucet, D., Suiro, S., and Maré, T., Temperature dependent thermal conductivity evaluation of alumina based nanofluids, *Proceedings of the 13th International Heat Transfer Conference*, Sydney, 2006.
107. Hong, K. S., Hong, T., and Yang, H., Thermal conductivity of Fe nanofluids depending on the cluster size of nanoparticles, *Appl. Phys. Lett.*, 88(3), 031901-3, 2006.
108. Wang, X., Zhu, D., and Yang, S., Investigation of pH and SDBS on enhancement of thermal conductivity in nanofluids, *Chem. Phys. Lett.*, 470(1–3), 107–111, 2009.
109. Masuda, H., Ebata, A., Teramae, K., and Hishinuma, N., Alteration of thermal conductivity and viscosity of liquid by dispersing ultra-fine particles (Dispersion of γ-Al_2O_3, SiO_2, and TiO_2 ultra-fine particles), *Netsu Bussei*, 4(4), 227–233, 1993.
110. Xie, H., Wang, J., Xi, T., and Liu, Y., Thermal conductivity of suspensions containing nanosized SiC particles, *Int. J. Thermophys.*, 23(2), 571–580, 2002.
111. Xie, H., Wang, J., Xi, T., Liu, Y., and Ai, F., Dependence of the thermal conductivity of nanoparticle-fluid mixture on the base fluid, *J. Mater. Sci. Lett.*, 21(19), 1469–1471, 2002.
112. Li, C. H. and Peterson, G. P., Experimental investigation of temperature and volume fraction variations on the effective thermal conductivity of nanoparticle suspensions (Nanofluids), *J. Appl. Phys.*, 99(8), 084314-1–084314-8, 2006.
113. Mintsa, H. A., Roy, G., Nguyen, C. T., and Doucet, D., New temperature dependent thermal conductivity data for water-based nanofluids, *Int. J. Therm. Sci.*, 48(2), 363–371, 2009.
114. Turgut, A., Tavman, I., Chirtoc, M., Schuchmann, H., Sauter, C., and Tavman, S., Thermal conductivity and viscosity measurements of water-based TiO_2 nanofluids, *Int. J. Thermophys.*, 30(4), 1213–1226, 2009.
115. Assael, M. J., Metaxa, I. N., Arvanitidis, J., Christofilos, D., and Lioutas, C., Thermal conductivity enhancement in aqueous suspensions of carbon multi-walled and double-walled nanotubes in the presence of two different dispersants, *Int. J. Thermophys.*, 26(3), 647–664, 2005.
116. Liu, M., Lin, M. C., Huang, I., and Wang, C., Enhancement of thermal conductivity with carbon nanotube for nanofluids, *Int. Commun. Heat Mass*, 32(9), 1202–1210, 2005.
117. Pak, B. C. and Cho, Y. I., Hydrodynamic and heat transfer study of dispersed fluids with submicron metallic oxide particles, *Exp. Heat Transfer*, 11(2), 151–170, 1998.
118. Xuan, Y. and Roetzel, W., Conceptions for heat transfer correlation of nanofluids, *Int. J. Heat Mass Transfer*, 43(19), 3701–3707, 2000.
119. Wang, B., Zhou, L., and Peng, X., Surface and size effects on the specific heat capacity of nanoparticles, *Int. J. Thermophys.*, 27(1), 139–151, 2006.

120. Einstein, A., A new determination of the molecular dimensions, *Ann. Phys.*, 324(2), 289–306, 1906.
121. Brinkman, H. C., The viscosity of concentrated suspensions and solutions, *J. Chem. Phys.*, 20(4), 571, 1952.
122. Tseng, W. J. and Lin, K., Rheology and colloidal structure of aqueous TiO_2 nanoparticle suspensions, *Mater. Sci. Eng. A*, 355(1–2), 186–192, 2003.
123. Maïga, S. E. B., Nguyen, C. T., Galanis, N., and Roy, G., Heat transfer behaviours of nanofluids in a uniformly heated tube, *Superlatt. Microstruct.*, 35(3–6), 543–557, 2004.
124. Kulkarni, D. P., Das, D. K., and Chukwu, G. A., Temperature dependent rheological property of copper oxide nanoparticles suspension (Nanofluid), *J. Nanosci. Nanotechnol.*, 6, 1150–1154, 2006.
125. Shah, R. K. and London, A. L., *Laminar Flow Forced Convection in Ducts, Supplement 1 to Advances in Heat Transfer*, Academic Press, New York, 1978.
126. Sieder, E. N. and Tate, G. E., Heat transfer and pressure drop of liquids in tubes, *Ind. Eng. Chem.*, 28(12), 1429–1435, 1936.
127. Kakaç, S. and Pramuanjaroenkij, A., Review of convective heat transfer enhancement with nanofluids, *Int. J. Heat Mass Transfer*, 52(13–14), 3187–3196, 2009.
128. Özerinç, S., Heat transfer enhancement with nanofluids, M.S. Thesis, Middle East Technical University, Ankara 2010.
129. Godson, L., Raja, B., Mohan Lal, D., and Wongwises, S., Enhancement of heat transfer using nanofluids-An overview, *Renew. Sust. Energy Rev.*, 14(2), 629–641, 2010.
130. Xuan, Y. and Li, Q., Investigation on convective heat transfer and flow features of nanofluids, *J. Heat Transfer*, 125(1), 151–155, 2003.
131. Yang, Y., Zhang Z. G., Grulke E. A., Anderson W. B., and Wu G., Heat transfer properties of nanoparticle-in-fluid dispersions (nanofluids) in laminar flow, *Int. J. Heat Mass Transfer*, 48(6), 1107–1116, 2005.
132. Maïga, S. E. B., Nguyen C. T., Galanis N., Roy G., Mare T., and Coqueux M., Heat transfer enhancement in turbulent tube flow using Al_2O_3 nanoparticle suspension, *Int. J. Numer. Method. H.*, 16(3), 275–292, 2006.
133. Vasu, V., Rama Krishna, K., and Kumar, A., Analytical prediction of forced convective heat transfer of fluids embedded with nanostructured materials (nanofluids), *Pramana*, 69(3), 411–421, 2007.
134. Putra, N., Roetzel, W., and Das, S., Natural convection of nano-fluids, *Heat Mass Transfer*, 39(8), 775–784, 2003.
135. Chun, B., Kang, H., and Kim, S., Effect of alumina nanoparticles in the fluid on heat transfer in double-pipe heat exchanger system, *Korean J. Chem. Eng.*, 25(5), 966–971, 2008.
136. Anoop, K., Sundararajan, T., and Das, S. K., Effect of particle size on the convective heat transfer in nanofluid in the developing region, *Int. J. Heat Mass Transfer*, 52(9–10), 2189–2195, 2009.
137. Asirvatham, L. G., Vishal, N., Gangatharan, S. K., and Lal, D. M., Experimental study on forced convective heat transfer with low volume fraction of CuO/water nanofluid, *Energies*, 2(1), 97–110, 2009.
138. Jung, J., Oh, H., and Kwak, H., Forced convective heat transfer of nanofluids in microchannels, *Int. J. Heat Mass Transfer*, 52(1–2), 466–472, 2009.

139. Chandrasekar, M., Suresh, S., and Chandra Bose, A., Experimental studies on heat transfer and friction factor characteristics of Al_2O_3/water nanofluid in a circular pipe under laminar flow with wire coil inserts, *Exp. Therm. Fluid Sci.*, 34(2), 122–130, 2010.
140. Duangthongsuk, W. and Wongwises, S., An experimental study on the heat transfer performance and pressure drop of TiO_2-water nanofluids flowing under a turbulent flow regime, *Int. J. Heat Mass Transfer*, 53(1–3), 334–344, 2010.
141. Wu, X., Wu, H., and Cheng, P., Pressure drop and heat transfer of Al_2O_3-H_2O nanofluids through silicon microchannels, *J. Micromech. Microeng.*, 19(10), 105020, 2009.

6

Fouling of Heat Exchangers

6.1 Introduction

Fouling can be defined as the accumulation of undesirable substances on a surface. In general, it refers to the collection and growth of unwanted material, which results in inferior performance of the surface. Fouling occurs in natural as well as synthetic systems. Arteriosclerosis serves as an example of fouling in the human body wherein the deposit of cholesterol and the proliferation of connective tissues in an artery wall form plaque that grows inward. The resulting blockage or narrowing of arteries places increased demand on the heart.

In the present context, the term fouling is used to specifically refer to undesirable deposits on the heat exchanger surfaces. A heat exchanger must effect a desired change in the thermal conditions of the process streams within allowable pressure drops and continue to do so for a specified time period. During operation, the heat transfer surface fouls, resulting in increased thermal resistance and often an increase in the pressure drop and pumping power as well. Both of these effects complement each other in degrading the performance of the heat exchanger. The heat exchanger may deteriorate to the extent that it must be withdrawn from service for replacement or cleaning.

Fouling may significantly influence the overall design of a heat exchanger and may determine the amount of material employed for construction as well as performance between cleaning schedules. Consequently, fouling causes an enormous economic loss as it directly impacts the initial cost, operating cost, and heat exchanger performance. An up-to-date review of the design of heat exchangers with fouling is given by Kakaç and coworkers.[1,2]

Fouling reduces the effectiveness of a heat exchanger by reducing the heat transfer and by affecting the pressure drop (Figure 6.1).

6.2 Basic Considerations

Thermal analysis of a heat exchanger is governed by the conservation of energy in that the heat release by the hot fluid stream equals the heat gain by

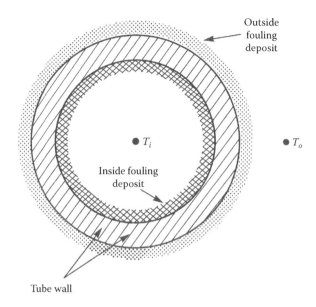

FIGURE 6.1
Fouled tube.

the cold fluid. The heat transfer rate, Q, is related to the exchanger geometry and flow parameters as

$$Q = UA_o \, \Delta T_m \tag{6.1}$$

where U is based on the outside heat transfer surface area of the exchanger. It is important to distinguish between overall heat transfer coefficient for clean (U_c) and fouled surfaces (U_f).

U_f can be related to the clean surface overall heat transfer coefficient, U_c, as

$$\frac{1}{U_f} = \frac{1}{U_c} + R_{ft} \tag{6.2}$$

where R_{ft} is the total fouling resistance, given as

$$R_{ft} = \frac{A_o R_{fi}}{A_i} + R_{fo} \tag{6.3}$$

The heat transfer rate under fouled conditions, Q_f, can be expressed as

$$Q_f = U_f \, A_f \, \Delta T_{mf} \tag{6.4}$$

Fouling of Heat Exchangers

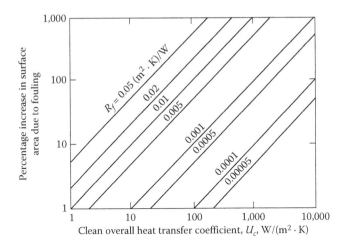

FIGURE 6.2
Effects of fouling on surface area. (From Chenoweth, J. M., *Heat Transfer in High Technology and Power Engineering*, Hemisphere, New York, pp. 406–419, 1987. With permission.)

where the subscript f refers to the fouled conditions. Process conditions usually set the heat duty and fluid temperatures at specified values, i.e., $Q_f = Q_c$ and $\Delta T_{mf} = \Delta T_{mc}$. Under these conditions, Equations 6.1, 6.2, and 6.4 show that

$$\frac{A_f}{A_c} = 1 + U_c R_{ft} \qquad (6.5)$$

where A_c is the required surface area if the heat exchanger remains clean. $U_c R_{ft}$ in Equation 6.5 represents the additional surface area required because of fouling of the heat exchanger. For a range of fouling resistances, Figure 6.2 shows the percentage of increase in the heat transfer surface area due to fouling. Obviously, the added surface is small if U_c is low (1–10 W/m² · K) even though the total fouling resistance may be high (R_{ft} = 50.0 m² K/kW). However, for high U_c (1,000–10,000 W/m² K), even a small fouling resistance (R_{ft} = 0.5 m² · K/kW) results in a substantial increase in the required heat transfer surface area. A 100% increase in the surface area due to fouling alone is not uncommon.

6.3 Effects of Fouling

Lower heat transfer and increased pressure drop resulted from fouling decrease the effectiveness of a heat exchanger. These effects and the basic thermohydraulic aspects of heat exchanger design are discussed in this section.

6.3.1 Effect of Fouling on Heat Transfer

A simple visualization of fouling, shown in Figure 6.1, depicts fouling buildup on the inside and outside of a circular tube. It is evident that fouling adds an insulating layer to the heat transfer surface. For a plain tubular heat exchanger, the overall heat transfer coefficient based on the outer surface area under fouled conditions, U_f, can be obtained by adding the inside and outside thermal resistances in Equation 6.2:

$$U_f = \frac{1}{\frac{A_o}{A_i h_i} + \frac{A_o}{A_i} R_{fi} + \frac{A_o \ln(d_o/d_i)}{2\pi k L} + R_{fo} + \frac{1}{h_o}} \quad (6.6)$$

The overall heat transfer coefficient for a finned tube (based on the outside surface area) is given by Equation 2.17. Fouling resistances R_{fi} or R_{fo}, also defined in Chapter 2, are sometimes referred to as "fouling factors." The heat transfer in the unwanted fouling material takes place by conduction.

Table 6.1 shows average total fouling resistance values specified in the design of some 750 shell-and-tube heat exchangers, by five different manufacturers.[3] Using the typical values of heat transfer coefficients for boiling, condensation, and gas flow and an average total fouling resistance given in Table 6.1 allows the preparation of Table 6.2, which shows the percentage increase in the heat transfer surface area for shell-and-tube boiler, evaporator, and condenser applications with single-phase flow in the tube side. It should be noted that if liquids are used for sensible heating/cooling, fouling may substantially increase the required surface area by a factor of 2 or even 3. For high heat transfer applications, fouling may even dictate the design of the heat exchange device.

TABLE 6.1

Specified Total Fouling Resistance.
$R_{ft} \times 10^4$, $m^2 \times K/W$

	Shell Side		
Tube Side	Vapor	Liquid	Two Phase
Vapor	3.7	5.1	4.8
Liquid	6.0	7.9	6.5
Two phase	5.1	6.7	5.1

Source: From Garrett-Price, B. A. et al., *Fouling of Heat Exchangers: Characteristics, Costs, Prevention, Control, and Removal*, Noyes Publications, Park Ridge, NJ, 1985. With permission.

Fouling of Heat Exchangers

TABLE 6.2
Added Surface Area for Typical Fluid Combinations

Shell Side (Boiling or Condensation)	h_o	$R_{ft} \times 10^4$	U_c	U_f	Increase in Area (%)
Tube side: gas at very low pressure; $h_i = 50$ W/m²·K					
Medium organics	1,000	4.8	47.6	46.5	2.3
Water, low pressure	5,000	4.8	49.5	48.4	2.4
Water, high pressure	10,000	4.8	49.8	48.6	2.4
Tube side: gas at high pressure; $h_i = 500$ W/m²·K					
Light organics	1,000	4.8	333.3	287.3	16.0
Medium organics	5,000	4.8	454.5	373.1	21.8
Steam	10,000	4.8	476.2	387.6	22.9
Tube side: medium organic liquids; $h_i = 1,000$ W/m²·K					
Medium organics	1,000	6.5	500.0	377.4	32.5
Water, low pressure	5,000	6.5	833.3	540.5	54.2
Water, high pressure	10,000	6.5	909.1	571.4	59.1
Tube side: water; $h_i = 5,000$ W/m²·K					
Light organics	1,000	6.5	833.3	287.3	54.2
Medium organics	5,000	6.5	2500.0	373.1	162.5
Water	10,000	6.5	3333.3	387.6	216.7

6.3.2 Effect of Fouling on Pressure Drop

Interestingly, more heat exchangers must be removed from service for cleaning due to excessive pressure drop than for an inability to meet the heat transfer requirements. As shown in Figure 6.1, fouling always has a finite, although sometimes small, thickness. The change in the flow geometry due to fouling affects the flow field and the pressure drop (or pumping power). For example, in a tubular heat exchanger, a fouling layer roughens the surface, decreases the inside diameter, and increases the outside diameter of the tubes.

The frictional pressure drop in the tube for a single-phase flow can be calculated by Equation 4.3:

$$\Delta p = 4f \left(\frac{L}{d_i} \right) \frac{\rho u_m^2}{2} \qquad (6.7)$$

where f is the Fanning friction factor. Various graphs and correlations to determine the friction factor for single-phase flow are available in the literature.[1,2]

The fouling layer decreases the inside diameter and roughens the surface, thus causing an increase in the pressure drop given by Equation 6.7. Pressure drop under fouled and clean conditions can be related as

$$\frac{\Delta p_f}{\Delta p_c} = \frac{f_f}{f_c} \frac{d_c}{d_f} \left(\frac{u_f}{u_c}\right)^2 \tag{6.8}$$

By assuming that the mass flow rate ($\dot{m} = u_m \rho A$) under clean and fouled conditions is the same, Equation 6.8 can be modified as

$$\frac{\Delta p_f}{\Delta p_c} = \frac{f_f}{f_c} \left(\frac{d_c}{d_f}\right)^5 \tag{6.9}$$

Even if the fouling does not affect the friction coefficient, a small reduction in tube inside diameter increase the pressure drop (Table 6.3). The fouling factor can be related to the thermal conductivity k_f of the fouling material and the fouling thickness t_f as

$$R_f = \frac{t_f}{k_f} \quad \text{(for a plane wall)} \tag{6.10a}$$

$$R_f = \frac{\ln(d_c/d_f)}{2k_f} \quad \text{(for a cylindrical tube wall)} \tag{6.10b}$$

TABLE 6.3

Added Pressure Drop for Typical Fouling Materials

Material	Thermal Conductivity (W/m · K)	Fouling[a] Thickness, t (mm)	Area Remaining (%)	Increase in Pressure Drop (%)
Hematite	0.6055	0.24	95.7	11.6
Biofilm	0.7093	0.28	95.0	13.7
Calcite	0.9342	0.37	93.5	18.4
Serpentine	1.0380	0.41	92.8	20.7
Gypsum	1.3148	0.51	90.9	26.9
Magnesium phosphate	2.1625	0.83	85.5	47.9
Calcium sulphate	2.3355	0.90	84.4	52.6
Calcium phosphate	2.5950	0.99	82.9	59.9
Magnetic iron oxide	2.8718	1.09	81.2	68.2
Calcium carbonate	2.9410	1.12	80.8	70.3

[a] Assuming fouling resistance of 4×10^{-4} m² · K/W, OD = 25.4 mm, ID = 22.1 mm tube.

ID under fouled conditions, d_f, can be obtained by rearranging Equation 6.10b:

$$d_f = d_c \exp\left(-\frac{2k_f R_f}{d_c}\right) \tag{6.11}$$

and the fouling thickness, t_f, is expressed as

$$t_f = 0.5 d_c \left[1 - \exp\left(-\frac{2k_f R_f}{d_c}\right)\right] \tag{6.12}$$

In general, the fouling layer consists of several materials. Thermal conductivity may also be nonuniform. Approximate thermal conductivities of pure materials that can constitute fouling deposits are given in the second column of Table 6.3.[3] These values have been used to estimate fouling thickness in a 25.4 mm OD, 16 BWG (22.1 mm ID) tube with a tube-side fouling resistance of 0.0004 m^2 K/W. It is assumed that the fouling layer is composed solely of one material. The percentage of remaining flow area and the percentage increase in the pressure drop are given in columns 3 and 4 of Table 6.3, respectively. As can be seen, the pressure drop increases by about 70% in some instances for the assumed fouling. In these calculations, it is assumed that the fouling does not affect the friction factor (i.e., $f_f = f_c$). Moreover, an increase in the pressure drop due to excess surface area (required for heat transfer) has not been taken into account. Aging of the deposit starts soon after it is laid on the surface. Mechanical properties of the deposit can change during this phase, for example, due to the changes in crystal or chemical structure. Slow poisoning of microorganisms due to corrosion at the surface may weaken the biofouling layer. A chemical reaction taking place at the deposit surface may alter the chemical composition of the deposit and thereby change its mechanical strength.

6.3.3 Cost of Fouling

Fouling of heat transfer equipment introduces an additional cost to the industrial sector. The added cost is in the form of (1) increased capital expenditure, (2) increased maintenance cost, (3) loss of production, and (4) energy losses. In order to compensate for fouling, the heat transfer area of a heat exchanger is increased. Pumps and fans are oversized to compensate for over-surfacing and the increased pressure drop resulting from a reduction in the flow area. Duplicate heat exchangers may have to be installed in order to ensure continuous operation while a fouled heat exchanger is cleaned. High-cost materials such as titanium, stainless steel, or graphite may be required for certain fouling situations. Cleaning equipment may be required for online cleaning. All of these items contribute to increasing the capital expenditure.

Online and offline cleaning add to the maintenance costs. Fouling increases the normally scheduled time incurred in maintaining and repairing equipment. Loss of production because of operation at reduced capacity or downtime can be costly. Finally, energy losses due to reduction in heat transfer and increase in pumping-power requirements can be a major contributor to the cost of fouling.

The annual costs of fouling and corrosion in U.S. industries, excluding electric utilities, were placed between $3 and $10 billion in 1982.[4] It is clear that the deleterious effects of fouling are extremely costly.

6.4 Aspects of Fouling

A landmark paper by Taborek et al.[5] cited fouling as the major unresolved problem in heat transfer. Since then, the great financial burden imposed by fouling on the industrial sector has been recognized. This has resulted in a significant increase in the literature on fouling, and various aspects of the problem have been resolved. The major unresolved problem of 1972 is still the major unsolved problem now. This is because the large amount of fouling research has not brought about a significant solution to the prediction and mitigation of fouling.

The next section discusses some fundamental aspects that help in understanding the types and mechanisms of fouling. The commonly used methods that aid in developing models to predict fouling are also outlined.

6.4.1 Categories of Fouling

Fouling can be classified in a number of different ways. These may include the type of heat transfer service (boiling, condensation), the type of fluid stream (liquid, gas), or the kind of application (refrigeration, power generation). Because of the diversity of process conditions, most fouling situations are virtually unique. However, in order to develop a scientific understanding, it is best to classify fouling according to the principal process that causes it. Such a classification, developed by Epstein,[6] has received wide acceptance. Accordingly, fouling is classified into the following categories: particulate, crystallization, corrosion, biofouling, and chemical reaction.

6.4.1.1 Particulate Fouling

The accumulation of solid particles suspended in the process stream onto the heat transfer surface results in particulate fouling. In boilers, this may occur when unburnt fuel or ashes are carried over by the combustion gases.

Air-cooled condensers are often fouled because of dust deposition. Particles are virtually present in any condenser cooling water. The matter involved may cover a wide range of materials (organic, inorganic) and sizes and shapes (from the submicron to a few millimeters in diameter). Heavy particles settle on a horizontal surface because of gravity. Other mechanisms may be involved, however, for fine particles to settle onto a heat transfer surface at an inclination.

6.4.1.2 Crystallization Fouling

A common way in which heat exchangers become fouled is through the process of crystallization. Crystallization arises primarily from the presence of dissolved inorganic salts in the process stream, which exhibits supersaturation during heating or cooling. Cooling-water systems are often prone to crystal deposition because of the presence of salts such as calcium and magnesium carbonates, silicates, and phosphates. These are inverse solubility salts that precipitate as the cooling water passes through the condenser (i.e., as the water temperature increases). The problem becomes serious if the salt concentration is high. Such a situation may arise, for example, because of accumulation in cooling-water systems with an evaporative cooling tower. The deposits may result in a dense, well-bonded layer referred to as a scale, or a porous, soft layer described as a soft scale, sludge, or powdery deposit.

6.4.1.3 Corrosion Fouling

A heat transfer surface exposed to a corrosive fluid may react, producing corrosion products. These corrosion products can foul the surface, provided that the pH value of the fluid does not dissolve the corrosion products as they are formed. For example, impurities in fuel like alkali metals, sulfur, and vanadium can cause corrosion in oil-fired boilers. Corrosion is particularly serious on the liquid side. Corrosion products may also be swept away from the surface where they are produced and transported to other parts of the system.

6.4.1.4 Biofouling

Deposition and/or growth of material of a biological origin on a heat transfer surface results in biofouling. Such material may include microorganisms (e.g., bacteria, algae, and molds) and their products, and the resulting fouling is known as microbial fouling. In other instances, organisms such as seaweed, water weeds, and barnacles form deposits known as macrobial fouling. Both types of biofouling may occur simultaneously. Marine or power plant condensers using seawater are prone to biofouling.

6.4.1.5 Chemical Reaction Fouling

Fouling deposits are formed as a result of chemical reaction(s) within the process stream. Unlike corrosion fouling, the heat transfer surface does not participate in the reaction, although it may act as catalyst. Polymerization, cracking, and coking of hydrocarbons are prime examples.

It must be recognized that most fouling situations involve a number of different types of fouling. Moreover, some of the fouling processes may complement each other. For example, corrosion of a heat transfer surface promotes particulate fouling. Significant details about each type of fouling are available in the literature. Somerscales and Knudsen[7] and Melo et al.[8] are excellent sources of such information.

6.4.2 Fundamental Processes of Fouling

Even without complications arising from the interaction of two or more categories, fouling is an extremely complex phenomenon. This is primarily due to the large number of variables that affect fouling. To organize thinking about the topic, it will therefore be extremely useful to approach fouling from a fundamental point of view. Accordingly, the fouling mechanisms, referred to as sequential events by Epstein,[9] are initiation, transport, attachment, removal, and aging and are described in the following sections.

6.4.2.1 Initiation

During initiation, the surface is conditioned for the fouling that will take place later. Surface temperature, material, finish, roughness, and coatings strongly influence the initial delay, induction, or incubation period. For example, in crystallization fouling, the induction period tends to decrease as the degree of supersaturation increases with respect to the heat transfer surface temperature. For chemical reaction fouling, the delay period decreases with increasing temperature because of the acceleration of induction reactions. Surface roughness tends to decrease the delay period.[10] Roughness projections provide additional sites for crystal nucleation, thereby promoting crystallization, while grooves provide regions for particulate deposition.

6.4.2.2 Transport

During this phase, fouling substances from the bulk fluid are transported to the heat transfer surface. Transport is accomplished by a number of phenomena including diffusion, sedimentation, and thermophoresis. A great deal of information available for each of these phenomena has been applied to study the transport mechanism for various fouling categories.[7,8]

The difference between fouling species, oxygen or reactant concentration in the bulk fluid, C_b, and that in the fluid adjacent to the heat transfer surface,

Fouling of Heat Exchangers

C_s, results in transport by diffusion. The local deposition flux, $\dot{m}_d$, can be written as

$$\dot{m}_d = h_D(C_b - C_s) \tag{6.13}$$

where h_D is the convective mass transfer coefficient. h_D is obtained from the Sherwood number ($Sh = h_D d/D$) which, in turn, depends on the flow and geometric parameters.

Because of gravity, particulate matter in a fluid is transported to the inclined or horizontal surface. This phenomenon, known as sedimentation, is important in applications where particles are heavy and fluid velocities are low.

Thermophoresis is the movement of small particles in a fluid stream when a temperature gradient is present. Cold walls attract colloidal particles, while hot walls repel these particles. Thermophoresis is important for particles below 5 μm in diameter and becomes dominant at about 0.1 μm.

A number of other processes such as electrophoresis, inertial impaction, and turbulent downsweeps may be present. Theoretical models to study these processes are available in the literature.[11–13] However, application of these models for fouling prediction is often limited by the fact that several of the preceding processes may be involved simultaneously in a particular fouling situation.

6.4.2.3 Attachment

Part of the fouling material transported attaches to the surface. Considerable uncertainty about this process exists. Probabilistic techniques are often used to determine the degree of adherence. Forces acting on the particles as they approach the surface are important in determining attachment. Additionally, properties of the material such as density, size, and surface conditions are important.

6.4.2.4 Removal

Some material is removed from the surface immediately after deposition and some is removed later. In general, shear forces at the interface between the fluid and deposited fouling layer are considered responsible for removal. Shear forces, in turn, depend on the velocity gradients at the surface, the viscosity of the fluid, and surface roughness. Dissolution, erosion, and spalling have been proposed as plausible mechanisms for removal. In dissolution, the material exits in ionic form. Erosion, whereby the material exits in particulate form, is affected by fluid velocity, particle size, surface roughness, and bonding of the material. In spalling, the material exits as a large mass. Spalling is affected by thermal stress set up in the deposit by the heat transfer process.

6.4.2.5 Aging

Once deposits are laid on the surface, aging begins. The mechanical properties of the deposit can change during this phase because of changes in the crystal or chemical structure, for example. Slow poisoning of microorganisms due to corrosion at the surface may weaken the biofouling layer. A chemical reaction taking place at the deposit surface may alter the chemical composition of the deposit and, thereby, change its mechanical strength.

6.4.3 Prediction of Fouling

The overall result of the processes listed above is the net deposition of material on the heat transfer surface. Clearly, the deposit thickness is time dependent. In the design of heat exchangers, a constant value of fouling resistance, R_f, interpreted as the value reached in a time period after which the heat exchanger will be cleaned, is used. To determine the cleaning cycle, one should be able to predict how fouling progresses with time. Such information is also required for proper operation of the heat exchanger. Variation of fouling with time can be expressed as the difference between deposit rate, ϕ_d, and removal rate, ϕ_r, functions.[14]

$$\frac{dR_f}{dt} = \phi_d - \phi_r \tag{6.14}$$

The behavior of ϕ_d and ϕ_r depend upon a large number of parameters. The resulting fouling behavior can be represented as a fouling factor–time curve, as shown in Figure 6.3. The shape of the curve relates to the phenomena occurring during the fouling process.

If either deposition rate is constant and removal rate is negligible or the difference between deposition and removal rates is constant, the fouling–time curve will assume a straight line function, as shown by curve A in Figure 6.3. Linear fouling is generally represented by tough, hard, adherent deposits. Fouling in such cases will continue to increase unless some type of cleaning is employed. If the deposit rate is constant and the removal rate is ignored, Equation 6.14 can be integrated to yield

$$R_f = \phi_f t \tag{6.15}$$

which was first developed by McCabe and Robinson.[15]

Asymptotic fouling curve C is obtained if the deposit rate is constant and the removal rate is proportional to the fouling layer thickness, suggesting that the shear strength of the layer is decreasing or other mechanisms deteriorating the stability of the layer are taking place. Such a situation will generally occur if the deposits are soft since they can flake off easily. Fouling in such cases reaches an asymptotic value.

Fouling of Heat Exchangers

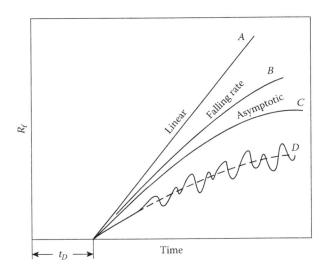

FIGURE 6.3
Typical fouling factor-time curve. (From Chenoweth, J. M., *Heat Transfer in High Technology and Power Engineering*, Hemisphere, New York, pp. 406–419, 1987. With permission.)

Falling-rate fouling, shown by curve B, lies between the linear and asymptotic fouling curves. Such behavior may result if the deposition rate is inversely proportional to the fouling thickness. A periodic change in operating conditions results in the sawtooth configuration shown in curve D. This situation is typical of commercial cooling-tower water. Assuming the removal function to be proportional to the fouling resistance and the deposition to be constant, Kern and Seaton[16] obtained the classical relation:

$$R_f = R_f^* \left(1 - e^{-t/\theta}\right) \qquad (6.16)$$

where θ is the time constant that indicates how quickly asymptotic fouling conditions are approached and R_f^* is the asymptotic fouling factor.

A number of semiempirical models have been developed over the years to predict the nature of the fouling curve in a given application. The general applicability of these models is limited since the various constants or coefficients involved are site dependent and would usually be unknown. Much of the current fouling research is directed toward establishing predictive models. Epstein[6] has tabulated a number of deposition and removal models developed over the past several years. The qualitative effects of increasing certain parameters on the deposition and removal rates and the asymptotic fouling factor are summarized in Table 6.4.[17] Velocity is the only parameter whose increase causes a reduction in the asymptotic fouling factor even though there may be some exceptions.

TABLE 6.4

Effect of Parameters on Fouling

Parameter Increased	Deposition Rate	Removal Rate	Asymptotic Fouling
Stickness	Increases	Decreases	Increases
Surface temperature	Increases	Questionable	Increases
Toughness	Questionable	Decreases	Increases
Roughness	Increases (?)	Increases	Questionable
In situ corrosion	Increases	Questionable	Increases
Ex situ corrosion	Increases	Questionable	Increases
Velocity	Decreases	Increases	Decreases

6.5 Design of Heat Exchangers Subject to Fouling

Although fouling is time dependent, only a fixed value can be prescribed during the design stage. Therefore, the operating characteristics and cleaning schedules of the heat exchanger depend on the design fouling factor. Many heat exchangers operate for long periods without being cleaned, while others might require frequent cleaning. If fouling is anticipated, provisions should be made during the design stage. A number of different approaches are used to provide an allowance for fouling, all of which result in an excess surface area for heat transfer. Current methods include specifying the fouling resistances, the cleanliness factor, or the percentage over surface.

6.5.1 Fouling Resistance

A common practice is to prescribe a fouling resistance or fouling factor (as in Equation 6.6) on each side of the surface where fouling is anticipated. The result is a lower overall heat transfer coefficient. Consequently, excess surface area is provided to achieve the specified heat transfer. The heat exchanger will perform satisfactorily until the specified value of the fouling resistance is reached, after which it must be cleaned. The cleaning interval is expected to coincide with the plant's regular maintenance schedule so that additional shutdowns can be avoided.

It is extremely difficult to predict a specific fouling behavior for most cases since a large number of variables can materially alter the type of fouling and its rate of buildup. Sources of fouling resistance in the literature are rather limited, in part, because of the relatively recent interest in fouling research. Tables found in the standards of the Tubular Exchanger Manufacturers Association (TEMA),[19] reproduced here as Tables 6.5 to 6.11, are probably the most referenced source of fouling factors used in the design of heat exchangers. Unfortunately, the TEMA tables do not cover the large variety of possible

TABLE 6.5

TEMA Design Fouling Resistances for Industrial Fluids

Industrial Fluids	R_f (m² · K/W)
Oils	
Fuel oil no. 2	0.000352
Fuel oil no. 6	0.000881
Transformer oil	0.000176
Engine lube oil	0.000176
Quench oil	0.000705
Gases and Vapors	
Manufactured gas	0.001761
Engine exhaust gas	0.001761
Steam (nonoil bearing)	0.000088
Exhaust steam (oil bearing)	0.000264–0.000352
Refrigerant vapors (oil bearing)	0.000352
Compressed air	0.000176
Ammonia vapor	0.000176
CO_2 vapor	0.000176
Chlorine vapor	0.000352
Coal flue gas	0.001761
Natural gas flue gas	0.000881
Liquids	
Molten heat transfer salts	0.000088
Refrigerant liquids	0.000176
Hydraulic fluid	0.000176
Industrial organic heat transfer media	0.000352
Ammonia liquid	0.000176
Ammonia liquid (oil bearing)	0.000528
Calcium chloride solutions	0.000528
Sodium chloride solutions	0.000528
CO_2 liquid	0.000176
Chlorine liquid	0.000352
Methanol solutions	0.000352
Ethanol solutions	0.000352
Ethylene glycol solutions	0.000352

Source: From, *Standards of the Tubular Exchanger Manufacturers Association*, 1988. With permission. ©1988 Tubular Exchanger Manufacturers Association.

TABLE 6.6

Fouling Resistances for Chemical Processing Streams

Streams	R_f (m² · K/W)
Gases and vapors	
Acid gases	0.000352–0.000528
Solvent vapors	0.000176
Stable overhead products	0.000176
Liquids	
MEA and DEA solutions	0.000352
DEG and TEG solutions	0.000352
Stable side draw and bottom product	0.000176–0.000352
Caustic solutions	0.000352
Vegetable oils	0.000528

Source: From *Standards of the Tubular Exchanger Manufacturers Association,* 1988. With permission. ©1988 Tubular Exchanger Manufacturers Association.

TABLE 6.7

Fouling Resistances for Natural Gas–Gasoline Processing Streams

Streams	R_f (m² · K/W)
Gases and vapors	
Natural gas	0.000176–0.000352
Overhead products	0.000176–0.000352
Liquids	
Lean oil	0.000352
Rich oil	0.000176–0.000352
Natural gasoline and liquefied petroleum gases	0.000176–0.000352

Source: From *Standards of the Tubular Exchanger Manufacturers Association,* 1988. With permission. ©1988 Tubular Exchanger Manufacturers Association.

process fluids, flow conditions, and heat exchanger configurations. These values allow the exchanger to perform satisfactorily in a designated service for a "reasonable time" between cleanings. The interval between cleanings is not known *a priori* since it depends on the performance of the heat exchanger while it is in service. Quite often, sufficient excess area is provided for the exchanger to perform satisfactorily under fouled conditions. Proprietary research data, plant data, and personal or company experience are other sources of information on fouling resistance.

TABLE 6.8

Fouling Resistances for Oil Refinery Streams

Streams	R_f (m² · K/W)		
Crude and vacuum unit gases and vapors			
Atmospheric tower overhead vapors	0.000176		
Light naphthas	0.000176		
Vacuum overhead vapors	0.000352		
Crude and vacuum liquids			
Crude oil			
	−30°C to 120°C Velocity m/s		
	<0.6	0.6–1.2	>1.2
Dry	0.000528	0.000352	0.000352
Salt[a]	0.000528	0.000352	0.000352
	120°C to 175°C Velocity m/s		
	<0.6	0.6–1.2	>1.2
Dry	0.000528	0.000352	0.000352
Salt[a]	0.000881	0.000705	0.000705
	175°C to 230°C Velocity m/s		
	<1.5	0.6–1.2	>1.2
Dry	0.000705	0.000528	0.000528
Salt[a]	0.001057	0.000881	0.000881
	230°C and over Velocity m/s		
	<1.5	0.6–1.2	>1.2
Dry	0.000881	0.000705	0.000705
Salt[a]	0.001233	0.001057	0.001057
Gasoline	0.000352		
Naphtha and light distillates	0.000352–0.000528		
Kerosene	0.000352–0.000528		
Light gas oil	0.000352–0.000528		
Heavy gas oil	0.000528–0.000881		
Heavy fuel oils	0.000881–0.001233		

(*continued*)

TABLE 6.8 (CONTINUED)
Fouling Resistances for Oil Refinery Streams

Streams	R_f (m² · K/W)
Asphalt and Residuum	
Vacuum tower bottoms	0.001761
Atmosphere tower bottoms	0.001233
Cracking and coking unit streams	
Overhead vapors	0.000352
Light cycle oil	0.000352–0.000528
Heavy cycle oil	0.000528–0.000705
Light coker gas oil	0.000528–0.000705
Heavy coker gas oil	0.000705–0.000881
Bottoms slurry oil (1.4 m/s minimum)	0.000528
Light liquid products	0.000176

Source: From, *Standards of the Tubular Exchanger Manufacturers Association*, 1988. With permission. ©1988 Tubular Exchanger Manufacturers Association.

[a] Assumes desalting at approximately 120°C.

TABLE 6.9
Fouling Resistances for Oil Refinery Streams

Streams	R_f (m² · K/W)
Catalytic reforming, hydrocracking, and hydrodesulfurization	
Reformer charge	0.000264
Reformer effluent	0.000264
Hydrocracker charge and effluent[a]	0.000352
Recycle gas	0.000176
Hydrodesulphurization charge and effluent[a]	0.000352
Overhead vapors	0.000176
Liquid product over 50° A.P.I.	0.000176
Liquid product 30–50° A.P.I.	0.000352
Light ends processing	
Overhead vapors and gases	0.000176
Liquid products	0.000176
Absorption oils	0.000352–0.000528
Alkylation trace acid streams	0.000352
Reboiler streams	0.000352–0.000528
Lube oil processing	
Feed stock	0.000352
Solvent feed mix	0.000352

TABLE 6.9 (CONTINUED)
Fouling Resistances for Oil Refinery Streams

Streams	R_f (m² · K/W)
Solvent	0.000176
Extract[b]	0.000528
Raffinate	0.000176
Asphalt	0.000881
Wax slurries[b]	0.000528
Refined lube oil	0.000176
Visbreaker	
Overhead vapor	0.000528
Visbreaker bottoms	0.001761
Naphtha hydrotreater	
Feed	0.000528
Effluent	0.000352
Naphthas	0.000352
Overhead vapors	0.000264

Source: From *Standards of the Tubular Exchanger Manufacturers Association*, 1988. With permission. © 1988 Tubular Exchanger Manufacturers Association.

[a] Depending on charge, characteristics, and storage history, charge resistance may be many times this value.

[b] Precautions must be taken to prevent wax deposition on cold tube walls.

TABLE 6.10
Fouling Resistances for Oil Refinery Streams

Streams	R_f (m² · K/W)
Catalytic hydrodesulphurizer	
Charge	0.000705–0.000881
Effluent	0.000352
H.T. separator overhead	0.000352
Stripper charge	0.000528
Liquid products	0.000352
HF alky unit	
Alkylate, deprop. bottoms, main fract. overhead main fract. Feed	0.000528
All other process streams	0.000352

Source: From, *Standards of the Tubular Exchanger Manufacturers Association*, 1988. With permission. ©1988 Tubular Exchanger Manufacturers Association.

TABLE 6.11

Fouling Resistances for Water

Temperature of Heating Medium	Up to 115°C		R_f (m² · K/W) 115°C to 205°C	
Temperature of Water Water Velocity (m/s)	50°C		Over 50°C	
	0.9 and Less	Over 0.9	0.9 and Less	Over 0.9
350.000176				
	0.000528	0.000352	Cooling tower and artificial spray pond	Treated make up 0.0.52
Cooling tower and artificial spray pond				
Treated make up	0.000176	0.000176	0.000352	0.000352
Untreated	0.000528	0.000528	0.000881	0.000705
City or well water	0.000176	0.000176	0.000352	0.000352
River water				
Minimum	0.000352	0.000176	0.000528	0.000352
Average	0.000528	0.000352	0.000705	0.000528
Muddy or silty	0.000528	0.000352	0.000705	0.000528
Hard (over 15 grains/gal)	0.000528	0.000528	0.000881	0.000881
Engine jacket	0.000176	0.000176	0.000176	0.000176
Distilled or closed cycle				
Condensate	0.000088	0.000088	0.000088	0.000088
Treated boiler feedwater	0.000176	0.000088	0.000176	0.000176
Boiler blowdown	0.000352	0.000352	0.000352	0.000352

Source: From *Standards of the Tubular Exchanger Manufacturers Association,* 1988. With permission. ©1988 Tubular Exchanger Manufacturers Association.

6.5.2 Cleanliness Factor

Another approach for providing an allowance for fouling is by specifying cleanliness factor (CF), a term developed for the steam power industry. CF relates the overall heat transfer coefficient when the heat exchanger is fouled to when it is clean:

$$CF = \frac{U_f}{U_c} \tag{6.17}$$

It is apparent from Equation 6.17 that this approach provides a fouling allowance that varies directly with the clean surface overall heat transfer

Fouling of Heat Exchangers

coefficient. From Equations 6.2 and 6.17, the cleanliness factor can be related to the total fouling resistance:

$$R_{ft} = \frac{1-CF}{U_c CF} \tag{6.18}$$

or

$$CF = \frac{1}{1+R_{ft}U_c} \tag{6.19}$$

Equation 6.18 is used to obtain Figure 6.4, which shows fouling resistance vs. clean surface overall heat transfer coefficient curves for various cleanliness factors. Figure 6.4 illustrates that a given CF will respond to higher fouling resistance, R_{ft}, if the overall heat transfer coefficient, U_c, is low. Such a trend is desirable for the design of steam condensers, where U_c is proportional to the velocity. As shown in Table 6.4, lower velocities (hence low U_c) result in increased fouling. Although the cleanliness factor results in favorable trends, the designer is still left with the problem of selecting the appropriate CF for his application. Typical designs are based upon a cleanliness factor of 0.85.

6.5.3 Percent over Surface

In this approach, the designer simply adds a certain percentage of clean surface area to account for fouling. The added surface implicitly fixes the total fouling resistance depending upon the clean surface overall heat transfer

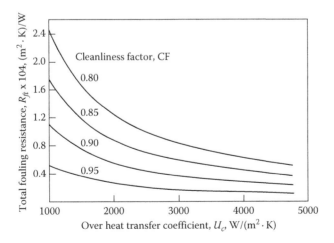

FIGURE 6.4
Calculated fouling resistance based on cleanliness factor.

coefficient. If heat transfer rate and fluid temperatures under clean and fouled conditions are the same ($Q_f = Q_c$ and $\Delta T_{mf} = \Delta T_{mc}$), the percent over surface (OS) can be obtained from Equation 6.5 as

$$OS = 100\left(\frac{A_f}{A_c} - 1\right) = 100\, U_c R_{ft} \qquad (6.20)$$

In a shell-and-tube heat exchanger, the additional surface can be provided either by increasing the length of tubes or by increasing the number of tubes (hence the shell diameter). Such a change will affect the design conditions such as flow velocities, number of cross passes, or baffle spacing. Therefore, the new design with the increased surface should be related to achieve optimum conditions.

Example 6.1

A double-pipe heat exchanger is used to condense steam at a rate of 120 kg/hr at 45°C. Cooling water (seawater) enters through the inner tube at a rate of 1.2 kg/s at 15°C. The tube with 25.4 mm OD, 22.1 mm ID is made of mild steel, $k = 45$ W/m · K. The heat transfer coefficient on the steam side, h_o, is 7000 W/m² · K. Calculate the overall heat transfer coefficient under clean and fouled conditions.

Solution

At 45°C, c_p (water) = 4.18 kJ/kg · K and h_{fg} = 2392 kJ/kg from the appendix (Table B2). The exit temperature of the water can be obtained from the heat balance:

$$Q = \dot{m}_o h_{fg} = \dot{m}_i c_p (T_{out} - T_{in})$$

$$Q = \frac{120}{3600}(2392) = 79.73 \text{ kW}$$

$$T_{out} = T_{in} + \frac{Q}{\dot{m}_i c_p} = 15 + \frac{79.73 \times 10^3}{1.2 \times 4.18 \times 10^3} = 30.89°C$$

At the mean temperature of 23°C, the properties of water are

$\rho = 997.207$ kg/m³, $k = 0.605$ W/m · K, $\mu = 9.09 \times 10^{-4}$ Pa s, $Pr = 6.29$

$$Re = \frac{u_m d_i \rho}{\mu} = \frac{4\dot{m}_i}{\mu \pi d_i} = \frac{4 \times 1.2}{9.09 \times 10^{-4} \pi (0.0221)} = 76{,}056$$

The flow is turbulent.
The heat transfer coefficient inside the tube can be calculated by use of the correlation given for turbulent flow in Chapter 3. The Gnielinski correlation for constant properties, Equation 3.31, is used here:

$$f = (1.58 \ln Re - 3.28)^{-2}$$
$$f = (1.58 \ln 76056 - 3.28)^{-2} = 0.0047$$

$$Nu_b = \frac{(f/2)(Re - 1000)Pr}{1 + 12.7(f/2)^{1/2}(Pr^{2/3} - 1)}$$

$$Nu_b = \frac{0.0047/2\,(76056 - 1000)6.29}{1 + 12.7\,(0.0047/2)^{1/2}\,(6.29^{2/3} - 1)} = 447$$

$$h_i = \frac{Nu\,k}{d_i} = \frac{447 \times 0.605}{0.0221} = 12236.9\ \text{W/m}^2 \cdot \text{K}$$

The overall heat transfer coefficients for the clean and fouled surfaces based on the outside surface area of the tube are

$$U_c = \frac{1}{\dfrac{d_o}{d_i}\dfrac{1}{h_i} + \dfrac{d_o \ln (d_o/d_i)}{2k} + \dfrac{1}{h_o}}$$

and

$$U_f = \frac{1}{\dfrac{d_o}{d_i}\dfrac{1}{h_i} + \dfrac{d_o}{d_i} R_{fi} + \dfrac{d_o \ln (d_o/d_i)}{2k} + R_{fo} + \dfrac{1}{h_o}}$$

The inside and the outside fouling resistances of the inner tube are obtained from Table 6.11 as

$$R_{fi} = 0.088\ \text{m}^2 \cdot \text{K/kW} \quad \text{for sea water}$$

$$R_{fo} = 0.088\ \text{m}^2 \cdot \text{K/kW} \quad \text{for condensate}$$

The water side velocity is

$$u_m = \frac{\dot{m}_i}{\rho_i A_i} = \frac{1.2}{(997.207)\dfrac{\pi}{4}(0.022)^2} = 3.17\ \text{m/s}$$

and the mean temperature difference is

$$\Delta T_{lm} = \frac{\Delta T_1 - \Delta T_2}{\ln \dfrac{\Delta T_1}{\Delta T_2}} = \frac{(45-15)-(45-30.89)}{\ln \dfrac{(45-15)}{(45-30.89)}} = 21.06°\ \text{C}$$

The surface area is calculated from Equations 6.1 and 6.4. The distribution of resistances, the outside heat transfer coefficients, and the surface area can be tabulated as

Distribution of Resistances	Clean m² · K/W	%	Fouled m² · K/W	%
Water side, h_i = 12236.9 W/m² · K	0.939×10^{-4}	34.01	0.939×10^{-4}	20.18
Inside fouling	0		1.011×10^{-4}	21.73
Tube wall	0.393×10^{-4}	14.23	0.393×10^{-4}	8.45
Outside fouling	0		0.88×10^{-4}	18.92
Steam side, h_o = 7000 W/m² · K	1.429×10^{-4}	51.76	1.429×10^{-4}	30.72
Total resistance, m² · K/W	2.761×10^{-4}		4.652×10^{-4}	
Overall heat transfer coefficient, W/m² · K	3621.9		2149.6	
Mean temperature difference, K	21.06		21.06	
Surface area, m²	1.045		1.761	

For assumed values of fouling resistances, the overall heat transfer coefficient, under fouled conditions, is about 60% of that under clean conditions. In this example, the heat transfer surface area should be increased by 76% to obtain the desired heat transfer rate under fouled conditions. Increasing the surface area by such a magnitude may be expensive. An alternate solution would be to design with lower fouling resistances and arrange for cleaning and/or mitigation techniques to control fouling.

6.5.3.1 Cleanliness Factor

If a cleanliness factor of 0.85 (typical value) is used, U_f can be obtained from Equation 6.17:

$$U_f = 0.85 \times 3621.9 = 3079 \text{ W/m}^2 \cdot \text{K}$$

From Equation 6.2, this is equivalent to a total fouling resistance of 4.867×10^{-5} m² · K/W, which is rather small compared to the value obtained from Table 6.11. The fouled area is obtained as resulting in an excess area of only 23%.

6.5.3.2 Percent over Surface

If 25% over surface (typical value) is used, using Equation 6.20, the total fouling resistance is obtained as 6.895×10^{-5} m² · K/W, which is about 39% of the TEMA value.

Seemingly different methods for providing an allowance for fouling essentially do the same thing. They result in an increase in the heat transfer surface area. Total fouling resistance, cleanliness factor, and percent over surface can be related to each other through Equations 6.3 and 6.17–6.20. Such a relationship for the above example is presented in Table 6.12.

Example 6.2

A double-pipe heat exchanger is used to condense steam at a rate of 113.68 kg/hr at 50°C. Cooling water enters through the tubes at a rate of 0.9 kg/s at 10°C. The tube with 25.4 mm OD, 22.1 mm ID is made of mild steel ($k = 45$ W/m · K). The heat transfer coefficient on the steam side, h_o is 10,000 W/m² · K, and on the inside, h_i, is 8000 W/m² · K.

Solution

To increase the heat transfer surface area, the plain tube is replaced by a low-finned tube L with fins either on the inside or the outside. Assume that the use of fins increases the available surface area by a factor of 2.9. Assume 100% fin efficiency, an identical wall resistance, and inside fouling for both plain and finned tubes of 0.000176 m² · K/W. Fouling on the shell side is negligible.

Table 6.13 shows various conditions and results of this example. The heat transfer rate is reflected by the product U × A. It is evident that the finned tube results in a higher heat transfer than the plain tube. For clean conditions, the heat transfer increases by 50% if fins are used on the inside and by 30% if fins are used on the outside. Fins are more effective on the inside because of the relatively smaller heat transfer coefficient (hence higher thermal resistance) on the inside.

Under fouled conditions, the increase in heat transfer is 88% and 16% for fins on the inside and the outside, respectively. In this example, enhancement is favorable if done on the inside of the tube. External fins are not very effective since the heat transfer increases by only 16%, even though the surface area is increased by 290%. As a general rule, enhancement is effective if done

TABLE 6.12
Relationship between R_{ft}, CF, and OS

R_{ft} (m²·K/W)	CF	OS (%)
0.00005	0.85	17.7
0.00010	0.74	35.5
0.00015	0.65	53.0
0.00020	0.59	70.7
0.00025	0.53	88.4
0.00030	0.49	106.0
0.00035	0.45	123.7
0.00040	0.41	141.4

TABLE 6.13

Fouling in Plain and Finned Tubes

Quantity	Plain	Finned Inside	Finned Outside
d_o	25.4 mm	25.4 mm	25.4 mm
d_i	22.1 mm	22.1 mm	22.1 mm
h_i	8,000 W/m²·K	8,000 W/m²·K	8,000 W/m²·K
h_o	10,000 W/m²·K	10,000 W/m²·K	10,000 W/m²·K
k	45 w/mK	45 W/mK	45 W/mK
Area ratio			
A_o/A_{op}	1.0	1.0	2.9
A_i/A_{ip}	1.0	2.9	1.0
A_o/A_i	1.1493	0.3963	3.3333
U_c	3,534 W/m²·K	5,296 W/m²·K	1,586 W/m²·K
$U_c A_o/L$	282 w/K/m	423 W/K/m	367 W/K/m
% Increase	0	50	30
R_{fi}	0.000176 m²·K/W	0.000176 m²·K/W	0.000176 m²·K/W
U_f	2,061 W/m²·K	3,867 W/m²·K	822 W/m²·K
$U_f A_o/L$	164 W/K/m	309 W/K/m	190 W/K/m
% Increase	0	88	16

on the side with dominant thermal resistance. Identical fouling resistance has been used in this example. Such an assumption may not be realistic.

6.6 Operations of Heat Exchangers Subject to Fouling

The specifications of excess area in the heat exchanger, due to the anticipated fouling, leads to operational problems and may be the cause of fouling of the surface beyond that which was used in the design. Due to the time-dependent nature of fouling, the operating conditions will usually be different than the design conditions. The following example illustrates the effect of fouling on operation of the heat exchanger for two different situations.

Example 6.3

The condenser in Example 6.2 is designed with a total fouling resistance of 0.000176. The tube is made of copper. Consider its operation if (1) the water flow rate is maintained at the design value of 0.9 kg/s, or (2) the heat transfer rate is maintained at design value of 75,240 W.

Solution

For simplicity, the wall resistance and thickness of the tube are ignored. Therefore, the clean surface overall heat transfer coefficient, U_c, is

Fouling of Heat Exchangers

$$\frac{1}{U_c} = \frac{1}{8000} + \frac{1}{10,000}$$

or

$$U_c = 4444.4 \text{ W/m}^2 \cdot \text{K}$$

$$\frac{1}{U_f} = \frac{1}{U_c} + R_{ft} = \frac{1}{4444.4} + 0.000176 = 0.000401$$

The design surface area can be obtained from

$$Q = U_f A_f \Delta T_{mf}$$
$$Q = 75240 \text{ W}, \ U_f = 2493.8 \text{ W/m}^2 \cdot \text{K}, \ \Delta T_{mf} = 28.85° \text{ C}$$

Therefore, $A_f = 1.0456$ m² ($A_c = 0.5868$ m², over surface = 78.2%). The heat exchanger area is fixed at 1.0456 m².

If the water flow rate is maintained at the design value, the heat transfer will be highest when the surface is clean. A higher heat transfer results in a higher water outlet temperature. However, the overall heat transfer coefficient will decrease as the fouling builds up, and the outlet water temperature decreases toward the design outline.

Fouling resistance used in the design brings about a 78% increase in the surface area over that required if there was no fouling resistance. Therefore, to achieve the design heat transfer rate when the heat exchanger is placed in operation ($R_f = 0$), the water flow rate needs to be adjusted. A common practice (though not recommended) adapted by plant operators is to reduce the mass flow rate. The resulting change in velocity affects the tube-side heat transfer coefficient, h_i, and the water outlet temperature, T_{out}. The water flow rate and the exit temperature to achieve the design heat transfer rate can be computed from Equations 6.21 and 6.22. However, the tube-side heat transfer coefficient, h_i, should be related to the design heat transfer coefficient, h_d (=8000 W/m² · K), as

$$Q = U_f A_f \Delta T_m = \dot{m}_i c_p (T_o - T_i) \tag{6.21}$$

$$\frac{1}{U_f} = \frac{1}{U_c} + R_{ft} \tag{6.22}$$

Both higher water temperatures and lower water flow rates increase the tendency of the surface exposed to water to foul. This demonstrates that the adding of heat transfer surface area to allow for anticipated fouling tends to accelerate the rate of fouling.

It would be desirable if, during the initial operating period, the water velocity could be maintained at the design value. This can be accomplished by recirculating some of the water or flooding the condenser. Additional capital costs of pumps and piping required for recirculation should be balanced against the advantage of reducing fouling in the heat exchanger.

Performance of the heat exchanger should be monitored during operation. Based on these observations, a proper cleaning schedule can be established. If proper fouling data are available, the cleaning schedule can be based on a rational economic criterion. Somerscales[20] provides a discussion of some of these techniques.

6.7 Techniques to Control Fouling

There are a number of strategies to control fouling. Additives that act as fouling inhibitors can be used while the heat exchanger is in operation. If it is not possible to stop fouling, it becomes a practical matter to remove it. Surface cleaning can be done either online or offline. Table 6.14 (Ref. 21) provides a summary of the various techniques used to control fouling. The following sections provide a discussion of some of these techniques.

6.7.1 Surface Cleaning Techniques

If prior arrangement is made, cleaning can be done online. At other times, offline cleaning must be used. Cleaning methods can be classified as continuous cleaning or periodic cleaning.

6.7.1.1 Continuous Cleaning

Two of the most common techniques are the sponge-ball and brush systems. The sponge-ball system recirculates rubber balls through a separate loop feeding into the upstream end of the heat exchanger. The system requires extensive installation and is therefore limited to large facilities. The brush system has capture cages at the ends of each tube. It requires a flow-reversal valve, which may be expensive.

6.7.1.2 Periodic Cleaning

Fouling deposits may be removed by mechanical or chemical means. The mechanical methods of cleaning include high-pressure water jets, steam, brushes, and water guns. High-pressure water works well for most deposits, but, frequently, a thin layer of the deposit is not removed, resulting in

TABLE 6.14
Various Techniques to Control Fouling

Online Techniques	Offline Techniques
Use and control of appropriate additives	Disassembly and manual cleaning
Inhibitors	
Antiscalants	Lances
Dispersants	Liquid jet
Acids	Steam
	Air jet
Online cleaning	Mechanical cleaning
Sponge balls	Drills
Brushes	Scrapers
Sonic horns	
Soot blowers	
Chains and scrapers	Chemical cleaning
Thermal shock	
Air bumping	

Source: From *Chenoweth, J. M. Fouling Science and Technology*, Kluwer, Dordrecht, The Netherlands, pp. 477–494, 1988. With permission.

greater affinity for fouling when the bundle is returned to service.[5,22] High-temperature steam is useful for the removal of hydrocarbon deposits. Brushes or lances are scraping devices attached to long rods and sometimes include a water or steam jet for flushing and removing the deposit. Chemical cleaning is designed to dissolve deposits by a chemical reaction with the cleaning fluid. The advantage of chemical cleaning is that a hard-to-reach area can be cleaned (e.g., finned tubes). However, the solvent selected for chemical cleaning should not corrode the surface.

6.7.2 Additives

Chemical additives are commonly used to minimize fouling. The effect of additives is best understood for water. For various types of fouling, Strauss and Puckorious[23] provide the following observations.

6.7.2.1 Crystallization Fouling

Minerals from the water are removed by softening. The solubility of the fouling compounds is increased by using chemicals such as acids and polyphosphates. Crystal modification by chemical additives is used to make deposits easier to remove. Preventive techniques are to keep a surface temperature above the freezing value of vapors if it is possible and to avoid the fouling, preheating should be done.[24]

6.7.2.2 Particulate Fouling

Particles are removed mechanically by filtration. Flocculants are used to aid filtration. Dispersants are used to maintain particles in suspension. Increasing an outlet temperature value of the exhaust gases, above the melting point of the particulates, decrease particles fouling done.[24]

6.7.2.3 Biological Fouling

Chemical removal using continuous or periodic injections of chlorine and other biocides is most common.

6.7.2.4 Corrosion Fouling

Additives are used to produce protective films on the metal surface. Corrosion pollution in a heat exchanger depends on the temperature of the exhaust stream. Since there are electrochemical reactions on active metal surfaces; increasing stream velocity up to the maximum value will also increase the corrosion rate. However, there is no effect observed for a passive surface. pH value plays an important role on this fouling, where steel surfaces have a pH value in the range from 11 to 12, there is corrosion occured.[24]

A number of additive options may be available depending on the type of fluid and application. Additional information on additives is given by Marner and Suitor.[22] Some techniques should be considered parallel to or before the above techniques such as removal of potential residues from inlet gas and adjusting design up front to minimize fouling.[24]

Nomenclature

A	heat transfer surface area, m²
A_{cr}	cross-sectional area, m²
C_b	concentration in the fluid, kg/m³
c_p	specific heat at constant pressure, J/kg · K
C_s	concentration at the heat transfer surface, kg/m³
CF	cleanliness factor defined by Equation 6.17
D	diffusion coefficient, m²/s
d	tube diameter, m
f	Fanning friction factor
h	heat transfer coefficient, W/m² · K
h_D	mass transfer coefficient, m/s
h_{fg}	heat of vaporization, J/kg

k	thermal conductivity, W/m · K
L	length of the heat transfer surface, m
$\dot{m}$	fluid mass flow rate, kg/s
Nu	Nusselt number, hd/k
OS	over surface defined in Equation 6.20, %
ΔP	pressure drop, Pa
Pr	Prandtl number, $\mu c_p/k$
Q	heat transfer rate, W
Re	Reynolds number, $\rho u_m d/\mu$
R_f	fouling resistance, m² · K/kW
R_{ft}	total fouling resistance, m² · K/kW
R_f^*	asymptotic fouling resistance, m² · K/kW
Sh	Sherwood number, $h_D d/D$
T	temperature, °C, K
ΔT_m	effective mean temperature difference, °C, K
t	time, s
t_f	fouling thickness, m
t_D	delay time, s
U	overall heat transfer coefficient, W/m² · K
u_m	average fluid velocity, m/s

Greek Symbols

ρ	fluid density, kg/m³
μ	viscosity, kg/m · s
ϕ_d	fouling deposition function, m² · K/kJ
ϕ_r	fouling removal function, m² · K/kJ
θ	time constant, s

PROBLEMS

6.1. The heat transfer coefficient of a steel ($k = 43$ W/m · K) tube (1.9 cm ID and 2.3 cm OD) in a shell-and-tube heat exchanger is 500 W/m² · K on the inside and 120 W/m² · K on the shell side, and it has a deposit with a total fouling factor of 0.000176 m² · K/W. Calculate

   a. The overall heat transfer coefficient

   b. The cleanliness factor, and percent over surface.

6.2. Consider a shell-and-tube heat exchanger having one shell and four tube passes. The fluid in the tubes enters at 200°C and leaves at 100°C. The temperature of the fluid is 20°C entering the shell and 90°C leaving the shell. The overall heat transfer coefficient based on the clean surface area of 12 m² is 300 W/m² · K. During the operation of this heat exchanger for six months, assume that a deposit with a fouling factor of 0.000528 m² · K/W has built up. Estimate the percent decrease in the heat transfer rate between the fluids.

6.3. Assume the water for a boiler is preheated using flue gases from the boiler stack. The flue gases are available at a rate of 0.25 kg/s at 150°C, with a specific heat of 1000 J/kg · K. The water entering the exchanger at 15°C at the rate of 0.05 kg/s is to be heated to 90°C. The heat exchanger is to be of the type with one shell pass and four tube passes. The water flows inside the tubes, which are made of copper (2.5 cm ID, 3.0 cm OD). The heat transfer coefficient on the gas side is 115 W/m² · K, while the heat transfer coefficient on the water side is 1150 W/m² · K. A scale on the water side and gas side offer an additional total thermal resistance of 0.000176 m² · K/W.

  a. Determine the overall heat transfer coefficient based on the outer tube diameter.
  b. Determine the appropriate mean temperature difference for the heat exchanger.
  c. Estimate the required total tube length.
  d. Calculate the percent over surface design and the cleanliness factor.

6.4. A horizontal shell-and-tube heat exchanger is used to condense organic vapors. The organic vapors condense on the outside of the tubes, while water is used as the cooling medium on the inside of the tubes. The condenser tubes are 1.9 cm OD, 1.6 cm ID copper tubes that are 2.4 m in length. There are a total of 768 tubes. The water makes four passes through the exchanger. Test data obtained when the unit was first placed into service are as follows:

  Water rate = 3700 liters/min
  Inlet water temperature = 29°C
  Outlet water temperature = 49°C
  Organic-vapor condensation temperature = 118°C

After three months of operation, another test made under the same conditions as the first (i.e., same water rate and inlet temperature and same condensation temperature) showed that the exit water temperature was 46°C.

  a. What is the tube-side fluid (water) velocity?
  b. Assuming no changes in either the inside heat transfer coefficient or the condensing coefficient, negligible shell-side

fouling, and no fouling at the time of the first test, estimate the tube-side fouling coefficient at the time of the second test.

6.5. In a double-pipe heat exchanger, deposits of calcium carbonate with a thickness of 1.12 mm and magnesium phosphate with a thickness of 0.88 mm on the inside and outside of the inner tube, respectively, have formed over time. Tubes (ID = 1.9 cm, OD = 2.3 cm) are made of carbon steel (k = 43 W/m · K). Calculate the total fouling resistance based on the outside surface area of the heat exchanger.

6.6. In a shell-and-tube heat exchanger, water at a flow rate of 3 kg/s is heated from 20°C to 90°C. On the shell side, steam condenses at 100°C to heat the water, resulting in an overall heat transfer coefficient of 2100 W/m² · K. After a period of six months of continuous operation, the temperature of tube inside water drops to 80°C. Calculate

   a. The overall heat transfer coefficient under six months of operating conditions
   b. The total fouling resistance under operating conditions.

6.7. In problem 6.1, if shell-side fluid is refined lube oil and tube-side fluid is sea water with a velocity of 2 m/s, what value of the overall heat transfer coefficient should be used for design proposals if the shell-side heat transfer coefficient remains unchanged? Calculate the over surface design. Is this acceptable?

6.8. A counterflow double-pipe heat exchanger is designed to cool lubricating oil for a large industrial gas turbine engine. The flow rate of cooling water through the inner tube is 0.2 kg/s, and the water enters the tubes at 20°C, while the flow rate of oil through the outlet annulus is 0.4 kg/s. The oil and water enter the heat exchanger at temperatures of 60°C and 30°C, respectively. The heat transfer coefficients in the annulus and in the inner tube have been estimated as 8 W/m² · K and 2445 W/m² · K, respectively. The outer diameter of the inner tube is 25 mm, and the inner diameter of the outer tube is 45 mm. The total length of the double-pipe heat exchanger (total length per hairpin) is 15 m and total heat exchanger area is 325 m². In the analysis, the tube wall resistance and the curvature of the wall are neglected. What is the total value of the fouling resistance used in this design?

6.9. A shell-and-tube type condenser with one shell pass and four tube passes is used to condense organic vapor. The condensation occurs on the shell side, while the coolant water flows inside the tubes, which are 1.9 cm OD, 1.6 cm ID copper tubes. The length of the heat exchanger is 3 m. The total number of tubes is 840. The steam condenses at 105°C. The initial data of the condenser are recorded as

water rate: 70 kg/s

water inlet temperature: 20°C

water outlet temperature: 45°C.

After four months of operation under the same conditions, the exit temperature of water drops to 40°C. Assuming that shell-side fouling is negligible, there is no fouling at the time of the first operation, and the inside and outside heat transfer coefficients are unchanged, estimate the tube-side fouling factor after operation for four months.

6.10. In a single-phase double-pipe heat exchanger, water is to be heated from 35°C to 95°C. The water flow rate is 4 kg/s. Condensing steam at 200°C in the annulus is used to heat the water. The overall heat transfer coefficient used in the design of this heat exchanger is 3500 W/m^2 · K. After an operation of six months, the outlet temperature of the hot water drops to 90°C. Maintaining the outlet temperature of the water is essential for the purpose of this heat exchanger; therefore, fouling is not acceptable. Calculate the total fouling factor under these operations and comment on whether the cleaning cycle must be extended.

6.11. Distilled water enters the tubes of a shell-and-tube heat exchanger at 200°C and leaves at 100°C. The objective is to design a heat exchanger to heat city water from 20°C to 90°C on the shell side. The overall heat transfer coefficient based on the outside clean surface area of 12 m^2 is 1500 W/m^2 · K. Tubes are carbon steel ID = 16 mm and OD = 19 mm. During the operation of this heat exchanger for six months, a deposit builds up on the shell-side only. Estimate the percent decrease in the heat transfer rate between the fluids.

References

1. Kakaç, S., Ed., *Boilers, Evaporators and Condensers*, John Wiley & Sons, New York, 1991.
2. Kakaç, S., Shah, R. K., and Aung, W., Eds., *Handbook of Single-Phase Convective Heat Transfer*, John Wiley & Sons, New York, 1987.
3. Chenoweth, J. M., Fouling problems in heat exchangers, in *Heat Transfer in High Technology and Power Engineering*, Yang, W. K. and Mori, Y., Eds., Hemisphere, Washington, D.C., 1987, 406.

4. Garrett-Price, B. A., Smith, S. A., Watts, R. L., Knudsen, J. G., Marner, W. J., and Suitor, J. W., *Fouling of Heat Exchangers: Characteristics, Costs, Prevention, Control, and Removal*, Noyes, Park Ridge, N.J., 1985.
5. Taborek, J., Akoi, T., Ritter, R. B., and Palen, J. W., Fouling: the major unresolved problem in heat transfer, *Chem. Eng. Prog.*, 68, 59, 1972.
6. Epstein, N., Fouling in heat exchangers, in *Heat Transfer 1978*, Vol. 6, Hemisphere, Washington, D.C., 1978, 235.
7. Somerscales, E. F. C. and Knudsen, J. G., Eds., *Fouling of Heat Transfer Equipment*, Hemisphere, Washington, D.C., 1981.
8. Melo, L. F., Bott, T. R., and Bernardo, C. A., Eds., *Fouling Science and Technology*, Kluwer, The Netherlands, 1988.
9. Epstein, N., Thinking about heat transfer fouling: a 5 × 5 matrix, *Heat Transfer Eng.*, 4, 43, 1983.
10. Epstein, N., Fouling in heat exchangers, in *Low Reynolds Number Flow Heat Exchangers*, Kakaç, S., Shah, R. K., and Bergles, A. E., Eds., Hemisphere, Washington, D.C., 1981.
11. Friedlander, S. K., *Smoke, Dust and Haze*, John Wiley & Sons, New York, 1977.
12. Whitmore, P. J. and Meisen, A., Estimation of thermo- and diffusiophoretic particle deposition, *Can. J. Chem. Eng.*, 55, 279, 1977.
13. Nishio, G., Kitani, S., and Tadahashi, K., Thermophoretic deposition of aerosol particles in a heat-exchanger pipe, *Ind. Eng. Chem. Proc. Design Dev.*, 13, 408, 1974.
14. Taborek, J., Aoki, T., Ritter, R. B., and Palen, J. W., Predictive methods for fouling behavior, *Chem. Eng. Progr.*, 68, 69, 1972.
15. McCabe, W. L. and Robinson, C. S., Evaporator scale formation, *Ind. Eng. Chem.*, 16, 478, 1924.
16. Kerm, D.Q. and Seaton, R. E., A theoretical analysis of thermal surface fouling, *Brit. Chem. Eng.*, 4, 258, 1959.
17. Knudson, J. G., Fouling of heat exchangers: are we solving the problem?, *Chem. Eng. Prog.*, 80, 63, 1984.
18. Garret-Price, B. A., Smith, S. A., Watts, R. L., and Knudsen, J. G., Industrial Fouling: Problem Characterization, Economic Assessment, and Review of Prevention, Mitigation, and Accomodation Techniques, Report PNL-4883, Pacific Northwest Laboratory, Richland, Washington, 1984.
19. Tubular Exchanger Manufacturers Association (TEMA), *Standards of the Tubular Exchanger Manufacturers Association*, 7th ed., New York, 1988.
20. Somerscales, E. F. C., Fouling, in *Two-Phase Flow Heat Exchangers: Thermo-Hydraulic Fundamentals and Design*, Kakaç, S., Bergles, A. E., and Fernandes, E. O., Eds., John Wiley & Sons, New York, 1988, ch. 21.
21. Chenoweth, J. M., General design of heat exchangers for fouling conditions, in *Fouling Science and Technology*, Melo, L. F., Bott, T. R., and Bernardo, C. A., Eds., Kluwer, The Netherlands, 1988, 477.
22. Marner, W. J. and Suitor, J. W., Fouling with convective heat transfer, in *Handbook of Single-Phase Convective Heat Transfer*, Kakaç, S., Shah, R. K., and Aung, W., Eds., John Wiley & Sons, New York, 1987, ch. 21.
23. Strauss, S. D. and Puckorious, P. R., Cooling-water treatment for control of scaling, fouling and corrosion, *Power*, 128, S1, 1984.
24. Shah, R. K. and Sekulić, D. P., *Fundamentals of Heat Exchanger Design*, John Wiley & Sons, New York, 2003.

7

Double-Pipe Heat Exchangers

7.1 Introduction

A typical double-pipe heat transfer exchanger consists of one pipe placed concentrically inside another pipe of a larger diameter with appropriate fittings to direct the flow from one section to the next, as shown in Figures 1.8 and 7.1. One fluid flows through the inner pipe (tube side), and the other flows through the annular space (annulus). The inner pipe is connected by U-shaped return bends enclosed in a return-bend housing. Double-pipe heat exchangers can be arranged in various series and parallel arrangements to meet pressure drop and MTD requirements. The major use of the double-pipe heat exchanger is for sensible heating or cooling of process fluids where small heat transfer areas (up to 50 m^2) are required. This configuration is also very suitable for one or both of the fluids at high pressure because of the smaller diameter of the pipes. The major disadvantage is that they are bulky and expensive per unit of heat transfer surface area.

These double-pipe heat exchangers are also called hairpin heat exchangers, and they can be used when one stream is a gas, viscous liquid, or small in volume. These heat exchangers can be used under severe fouling conditions because of the ease of cleaning and maintenance. The outer surface of the inner tube can be finned and then the tube can be placed concentrically inside a large pipe (Figures 1.8 and 7.2). In another type, there are multitubes, finned or bare, inside a larger pipe (Figure 7.3). The fins increase the heat transfer surface per unit length and reduce the size and, therefore, the number of hairpins required for a given heat duty. Integrally, resistance-welded longitudinal fins (usually fabricated in a U-shape) have proven to be the most efficient for double-pipe heat exchangers. Fins are most efficient when the film coefficient is low.

Double-pipe heat exchangers are based on modular principles, individual hairpin sections arranged in proper series (Figure 7.4) and parallel or series-parallel combinations (Figures 7.5 and 7.6) to achieve the required duty. When heat duties change, more hairpins can be easily added to satisfy the new requirements. For fluids with a lower heat transfer coefficient such as oil

Multi-tube hairpin sections—Available in both finned and bare tube designs. Shell sizes range from 3″ to 16″ IPS. Applications: Designs are available to meet most heat exchanger requirements; from heating heavy oils and asphalts to light chemicals, gasoline, butanes, and other hydrocarbons. Also available are designs for high-pressure-gas heat transfer and for processing lethal or hard-to-contain fluids and gases such as hydrogen, dowtherm and acids.

Double-pipe hairpin sections—Available in both finned and bare tube designs. Shell sizes range from 2″ to 6″ IPS. Applications: Ideal for all severe operating conditions. A simple, low-cost section is also available for standard petroleum, petrochemical or chemical service.

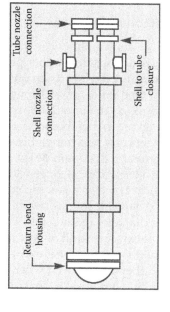

FIGURE 7.1
Double-pipe hairpin heat exchanger. (Courtesy of Brown Fintube, Inc.)

Double-Pipe Heat Exchangers

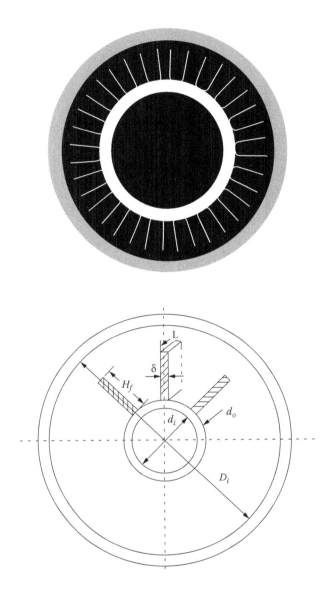

FIGURE 7.2
Cross section of a longitudinally finned inner tube heat exchanger.

or gas, outside-finned inner tubes are justified, and this type of fluid stream will flow through the annulus.

Double-pipe heat exchangers are used as counterflow heat exchangers and can be considered as an alternative to shell-and-tube heat exchangers. Double-pipe heat exchangers have outer pipe inside diameters of 50–400 mm at a nominal length of 1.5–12.0 m per hairpin. The outer diameter of the inner tube may vary between 19 and 100 mm.

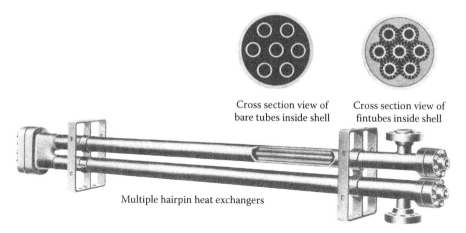

FIGURE 7.3
Multitube hairpin heat exchanger. (Courtesy of Brown Fintube, Inc.)

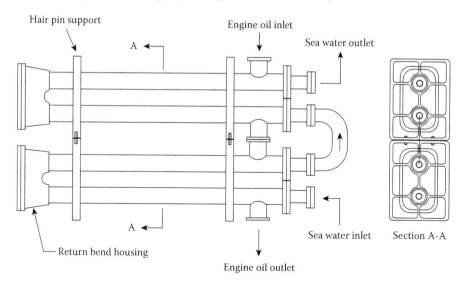

FIGURE 7.4
Two hairpin sections arranged in series.

7.2 Thermal and Hydraulic Design of Inner Tube

Correlations available in the literature are used to calculate heat transfer coefficients inside the inner tube. Important correlations are given in Chapter 3. The frictional pressure drop for flow inside circular tube is calculated using the Fanning Equation (Equation 4.3):

Double-Pipe Heat Exchangers

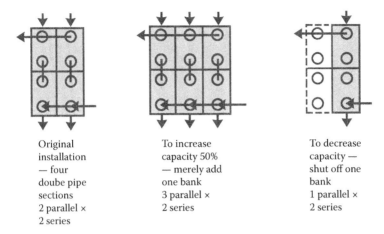

FIGURE 7.5
Parallel-series combinations. (Courtesy of Brown Fintube, Inc.)

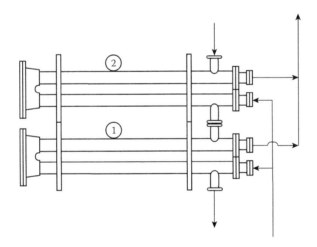

FIGURE 7.6
Two double-pipe units in series on the annulus (shell) side and parallel on the tube side.

$$\Delta p = 4f \frac{L}{d_i} \rho \frac{u_m^2}{2} \tag{7.1}$$

or

$$\Delta p = 4f \frac{L}{d_i} \frac{G^2}{2\rho} \tag{7.2}$$

Hairpin exchangers are used both for single- and two-phase flows; our analysis in this chapter is limited to single-phase convective heat transfer. We will also consider the case where the outside of the inner tubes are longitudinally finned to facilitate mechanical cleaning.

7.3 Thermal and Hydraulic Analysis of Annulus

Heat transfer and pressure calculations for flow inside the annulus shell are similar to those for the tube-side flow, provided that the hydraulic diameter of the annulus is used instead of the tube inner diameter. The hydraulic (equivalent) diameter is given by Equation 3.14.

$$D_h = 4 \left\{ \frac{\text{Netfree - flow area}}{\text{wetted perometter}} \right\} \tag{7.3}$$

It should be noted that only a portion of the wetted perimeter is heated or cooled. Therefore, the hydraulic (equivalent) diameter for the heat transfer calculations is not the same as that used in pressure-drop calculations, given by Equation 3.18.

$$D_h = \frac{4A_c}{P_w} \tag{7.4}$$

Equation 7.4 is used for the calculation of the Reynolds number, and the equivalent diameter based on heat transfer is given by Equation 3.19:

$$D_e = \frac{4A_c}{P_h} \tag{7.5}$$

where A_c is the net cross-sectional area for flow and P_w and P_h are the wetted perimeters for the pressure-drop and heat transfer calculations, respectively. Equation 7.5 will be used to calculate the heat transfer coefficient from the Nusselt number.

7.3.1 Hairpin Heat Exchanger with Bare Inner Tube

In the case of double pipe with a bare inner tube of outside diameter d_o and an outer tube of the inside diameter D_i, we get

$$D_h = \frac{4 \left(\frac{\pi D_i^2}{4} - \frac{\pi d_o^2}{4} \right)}{\pi D_i + \pi d_o} = D_i - d_o \tag{7.6}$$

Double-Pipe Heat Exchangers

$$D_e = \frac{4\left(\frac{\pi D_i^2}{4} - \frac{\pi d_o^2}{4}\right)}{\pi d_o} = \frac{D_i^2 - d_o^2}{d_o} \quad (7.7)$$

Example 7.1

Water at a flow rate of 5,000 kg/h will be heated from 20°C to 35°C by hot water at 140°C. A 15°C hot water temperature drop is allowed. A number of 3.5 m hairpins of 3 in. (ID = 0.0779 m) by 2 in. (ID = 0.0525 m, OD = 0.0603 m) counterflow double-pipe heat exchangers with annuli and pipes, each connected in series, will be used. Hot water flows through the inner tube. Fouling factors are: $R_{fi} = 0.000176$ m² · K/W, $R_{fo} = 0.000352$ m² · K/W. Assume that the pipe is made of carbon steel ($k = 54$ W/m · K). The heat exchanger is insulated against heat losses.

1. Calculate the number of hairpins.
2. Calculate the pressure drops.

Solution

Inner tube side heat transfer coefficient—First calculate the Reynolds number to determine if the flow is laminar or turbulent.
From the appendix, the properties of hot water at $T_b = 132.5°C$ are

$$\rho = 932.53 \text{ kg/m}^3, \quad c_p = 4.268 \text{ kJ/kg·K}$$
$$k = 0.687 \text{ W/m·K}, \quad \mu = 0.207 \times 10^{-3} \text{ Pa·s}$$
$$Pr = 1.28$$

The hot water mass flow rate is calculated as follows:

$$Q = (\dot{m}c_p)_c \Delta T_c = (\dot{m}c_p)_h \Delta T_h$$
$$\dot{m}_h = \frac{(\dot{m}c_p)_c \Delta T_c}{c_p \Delta T_h} = \frac{(5,000/3,600) \times 4.179 \times (35-20)}{4.268 \times (15)} = 1.36 \text{ kg/s}$$

where $c_p = 4.179$ kJ/kg · K for cold water at $T_b = 27.5°C$.
The velocity and Reynolds number are calculated as follows:

$$u_m = \frac{\dot{m}_h}{\rho_h A_c} = \frac{1.36}{(932.53)\left(\frac{\pi}{4}\right)(0.0525)^2} = 0.673 \text{ m/s}$$

$$Re = \frac{\rho u_m d_i}{\mu} = \frac{4\dot{m}_h}{\pi \mu d_i} = \frac{4 \times 1.36}{\pi \times 0.207 \times 10^{-3} \times 0.0525} = 159,343$$

Hence, the flow is turbulent; a correlation can be selected from Chapter 3. Prandtl's correlation is used here with constant properties:

where

$$Nu_b = \frac{(f/2)(Re_b)Pr_b}{1+8.7(f/2)^{1/2}(Pr_b-1)}$$

$$f = (1.58 \ln Re - 3.28)^{-2}$$
$$= [1.58 \ln(1,59,343) - 3.28]^{-2}$$
$$= 4.085 \times 10^{-3}$$

$$Nu_b = \frac{(0.004085/2)(1,59,343)(1.28)}{1+8.7\left(\frac{0.004085}{2}\right)^{1/2}(1.28-1)} = 375.3$$

$$h_i = \frac{Nu_b k}{d_i} = \frac{375.3 \times 0.687}{0.0525} = 4911 \text{ W/m}^2 \cdot \text{K}$$

To determine the heat transfer coefficient in the annulus, the following properties at $T_b = 27.5°C$ from the appendix are used:

$$\rho = 996.4 \text{ kg/m}^3, \quad c_p = 4.179 \text{ kJ/kg} \cdot \text{K}$$
$$k = 0.609 \text{ W/m} \cdot \text{K}, \quad \mu = 8.41 \times 10^{-6} \text{ Pa} \cdot \text{s}$$
$$Pr = 5.77$$

The velocity of cold water, u_m, through the annulus is calculated:

$$u_m = \frac{\dot{m}}{A_c \rho} = \frac{5,000/3,600}{\frac{\pi}{4}(0.0779^2 - 0.0603^2)(996.4)} = 0.729 \text{ m/s}$$

$$D_h = \frac{4A_c}{P_w} = D_i - d_o = 0.0779 - 0.0603 = 0.0176 \text{ m}$$

$$Re = \frac{\rho u_m D_h}{\mu} = \frac{996.4 \times 0.729 \times 0.0176}{0.841 \times 10^{-3}} = 15,201$$

Therefore, the flow is turbulent. Prandtl's correlation is also used for the annulus:

$$f = (3.64 \log_{10} Re_b - 3.28)^{-2} = [3.64 \log_{10}(15,201) - 3.28]^{-2} = 7.021 \times 10^{-3}$$

$$f/2 = 3.51 \times 10^{-3}$$

$$Nu_b = \frac{(f/2)(Re_b)Pr_b}{1+8.7(f/2)^{1/2}(Pr_b-1)}$$

$$Nu_b = \frac{(3.51 \times 10^{-3})(15,201)(5.77)}{1+8.7(3.51 \times 10^{-3})^{1/2}(5.77-1)} = 89.0$$

Double-Pipe Heat Exchangers

The equivalent diameter for heat transfer, from Equation 7.7, is

$$D_e = \frac{D_i^2 - d_o^2}{d_o} = \frac{0.0779^2 - 0.0603^2}{0.0603} = 0.0403 \text{ m}$$

and

$$h_o = \frac{Nu_b k}{D_e} = \frac{89.0 \times 0.609}{0.0403} = 1345 \text{ W/m}^2 \cdot \text{K}$$

The overall heat transfer coefficient based on the outside area of the inner tube is

$$\frac{1}{U_f} = \frac{d_o}{d_i h_i} + \frac{d_o R_{fi}}{d_i} + \frac{d_o \ln(d_o/d_i)}{2k} + R_{fo} + \frac{1}{h_o}$$

$$= \frac{0.0603}{0.0525 \times 4{,}911} + \frac{0.0603 \times 1.76 \times 10^{-4}}{0.0525}$$

$$+ \frac{0.0603 \ln\left(\frac{603}{525}\right)}{2 \times 54} + 3.52 \times 10^{-4} + \frac{1}{1345}$$

$$U_f = 622 \text{ W/m}^2 \cdot \text{K}$$

The heat transfer surface area can be calculated as follows:

$$A_o = \frac{Q}{U_o \Delta T_m}$$

$$\Delta T_m = \Delta T_1 = \Delta T_2 = 105°\text{C}$$

The heat duty of the heat exchanger is

$$Q = (\dot{m} c_p)_c \Delta T_c = 1.389 \times 4.179 \times 15 = 87.1 \text{ kW}$$

So,

$$A_o = \frac{87.1 \times 1{,}000}{622 \times 105} = 1.33 \text{ m}^2$$

The heat transfer area per hairpin is

$$A_{hp} = 2\pi d_o L = 2\pi \times 0.0603 \times 3.5 = 1.325 \text{ m}^2$$

$$\frac{A_o}{A_{hp}} = \frac{1.33}{1.325} = 1$$

Therefore, the number of the hairpins, N_{hp}, equals 1.

The clean heat transfer coefficient based on the outside heat transfer area is

$$\frac{1}{U_c} = \frac{d_o}{d_i h_i} + \frac{d_o \ln(d_o/d_i)}{2k} + \frac{1}{h_o}$$

$$= \frac{0.0603}{0.0525 \times 4{,}911} + \frac{0.0603 \ln(603/525)}{2 \times 54} + \frac{1}{1345}$$

$$U_c = 948 \text{ W/m}^2 \cdot \text{K}$$

The cleanliness factor can also be calculated

$$CF = \frac{U_f}{U_c} = \frac{622}{948} = 0.66$$

as can be the percentage over surface:

$$OS = 100 U_c R_{ft}$$

$$R_{ft} = \frac{1-CF}{U_c CF} = \frac{1-0.66}{948 \times 0.66} = 0.543 \times 10^{-3} \text{ m}^2 \cdot \text{K/W}$$

$$OS = 100 \times 948 \times 0.543 \times 10^{-3} = 51.5\%$$

To determine the pressure drop in the tube side, the frictional pressure drop can be calculated from Equation 4.16:

$$\Delta p_t = 4f \frac{2L}{d_i} N_{hp} \frac{\rho u_m^2}{2}$$

$$\Delta p_i = 4 \times 4.085 \times 10^{-3} \frac{2 \times 3.5}{0.0525} \times 1 \times 932.53 \frac{(0.673)^2}{2} = 460.1 \text{ Pa}$$

To calculate the pumping power for the tube stream, Equation 4.19 for pumping power is used:

$$P_t = \frac{\Delta p_t \cdot \dot{m}_h}{\eta_p \rho_h} = \frac{460.1 \times 1.36}{0.80 \times 932.33} = 0.84 \text{ W}$$

For pressure drop in the annulus side,

$$\Delta p_a = 4f \frac{2L}{D_h} \rho \frac{u_m^2}{2} N_{hp}$$

$$\Delta p_a = 4 \times 7.02 \times 10^{-3} \frac{2 \times 3.5}{0.0176} \times 996.41 \frac{(0.719)^2}{2} \times 1 = 2876.4 \text{ Pa}$$

Double-Pipe Heat Exchangers

and to calculate pumping power for the annulus stream,

$$P_a = \frac{\Delta p_a \dot{m}_c}{\eta_p \rho_c} = \frac{2876.4 \times 1.389}{0.80 \times 996.41} = 5.0 \text{ W}$$

In practice, additional pressure drops in the bend, inlet and outlet should be added.

7.3.2 Hairpin Heat Exchangers with Multitube Finned Inner Tubes

The total wetted perimeter of the annulus with longitudinally finned inner tubes can be written as (see Figure 7.2)

$$P_w = \pi(D_i + d_o N_t) + 2H_f N_f N_t \tag{7.8}$$

and the heat transfer perimeter of the annulus can be calculated from

$$P_h = (\pi d_o + 2H_f N_f)N_t \tag{7.9}$$

It is assumed that the outside of the annulus is insulated against heat losses and the heat transfer occurs through the wall and the fins of the inner tube.

The net cross-sectional freeflow area in the annulus with longitudinal finned inner tubes is given by

$$A_c = \frac{\pi}{4}(D_i^2 - d_o^2 N_t) - \delta H_f N_t N_f \tag{7.10}$$

From Equations 7.4, 7.5, 7.8, 7.9, and 7.10, the hydraulic diameter for the Reynolds number and the pressure drop is

$$D_h = \frac{4A_c}{P_w}$$

and the equivalent diameter based on heat transfer,

$$D_e = \frac{4A_c}{P_h}$$

can be calculated. The validity of the equivalent diameter/hydraulic diameter approach has been substantiated by the results of experiments with finned annuli.[1]

For a double-pipe heat exchanger of length L, the unfinned (bare) and finned areas are, respectively,

$$A_u = 2N_t\left(\pi d_o L - N_f L \delta\right) \tag{7.11}$$

$$A_f = 2N_t N_f L(2H_f + \delta) \tag{7.12}$$

The total outside heat transfer surface area of a hairpin is the sum of the unfinned and finned surfaces, given by Equations 7.11 and 7.12 (see Figure 7.2):

$$A_t = A_u + A_f = 2N_t L\left(\pi d_o + 2N_f H_f\right) \tag{7.13}$$

The overall heat transfer coefficient based on the outside area of the inner tube is given by Equation 2.17. It should be noticed that A_o is replaced by A_t:

$$U_o = \frac{1}{\dfrac{A_t}{A_i}\dfrac{1}{h_i} + \dfrac{A_t}{A_i}R_{fi} + A_t R_w + \dfrac{R_{fo}}{\eta_o} + \dfrac{1}{\eta_o h_o}} \tag{7.14}$$

where

$$\eta_o = \left[1 - \left(1 - \eta_f\right)\dfrac{A_f}{A_t}\right]$$

which is defined by Equation 2.15. As can be seen from Equation 7.14, area ratios A_t/A_i and A_f/A_t must be calculated to find the overall heat transfer coefficient based on the outside surface area of the inner tube. It should be noted that R_w is calculated for bare tubes using Equation 2.12b.

The fin efficiency is given by Equation 2.18:

$$\eta_f = \dfrac{\tanh(mH_f)}{mH_f} \tag{7.15}$$

where

$$m = \sqrt{\dfrac{2h}{\delta k_f}}$$

Equation 7.15 gives the efficiency of a rectangular continuous longitudinal fin[2] with a negligible heat loss from the tip of the fin. In practice, fins generally have no uniform cross sections, and therefore, the analysis of fin thermal behavior is more complex. For complex geometry, fin efficiencies are available in graphical forms.[3,4,5]

Double-Pipe Heat Exchangers

As can be seen from Equation 7.15, the finned area efficiency compared to the bare surface is determined by fin thickness, height, thermal conductivity, and, most importantly, fluid heat transfer coefficient. Therefore, having the fluid with the poorest heat transfer properties on the finned side is the best way to utilize the maximum area of the fins.

If the hairpin heat exchanger units operate as purely countercurrent or parallel flow, then the correction factor F is unity. But for series/parallel systems, a correction must be included in the calculations given by Equation 7.18.

In a design problem, the heat duty, Q, of the heat exchanger will be known and the overall heat transfer coefficient and ΔT_m are determined as outlined above. The surface area in this case will be the external total surface area of the inner tubes and is calculated from Equation 2.32:

$$Q = UN_{hp}A_t\Delta T_m \tag{7.16}$$

where N_{hp} is the number of hairpins and A_t is given by Equation 7.13. If the length of the hairpin is fixed, then the number of hairpins, N_{hp}, can be calculated for a specific heat duty.

The overall heat transfer coefficients can also be based on the internal area of the inner tube, A_i. In this case, the heat transfer surface area will be the inner surface area of the inner tube, which is

$$A_i = 2(\pi d_i L N_t)$$

Then, the heat duty is

$$Q = U_i N_{hp} A_i \Delta T_m \tag{7.17}$$

In design calculations, the length of the hairpin, L, can first be assumed, and then the number of hairpins is calculated.

Example 7.2

The objective of this example is to design an oil cooler with sea water using a finned tube double-pipe heat exchanger.

Engine oil at a rate of 3 kg/s will be cooled from 65°C to 55°C by sea water at 20°C. The sea water outlet temperature is 30°C, and it flows through the inner tube. The properties are given as

Fluid	Annulus Fluid, Oil	Tube-Side Fluid, Sea Water
Density, kg/m²	885.27	1013.4
Specific heat, kJ/kg · K	1.902	4.004
Viscosity, kg/m · s	0.075	9.64×10^{-4}
Thermal conductivity, W/m · K	0.1442	0.639
Prandtl number, Pr	1,050	6.29

The following design data are selected:

Length of hairpin = 4.5 m
Annulus nominal diameter = 2 in.
Nominal diameter of the inner tube = 3/4 in.
Fin height: $H_f = 0.0127$ m
Fin thickness: $\delta = 0.9$ mm
Number of fins per tube = 30
Material throughout = carbon steel (k = 52 W/m · K)
Number of tubes inside the annulus, $N_t = 1$.

Proper fouling factors are to be selected, and the surface area of the heat exchanger as well as the number of hairpins including pressure drops and pumping powers for both streams are to be calculated.

Solution

From Table 9.2 for the tube of nominal diameter 3/4 in. schedule 40, $d_o = 0.02667$ m, $d_i = 0.02093$ m. The annulus nominal diameter is 2 in., and $D_i = 0.0525$ m.

The sea water mass flow rate can be calculated from the heat balance as

$$\dot{m}_c = \frac{(\dot{m}c_p)_h \Delta T_h}{c_{pc} \Delta T_c}$$

$$= \frac{3 \times 1.902 \times (10)}{4.004 \times (10)} = 1.425 \text{ kg/s}$$

Annulus—The net cross-sectional area in the annulus with longitudinal finned tubes is given by Equation 7.10:

$$A_c = \frac{\pi}{4}(D_i^2 - d_o^2 N_t) - (\delta H_f N_f N_t)$$

$$= \frac{\pi}{4}\left[(0.0525)^2 - (0.0266)^2\right] - (0.9 \times 10^{-3})(0.0127)(30)(1)$$

$$= 1.263 \times 10^{-3} \text{ m}^2$$

The hydraulic diameter is

$$D_h = \frac{4A_c}{P_w}$$

where P_w is given by Equation 7.8:

Double-Pipe Heat Exchangers

$$P_w = \pi(D_i + d_o N_t) + 2H_f N_f N_t$$

$$= \pi(0.0525 + 0.0266) + 2(0.0127)(30)$$

$$= 1.011 \text{ m}$$

$$D_h = \frac{4A_c}{P_w} = \frac{4(1.263 \times 10^{-3})}{1.011} = 5.0 \times 10^{-3} \text{ m}$$

The wetted perimeter for heat transfer can be calculated using Equation 7.9:

$$P_h = \pi d_o N_t + 2N_f H_f N_t$$
$$= \pi(0.0266)(1) + 2(30)(0.0127)(1) = 0.845 \text{ m}$$

and the equivalent diameter for heat transfer can be determined as follows:

$$D_e = \frac{4A_c}{P_h} = \frac{4(1.263 \times 10^{-3})}{0.845} = 5.98 \times 10^{-3} \text{ m}$$

Inner Tube, Sea Water—The velocity in this case is

$$u_m = \frac{\dot{m}_c}{\rho \pi \frac{d_i^2}{4}} = \frac{4 \times 1.425}{1013.4 \times \pi(0.0209)^2} = 4.1 \text{ m/s}$$

which is a rather high velocity in practical applications.
The Reynolds number is

$$Re = \frac{\rho u_m d_i}{\mu} = \frac{1013.4 \times 4.1 \times 0.02093}{9.64 \times 10^{-4}} = 90{,}082$$

Hence, the flow is turbulent. One of the turbulent forced convection corrections given in Chapter 3 can be selected.
The Petukhov–Kirillov correlation is used here:

$$f = (1.58 \ln Re - 3.28)^{-2} = (1.58 \ln 90{,}082 - 3.28)^{-2} = 0.0046$$

$$Nu_b = \frac{(f/2)(Re_b)Pr_b}{1.07 + 12.7(f/2)^{1/2}(Pr_b^{2/3} - 1)}$$

$$= \frac{(0.0023)(90{,}082)(6.29)}{1.07 + 12.7(0.0023)^{1/2}(6.29^{2/3} - 1)} = 513.8$$

$$h_i = \frac{Nu \cdot k}{d_i} = \frac{(513.8 \times 0.639)}{0.02093} = 15{,}685.9 \text{ W/m}^2 \cdot \text{K}$$

Annulus, Oil—The velocity in this case is calculated as follows:

$$u_m = \frac{\dot{m}_h}{\rho A_c} = \frac{3}{885.27 \times 1.263 \times 10^{-3}} = 2.68 \text{ m/s}$$

$$Re = \frac{\rho u_m D_h}{\mu} = \frac{885.27 \times 2.68 \times 5 \times 10^{-3}}{0.075} = 158.17$$

It is a laminar flow.

The Sieder–Tate correlation (Equation 3.24) can be used to calculate the heat transfer coefficient in the annulus:

$$Nu_b = 1.86 \left(Re_b \, Pr_b \, \frac{D_h}{L} \right)^{1/3} \left(\frac{\mu_b}{\mu_w} \right)^{0.14}$$

$$Re_b \, Pr_b \, \frac{D_h}{L} = (158.17)(1050) \frac{5 \times 10^{-3}}{4.5} = 184.5$$

$$T_w \approx \frac{1}{2} \left(\frac{65+55}{2} + \frac{20+30}{2} \right) = 42.5° \text{ C}$$

$$\mu_w = 0.197 \text{ Pa·s}$$

$$\left(Re_b Pr_b \frac{D_h}{L} \right)^{1/3} \left(\frac{\mu_b}{\mu_w} \right)^{0.14} = (184.5)^{1/3} \left(\frac{0.075}{0.197} \right)^{0.14} = 4.97 > 2$$

Therefore, the Sieder–Tate correlation is applicable.

$$Nu_b = 1.86 (184.5)^{1/3} \left(\frac{0.075}{0.197} \right)^{0.14} = 9.25$$

$$h_o = \frac{Nu \cdot k}{D_e} = \frac{9.15 \times 0.1442}{5.98 \times 10^{-3}} = 223 \text{ W/m}^2 \cdot \text{K}$$

Note that Equation 3.25 is satisfied using Equation 3.24. Finned and unfinned heat transfer areas are given by Equations 7.11 and 7.12.

$$A_f = 2N_t N_f L (2H_f + \delta) = (2)(1)(30)(4.5)(2 \times 0.0127 + 0.9 \times 10^{-3})$$
$$A_f = 7.101 \text{ m}^2$$
$$A_u = 2N_t (\pi d_o L - N_f L \delta) = (2)(1)[\pi(0.0266)(4.5) - (30)(4.5)(0.9 \times 10^{-3})]$$
$$A_u = 0.509 \text{ m}^2$$

The total area of a hairpin is

$$A_t = A_u + A_f = 7.101 + 0.509 = 7.61 \text{ m}^2$$

From Equation 7.15, the fin efficiency is

$$\eta_f = \frac{\tanh(mH_f)}{mH_f}$$

$$m = \sqrt{\frac{2h_o}{\delta k_f}} = \sqrt{\frac{2(223.05)}{(0.9\times 10^{-3})(52)}} = 97.63$$

$$\eta_f = \frac{\tanh(97.63\times 0.0127)}{97.63\times 0.0127} = 0.682$$

The overall surface efficiency, from Equation 2.14, is

$$\eta_o = \left[1-(1-\eta_f)\frac{A_f}{A_t}\right] = \left[1-(1-0.682)\frac{7.101}{7.610}\right] = 0.703$$

and the overall heat transfer coefficient (fouled) is

$$U_{of} = \cfrac{1}{\cfrac{A_t}{A_i}\cfrac{1}{h_i} + \cfrac{A_t}{A_i}R_{fi} + A_t R_w + \cfrac{R_{fo}}{\eta_o} + \cfrac{1}{\eta_o h_o}}$$

$$= \cfrac{1}{\cfrac{A_t}{A_i h_i} + \cfrac{A_t}{A_i}R_{fi} + \cfrac{A_t \ln(d_o/d_i)}{2\pi k \cdot 2L} + \cfrac{R_{fo}}{\eta_o} + \cfrac{1}{\eta_o h_o}}$$

$$A_i = 2\pi d_i L = 2\pi \times 0.02093 \times 4.5 = 0.592 \text{ m}^2$$

From Tables 5.5 and 5.11, the fouling resistances are

$$R_{fo} = 0.176\times 10^{-3} \text{ m}^2\cdot\text{K/W} \text{ (engine oil)}$$
$$R_{fi} = 0.088\times 10^{-3} \text{ m}^2\cdot\text{K/W} \text{ (sea water)}$$

So,

$$U_{of} = \left[\frac{7.610}{(0.592)(15685.9)} + \frac{7.610}{0.592}(0.088\times 10^{-3}) + \frac{7.610\ln\left(\frac{0.0266}{0.0209}\right)}{2(52)\times\pi\times(2\times 4.5)}\right.$$

$$\left. + \frac{0.176\times 10^{-3}}{0.703} + \frac{1}{(0.703)(223)}\right]^{-1}$$

$$= 108.6 \text{ W/m}^2\cdot\text{K}$$

Under clean conditions,

$$U_{oc} = \cfrac{1}{\cfrac{A_t}{A_i h_i} + \cfrac{A_t \ln(d_o/d_i)}{2\pi k \cdot 2L} + \cfrac{1}{\eta_o h_o}}$$

$$= \left[\frac{7.610}{(0.592)(15685.9)} + \frac{7.610 \ln\left(\frac{0.0266}{0.0209}\right)}{2(52) \times \pi(2 \times 4.5)} + \frac{1}{(0.703)(223)} \right]^{-1}$$

$$= 127.6 \text{ W/m}^2 \cdot \text{K}$$

The cleanliness factor, CF, is determined as follows:

$$CF = \frac{U_{of}}{U_{oc}} = \frac{108.6}{127.6} = 0.85$$

The total heat transfer surface area is

$$A_o = \frac{Q}{U_o \Delta T_m}$$

$$Q = (\dot{m} c_p)_h (T_{h_1} - T_{h_2}) = (3)(1.902 \times 10^3)(65 - 55) = 57{,}060 \text{ W}$$

$$\Delta T_m = \Delta T_1 = \Delta T_2 = 35° \text{ C}$$

Without fouling,

$$A_{oc} = \frac{Q}{U_{oc} \Delta T_m} = \frac{57{,}060}{(127.6)(35)} = 12.78 \text{ m}^2$$

With fouling,

$$A_{of} = \frac{Q}{U_{of} \Delta T_m} = \frac{57060}{(108.6)(35)} = 15.01 \text{ m}^2$$

The hairpin heat transfer surface area is

$$A_{hp} = A_t = 7.61 \text{ m}^2$$

and the number of hairpins can be determined:

$$N_{hp} = \frac{A_{of}}{A_{hp}} = \frac{15.01}{7.61} = 1.97$$

One may select two hairpins.
Pressure Drops and Pumping Powers—For the inner tube,

$$\Delta p_t = 4f \frac{2L}{d_i} \rho \frac{u_m^2}{2} N_{hp}$$

$$= 4(0.0046)\frac{4.5 \times 2}{0.0209}(1013.4)\frac{(4.1)^2}{2}(2) = 135 \text{ kPa}$$

Pumping Power

$$W_a = \frac{\Delta p_a \dot{m}_a}{\eta_a \rho_a} = \frac{(135 \times 10^3)(1.425)}{(0.80)(1013.4)} = 237.3 \text{ W}$$

In the annulus,

$$f_{cp} = \frac{16}{Re} = \frac{16}{158.17} = 0.1011$$

Using Equation 3.21b with Table 3.2, we have

$$f = f_{cp}\left(\frac{\mu_b}{\mu_w}\right)^{-0.5} = 0.1011\left(\frac{0.075}{0.197}\right)^{-0.5} = 0.164$$

$$\Delta p_a = 4f \frac{2L}{D_h} \rho \frac{u_m^2}{2} N_{hp}$$

$$= 4(0.164)\frac{4.5 \times 2}{(5 \times 10^{-3})}(885.27)\frac{(2.68)^2}{2}(2) = 7.5 \text{ MPa}$$

$$w_a = \frac{\Delta p_a \dot{m}_a}{\eta_a \rho_a} = \frac{(7.5 \times 10^6)(3)}{(0.80)(885.27)} = 31.8 \text{ kW}$$

7.4 Parallel–Series Arrangements of Hairpins

When the design dictates a large number of hairpins in a double-pipe heat exchanger designed for a given service, it may not always be possible to connect both the annulus and the tubes in series for a pure counterflow arrangement. A large quantity of fluid through the tube or annulus may result in high pressure drops caused by high velocities, which may exceed available pressure drop. In such circumstances, the mass flow may be separated into a number of parallel streams, and the smaller mass flow rate side can be connected in series. The system then results in a parallel–series arrangement (Figure 7.5). Two double-pipe units in series on the annulus (shell) side and in parallel on the tube side are shown in Figure 7.6. The inner-pipe fluid is divided into two streams, one facing the hotter fluid in unit 2 and the other facing the colder fluid in unit 1. For the inner pipe, a new heat transfer coefficient must be calculated. Therefore, in each hairpin section, different amounts of heat are transferred; the true mean temperature difference, ΔT_m, will be significantly different from the LMTD. In this case, the true mean temperature difference in Equation 7.16 is given by[3,6]

$$\Delta T_m = S(T_{h_1} - T_{c_1}) \tag{7.18}$$

where the dimensionless quantity S is defined as

$$S = \frac{(\dot{m}c_p)_h (T_{h_1} - T_{h_2})}{UA(T_{h_1} - T_{c_1})} \tag{7.19}$$

For n hairpins, the value of S depends on the number of hot and cold streams and their series–parallel arrangement. In the simplest form, either the cold fluid is divided between the n hairpins in parallel or the mass flow rate of the hot fluid is equally divided between the n hairpins.

For one-series hot fluid and n_1-parallel cold streams,

$$S = \frac{1 - P_1}{\left(\dfrac{n_1 R_1}{R_1 - 1}\right) \ln\left[\left(\dfrac{R_1 - 1}{R_1}\right)\left(\dfrac{1}{P_1}\right)^{1/n_1} + \dfrac{1}{R_1}\right]} \tag{7.20}$$

where

$$P_1 = \frac{T_{h_2} - T_{c_1}}{T_{h_1} - T_{c_1}}, \quad R_1 = \frac{T_{h_1} - T_{h_2}}{n_1 (T_{c_2} - T_{c_1})} \tag{7.21}$$

For one-series cold stream and n_2-parallel hot streams,

$$S = \frac{1 - P_2}{\left(\dfrac{n_2}{1 - R_2}\right) \ln\left[(1 - R_2)\left(\dfrac{1}{P_2}\right)^{1/n_2} + R_2\right]} \tag{7.22}$$

where

$$P_2 = \frac{T_{h_1} - T_{c_2}}{T_{h_1} - T_{c_1}}, \quad R_2 = \frac{n_2 (T_{h_1} - T_{h_2})}{T_{c_2} - T_{c_1}} \tag{7.23}$$

The above relations are derived by Kern.[3]

The total rate of heat transfer can be expressed as

$$Q = UAS(T_{h_1} - T_{c_1}) \quad (7.24)$$

It is assumed that the parallel fluid stream mass flow is split equally between the n hairpins and that the overall heat transfer coefficient, U, the specific heat of two fluids are constant, and the heat transfer areas of the units are equal. One can calculate the LMTD and apply a correction factor, F, depending on the number of series–parallel combinations; for these, graphs are available.[3,5,10] But when the number of tube-side parallel paths is equal to the number of shell-side parallel paths, then the normal LMTD should be used.[10]

In the design of a longitudinal-flow finned bundle-type heat exchanger (Figure 7.7), the correction factor, F, must be applied to the LMTD. Since the flow is not purely countercurrent, the F factor will be obtained from Figures 2.7–2.11.[7,8]

(a)

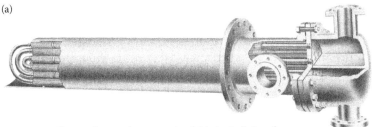

Process heat exchangers—Available in shell sizes from 8⅝" to 36" O.D. Applications: Used in a wide variety of services on both light and heavy fluids ranging from higher heat transfer materials such as gases and light gasolines to heavy lube oil stocks and viscous chemicals.

(b)

Tank suction and line heaters—Available in shell sizes from 8⅝" to 48" O.D. Applications: For heating viscous fluids as they are removed from storage or pumped over long distances. The suction heater is installed with its open and extending into the tank, and only the fluid removed is raised to the desired temperature. Line heaters can be used on either the suction or the pressue side of pumps depending upon design criteria.

FIGURE 7.7
The longitudinally finned bundle exchanger. (Courtesy of Brown Fintube, Inc.)

7.5 Total Pressure Drop

The total pressure drop in a heat exchanger includes fluid-frictional pressure drop, entrance and exit pressure drops, and static head and momentum-change pressure drop. The frictional pressure drop is calculated from Equations 7.1 or 7.2:

$$\Delta p_f = 4f \left(\frac{2L}{D_h} \right) \rho \left(\frac{u_m^2}{2} \right) N_p \qquad (7.25)$$

or

$$\Delta p_f = 4f \left(\frac{2L}{D_h} \right) \left(\frac{G^2}{2\rho} \right) N_p \qquad (7.26)$$

If the equivalent straight length of the U-bend is known, this should be added to L in tube-side pressure drop ($D_h = d_i$) calculations. The correlations for friction factor for smooth tubes are given in Table 4.1, and those for commercial tubes are shown in Figure 4.1. It should be noted that if there are large temperature differences between the wall and the fluid, the effect of property variations on friction coefficient should be included using Equations 3.21b or 3.22b.[7,9] Friction factors for helicals, spirals, and bends are discussed in Chapter 4.

There will be pressure losses through the inlet and outlet nozzles, which can be estimated from

$$\Delta p_n = K_c \frac{\rho u_m^2}{2} \qquad (7.27)$$

where $\rho u_m^2 / 2$, is the dynamic head, $K_c = 1.0$ at the inlet, and $K_c = 0.5$ at the outlet nozzle.[4]

Static head is calculated as

$$\Delta p_f = \rho \Delta H \qquad (7.28)$$

For horizontal hairpins, ΔH is the total difference in elevation between the inlet and outlet nozzles.

For fully developed conditions, the momentum-change pressure drop is calculated from

$$\Delta p_m = G^2 \left(\frac{1}{\rho_{out}} - \frac{1}{\rho_{in}} \right) \qquad (7.29)$$

It is important to note that available (or permissible) pressure drop is usually a dictating factor in heat exchanger design.

Double-Pipe Heat Exchangers

In longitudinally finned-tube heat exchangers, the heat transfer coefficient can be enhanced by a "cut-and-twist" at regular intervals. The fins are cut radially from the edge of the root, and the leading edge is twisted. This has a mixing effect and is normally accounted for by using the cut-and-twist pitch as the developed length in the film coefficient calculation.[10] The optimum cut-and-twist pitch lies between 300 and 1,000 mm. In this case, the pressure drop will increase. The cut-and-twist loss is accounted for by multiplying the frictional pressure drop by a factor, F_{ct}, defined by the following empirical relationship:

$$F_{ct} = 1.58 - 0.525 L_{ct} \tag{7.30}$$

where L_{ct} is the cut-and-twist length in meters. The optimum value of L_{ct} is about 0.3–1 m. If cut-and-twist is provided, then, in calculating heat transfer coefficient in laminar flow, $L = L_{ct}$, which results in the enhancement of heat transfer.

7.6 Design and Operational Features

As explained in previous sections, in hairpin exchangers, two double pipes are joined at one end by a U bend welded to the inner pipes and a return bend housing on the annulus (shell) side, which has a removable cover to allow withdrawal of the inner pipes (Figures 7.3 and 7.8).

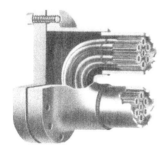

Return bend housing and cover plate

4-bolt standard shell to tube closure and tubeside joint with ASA lap joint shell flange (pressure to 500 psig)

FIGURE 7.8
(a) Return-bend housing and the cover plate; (b) shell-to-tube closure and tube-side joint. (Courtesy of Brown Fintube, Inc.)

TABLE 7.1

Typical Double-Pipe Geometries with One Finned Tube[a]

Shell IPS (mm OD)	Tube OD (mm)	Tube Wall (mm)	Fin Height (mm)	No. of Fins (max.)	NFA (mm²)	D_h (mm)	$\dfrac{A_f}{A_o}$	$\dfrac{A_o}{A_i}$	Surface Area (m²/m)
Sch. 40 Shell (standard pressure units)									
2 in. (60)	25.4	2.77	11.10	20	1 445	8.41	0.85	8.40	1.05
3 in. (89)	25.4	2.77	23.80	20	3 826	12.0	0.92	16.50	2.07
3 in. (89)	48.3	3.68	12.70	36	2 510	7.65	0.86	8.30	2.13
3.5 in. (102)	48.3	3.68	19.05	36	3 916	8.66	0.90	11.86	3.05
3.5 in. (102)	60.3	3.91	12.70	40	3 039	8.18	0.84	7.31	2.41
4 in. (114)	48.3	3.68	25.40	36	5 548	9.63	0.92	15.42	3.96
4 in. (114)	60.3	3.91	19.05	40	4 652	9.14	0.89	10.39	3.43
4 in. (114)	73.0	5.16	12.70	48	3 452	7.80	0.84	7.35	2.90
Sch. 80 Shell (high-pressure units)									
2 in. (60)	25.4	2.77	9.53	20	1 228	8.47	0.83	7.39	0.92
3 in. (89)	48.3	5.08	11.10	36	2 050	6.93	0.84	7.95	1.90
3.5 in. (102)	48.3	5.08	15.90	36	3 368	8.64	0.88	10.82	2.59
3.5 in. (102)	60.3	5.54	11.10	40	2 447	7.29	0.82	6.97	2.16
4 in. (114)	48.3	5.08	22.20	36	4 851	9.42	0.91	14.64	3.50
4 in. (114)	60.3	5.54	15.90	40	3 947	8.99	0.87	9.43	2.92
4 in. (114)	73.0	7.01	9.53	48	2 787	7.70	0.80	6.17	2.29

Source: From Guy A. R., *Heat Exchanger Design Handbook*, Hemisphere, Washington, DC, 1983.

[a] Fin thickness of 0.889 mm (i.e., weldable steels and alloys); surface area per meter is total for both legs.

Double-Pipe Heat Exchangers

Double-pipe heat exchangers have four key design components:

1. Shell nozzles
2. Tube nozzles
3. Shell-to-tube closure
4. Return-bend housing

The opposite site of the inner tubes has a closure arrangement. Return-bend housing, cover plate, and the shell to tube closure arrangements are shown in Figure 7.8. The inner tube can be removed from the tail end of the unit for cleaning and maintenance.

The longitudinally finned bundle exchanger shown in Figure 7.7 is another version of the double-pipe heat exchanger. It is composed of a bundle of finned tubes connected by U-bends and mounted in a shell. This configuration provides a single shell pass and two tube passes. It is used to heat viscous liquids such as asphalt and fuel oils to reduce viscosity for easier pumping.

The fins are welded onto the inner pipe using an electric resistance slam-welding technique. This technique is suitable for carbon steel and stainless steel. Combination of materials can also be used, such as a stainless steel inner pipe with carbon steel fins. Fins made from brass or similar materials can be soldered to copper, nickel, admiralty brass, or aluminum brass tubes. The fins are normally 0.5 mm thick for fin heights up to 12.7 mm and 0.8 mm thick for fin heights over 12.7 mm.

Double-pipe heat exchangers are light and easily handled with a minimum of lifting equipment. Multiple units can be easily bolted together. Annulus sides connected in series only need a connecting gasket, and the tube side uses a simple return-bend. The typical geometries of double-pipe heat exchangers are given in Table 7.1.[10] Double-pipe heat exchangers are easier to clean in case of fouling build-up. They are flexible, with low installation costs. Simple constructions, easily reached bolting, lightweight fin tube elements, and a minimum number of parts all contribute to minimizing costs.

Nomenclature

A heat transfer surface area, m²
A_c net freeflow area, m²
c_p specific heat at constant pressure, J/kg · K
D_e equivalent diameter for heat transfer, m
D_h hydraulic diameter for pressure drop, m
d diameter of the inner tube, m
D diameter of the annulus, m

F	correction factor for LMTD
f	Fanning friction factor
G	mean mass velocity, kg/m² · s
g	acceleration due to gravity, 9.81 m/s²
H	difference in elevation between inlet and outlet nozzles, m
h	heat transfer coefficient, W/m² · K
H_f	height of the fin, m
K_c	velocity-head coefficient
k	thermal conductivity, W/m · K
L	nominal length of the exchanger section, also length of longitudinal fin, m
L_{ct}	cut-and-twist length, m
$\dot{m}$	mass flow rate, kg/s
N_f	number of fins per tube
N_{hp}	number of exchanger section hairpins
N_t	number of tubes in one leg of hairpin exchanger; number of tube holes in the tubesheet of a bundle exchanger (twice the number of U-bends), or the number of U-tubes in a bundle exchanger
n_1, n_2	number of parallel cold and hot streams, Equations 7.20 and 7.22
Nu	Nusselt number, hd_i/k or hD_e/k
P	perimeter, m
P_1, P_2	temperature ratios defined by Equations 7.21 and 7.23
Δp	pressure drop, kPa
Pr	Prandtl number, $\mu c_p/k$
R	thermal resistance, m² · K/W
R_1, R_2	temperature ratios defined by Equations 7.21 and 7.23
Re	Reynolds number, GD_h/μ or Gd_i/μ
Q	heat load or duty, W
S	dimensionless parameter for true temperature difference of a series–parallel arrangement, defined by Equations 7.18 and 7.19
T	absolute temperature, °C, K
ΔT	temperature difference, °C, K
tanh	hyperbolic tangent
U_o	overall heat transfer coefficient based on total external surface area, W/m² · K
u_m	mean stream velocity, m/s

Greek Symbols

η_f	fin efficiency
η_o	overall surface efficiency
η_p	pump efficiency
μ	dynamic viscosity, Pa
ρ	mass density, kg/m³
δ	fin thickness, m

Double-Pipe Heat Exchangers

Subscripts

b	bulk
c	cold, cross-section
e	equivalent
f	fins, fin-side, or frictional
h	hot, heat transfer
h	hydraulic
hp	hairpins
i	inside of tubes
in	inlet
m	momentum
n	nozzles
o	outside of bare tubes
out	outlet
s	shell side or fin side
s-p	series–parallel arrangement
t	total
u	unfinned
w	wall, wetted

PROBLEMS

7.1. A counterflow double-pipe heat exchanger is used to cool the lubricating oil for a large industrial gas turbine engine. The flow rate of cooling water through the inner tube is $\dot{m}_c = 0.2$ kg/s, while the flow rate of oil through the outer annulus is $\dot{m}_h = 0.4$ kg/s. The oil and water enter at temperatures of 60°C and 30°C, respectively. The heat transfer coefficient in the annulus is calculated to be 15 W/m² · K. The inner tube diameter is 25 mm and the inside diameter of the outer annulus is 45 mm. The outlet temperature of the oil is 40°C. Assume $c_p = 4{,}178$ J/kg · K for water and $c_p = 2{,}006$ J/kg · K for oil. The tube wall resistance and the curvature of the wall are neglected. Calculate the length of the heat exchanger if fouling is neglected.

7.2. In Problem 7.1, the heat transfer coefficient in the annulus is given as 15 W/m² · K. Is this value acceptable? Assume that the water used is city water with a fouling resistance of 0.000176 m² · K/W inside the tube. Oil side deposit is neglected. If the length of the hairpin used is 4 m, calculate the heat transfer area and the number of hairpins. If the space to accommodate this heat exchanger is limited, what are the alternatives?

7.3. Engine oil (raffinate) with a flow rate of 5 kg/s will be cooled from 60°C to 40°C by sea water at 20°C in a double-pipe heat exchanger. The water flows through the inner tube, whose outlet is heated to 30°C. The inner tube outside and inside diameters are $d_o = 1.315$ in. (= 0.0334 m) and $d_i = 1.049$ in. (= 0.02664 m), respectively. For the annulus, $D_o = 4.5$ in. (= 0.1143 m) and $D_i = 4.206$ in. (= 0.10226 m). The length of the hairpin is fixed at 3 m. The wall temperature is 35°C. The number of the tubes in the annulus is 3. The thermal conductivity of the tube wall is 43 W/m · K. Calculate:

a. The heat transfer coefficient in the annulus
b. The overall heat transfer coefficient
c. The pressure drop in the annulus and inner tube (only straight sections will be considered)
d. What is your decision as an engineer? How can you improve the design?

7.4. The objective of this problem is to design an oil cooler with sea water. The decision was made to use a hairpin heat exchanger.

Fluid	Annulus Fluid, Engine Oil	Tube-Side Fluid, Sea Water
Flow rate (in) kg/s	4	—
Inlet temperature, °C	65	20
Outlet temperature, °C	55	30
Density, kg/m³	885.27	1013.4
Specific heat, kJ/kg °C	1.902	4.004
Viscosity, kg/m · s	0.075	9.64×10^{-4}
Prandtl number (Pr)	1050	6.29
Thermal conductivity, W/m · K	0.1442	0.6389

Length of hairpin = 3 m
Annulus nominal diameter = 2 in.
Nominal diameter of inner tube (schedule 40) = 3/4 in.
Fin height, $H = 0.00127$ m
Fin thickness, $\delta = 0.9$ mm
Number of fins/tubes = 18
Material throughout = carbon steel
Thermal conductivity, $k = 52$ W/m · K
Number of tubes inside annulus = 3
Select the proper fouling factors. Calculate:

a. The velocities in the tube and in the annulus
b. The overall heat transfer coefficient for a clean and fouled heat exchanger

c. The total heat transfer area of the heat exchanger with and without fouling (OS design)
d. The surface area of a hairpin and the number of hairpins
e. Pressure drop inside the tube and in the annulus
f. Pumping powers for both streams.

7.5. Repeat Problem 7.4 with unfinned inner tubes. Write down your comments.

7.6. A double-pipe heat exchanger is designed as an engine oil cooler. The flow rate of oil is 5 kg/s, and it will be cooled from 60°C to 40°C through annulus (ID = 0.10226 m, OD = 0.1143 m). Sea water flows through the tubes (ID = 0.02664 m, OD = 0.03340 m) and is heated from 10°C to 30°C. The number of bare tubes in the annulus is 3, and the length of the hairpin is 3 m. Assume that the tube wall temperature is 35°C. Design calculations give the number of hairpins as 85. Allowable pressure drop in the heat exchanger for both streams is 20 psi. Is this design acceptable? Outline your comments.

7.7. Assume that the mass flow rate of oil in Problem 7.6 is doubled (10 kg/s), which may result in an unacceptable pressure drop. It is decided that two hairpins will be used in which the hot fluid flows in parallel between two units and the cold fluid flows in series. Calculate:
a. The pressure drop in the annulus for series arrangements for both streams
b. The pressure drop in the annulus if the hot fluid is split equally and flows in parallel between these two units.

7.8. Sea water at 30°C flows on the inside of a 25 mm ID steel tube with a 0.8 mm wall thickness at a flow rate of 0.4 kg/s. The tube forms the inside of a double-pipe heat exchanger. Engine oil (refined lube oil) at 100°C flows in the annular space with a flow rate of 0.1 kg/s. The outlet temperature of the oil is 60°C. The material is carbon steel. The inner diameter of the outer tube is 45 in. Calculate:
a. The heat transfer coefficient in the annulus
b. The heat transfer coefficient inside the tube
c. The overall heat transfer coefficient with fouling
d. The area of the heat exchanger; assuming the length of a hairpin to be 4 m, calculate the number of hairpins
e. The pressure drops and pumping powers for both streams.

7.9. A counterflow double-pipe heat exchanger is used to cool the lubricating oil for a large industrial gas turbine engine. The flow rate of oil through the annulus is $\dot{m}_h = 0.4$ kg/s. The oil and water enter at temperatures of 60°C and 30°C, respectively. The heat transfer coefficient in the inner tube is calculated to be 3,000 W/m²·K. The inner tube diameter is 25 mm and the inside diameter of the outer tube is 45 mm. The outlet temperature of oil and water are 40°C and 50°C, respectively. Assume $c_p = 4{,}178$ J/kg·K for water and $c_p = 2{,}000$ J/kg·K for oil. The tube wall resistance and the curvature of the wall are neglected. Assume the length of the double-pipe heat exchanger is 4 m. Calculate:

a. The heat transfer coefficient in the annulus
b. The heat transfer area and the number of hairpins.

Because of the space limitation, assume that the maximum number of hairpins should not be more than 10. Is this condition satisfied? If not, suggest a new design for a solution for the use of a double-pipe heat exchanger.

7.10. In Example 7.1, assume that the mass flow rate of cold fluid is increased to 20,000 kg/h. The remaining geometrical parameters and the process specifications remain the same. In this case, pressure drop will increase. Assume that the allowable pressure drop in the annulus is 10 kPa. The velocity of cold water through the annulus will be around 3 m/s, which would require a large pressure drop to drive the cold fluid through the inner tube. The pressure drop could be reduced by using several units with cold water flowing in four parallel units and hot water flowing through the inner tube in series. Calculate:

a. The overall heat transfer coefficient
b. Pressure drop in the annulus
c. The appropriate mean temperature difference.

7.11. A sugar solution ($\rho = 1{,}080$ kg/m³, $c_p = 3{,}601$ J/kg·K, $k_f = 0.5764$ W/m·K, $\mu = 1.3 \times 10^{-3}$ N·s/m²) flows at rate of 2 kg/s and is to be heated from 25°C to 50°C. Water at 95°C is available at a flow rate of 1.5 kg/s ($c_p = 4{,}004$ J/kg·K). Sugar solution flows through the annulus. Designing a double-pipe heat exchanger for this process is proposed. The following geometrical parameters are assumed:

Length of the hairpin = 3m
Do (annulus) = 6 cm
Di (annulus) = 5.25 cm

do (Inner tube) = 2.6 cm
di (Inner tube) = 2.09 cm
Number of tubes inside the annulus = 1
Number of fins = 20
Fin height = 0.0127 m
Fin thickness = 0.0009 m
Thermal conductivity of the material, k = 52 W/m · K.
a. Calculate the hydraulic diameter of the annulus for pressure-drop analysis.
b. Calculate the equivalent diameter for heat transfer analysis.
c. Calculate the heat transfer surface area of this heat exchanger and the number of hairpins.

DESIGN PROJECT 7.1
Oil Cooler: Finned-Tube Double-Pipe Heat Exchanger

The objective of this example is to design an oil cooler with sea water. The decision was made to use a hairpin heat exchanger.

Fluid	Annulus Fluid, Engine Oil	Tube-Side Fluid, Sea Water
Flow rate, kg/s	5	—
Inlet temperature, °C	65	20
Outlet temperature, °C	55	30
Density, kg/m^3	885.27	1013.4
Specific heat, kJ/kg · °C	1.902	4.004
Viscosity, kg/m · s	0.075	9.64 × 10^{-4}
Thermal conductivity, W/m · K	0.1442	0.639
Prandtl number (Pr)	1050	6.29

Length of the hairpin = 3 m
Annulus nominal diameter = 2 in.
diameter of the inner tube = 3/4 in.
Fin height H_f = 0.0127 m
Fin thickness δ = 0.9 mm
Number of fins per tube = 30
Material throughout = carbon steel
Thermal conductivity, k = 52 W/m · K
Number of tubes inside the annulus (varied, starting with N_t = 1).

Select the proper fouling factors. The geometrical information is provided to initiate the analysis. These will be taken as variable parameters to come up with a suitable design; one can start with one inner tube and complete the hand calculations by then checking geometrical parameters to determine the effects of these changes on the design. Cost analysis of the selected design will be made. The final report will include material selection, mechanical design, and technical drawings of the components and the assembly. Calculate:

a. The velocities in the tube and in the annulus
b. The overall heat transfer coefficient for a clean and fouled heat exchanger
c. The total heat transfer area of the heat exchanger with and without fouling (OS design)
d. The surface area of a hairpin and the number of hairpins
e. Pressure drop inside the tube and in the annulus
f. Pumping powers for both streams
g. Technical drawings.

DESIGN PROJECT 7.2
Oil Cooler with Bare Inner Tube

The following parameters are fixed:

Fluid	SAE-30 Oil	Sea Water
Inlet temperature, °C	65	20
Outlet temperature, °C	55	30
Pressure drop limitations, kPa	140	5
Total mass flow rate, kg/s	2.5	—

For hand calculations, assume the length of heat exchanger is 3 m. Assume nominal pipe sizes of 3.5" and 2" as outer and inner tubes, respectively. In the design, the geometric parameters must be taken as variable to come up with an acceptable design. In the hand calculations for the given data, the steps indicated in Design Project 7.1 may be followed.

References

1. Delorenzo, N. and Anderson, E. D., Heat transfer and pressure drop of liquids in double pipe fintube exchangers, *Trans. ASME*, 67, 697, 1945.
2. Kakaç, S. and Yener, Y., *Heat Conduction*, 3rd ed., Taylor and Francis, New York, 1993.

3. Kern, D. Q., *Process Heat Transfer*, McGraw-Hill, New York, 1950.
4. Fraas, A. P., *Heat Exchanger Design*, 2nd ed., John Wiley & Sons, New York, 1989.
5. Hewitt, G. F., Shires, G. L., and Bott, T. R., *Process Heat Transfer*, CRC Press, Boca Raton, FL, 1994.
6. Cherrmisinoff, N. P., Ed., *Handbook of Heat and Mass Transfer*, Vol. 1, McGraw-Hill, New York, 1986.
7. Kakaç, S., Ed., *Boilers, Evaporators and Condensers*, John Wiley & Sons, New York, 1991.
8. Tubular Exchanger Manufacturers Association (TEMA), *Standards of the Tubular Exchanger Manufacturers Association*, New York, 1988.
9. Kakaç, S., Shah, R. K., and Aung, W., *Handbook of Single-Phase Convective Heat Transfer*, John Wiley & Sons, New York, 1987.
10. Guy, A. R., Double-pipe heat exchangers, in *Heat Exchanger Design Handbook*, Schlünder, E. U., Ed., Hemisphere, Washington, DC, 1983, ch. 3.

8

Design Correlations for Condensers and Evaporators

8.1 Introduction

Condensation and evaporation heat transfer occurs in many engineering applications, such as in power condensers, boilers, and steam generators, which are all important components in conventional and nuclear power stations. Evaporators and condensers are also essential parts of the vapor-compression refrigeration cycles and in process industry. In the design of these condensers and evaporators, proper correlations must be selected to calculate the condensing and boiling heat transfer coefficients.

When a saturated vapor comes in contact with a surface at a lower temperature, condensation occurs. When evaporation occurs at a solid–liquid interface, it is called boiling. The boiling occurs when the temperature of the surface, T_w, exceeds the saturation temperature, T_{sat}, corresponding to the liquid pressure.

8.2 Condensation

The most common type of condensation involved in heat exchangers is surface condensation, where a cold wall, at a temperature lower than the local saturation temperature of the vapor, is placed in contact with the vapor. In this situation, the number of vapor molecules that strike the cold surface and stick to it is greater than the number of liquid molecules jump into the vapor phase. The resulting net liquid (i.e., condensate) will accumulate in one of two ways. If the liquid wets the cold surface, the condensate will form a continuous film, and this mode of condensation is referred to as filmwise condensation. If the liquid does not wet the cold surface, it will form into numerous microscopic droplets. This mode of condensation is referred to as dropwise condensation and results in much larger heat transfer coefficients than during filmwise conditions. Since long-term dropwise condensation conditions are very difficult to sustain, all surface condensers today are designed to operate in the filmwise mode. More information on the fundamentals of condensation is given by Marto.[1,2]

8.3 Film Condensation on a Single Horizontal Tube

8.3.1 Laminar Film Condensation

Nusselt[3] treated the case of laminar film condensation of a quiescent vapor on an isothermal horizontal tube, as depicted in Figure 8.1. In this situation, the motion of the condensate is determined by a balance of gravitational and viscous forces. In the Nusselt analysis, convection terms in the energy equation are neglected, so the local heat transfer coefficient around the tube can be written simply as:

$$h(\phi) = \frac{k_l}{\delta(\phi)} \tag{8.1}$$

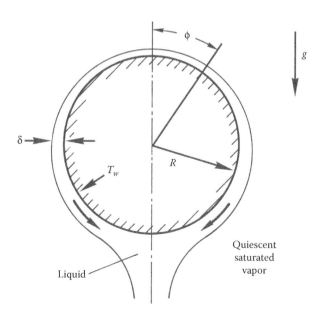

FIGURE 8.1
Film condensation profile on a horizontal tube. (Adapted From Marto, P. J., *Boilers, Evaporators and Condensers*, John Wiley & Sons, New York, 1991.)

Design Correlations for Condensers and Evaporators

Here δ is the film thickness, which is a function of the circumferential angle ϕ. Clearly, at the top of the tube, where the film thickness is a minimum, the heat transfer coefficient is a maximum, and the heat transfer coefficient falls to zero as the film thickness increases to infinity. The Nusselt theory yields the following average heat transfer coefficient[3]:

$$\frac{h_m d}{k_l} = 0.728 \left[\frac{\rho_l (\rho_l - \rho_g) g i_{lg} d^3}{\mu_l (T_{sat} - T_w) k_l} \right]^{1/4} \tag{8.2}$$

Example 8.1

Quiescent refrigerant 134-A vapor at a saturation temperature of 47°C is condensing upon a horizontal smooth copper tube whose outside wall temperature is maintained constant at 40°C.

The outside tube diameter is 19 mm. Calculate the average condensation heat transfer coefficient on the tube.

Solution

The average heat transfer coefficient can be calculated using the Nusselt expression, Equation 8.2. From the appendix, the thermophysical properties of R-134a at 47°C are

$$\rho_l = 1117.3 \text{ kg/m}^3$$

$$\rho_g = 62.5 \text{ kg/m}^3$$

$$k_l = 0.068 \text{ W/m} \cdot \text{K}$$

$$\mu_l = 1.72 \times 10^{-4} \text{ Pa} \cdot \text{s}$$

$$i_g = 154.6 \text{ kJ/kg}.$$

Substituting in Equation 8.2 yields

$$h_m = 0.728 \frac{(0.068)}{(0.019)} \left[\frac{(1117.3)(1117.3 - 62.5)(9.81)(154.6 \times 10^3)(0.019)^3}{(1.72 \times 10^{-4})(47 - 40)(0.068)} \right]^{1/4}$$

$$h_m = 1620.8 \text{ W/m}^2 \cdot \text{K}$$

8.3.2 Forced Convection

When the vapor surrounding a horizontal tube is moving at a high velocity, the analysis for film condensation is affected in two important ways: (1) the surface shear stress between the vapor and the condensate and (2) the effect of vapor separation.

The early analytical investigations of this problem were extensions of Nusselt's analysis to include the interfacial shear boundary condition at the edge of the condensate film. Shekriladze and Gomelauri[4] assumed that the primary contribution to the surface shear was due to the change in momentum across the interface. Their simplified solution for an isothermal cylinder without separation and with no body forces is

$$Nu_m = \frac{h_m d}{k_l} = 0.9\, \tilde{Re}^{1/2} \tag{8.3}$$

where $\tilde{Re}$ is defined as a two-phase Reynolds number involving the vapor velocity and condensate properties, $u_g d/v_l$. When both gravity and velocity are included they recommend the relationship

$$\frac{Nu_m}{\tilde{Re}^{1/2}} = 0.64\left(1+\left(1+1.69 F\right)^{1/2}\right)^{1/2} \tag{8.4}$$

where

$$F = \frac{g d \mu_l i_{lg}}{u_g^2 k_l \Delta T} \tag{8.5}$$

Equation 8.4 neglects vapor separation, which occurs somewhere between 82° and 180° from the stagnation point of the cylinder. After the separation point, the condensate film rapidly thickens and, as a result, heat transfer is deteriorated. A conservative approach, suggested by Shekriladze and Gomelauri,[4] is to assume that there is no heat transferred beyond the separation point. If the minimum separation angle of 82° is then chosen, a most conservative equation results and the heat transfer decreases by approximately 35%. Therefore, Equation 8.3 reduces to

$$Nu_m = 0.59\, \tilde{Re}^{1/2} \tag{8.6}$$

An interpolation formula based on this conservative approach, which satisfies the extremes of gravity-controlled and shear-controlled condensation, was proposed by Butterworth:

$$\frac{Nu_m}{\tilde{Re}^{1/2}} = 0.416\left(1+\left(1+9.47 F\right)^{1/2}\right)^{1/2} \tag{8.7}$$

Vapor boundary layer effects, especially separation, and the effect of the pressure gradient around the lower part of the tube provide significant difficulties in arriving at an accurate analytical solution. As a result, approximate conservative expressions are used. Figure 8.2 compares the predictions

Design Correlations for Condensers and Evaporators

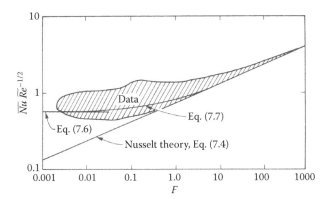

FIGURE 8.2
Condensation in downflow over horizontal tubes. (From Marto, P. J., *Boilers, Evaporators and Condensers*, John Wiley & Sons, New York, 1991. With permission.)

of Equations 8.6 and 8.7 along with the Nusselt equation, Equation 8.2, to experimental data from Rose.[5] In general, Equation 8.7 is conservative and can be used with reasonable confidence.

Example 8.2

Suppose that the refrigerant-134A vapor used in Example 8.1 were moving downward over the tube at a velocity of 10 m/s instead of being quiescent as originally stated. Calculate the average heat transfer coefficient in this situation.

Solution

For downward moving vapor over a horizontal tube, the average heat transfer coefficient can be calculated using Equations 8.3, 8.4, and 8.7. First, using Equation 8.7,

$$\frac{h_m d}{k_l} = 0.416\left(1+\left(1+9.47F\right)^{1/2}\right)^{1/2} \tilde{Re}^{1/2}$$

where F is given by Equation 8.5:

$$F = \frac{g d \mu_l i_{lg}}{u_g^2 k_l \Delta T}$$

On substitution of R-134A properties into Equation 8.5, as listed in Example 8.1, yields

$$F = \frac{(9.81)(0.019)(1.72 \times 10^{-4})(1.546 \times 10^5)}{(10^2)(0.068)(47-40)} = 0.104$$

Then, the two-phase Reynolds number is

$$\tilde{Re} = \frac{\rho_l u_g d}{\mu_l} = \frac{(1117.3)(10)(0.019)}{(1.72 \times 10^{-4})} = 1.23 \times 10^6$$

When these values for F and $\tilde{Re}$ are substituted into Equation 8.7, this yields

$$h_m = 0.416 \frac{(0.068)}{(0.019)} \left[1 + \left(1 + 9.47 \times 0.104\right)^{1/2} \right]^{1/2} \left(1.23 \times 10^6\right)^{1/2}$$

$$= 2563 \text{ W/m}^2 \cdot \text{K}$$

which represents a 58% increase over the quiescent vapor case of Example 8.1.

Recalculating using Equation 8.3,

$$h_m = 0.9 \frac{k_l}{d} \tilde{Re}^{1/2}$$

$$= 0.9 \frac{(0.068)}{(0.019)} \left(1.23 \times 10^6\right)^{1/2}$$

$$= 3572 \text{ W/m}^2 \cdot \text{K}$$

which represents a 120% increase over the quiescent vapor case of Example 8.1 and is 39% higher than the value from Equation 8.7.

Once again, recalculating using Equation 8.4,

$$h_m = 0.64 \frac{k_l}{d} \left[1 + \left(1 + 1.69 \, F^{1/2}\right) \right]^{1/2} \tilde{Re}^{1/2}$$

$$= 0.64 \frac{(0.068)}{(0.019)} \left[1 + \left(1 + 1.69 \times 0.104\right)^{1/2} \right]^{1/2} \left(1.23 \times 10^6\right)^{1/2}$$

$$= 3629 \text{ W/m}^2 \cdot \text{K}$$

which represents a 124% increase over the quiescent vapor case of Example 8.1 and is 42% higher than the value from Equation 8.7 and 1.4% higher than the value from Equation 8.3.

8.4 Film Condensation in Tube Bundles

During film condensation in tube bundles, the conditions are much different than for a single tube. The presence of neighboring tubes creates several added complexities, as depicted schematically in Figure 8.3.[1,2] In the idealized case, Figure 8.3a, the condensate from a given tube is assumed

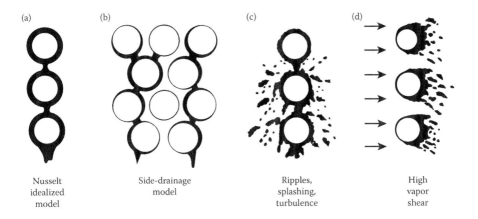

FIGURE 8.3
Schematic representation of condensate flow. (Adapted From Marto, P. J., *Two-Phase Flow Heat Exchangers: Thermal-Hydraulic Fundamentals and Designs,* Kluwer, The Netherlands, pp. 221–291, 1988.)

to drain by gravity to the lower tubes in a continuous, laminar sheet. In reality, depending upon the spacing-to-diameter ratio of the tubes and depending upon whether they are arranged in a staggered or inline configuration, the condensate from one tube may not fall on the tube directly below it but instead may flow sideways, as shown in Figure 8.3b. Also, it is well-known experimentally that condensate does not drain from a horizontal tube in a continuous sheet but in discrete droplets along the tube axis. When these droplets strike the lower tube, considerable splashing can occur, as in Figure 8.3c, causing ripples and turbulence in the condensate film. Perhaps most important of all, large vapor velocities can create significant shear forces on the condensate, stripping it away, independent of gravity, as depicted by Figure 8.3d.

8.4.1 Effect of Condensate Inundation

In the absence of vapor velocity, as condensate flows by gravity onto lower tubes in a bundle, the condensate thickness around the lower tubes should increase and the condensation heat transfer coefficient should, therefore, decrease.

Nusselt[3] extended his analysis to include film condensation on a vertical in-line column of horizontal tubes. He assumed that all the condensate from a given tube drains as a continuous laminar sheet directly onto the top of the tube below it. With this assumption, together with the assumption that $(T_{sat} - T_w)$ remains the same for all tubes, he showed that the average coefficient for a vertical column of N tubes, compared to the coefficient for the first tube (i.e., the top tube in the row), is

$$\frac{h_{m,N}}{h_1} = N^{-1/4} \tag{8.8}$$

In Equation 8.8, h_1 is calculated using Equation 8.2. In terms of the local coefficient for the Nth tube, the Nusselt theory gives

$$\frac{h_N}{h_1} = N^{3/4} - (N-1)^{3/4} \tag{8.9}$$

Kern[6] proposed a less conservative relationship

$$\frac{h_{m,N}}{h_1} = N^{-1/6} \tag{8.10}$$

or, in terms of the local value,

$$\frac{h_N}{h_1} = N^{5/6} - (N-1)^{5/6} \tag{8.11}$$

Eissenberg[7] experimentally investigated the effects of condensate inundation by using a staggered tube bundle. He postulated a side-drainage model that predicts a less severe effect of inundation:

$$\frac{h_N}{h_1} = 0.60 + 0.42\, N^{-1/4} \tag{8.12}$$

Numerous experimental measurements have been made in studying the effect of condensate inundation. The data, however, are very scattered. As a result, it is not too surprising that there is no successful theoretical model today that can accurately predict the effect of condensate inundation on the condensation performance for a variety of operating conditions. For design purposes, the Kern expressions, either Equation 8.10 or 8.11, are conservative and have been recommended by Butterworth.[8]

Example 8.3

Suppose that refrigerant-134A, used in Example 8.1, is condensing under quiescent conditions on the shell side of a bundle of 41 tubes. The bundle can be configured in a square, inline arrangement or in a triangular, staggered arrangement, as shown in Figure 8.4. Find the average shell-side coefficient for each of the configurations.

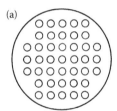

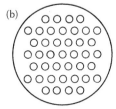

FIGURE 8.4
Tube bundle layout: (a) square, inline arrangement, (b) triangular, staggered arrangement. (Adapted From Marto, P. J., *Boilers, Evaporators and Condensers*, John Wiley & Sons, New York, 1991.)

Solution

In order to find the average heat transfer coefficient for the bundle, correct the Nusselt expression for a single tube, Equation 8.2, using the Kern relationship, Equation 8.10, then Equation 8.8 and, finally, Equation 8.12.

In a square, inline arrangement, there are five columns of seven tubes each and two columns of three tubes each. This arrangement would be approximately equivalent to seven columns of 6 tubes each. Therefore, $N = 6$. Using $h_1 = 1620.8$ W/m² · K from Equation 8.2 in Example 8.1, along with Equation 8.10 gives

$$h_{m,N} = h_1 (N)^{-1/6}$$
$$h_{m,6} = h_1 (6)^{-1/6} = 1620.8(0.7418)$$
$$= 1202 \text{ W/m}^2 \cdot \text{K}$$

Now, using Equation 8.8 yields

$$h_{m,N} = h_1 (N)^{-1/4}$$
$$h_{m,6} = h_1 (6)^{-1/4} = 1620.8(0.6389)$$
$$h_{m,6} = 1036 \text{ W/m}^2 \cdot \text{K}$$

Recalculating using Equation 8.12 yields

$$h_{m,N} = h_1(0.60 + 0.42\ N^{-1/4})$$
$$h_{m,6} = 1620.8(0.60 + 0.42(6)^{-1/4})$$
$$= 1407\ \text{W/m}^2 \cdot \text{K}$$

Therefore, Equations 8.8 and 8.12 provide the most and least conservative values of $h_{m,6}$, respectively. The value from Equation 8.12 is 36% higher than that of Equation 8.8. The Kern relation gives an intermediate value of $h_{m,6}$.

In a triangular, staggered arrangement, assuming that the condensate falls straight down and not sideways, there are seven columns of 3 tubes, four columns of 4 tubes, and two columns of 2 tubes. This arrangement would be equivalent to approximately thirteen columns of 3 tubes each. Therefore, $N = 3$.

Once again, using $h_1 = 1620.8$ W/m² · K from Equation 8.2 in Example 8.1 along with Equation 8.10 gives

$$h_{m,N} = h_1(N)^{-1/6}$$
$$h_{m,3} = h_1(3)^{-1/6} = 1620.8(0.8327)$$
$$= 1350\ \text{W/m}^2 \cdot \text{K}$$

Now, again using Equation 8.8,

$$h_{m,N} = h_1(N)^{-1/4}$$
$$h_{m,3} = h_1(3)^{-1/4} = 1620.8(0.7598)$$
$$= 1232\ \text{W/m}^2 \cdot \text{K}$$

Equation 8.12 gives

$$h_{m,N} = h_1(0.60 + 0.42\ N^{-1/4})$$
$$h_{m,3} = 1620.8\left[0.60 + 0.42(3)^{-1/4}\right]$$
$$= 1490\ \text{W/m}^2 \cdot \text{K}$$

Here, Equation 8.12 gives a value of $h_{m,3}$ that is 21% greater than Equation 8.8, and, once again, the Kern relation Equation 8.10 yields an intermediate value of $h_{m,3}$. Therefore, the results of this example show that the staggered arrangement of tubes yields an average heat transfer coefficient that is generally larger than that for an inline arrangement.

One way of preventing the inundation of condensate on lower tubes is to incline the tube bundle with respect to the horizontal. Shklover and Buevich[9] conducted an experimental investigation of steam condensation in an inclined bundle of tubes. They found that inclination of the bundle increases the average heat transfer coefficient over the horizontal bundle result by as much as 25%. These favorable results led to the design of an inclined bundle condenser with an inclination angle of 5°.

8.4.2 Effect of Vapor Shear

In tube bundles, the influence of vapor shear has been measured by Kutateladze et al.,[10] Fujii et al.,[11] and Cavallini et al.[12] Fujii et al.[11] found that there was little difference between the downward-flow and horizontal-flow data obtained, but the upward-flow data were as much as 50% lower in the range $0.1 < F < 0.5$. The following empirical expression was derived, which correlates the downward-flow and horizontal-flow data reasonably well:

$$\frac{Nu_m}{\tilde{Re}^{1/2}} = 0.96 \, F^{1/5} \tag{8.13}$$

for $0.03 < F < 600$. Cavallini et al.[12] compared their data with the prediction of Shekriladze and Gomelauri[4] (Equation 8.4), and found the prediction to be conservative.

In a tube bundle, it is not clear which local velocity should be used to calculate vapor shear effects. Butteworth[8] points out that the use of the maximum cross-sectional area gives a conservative prediction. Nobbs and Mayhem[9] have used the mean local velocity through the bundle. They calculated this velocity based on a mean flow width, which is given by

$$w = \frac{p_L p_t - \pi d^2/4}{p_L} \tag{8.14}$$

where p_L and p_t are the tube pitches (i.e., centerline-to-centerline distance) in the longitudinal and transverse directions, respectively.

8.4.3 Combined Effects of Inundation and Vapor Shear

Initially, the effects of inundation and vapor shear were treated separately. The combined average heat transfer coefficient for condensation in a tube bundle was simply written as

$$h_{m,N} = h_1 C_N C_{ug} \tag{8.15}$$

where h_1 represents the average coefficient for a single tube from the Nusselt theory, Equation 8.2, and C_N and C_{ug} are correction factors to account for inundation and vapor shear, respectively.

However, in a tube bundle, a strong interaction exists between vapor shear and condensate inundation, and local heat transfer coefficients are very difficult to predict. Butterworth[8] proposed a relationship for the local heat transfer coefficient in the Nth tube row which separates out the effects of vapor shear and condensate inundation. A slightly modified form of his equation is

$$h_N = \left[\frac{1}{2} h_{sh}^2 + \left(\frac{1}{4} h_{sh}^4 + h_1^4 \right)^{1/2} \right]^{1/2} \times \left[N^{5/6} - (N-1)^{5/6} \right] \quad (8.16)$$

where h_{sh} is, from Equation 8.6,

$$h_{sh} = 0.59 \frac{k_l}{d} \tilde{Re}^{1/2}$$

and h_1 is obtained using Equation 8.2.

McNaught[13] has suggested that shell-side condensation may be treated as two-phase forced convection. He therefore proposed the following relationship for the local coefficient for the Nth tube row:

$$h_N = \left(h_{sh}^2 + h_G^2 \right)^{1/2} \quad (8.17)$$

where h_G is given by Equation 8.11:

$$h_G = h_l \left[N^{5/6} - (N-1)^{5/6} \right]$$

and h_{sh} is given as

$$h_{sh} = 1.26 \left[\frac{1}{X_{tt}} \right]^{0.78} h_l \quad (8.18)$$

In Equation 8.18, X_{tt} is the Lockhart–Martinelli parameter, defined as

$$X_{tt} = \left(\frac{1-x}{x} \right)^{0.9} \left(\frac{\rho_g}{\rho_l} \right)^{0.5} \left(\frac{\mu_l}{\mu_g} \right)^{0.1} \quad (8.19)$$

and h_l is the liquid-phase forced convection heat transfer coefficient across a bank of tubes. This is generally expressed as

$$h_l = C \frac{k_l}{d} Re_l^m Pr_l^n \quad (8.20)$$

Design Correlations for Condensers and Evaporators

where C, m, and n depend upon the flow conditions through the tube bank. One of the correlations given in Chapter 3 can be selected. The numerical values in Equation 8.18 were obtained for steam condensing in a bank of in-line or staggered tubes under the following conditions[9]:

$$\frac{p}{d} = 1.25$$

$$10 \leq G \leq 70 \text{ kg}/(m^2 \cdot s)$$

$$0.025 \leq x \leq 0.8$$

$$0.008 \leq X_{tt} \leq 0.8$$

The correlation includes the effect of condensate inundation. McNaught[13] found that Equations 8.17 and 8.18 correlated 90% of the steam data within ±25%. Care must be taken to avoid using this correlation when the operating conditions fall outside of the ranges indicated above.

Example 8.4

Steam at a saturation temperature of 100°C is condensing in a bundle of 320 tubes within a 0.56 m wide duct. The tubes are arranged in a square, in-line pitch (p = 35.0 mm), as shown in Figure 8.5. The bundle is made of up to 20 rows of tubes with 3 cm OD and with 16 tubes in each row. The tube wall temperature in each row is kept constant at 93°C. The steam flows downward in the bundle, and at the 6th row of tubes, the local

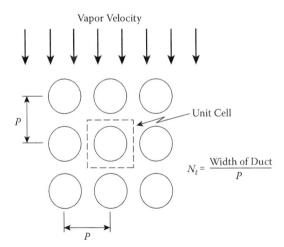

FIGURE 8.5
Schematic of square, inline tube arrangement. (From Marto, P. J., *Boilers, Evaporators and Condensers*, John Wiley & Sons, New York, 1991.)

mass flow rate of vapor is $m_g = 14.0$ kg/s. Find the local heat transfer coefficient for this 6th row of tubes using Butterworth's method.[7]

Solution

From the appendix, the thermophysical properties of steam at 100°C are

$$\rho_l = 957.9 \text{ kg/m}^3$$

$$\rho_g = 0.598 \text{ kg/m}^3$$

$$k_l = 0.681 \text{ W/m} \cdot \text{K}$$

$$\mu_l = 2.79 \times 10^{-4} \text{ kg/m} \cdot \text{s}$$

$$\mu_g = 1.2 \times 10^{-5} \text{ kg/m} \cdot \text{s}$$

$$i_{lg} = 2.257 \times 10^3 \text{ kJ/kg}$$

$$c_{pl} = 4.219 \text{ kJ/kg} \cdot \text{K}$$

$$Pr_l = 1.73.$$

To calculate the local steam velocity,

$$u_g = \frac{\dot{m}_g}{\rho_g A_m}$$

where A_m is the mean flow area. This mean flow area can be written in terms of the number of unit cells, N_t, and the mean width per cell, w,

$$A_m = w N_t L$$

where w is given by Equation 8.14

$$w = \frac{p_L p_t - \pi d^2/4}{p_L}$$

In this example, $p_L = p_t = p = 0.035$ m. Therefore,

$$w = \frac{(0.035)^2 - \pi (0.03)^2/4}{0.035} = 0.0148 \text{ m}$$

The mean flow area is then

$$A_m = (0.0148)(16)(4) = 0.947 \text{ m}^2$$

The local steam velocity is

$$u_g = \frac{(14.0)}{(0.598)(0.947)} = 24.7 \text{ m/s}$$

Using Butterworth's method, the local heat transfer coefficient is given by Equation 8.16:

$$h_N = \left[\frac{1}{2}h_{sh}^2 + \left(\frac{1}{4}h_{sh}^4 + h_1^4\right)^{1/2}\right]^{1/2} \times \left[N^{5/6} - (N-1)^{5/6}\right]$$

where

$$h_{sh} = 0.59\frac{k_l}{d}\tilde{Re}^{1/2}$$

and

$$\frac{h_m d}{k_l} = 0.728\left[\frac{\rho_l(\rho_l - \rho_g)gi_{lg}d^3}{\mu_l(T_{sat} - T_w)k_l}\right]^{1/4}$$

The two-phase Reynolds number is

$$\tilde{Re} = \frac{\rho_l u_g d}{\mu_l}$$

Therefore,

$$h_{sh} = 5.59\frac{(0.681)}{(0.03)}(2.54 \times 10^6)^{1/2} = 21{,}345 \text{ W/m}^2 \cdot \text{K}$$

From Equation 8.2,

$$h_1 = 0.728\frac{(0.681)}{(0.03)}\left[\frac{(957.9)(957.9 - 0.598)(9.81)(2.257 \times 10^6)(0.03)^3}{(2.79 \times 10^{-4})(100 - 93)(0.681)}\right]^{1/4}$$

$$= 13{,}241 \text{ W/m}^2 \cdot \text{K}$$

Therefore, substituting into Equation 8.16 with $N = 6$ yields

$$h_N = \left\{\frac{1}{2}(21345)^2 + \left[\frac{1}{4}(21345)^4 + (13241)^4\right]^{1/2}\right\}^{1/2} \times \left[6^{5/6} - (6-1)^{5/6}\right]$$

$$= 14{,}242 \text{ W/m}^2 \cdot \text{K}$$

Numerical modeling must be used to accurately predict the average shell-side heat transfer coefficient in a tube bundle. Marto[14] has reviewed computer methods to predict shell-side condensation. The vapor flow must be followed throughout the bundle and within the vapor flow lanes. Knowing the local vapor velocity, the local vapor pressure and temperature, and the distribution of condensate from other tubes (and the local concentration of any noncondensable gases), it is possible to predict a local coefficient in the bundle which can be integrated to arrive at the overall bundle performance. Early efforts to model the thermal performance of condensers were limited essentially to one-dimensional routines in the plane perpendicular to the tubes.[15–18] Today, more sophisticated two-dimensional models exist which have been utilized successfully to study the performance of complex tube bundle geometries.[19–24]

8.5 Condensation inside Tubes

8.5.1 Condensation inside Horizontal Tubes

During the film condensation inside the tubes, a variety of flow patterns can exist as the flow passes from the tube inlet to the exit. These different flow patterns can alter the heat transfer considerably so that local heat transfer coefficients must be calculated along the length of the tube. During horizontal in-tube condensation, the transition from annular to stratified flow is most important. Details of in-tube condensation can be found in the literature.[25]

Different heat transfer models are used to calculate the heat transfer coefficient depending on whether vapor shear or gravitational forces are more important.

The previously mentioned flow patterns have been studied by numerous investigators in recent years.[26–30] The transition from one flow pattern to another must be predicted in order to make the necessary heat transfer calculations. Breber et al.[29] have proposed a simple method of predicting flow pattern transitions that depend on the dimensionless mass velocity, j_g^*, defined as

$$j_g^* = \frac{xG}{\left[gd\rho_g(\rho_l - \rho_g)\right]^{1/2}} \tag{8.21}$$

and the Lockhart–Martinelli parameter, X_{tt}, given by Equation 8.19. Their flow pattern criteria are

$$j_g^* > 1.5 \quad X_{tt} < 1.0: \text{mist and annular} \tag{8.22a}$$

$$j_g^* < 0.5 \quad X_{tt} < 1.0: \text{wavy and stratified} \tag{8.22b}$$

$$j_g^* < 0.5 \quad X_{tt} > 1.5: \text{slug} \qquad (8.22c)$$

$$j_g^* > 1.5 \quad X_{tt} < 1.5: \text{bubble} \qquad (8.22d)$$

During condensation inside horizontal tubes, different heat transfer models are used, depending on whether vapor shear or gravitational forces are more important. When the vapor velocity is low (i.e., j_g^* is less than 0.5), gravitational forces dominate the flow and stratification of the condensate will occur. At high vapor velocities (i.e., $j_g^* > 1.5$) where interfacial shear forces are large, gravitational forces may be neglected, and the condensate flow will be annular. When the flow is stratified, the condensate forms as a thin film on the top portion of the tube walls. This condensate drains toward the bottom of the tube where a stratified layer exists, as shown schematically in Figure 8.6. The stratified layer flows axially due to vapor shear forces. In this circumstance, the Nusselt theory for laminar flow is generally valid over the top, thin film region of the tube. However, Butterworth[31] points out that if the axial vapor velocity is high, turbulence may occur in this thin film and the Nusselt analysis is no longer valid. In the stratified layer, heat transfer is generally negligible. For laminar flow, the average heat transfer coefficient over the entire perimeter may be expressed by a modified Nusselt result:

$$h_m = \Omega \left\{ \frac{k_l^3 \rho_l (\rho_l - \rho_g) g i_{fg}}{\mu_l d \Delta T} \right\}^{1/4}$$

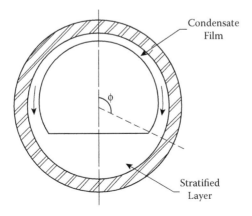

FIGURE 8.6
Idealized condensate profile inside a horizontal tube. (From Marto, P. J., *Boilers, Evaporators and Condensers*, John Wiley & Sons, New York, 1991.)

where the coefficient Ω depends on the fraction of the tube that is stratified. Jaster and Kosky[32] have shown that Ω is related to the void fraction of the vapor α_g:

$$\Omega = 0.728 \alpha_g^{3/4} \tag{8.23}$$

where

$$\alpha_g = \frac{1}{1+\left[(1-x)/x\right]\left(\rho_g/\rho_l\right)^{2/3}} \tag{8.24}$$

Generally, laminar flow models (based on a Nusselt analysis) predict heat transfer coefficients that are too low, and turbulent flow models must be used. These turbulent flow models are either empirically based, dimensionless correlations or are based on the heat transfer–momentum analogy. Some of the models are rather cumbersome to use and result in expressions that are inconvenient for design purposes.

Several correlations available in the literature have been verified for use with refrigerants. Some of these correlations are described in the following discussion, such as the correlations by Traviss et al.,[33] Cavallini and Zecchin,[34] and Shah.[35] For heat transfer in the superheated and subcooled regions of the condenser, the single-phase correlations discussed earlier during in-tube evaporation are applicable.

The correlation by Traviss et al.[33] was originally developed as part of an extensive study of condensation of R-12 and R-22. The correlation was also verified successfully by using these same R-12 and R-22 data. This in-tube condensation equation was derived by applying the momentum and heat transfer analogy to an annular flow model. The velocity distribution in the annular film was described by the von Karman universal velocity distribution. Radial temperature gradients in the vapor core were neglected, and a saturation temperature was assumed at the liquid–vapor interface. The resulting two-phase heat transfer coefficient is

$$Nu = \frac{h_{TP}d}{k_l} = Pr_l Re_l^{0.9} \frac{F_1(X_{tt})}{F_2(Re_l, Pr_l)} \tag{8.25}$$

where the liquid Reynolds number is

$$Re_l = \frac{G(1-x)d}{\mu_l} \tag{8.26}$$

and the nondimensional parameter, F_1, is

$$F_1 = 0.15\left[\frac{1}{X_{tt}} + \frac{2.85}{X_{tt}^{0.476}}\right] \tag{8.27}$$

Three functions are given for F_2, with the choice of which function to use in the correlation being dependent on the Reynolds number range. The three functions are

$$F_2 = 0.70 Pr_l Re_l^{0.5} \text{ for } Re_l \leq 50 \tag{8.28}$$

$$F_2 = 5 Pr_l + 5\ln\left[1 + Pr_l\left(0.09636 Re_l^{0.585} - 1\right)\right] \text{ for } 50 < Re_l \leq 1125 \tag{8.29}$$

$$F_2 = 5 Pr_l + 5\ln(1 + 5 Pr_l) + 2.5\ln\left(0.00313 Re_l^{0.812}\right) \text{ for } Re_l > 1125 \tag{8.30}$$

where the Reynolds number is defined in Equation 8.26.

The preceding correlation has been used extensively in the past. However, two more recent correlations, described next, are simpler to implement, and in addition, they have been shown to correlate experimental data just as well.[36]

Cavallini and Zecchin[34] developed a semiempirical equation that is simple in form and correlates refrigerant data quite well. Data for several refrigerants, including R-11, R-12, R-21, R-22, R-113, and R-114, were used to derive and verify the correlation. The basic form of the correlation was developed from a theoretical analysis similar to that used by Traviss et al.[33]

The working equation suggested by Cavallini and Zecchin is

$$h_{TP} = 0.05 Re_{eq}^{0.8} Pr_l^{0.33} \frac{k_l}{d} \tag{8.31}$$

where the equivalent Reynolds number, Re_{eq}, is defined by

$$Re_{eq} = Re_v\left(\frac{\mu_v}{\mu_l}\right)\left(\frac{\rho_l}{\rho_v}\right)^{0.5} + Re_l \tag{8.32}$$

The equation for Re_l was presented earlier in Equation 8.26, and Re_v is defined similarly as

$$Re_v = \frac{Gxd}{\mu_v} \tag{8.33}$$

The Cavallini–Zecchin correlation is very similar in form to any one of several single-phase turbulent correlations (e.g., the well-known Dittus–Boelter equation). Cavallini and Zecchin also suggest that their equation can be used to calculate the average heat transfer coefficients between the condenser inlet and outlet, provided that the thermophysical properties and the temperature difference between the wall and fluid do not vary considerably along the tube.

The Shah[35] correlation was developed from a larger group of fluids, including water, than the previous correlations. It was developed by establishing a connection between condensing and Shah's earlier correlations for boiling heat transfer without nucleate boiling. The resulting correlation in terms of h_l is

$$h_{TP} = h_l \left(1 + \frac{3.8}{Z^{0.95}}\right) \tag{8.34}$$

where

$$Z = \left(\frac{1-x}{x}\right)^{0.8} p_r^{0.4} \tag{8.35}$$

and h_l is the liquid-only heat transfer coefficient and can be found using the Dittus–Boelter equation:

$$h_l = 0.023 \left[\frac{G(1-x)d}{\mu_L}\right]^{0.8} \frac{Pr_l^{0.4} k_l}{d} \tag{8.36}$$

Shah also suggested integrating these equations over a length of tubing to obtain the mean heat transfer coefficient in the condensing region:

$$h_{TP_m} = \frac{1}{L} \int_0^L h_{TP} dL \tag{8.37}$$

For the case of a linear quality variation over a 100% to 0% range, the result is

$$h_{TP_m} = h_l \left(0.55 + \frac{2.09}{Pr^{0.38}}\right) \tag{8.38}$$

The results from this equation differ by only 5% from the value obtained when a mean quality 50% is used in the local heat transfer correlation, Equation 8.34.

The local heat transfer coefficients for the previous three correlations are compared in Figure 8.7.[37] As with the comparisons of in-tube evaporation correlations presented earlier, an ozone-safe refrigerant, namely R-134a, is used. Flow rate, temperature, tube diameter, and tube length conditions similar to a typical condenser were selected for this comparison. Over most of the quality range, the correlations agree with each other to within 20%. The local heat transfer coefficients decrease as the quality decreases, which is the result of the annular film thickness increasing as condensation proceeds from the high-quality inlet of the condenser to the low-quality exit.

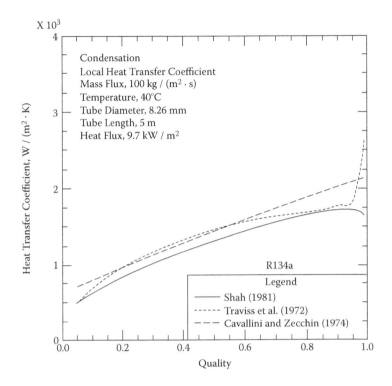

FIGURE 8.7
Local condensation heat transfer coefficient for in-tube flow of R-134A. (From Pate, M. B., *Boilers, Evaporators and Condensers*, John Wiley & Sons, New York, 1991. With permission.)

Average heat transfer coefficients, obtained by using Equation 8.37 to integrate the local value over tube length, are plotted as a function of mass flux in Figure 8.8.[37] All three correlations show good agreement; however, two of the correlations agree to within 10% of each other. Average coefficient data from several past experimental studies have been predicted to within +20% by these three correlations.[35–37]

8.5.2 Condensation inside Vertical Tubes

Condensation heat transfer in vertical tubes depends on the flow direction and its magnitude. For downward-flowing vapor at low velocities, the condensate flow is controlled by gravity and the heat transfer coefficient may be calculated by the Nusselt theory on a vertical surface:

$$Nu_m = \frac{h_m L}{k_l} = 0.943 \left\{ \frac{\rho_l (\rho_l - \rho_g) g i_{lg} L^3}{\mu_l \Delta T k_l} \right\} \tag{8.39}$$

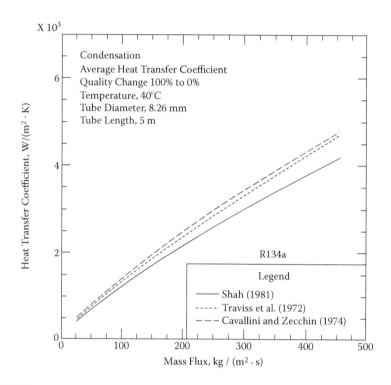

FIGURE 8.8
Average condensation heat transfer coefficient for in-tube flow of R-134A. (From Pate, M. B., *Boilers, Evaporators and Condensers*, John Wiley & Sons, New York, 1991. With permission.)

If the condensate film proceeds from laminar, wave-free to wavy conditions, a correction to Equation 8.39 can be applied:

$$\frac{h_{m,c}}{h_m} = 0.8 \left(\frac{Re_F}{4} \right)^{0.11} \tag{8.40}$$

where the film Reynolds number $Re_F > 30$. If turbulent conditions exist, then the average heat transfer coefficient can be calculated by one of the methods described by Marto.[1] If the vapor velocity is very high, then the flow is controlled by shear forces and the annular flow models may be used.[1,2]

For upward-flowing vapor, interfacial shear will retard the drainage of condensate. As a result, the condensate film will thicken and the heat transfer coefficient will decrease. In this case, Equation 8.39 may be used with a correction factor of 0.7 to 1.0 applied, depending on the magnitude of the vapor velocity. Care must be exercised to avoid vapor velocities that are high enough to cause flooding, which occurs when the vapor shear forces prevent the down-flow of condensate. One criterion to predict the onset of flooding is outlined by Wallis,[38] and is based on air-water systems:

$$\left(v_g^*\right)^{1/2} + \left(v_l^*\right)^{1/2} = C \tag{8.41}$$

where

$$v_g^* = \frac{v_g \rho_g^{1/2}}{\left[gd_i(\rho_l - \rho_g)\right]^{1/2}} \tag{8.42}$$

$$v_l^* = \frac{v_l \rho_l^{1/2}}{\left[gd_i(\rho_l - \rho_g)\right]^{1/2}} \tag{8.43}$$

The velocities v_g and v_l should be calculated at the bottom of the tube (where they are at their maximum values). Wallis[38] determined the parameter C to be 0.725 based on his measurements of air and water. Butterworth[31] suggests that C should be corrected for surface tension and for tube-end effects using the following relationship:

$$C_2 = 0.53 F_\sigma F_g \tag{8.44}$$

where F_σ is a correction factor for surface tension and F_g depends on the geometry of the end of the tube.

8.6 Flow Boiling

The study of flow boiling is of fundamental importance in the nuclear, chemical, and petrochemical industries. In the first case the main parameters of concern are the onset of nucleate boiling (ONB), the void fraction in the subcooled and saturated boiling regions which is required for pressure drop, transient reactor response, and instability prediction, as well as the heat transfer crises and postdryout heat transfer.

8.6.1 Subcooled Boiling

Consider a heated channel with flowing coolant, as shown in Figure 8.9; the subcooled boiling region commences at the point where bubbles begin to grow at the wall. The bubbles will form on the heated surface, become detached, and pass into the bulk of the liquid where they may condense if the bulk liquid temperature is below its boiling temperature. This process of heat transfer is called subcooled nucleate boiling. Otherwise, it is called

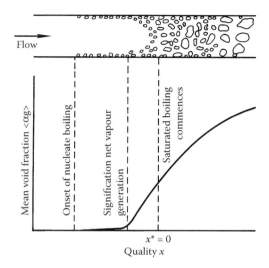

FIGURE 8.9
Variation of void fraction along the heated channel. (From Butterworth, D. and Shock, R. A., *Heat Transfer 1982, Proc. 7th Int. Heat Transfer Conf.*, Hemisphere, Washington, DC, 1982. With permission.)

saturated nucleate boiling or two-phase convective boiling. As the amount of superheat, $(T_w - T_s)$, increases, the number of nucleation sites where bubbles grow increases, and, consequently, there is a rapid increase in the rate of heat transfer. The ONB is indicated in Figure 8.9.[39]

If nucleate boiling has already begun in flow over any form of heated surface, two-phase boiling begins as soon as the bulk of the liquid reaches its boiling temperature. The formation of vapor increases the velocity of flow past the heated surface, thus facilitating the transfer of heat. Therefore, calculation of heat transfer in two-phase convection, in nucleate boiling, and in a combination of the two is important and will be explained by solved examples in the coming sections. It is important to note that at ONB, bubbles merely grow and collapse at, or very close to, the wall. There is, however, an increase in heat transfer coefficient either through simultaneous evaporation and condensation, i.e., the bubbles act as heat pipes, or through the increased bubble agitation from growth and collapse. Later, bubbles slide along the wall but do not survive in the bulk liquid. Not until some further distance downstream (see Figure 8.9), is there any significant net vapor generation (SNVG) resulting in an increase in void fraction as bubbles depart and survive in the bulk liquid.

The superheat for ONB and the corresponding radius are given by Davis and Anderson[40]:

$$\Delta T_{ONB} = \Delta T_{WONB} - T_s = \left(\frac{8\sigma T_s q''}{k \rho_v \Delta h_v} \right) \tag{8.45}$$

and the corresponding radius is

$$r_c = \left(\frac{2k\sigma T_s}{\Delta h_v \rho_v q''} \right)^{1/2} \tag{8.46}$$

Most developments of such equations use the Clausius–Clapeyron equation and differ only in the assumptions used to obtain the solution. Many data are available in the literature, and many of these have been compared with Equation 8.45; a detailed survey of the literature is given by Butterworth and Shock.[39] It is found that for water and other nonwetting fluids, Equations 8.45 and 8.46 are reliable, whereas ΔT_{ONB} is under-predicted and the corresponding r is over-predicted for well-wetting fluids, which include many organic substances.[40]

Bergles and Rohsenow[41] formulated Equation 8.45 with the properties of water at pressures of 1 bar and obtained the following equation:

$$q'' = 1120 \, p^{1.156} (1.8 \, \Delta T_{ONB})^{2.16/p^{0.0234}} \tag{8.47}$$

which can be used in design. Equation 8.47 has been extensively tested in the range of 1–138 bar.

8.6.2 Flow Pattern

The local hydrodynamic and heat transfer behavior is related to the distribution of liquid and vapor, referred to as the flow patterns or flow regimes. It is helpful to briefly discuss flow patterns, even though experience has shown that reasonably accurate correlations for pressure drop and heat transfer coefficient can be obtained without consideration of the flow pattern.

In a traditional once-through boiler, the flow enters as subcooled liquid and exits as superheated vapor. Subcooled boiling is observed before the fluid reaches a bulk saturated condition; the flow pattern is wall bubbly. The bubble boundary layer thickens because of the accumulation of uncondensed vapor, which is promoted by the decreasing condensation potential. The fluid is in a nonequilibrium state, with superheated liquid near the wall and subcooled liquid in the core, and the vapor would condense if the flow were brought to rest and mixed. At some point, the bulk enthalpy is at the saturated liquid condition ($x = 0$).

As the vapor volumetric fraction increases, the bubbles agglomerate and slug flow is observed (slug flow may also be observed in the subcooled region). Agglomeration of the slug flow bubbles leads to a transition regime termed churn flow, where the nominally liquid film and the nominally vapor core are in a highly agitated state. The subsequent flow pattern, annular flow, has the phases more clearly separated spacewise. However, the film may contain some vapor bubbles and the core may contain liquid in the form of drops or streamers. The film thickness usually varies with time with a distinct wave motion, and there is an interchange of liquid between the film and core.

There is a gradual depletion of the liquid due to evaporation. At some point before complete evaporation, the wall becomes nominally dry due to net entrainment of the liquid or abrupt disruption of the film. Beyond this dryout condition, drop flow prevails. A nonequilibrium condition again occurs, but in this case, the vapor becomes superheated to provide the temperature difference required to evaporate the vapor. Eventually, beyond the point where the bulk enthalpy is at the saturated vapor condition ($x = 1$), the liquid evaporates and normal superheated vapor is obtained.

The fluid and wall temperature profiles shown in Figure 8.10 (Ref. 42) pertain to uniform heat flux, as might be approximated in a fired boiler with complete vaporization. The temperature difference between the wall and the fluid is inversely proportional to the heat transfer coefficient. The normal single-phase coefficient is observed near the entrance of the tube, perhaps with an increase right at the inlet due to flow and/or thermal

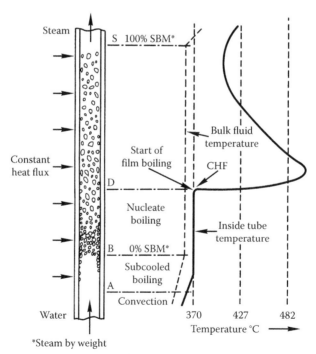

FIGURE 8.10
Temperature profiles in a vertical evaporator tube. (From Kitto, J. B. and Albrecht, M. J., *Two-Phase Flow Heat Exchangers: Thermal-Hydraulic Fundamentals and Designs*, Kluwer Publisher, The Netherlands, pp. 221–291, 1988. With permission.)

development. The coefficient increases rapidly as subcooled boiling is initiated because of the intense agitation of the bubbles, but levels off in established boiling over the subcooled and low-quality regions. The coefficient is usually expected to increase in annular flow because of the thinning of the liquid film. At the dryout point, the coefficient decreases rapidly as a result of the transition from, basically, a solid–liquid heat transfer to a solid–gas heat transfer. Droplets striking the surface elevate the heat transfer coefficient above what it would be for pure vapor.

Clearly, there are many similarities between flow boiling in vertical and horizontal tubes. Not surprisingly, therefore, these similarities are reflected in the literature. Indeed, some authors have correlated data for both horizontal and vertical tubes without remarking on the possible differences between the two. However, there is a very important difference between vertical and horizontal tubes, which is, of course, the stratifying effect of gravity. Many experimenters have observed the two-phase flow patterns and some have tried to interpret the heat transfer results in terms of these observations. Also, a few of the correlators of data have tried to include additional terms in their correlation to account for stratification.

Much of the literature on horizontal flow would, however, suggest to the casual reader that there are very great differences between boiling in vertical and horizontal tubes. This arises because the two topics have tended to be researched by different people with the separate aims of serving different industries. The research on vertical tubes has mainly been for the power industry, particularly the nuclear industry, while that on horizontal tubes has mainly been for the refrigeration and air conditioning industry. Collier[43] has suggested that the flow development in a horizontal evaporator channel is as given in Figure 8.11.

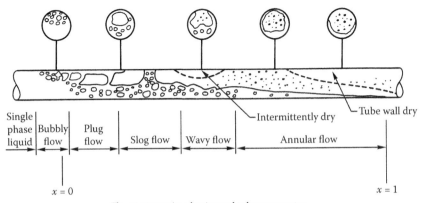

Flow patterns in a horizontal tube evaporator

FIGURE 8.11
Flow patterns in a horizontal tube evaporator. (From Collier, J. G., *Convective Boiling and Condensation*, 2nd ed., McGraw-Hill, London, 1981. With permission.)

Horizontal in-tube boiling stratification of the flow could occur, leaving a region at the top of the tube with a very low local heat transfer coefficient. The coefficient averaged around the circumference would be correspondingly low. The flow pattern inside the horizontal tube will depend on the velocity of flow. The figure shown by Collier would seem to represent a moderately high flow in the tube. With very high flows, one would expect less stratification, while with very low flows, stratified flow would be the predominant flow pattern, occurring almost directly beyond the point of initial vapor formation. The stratification will reduce the heat transfer coefficient. Therefore, the designer needs to know under what conditions heat transfer deterioration due to stratification may be expected. This topic will be discussed in the following sections with worked examples.

In recent years, several correlations have been developed to deal with the different stages of boiling and the effects of two-phase flow. Correlations for in-tube boiling developed by Chen,[44] Shah,[45,46] Güngör and Winterton,[47] and Kandlikar[48,49] will be discussed herein. The full range of boiling conditions for which the correlations apply include single-phase vapor entering the evaporator and two-phase flow occurring in sections within the evaporator coils. The flow is characterized by nucleate boiling at the inlet.

8.6.3 Flow Boiling Correlations

Established forced convection subcooled boiling is subject to the same surface and fluid variables as pool boiling. However, it was shown by Brown[50] that the surface effects become less pronounced as the levels of velocity and subcooling increase. Jens and Lottes[51] suggested the following equation for water boiling from stainless steel or nickel surfaces:

$$q'' = 3.91 \times 10^5 \, e^{0.065p} (T_w - T_s)^4 \qquad (8.48)$$

where p is the absolute pressure in bar, $(T_w - T_s)$ is in K, and q'' is in MW/m². No correlations based on a variety of surfaces are available for other fluids. The pressure effect has not been established for other fluids.

An extensive review of correlations for subcooled boiling of water over a wide range of pressures was carried out by Guglielmini et al.[52] For the partial boiling regions they examined several interpolation techniques and, without giving supporting evidence, recommended Pokhvalov et al.'s technique,[53] which is proposed by its authors for water and other fluids.

For the fully developed region, Guglielmini et al. compared 17 correlations, dating from 1955 to 1971. These include the often-used method of Thom et al.[54]:

$$\Delta T_s = 22.52 \, e^{-0.115p} (q''/10^6)^{1/2} \qquad (8.49)$$

(where p is the pressure in bar) resulting from data in flow boiling over a large range of pressures. Over a pressure range of 1–100 bar, there is a large scatter in predicted ΔT_s from the various correlations, which may have been due (1) to the differences in surface finish between tubes used by various workers, or (2) to differing degrees of accuracy in measuring ΔT_s, often in the range 5–10°C. Over the whole pressure range, Guglielmini et al.[52] recommend, after comparison with a database covering nine experimental programs with round, rectangular, and annulus test sections, the methods of Rassokhin et al.[55]:

$$\Delta T_s = m p^n q''^{1/3}$$
$$1 \leq p \leq 80 \text{ bar} \quad m = 0.304 \quad n = -0.25 \quad (8.50)$$
$$80 < p \leq 200 \text{ bar} \quad m = 34.7 \quad n = -1.25$$

and Labuntsov[56]:

$$\Delta T_s = \frac{1 - 0.0045\, p}{3.4\, p^{0.18}} q''^{1/3} \quad (8.51)$$

As shown in Figure 8.10, nucleate boiling prevails over the low-quality region and the heat transfer coefficient is essentially at the established subcooled boiling level. After annular flow is developed, however, the heat transfer coefficient frequently increases with increasing quality. Nucleation is suppressed because of the decreasing film thickness and the resultant increase in the heat transfer coefficient. Numerous correlations have been provided for this convective vaporization region, as noted by Collier.[43]

The more popular correlations incorporate both low quality and higher quality behavior in an additive fashion, for example, the widely used Chen correlation,[44] in which convective and nucleate boiling contributions are calculated separately and then added. The Chen method was originally devised for saturated boiling but can easily be extended to subcooled boiling by the superposition method:

$$h_{TP} = h_{cb} + h_{nb} = h_{LO} F + h_p S \quad (8.52)$$

where h_{LO} is the heat transfer coefficient if the liquid were flowing on its own at that point in the channel (single-phase convective correlation), and h_p is the pool boiling coefficient for the same wall superheat. F (boiling enhancement factor) allows for the increase in velocity of the liquid vapor and S (suppression factor) allows for the decrease in bubble activity due to the steepening of the temperature gradient.

The contributions of convective boiling and nucleate boiling are estimated separately. Chen's procedure for estimating the convective and nucleate boiling coefficients is described herein. His method for obtaining h_{nb} is based on the analysis of Forster and Zuber,[56] whereby Chen uses the pool boiling correlation with a suppression factor, S.

By using experimental data from different sources, Chen determined experimental values of the convective boiling enhancement factor, F, and based it on the Martinelli parameter, X_{tt}:

$$\frac{1}{X_{tt}} = \frac{x}{(1-x)^{0.9}} \left(\frac{\rho_l}{\rho_v}\right)^{0.5} \left(\frac{\mu_v}{\mu_l}\right)^{0.1} \tag{8.53}$$

The enhancement factor, therefore, can be determined by the following equation to fit the curve proposed by Chen[44]:

For $1/X_{tt} \leq 0.1, F = 1;$

For $1/X_{tt} > 0.1, F = 2.35(0.213 + 1/X_{tt})^{0.736}$ (8.54)

As vaporization takes place, the amount of liquid in the tube decreases, and then h_{LO} will also decrease. If one assumes that the heat transfer coefficient for the liquid flowing alone, h_{LO}, is proportional to the velocity raised to the power of 0.8 (in the Dittus–Boelter correlation), the value of F_o is related to F and quality x by[49]:

$$F_o = F(1-x) \tag{8.55}$$

The convection boiling heat transfer coefficient, h_{cb}, can then be found by multiplying the liquid heat transfer coefficient, h_{LO}, by the enhancement factor, F_o. To simplify the calculations of h_{LO}, a liquid heat transfer coefficient for zero vapor quality, h_{LO}, can be found to be constant throughout the length of the tube. Using the Dittus–Boelter correlation (one of the new correlations given in Chapter 3 may be used),

$$h_{LO} = 0.023 \, Re^{0.8} \, Pr^{0.4} \, k_l/d \tag{8.56}$$

The convection boiling coefficient can then be found by

$$h_{cb} = F_o h_{LO} \tag{8.57}$$

where F_o is the enhancement factor to account for the varying qualities given by Equation 8.55.

At low vapor qualities, the enhancement due to nucleate boiling must be considered. Chen's correlation for the nucleate component is as follows:

$$h_{vo} = h_{nb} = 0.00122 \frac{k_l^{0.079} \, C_{PL}^{0.45} \, \rho_L^{0.49} \, S\theta_B^{0.24} \, Dp_v^{0.75}}{s^{0.5} \, m_L^{0.29} \, (i_{lg}\rho_g)^{0.24}} \tag{8.58}$$

where S is a suppression factor dependent on the two-phase flow Reynolds number:

$$S = \frac{1}{1+2.53\times10^{-6}(Re_{TP})^{1.17}} \qquad (8.59)$$

and

$$\theta_B = \Delta T_{sat} = T_w - T_s \qquad (8.60)$$

where θ_B is the wall superheat, T_s is the saturation temperature, and T_w is the wall temperature. The two-phase Reynolds number is found by

$$Re_{TP} = (\dot{m}d/\mu_l)(1-x)F^{1.25} \qquad (8.61)$$

and Δp_v can be written from the Clapeyron equation as

$$\Delta p_v = \Delta T_s i_{lg} \rho_v / T_s \qquad (8.62)$$

The wall temperature is assumed at first to calculate an initial h_{NUCL}, then later checked as shown in Example 8.6. A reiterative process is used until the assumed wall superheat agrees with the final value.

The other class of correlation for mixed nucleation and convection boiling is that in which the multiplier h_{TP}/h_{LO} or h_{TP}/h_{LV} is a function of the Martinelli parameter, $1/X_{tt}$, to characterize convection and the boiling number, Bo,

$$\frac{h_{TP}}{h_{LO}} = C_1 \left\{ Bo + C_2 \left(\frac{1}{X_{tt}}\right)^{C_3} \right\}^{C_4} \qquad (8.63)$$

where, for example, Wright[58] gives $C_1 = 6{,}700$, $C_2 = 3.5 \times 10^{-4}$, $C_3 = 0.67$, and $C_4 = 1.0$. A relationship of the type of Equation 8.63 based on data for R-12, including entrance effects, was published by Mayinger and Ahrens[57]:

$$\frac{h_{TP}}{h_{LO}} = 0.85\times10^4 \left\{ Bo + 4.5\times10^{-4}\left(\frac{1}{X_{tt}}\right)^{0.35} \right\} \left(1+\frac{d}{z}\right)^{[1/X_{tt}]^{0.41}} \qquad (8.64)$$

Equation 8.64 suggests h_{TP} proportional to q'' in the nucleation-dominated region, although close examination of the data shows $h_{TP} \sim (q'')^{0.78}$. The Shah method[45] is valid for vertical and horizontal flow and is discussed in the following pages. Close analysis shows that, in the nucleate boiling region, it gives $h_{TP} \sim (q'')^{0.5}$.

Shah's correlation[45] to calculate the heat transfer coefficient of boiling flow through pipes is based on four dimensionless parameters, Fr_L, Co, Bo, and F_o which are the Froude number, convection number, boiling number, and enhancement factor, respectively. These dimensionless parameters characterize the flow and are employed to estimate the two-phase convective contribution to heat transfer in boiling.

The Froude number is defined as

$$Fr_L = \frac{G^2}{\rho_l^2 g d_i} \tag{8.65}$$

and determines whether stratification effects are important or not. A Froude number greater than 0.04 ($Fr_L > 0.04$) signifies that stratification effects are negligible and inertial forces are dominant compared with gravitational forces. For low Froude numbers, the Shah method is recommended because it allows for effects of stratification. A correction factor, K_{FR}, is used when $Fr_L < 0.04$[58]:

$$K_{FR} = (25\, Fr_L)^{-0.3} \tag{8.66}$$

When $Fr > 0.04$, $K_{FR} = 1$ for horizontal or inclined pipes whereas $K_{FR} = 1$ for vertical pipes at all rates as long as there is no liquid deficiency.

The convection number, Co, is defined as[58]

$$Co = [(1-x)/x]^{0.8} (\rho_v/\rho_l)^{0.5} K_{FR} \tag{8.67}$$

with the additional multiplication factor to Shah's original correlation to account for stratification effects for horizontal tubes when necessary.

The boiling number, Bo, defined as

$$Bo = q''/\dot{m} i_{lg} \tag{8.68}$$

determines the enhancement due to nucleate boiling. The heat flux, q'', is the initial estimated value and i_{lg} is the latent heat of vaporization. A boiling number less than 1.9×10^{-5} signifies that there is no enhancement due to nucleation.

The enhancement factor, F_o, is dependent on the characteristics of the boiling. It is the ratio of heat transfer for two-phase flow to liquid-only flow:

$$F_o = h_{TP}/h_{LO} \tag{8.69}$$

For pure convection boiling that occurs at high vapor qualities, and low boiling numbers, the convection boiling factor is defined by[58]

$$F_{cb} = 1.8\, Co^{-0.8} \quad Co < 1.0 \tag{8.70}$$

Design Correlations for Condensers and Evaporators

and for lower vapor qualities, where $Co > 1.0$,

$$F_{cb} = 1.0 + 0.8 \exp\left[1 - (Co)^{0.5}\right] \qquad Co > 1.0 \qquad (8.71)$$

where $F = F_{cb}$, and substituting it in Equation 8.55, F_o, the enhancement factor at varying vapor qualities, can be found.

The convection boiling heat transfer coefficient is given by Equation 8.57:

$$h_{cb} = F_o h_{LO} \qquad (8.72)$$

where h_{LO} is the liquid-only heat transfer coefficient and is found using the Dittus–Boelter correlation as recommended by Shah. One can use better correlations given in Chapter 9.

In the nucleate boiling regime, $Bo > 1.9 \times 10^{-5}$ for very low vapor qualities, and the nucleate boiling factor is[46]

$$F_{nb} = 231\, Bo^{0.5} \quad Co > 1.0 \qquad (8.73)$$

where the nucleate boiling effects are dominant and $F = F_{nb}$ in Equation 8.55.

At higher vapor qualities, and $0.02 < Co < 1.0$, combined nucleate and convective boiling effects are considered[58] and the enhancement factor is determined from

$$F_{cnb} = F_{nb}\left(0.77 + 0.13\, F_{cb}\right) \qquad (8.74)$$

which is developed empirically by Smith from Shah's experimental data.[58]

Whenever combined nucleate and convective boiling is applicable, the enhancement factor, F, is determined by choosing the greatest value of F_{nb} for $Co > 1.0$; alternatively, the greatest value of F_{cnb} and F_{cb} for $0.02 < Co < 1.0$ will be used for F in Equation 8.55.

The method of calculating the two-phase heat transfer coefficient, h_{TP}, for considering pure convection boiling is as follows:

1. Calculate the liquid only heat transfer coefficient, h_{LO}.
2. Determine the convection number, Co, at the desired vapor quality.
3. Find the convection boiling factor at the corresponding Co using Equation 8.70 or 8.71.
4. Determine the enhancement factor, F_o, by using a proper F in Equation 8.55.
5. The convection boiling heat transfer coefficient for tube-side flow is calculated by Equation 8.72.

To check the validity of the assumption that nucleate boiling is not present, the wall temperature for the ONB, T_{WONB} can be determined by the correlation developed by Davis and Anderson[40]:

$$T_{WONB} = \left(\frac{8\sigma q'' T_s}{k_l \Delta h_v \rho_v}\right)^{1/2} + T_s \qquad (8.75)$$

The heat flux, q'', is defined as

$$q'' = U(T_H - T_s) \qquad (8.76)$$

where T_H is the temperature of the heating fluid and T_s is the saturated fluid temperature. The overall heat transfer coefficient is U.

The wall temperature, T_w, can be calculated from the heat flux and two-phase heat transfer coefficient:

$$T_w = \frac{q''}{h_{cb}} + T_s \qquad (8.77)$$

where h_{cb} in the case of pure convection boiling is equal to the two-phase coefficient, h_{TP}.

If $T_{WONB} > T_W$, then nucleate boiling is not a concern and the assumption made is correct; but if $T_W > T_{WONB}$, then nucleate boiling should be considered and the calculations repeated for combined nucleate and convective boiling.

The method of calculating the boiling flow coefficient for combined convection and nucleation boiling is as follows:

1. Calculate the heat flux, q'', assuming no nucleate boiling as delineated previously.
2. To determine if enhancement due to nucleate boiling is significant, the boiling number, Bo, must first be calculated. If $Bo < 1.9 \times 10^{-5}$, the nucleate boiling is negligible and these steps are discontinued; otherwise, nucleate boiling should be considered.
3. Find the enhancement factor for nucleate boiling, F_{nb} or F_{cnb}, using Equation 8.73 or 8.74, respectively.
4. The boiling heat transfer coefficient at zero quality is then found by

$$h_{vo} = F_{nb} h_{LO} \qquad (8.78)$$

5. The enhancement factor at different vapor qualities, x, for $Co > 1.0$ can then be determined by making F in Equation 8.78 equal to the greater of F_{nb} or F_{cb} as given by Equations 8.73 and 8.71, respectively.

For $F_{cb} > F_{nb}$ the initial value of $q\delta$ which neglected nucleate boiling is correctly assumed and enhancement due to nucleation is negligible. If F_{nb}, however, is greater, then a more accurate value of heat flux must be calculated.

If $0.02 < Co < 1.0$, then the enhancement factor is the greater of F_{cnb} or F_{cb} where F_{cnb} is calculated by Equation 8.74 and F_{cb} is calculated by Equation 8.70. For $F_{cnb} > F_{cb}$, again, a more accurate value of heat flux, q'', must be calculated for nucleation enhancement effects with the new enhancement factor, $F = F_{cnb}$; otherwise, nucleation boiling enhancement is negligible.

The new boiling heat transfer coefficient at zero quality is defined as

$$h_{vo} = F h_{LO} \tag{8.79}$$

Shah's method of determining heat transfer during saturated boiling is applicable for saturated boiling inside pipes of all Newtonian fluids (except metallic fluids). Shah identified the Chart correlation[46] in which two regions are defined: a nucleate boiling region, $Co > 1.0$, where heat transfer is determined by the boiling number only, and a region of convection boiling, $Co < 1.0$, where nucleation bubbles are completely suppressed.

For the purpose of simplifying the use of the Chart correlation, Shah's method was introduced in equation form as outlined in Shah's second paper[46] and Smith's summary of Shah's method.

A comparative study of Shah's correlations to data points for 11 refrigerants revealed a mean deviation of ±14%.

The Güngör and Winterton[47] correlation was developed from a large database of saturated boiling points of halocarbon refrigerants. The form of the correlation is as follows:

$$h_{TP} = E h_{LO} + S h_p \tag{8.80}$$

where E is an enhancement factor defined as

$$E = 1 + 2.4 \times 10^4 \, Bo^{1.16} + 1.37 \left(1/X_{tt}\right)^{0.86} \tag{8.81}$$

and S is a suppression factor:

$$S = \left[1 + 1.15 \times 10^{-6} \, E^2 \, Re_l^{1.17}\right]^{-1} \tag{8.82}$$

The pool boiling term, proposed by Cooper,[59] is defined as

$$h_p = 55 \, Pr^{0.12} \left(-\log Pr\right)^{-0.55} M^{-0.5} \, q''^{0.67} \tag{8.83}$$

The values of h_L, the liquid-only heat transfer coefficient, and $1/X_{tt}$, the Martinelli parameter, are determined by Equations 8.55 and 8.53, respectively. The liquid Reynolds number is defined as

$$Re_{LO} = G(1-x)d_i/\mu_l \tag{8.84}$$

For $Fr < 0.05$, Güngör and Winterton[47] recommend that the enhancement factor be multiplied by the correction factors shown below:

$$E_2 = Fr_L^{(0.1-2Fr_L)} \tag{8.85}$$

$$S_2 = Fr_L^{1/2} \tag{8.86}$$

When the correlation was compared to actual experimental values, a mean deviation of ±21.3% was reported. The properties used to calculate the correlation should be determined at the saturated boiling temperature. Knowledge of q'', the heat flux, is also necessary for these equations.

Kandlikar[48,49] proposed a correlation based on Shah's database plus additional data. He introduced a fluid dependent parameter, F_{fl}, in the nucleate boiling term. The mean deviation for all the data sets considered is 17.1%. The correlation proposed is applicable to both horizontal and vertical tubes.

The resulting equations for two-phase flow as proposed by Kandlikar are in the form

$$h_{TP} = D1(Co)^{D2} h_{LO} + D3(Bo)^{D4} h_{LO} F_{fl} \quad \text{for vertical flow} \tag{8.87}$$

$$h_{TP} = D1(Co)^{D2}(25\,Fr_L)^{D5} h_{LO} + D3(Bo)^{D4}(25\,Fr_L)^{D6} h_{LO} F_{fl} \tag{8.88}$$
$$\text{for horizontal flow}$$

The single-phase liquid-only coefficient, h_{LO}, is calculated by the Dittus–Boelter equation. All the properties are evaluated at saturation temperature. The constants $D1$ through $D6$ were evaluated using experimental data from Mumm[60] and Wright[61] for water, Jallouk[62] for R-114, Chawla[63] for R-11, Steiner and Schlünder[64] for nitrogen, Mohr and Runge[65] for neon, and Bandel and Schlünder[66] for R-12. A reiterative process of using initial F_{fl} and recalculating the coefficients $D1$ through $D6$ from the experimental data was used to determine the values as given in Tables 8.1 and 8.2.

A general correlation for the flow boiling heat transfer coefficient was proposed by Kandlikar, which he reported yielded better results. Kandlikar's[67] general correlation was developed for both horizontal and vertical tube flow and a larger range of fluids. Its final form is as follows:

$$h_{TP} = C1(Co)^{C2}(25\,Fr_L)^{C5} h_{LO} + C3(Bo)^{C4} F_{fl} h_{LO} \tag{8.89}$$

Design Correlations for Condensers and Evaporators

TABLE 8.1
Values of Constants $D1$ Through $D6$

Constant	$Co < 0.65$	$Co > 0.65$
$D1$	1.091	0.809
$D2$	−0.948	−0.891
$D3$	887.46	387.53
$D4$	0.726	0.587
$D5$	0.333	0.096
$D6$	0.182	0.203

Source: Kandlikar, S. G., *J. Heat Transfer*, 113, 966, 1991; Kandlikar, S. S., *J. Heat Transfer*, 112, 219, 1990.

TABLE 8.2
Values of F_{fl} for Different Fluids

Fluid	F_{fl}
Water	1.0
R-11	1.35
R-114	2.15
R-12	2.10
Nitrogen	3.0
Neon	3.0

Source: Kandlikar, S. G., *J. Heat Transfer*, 113, 966, 1991; Kandlikar, S. S., *J. Heat Transfer*, 112, 219, 1990.

TABLE 8.3
Constants $C1$ Through $C5$. $C5 = 0$ for Vertical Tubes, and for Horizontal Tubes with $Fr_l > 0.04$

Constant	$Co < 0.65$	$Co > 0.65$
$C1$	1.136	0.6683
$C2$	−0.9	−0.2
$C3$	667.2	1058.0
$C4$	0.7	0.7
$C5$	0.3	0.3

Source: Kandlikar, S. G., *J. Heat Transfer*, 113, 966, 1991; Kandlikar, S. S., *J. Heat Transfer*, 112, 219, 1990.

For $Fr > 0.04$, the Froude number multiplier becomes unity. The values for the constants $C1$ through $C5$ are given in Table 8.3 as determined by Kandlikar, and the values for the fluid dependent parameter, F_{fl}, are listed in Table 8.4 for different fluids. The heat transfer coefficient can be evaluated at two different regions: the nucleate boiling region where $Co > 0.65$ and the convective

TABLE 8.4

Values of F_{fl} for Different Fluids

Fluid	F_{fl}
Water	1.00
R-11	1.30
R-12	1.50
R-13B1	1.31
R-22	2.20
R-113	1.10
R-114	1.24
R-152a	1.10
Nitrogen	4.70
Neon	3.50

Source: Kandlikar, S. G., *J. Heat Transfer*, 113, 966, 1991; Kandlikar, S. S., *J. Heat Transfer*, 112, 219, 1990.

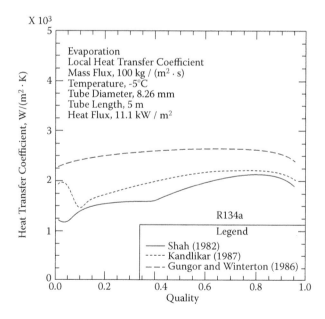

FIGURE 8.12
Local evaporation heat transfer coefficients for in-tube flow of R-134A. (From Pate, M. B., *Boilers, Evaporators and Condensers*, Wiley, New York, 1991. With permission.)

boiling region where $Co < 0.65$. These values of F_{fl} represent revised values from those originally presented. Any fluid not covered in the more recent listing should be retrieved from the previous listing. Three correlations given by Shah,[46] Kandlikar,[49] and Güngör and Winterton[47] are compared in Figures 8.12 and 8.13.

Design Correlations for Condensers and Evaporators

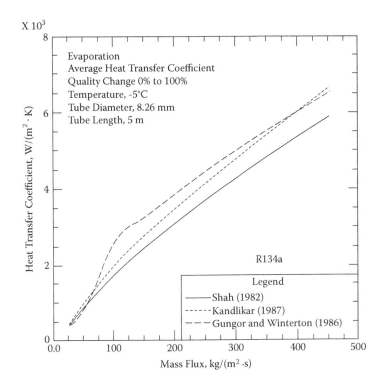

FIGURE 8.13
Average evaporation heat transfer coefficients for in-tube flow of R-134A. (From Pate, M. B., *Boilers, Evaporators and Condensers*, Wiley-Interscience, New York, 1991. With permission.)

Example 8.5

Refrigerant-22 (R-22) boiling at 250 K flows through a tubular heat exchanger. Tubes are arranged horizontally with an inner diameter of 0.0172 m. The mass velocity of the R-22 is 200 kg/m² · s. Heat is supplied by condensing the fluid outside at –12°C. The hot fluid-side heat transfer coefficient, h_h, is 5,400 W/m² · K. The tube is made of copper, and wall resistance and fouling are neglected. By the use of both Chen and Shah's methods, calculate the following:

  a. the convection heat transfer coefficient for R-22 for various values of quality assuming no nucleate boiling
  b. heat flux, q'' (W/m²)
  c. tube wall temperature (T_{WONB}) for the ONB.

Is the assumption of no nucleate boiling valid? It may be so at certain values of x_v.

Solution

The properties of refrigerant-22, from Appendix A, at 250 K (–23°C) are

$$P_s = 217.4 \text{ kPa } (2.174 \text{ bars})$$
$$\Delta h_v = 221900 \text{ J/kg}$$
$$\rho_v = 9.6432 \text{ kg/m}^3$$
$$\rho_l = 1360 \text{ kg/m}^3$$
$$c_{pl} = 1122 \text{ J/kg} \cdot \text{K}$$
$$k_l = 0.112 \text{ W/m} \cdot \text{K}$$
$$\sigma = 0.0155 \text{ N/m}$$
$$\mu_{gn} = 0.110 \times 10^{-4} \text{ Pa} \cdot \text{s}$$
$$\mu_l = 0.282 \times 10^{-3} \text{ Pa} \cdot \text{s}$$

The Prandtl number can be calculated as follows:

$$Pr = \frac{\mu_l c_{pl}}{k_l} = \frac{(0.282 \times 10^{-3})(1122)}{0.112} = 2.825$$

The Reynolds number for the liquid phase is

$$Re_{LO} = \frac{Gd_i}{\mu_l} = \frac{(200)(0.0172)}{0.000282} = 12.200$$

The Nusselt number at $x = 0$ can be calculated by Gnielinski's correlation:

$$Nu_{LO} = \frac{(f/2)(Re_{LO} - 1000) Pr_l}{1 + 12.7(f/2)^{1/2} (Pr_l^{2/3} - 1)}$$

where

$$f = (1.58 \ln Re_L - 3.28)^{-2}$$
$$f = [1.58 \ln(12200) - 3.28]^{-2}$$
$$Nu_{LO} = \frac{(0.00372)(12200 - 1000)(2.825)}{1 + 12.7(0.00372)^{1/2}(2.825^{2/3} - 1)} = 66.43$$

The heat transfer coefficient is then

$$h_{LO} = \frac{Nu_{LO} k_l}{d_i} = \frac{(66.43)(0.112)}{(0.0172)} = 432.55 \text{ W/m}^2 \cdot \text{K}$$

Design Correlations for Condensers and Evaporators

Using the Shah method,[45,46] define whether the effect of stratification is important or not. For this, calculate the Froude number:

$$Fr_{LO} = \frac{G^2}{\rho_l^2 g d_i} = \frac{(200)^2}{(1360)^2 (9.81)(0.0172)} = 0.1282$$

$Fr > 0.04$, therefore stratification effects are negligible and both Shah and Chen's methods may be used.

Chen's method, at $x = 0.05$, employs the Martinelli Parameter:

$$\frac{1}{X_{tt}} = \left(\frac{x}{1-x}\right)^{0.9} \left(\frac{\rho_l}{\rho_v}\right)^{0.5} \left(\frac{\mu_v}{\mu_l}\right)^{0.1}$$

$$= \left(\frac{0.05}{0.95}\right)^{0.9} \left(\frac{1360}{9.64}\right)^{0.5} \left(\frac{110}{2820}\right)^{0.1} = 0.6069$$

The enhancement factor can be calculated from Equation 8.55:

$$F_o = F(1-x_v)^{0.8}$$

where F, from Equation 8.54, is

$$F = 2.35\left(0.213 + \frac{1}{X_{tt}}\right)^{0.736} = 2.031$$

Therefore,

$$F_o = 2.031(1-0.05)^{0.8} = 1.949$$

and h_{cb} can be calculated from Equation 8.72:

$$h_{cb} = F_o(h_{LO}) = 1.949(432.55) = 842.7 \text{ W/m}^2 \cdot \text{K}$$

The overall heat transfer coefficient, with negligible wall resistance and fouling, is

$$U = \left(\frac{1}{h_{cb}} + \frac{1}{h_h}\right)^{-1} = \left(\frac{1}{842.7} + \frac{1}{5400}\right)^{-1} = 729.0 \text{ W/m}^2 \cdot \text{K}$$

To calculate heat flux,

$$q'' = U(T_H - T_s) = 729.0(-12-(-23)) = 8019 \text{ W/m}^2$$

To determine if nucleate boiling exists, the wall temperature for the ONB can be determined by Equation 8.45.

On the other hand, the wall temperature, T_w, can be calculated from the heat flux and heat transfer coefficient:

$$T_w = \frac{q''}{h_{cb}} + T_s = \frac{8109}{842.7} = 250 = 259.52 \text{ K}$$

For $x = 0.05$ and $T_w > T_{WONB}$, nucleate boiling is present and should be considered. This is demonstrated in the following solved problem, Example 8.6.

Shah's method accounts for stratification effects. At $x = 0.05$, the convection number, from Equation 8.67, can be determined:

$$Co = \left(\frac{1-x}{x}\right)^{0.8} \left(\frac{\rho_g}{\rho_l}\right)^{0.5} K_{FR}$$

where, $K_{FR} = 1$ since $Fr > 0.04$ and inertial forces are dominant compared with gravity forces.

$$Co = \left(\frac{1-0.05}{0.05}\right)^{0.8} \left(\frac{9.6432}{1360}\right)^{0.5} (1) = 0.8878$$

F_{cb} = convection boiling
$F_{cb} = 1.0 + 0.8\exp(1 - \sqrt{Co})$ $Co > 1.0$
$F_{cb} = 1.8\exp(1 - \sqrt{Co})$ $Co < 1.0$

Therefore,

$$F_{cb} = 1.8 \, Co^{-0.8}$$
$$= 1.8(0.8878)^{-0.8} = 1.9798$$

The enhancement factor, from Equation 8.55, is

$$F_o = F_{cb}(1-x)^{0.8}$$
$$= 1.9798(1 - 0.05)^{0.8} = 1.90$$
$$h_{cb} = F_o(h_{LO})$$
$$= 1.90(432.55) = 822.0 \text{ W/m}^2 \cdot \text{K}$$

Design Correlations for Condensers and Evaporators

The overall heat transfer coefficient, with negligible wall resistance and fouling, is

$$U = \left(\frac{1}{h_{cb}} + \frac{1}{h_h}\right)^{-1} = \left(\frac{1}{822} + \frac{1}{5400}\right)^{-1} = 713.0 \text{ W/m}^2 \cdot \text{K}$$

To calculate heat flux,

$$q'' = U(T_H - T_s) = 713.0(-12 - (-23)) = 7847 \text{ W/m}^2$$

The wall temperature for the ONB can be determined by Equation 8.45:

$$T_{WONB} = \left(\frac{8\sigma q'' T_s}{k_l \Delta h_v \rho_v}\right)^{1/2} + T_s = \left[\frac{8(0.0155)(7847)(250)}{(0.112)(221900)(9.64)}\right]^{1/2} + 250 = 251.01 \text{ K}$$

On the other hand, the wall temperature, T_w, can be calculated from the heat flux and heat transfer coefficient:

$$T_w = q''/h_{cb} + T_s = 7847/822 + 250 = 259.55 \text{ K}$$

For $x = 0.05$, $T_W > T_{WONB}$ and nucleate boiling is present and will be considered in Example 8.6. The calculations for each value of quality x can be performed in a similar way to calculate the various quantities as a function of quality.

Example 8.6

Repeat Example 8.5 with Refrigerant-22 at 250 K, now considering combined convection and nucleate boiling where necessary.

Solution

From Example 8.5,

$$T_H = -12° \text{ C}$$
$$T_s = -23° \text{ C}$$
$$G = 200 \text{ kg/m}^2 \cdot \text{s}$$
$$h_h = 5400 \text{ W/m}^2 \cdot \text{K}$$

Refrigerant-22 Properties at 250 K (–23°C) are, from the Appendix A,

$$P_{sat} = 217.4 \text{ kPA}(217.4 \text{ bars})$$
$$\Delta h_v = 221900 \text{ J/kg}$$
$$\rho_v = 9.6432 \text{ kg/m}^3$$
$$\rho_l = 1360 \text{ kg/m}^3$$
$$c_{pl} = 1122 \text{ J/kg} \cdot \text{K}$$
$$k_l = 0.112 \text{ W/m} \cdot \text{K}$$
$$\sigma = 0.0155 \text{ N/m}$$
$$\mu_v = 0.110 \times 10^{-4} \text{ Pa} \cdot \text{s}$$
$$\mu_l = 0.282 \times 10^{-3} \text{ Pa} \cdot \text{s}$$

At $x = 0.05$ (from Example 8.5),

$$h_{cb}(\text{Chen}) = 842.7 \text{ W/m}^2 \cdot \text{K}$$
$$h_{cb}(\text{Shah}) = 822.0 \text{ W/m}^2 \cdot \text{K}$$
$$q''(\text{Shah}) = 7847 \text{ W/m}^2$$

Using Chen's method at $x = 0.05$, the wall superheat is

$$\theta_B = \Delta T_{Sat} = T_w - T_s$$

where T_s is the saturation temperature and $T_W = -18.5°C$ (assumed wall temperature):

$$\theta_B = -18.5 - (-23) = 4.5° \text{ C}$$

The Reynolds number for two-phase flow, from Equation 8.61, is

$$Re_{TP} = \frac{\dot{m}d_i}{\mu_l}(1-x)F^{1.25}$$

where $F_{0.05} = 2.031$ is obtained in Example 8.5:

$$Re_{TP} = \frac{(200)(0.0172)}{(0.000282)}(0.95)(2.031)^{1.25} = 28{,}079$$

$$\Delta p_v = \Delta T_{Sat} \Delta h_v \rho_v / T_s = 4.5(221900)(9.643)/250 = 38.5 \text{ kPa}$$

Design Correlations for Condensers and Evaporators

The suppression factor, dependent on Re_{TP}, is, from Equation 8.59,

$$S = \left[1 + 2.53 \times 10^{-6} (Re_{TP})^{1.17}\right]^{-1}$$
$$= \left[1 + 2.53 \times 10^{-6} (28079)^{1.17}\right]^{-1} = 0.71$$

Chen's equation for the nucleate component, h_{vo}, is

$$h_{nb} = h_{vo} = 0.00122 \frac{k_l^{0.79} c_{pl} S\theta_B^{0.24} \Delta p_v^{0.75}}{\sigma^{0.5} \mu_l^{0.29} (\Delta h_v \rho_v)^{0.24}}$$

$$F_{cnb} = F_{nb}(0.77 + 0.13\, F_{cb}) = 3.06[0.77 + 0.13(1.98)] = 3.14$$

So, $F_{cb} < F_{cnb}$ and enhancement due to nucleate boiling is present. Therefore,

$$F = F_{cnb} = 3.14$$

Now go back and recalculate q'' with the new value of F_o for enhancement due to nucleate boiling:

$$h_{cb} = F_o h_{LO} = 3.14(432.55) = 1358 \text{ W/m}^2 \cdot \text{K}$$

The overall heat transfer coefficient, neglecting wall resistance and fouling, is

$$U = (1/h_{cb} + 1/h_H)^{-1} = (1/1345 + 1/5400)^{-1} = 1085.0 \text{ W/m}^2 \cdot \text{K}$$

and heat flux can be determined:

$$q'' = U(T_H - T_s) = 1085(-12 - (-23)) = 11.938 \text{ W/m}^2 \cdot \text{K}$$

Using Shah's method at $x = 0.05$, the boiling number, Bo, is

$$Bo = q''/G\Delta h_v = 7847/(200)(22,1900) = 1.77 \times 10^{-4}$$

This is the approximate value of the boiling number and $Bo > 1.9 \times 10^{-5}$; therefore, nucleation enhancement should be considered.
Calculate the value of the nucleate boiling factor F for nucleation:

$$F_{nb} = 231\, Bo^{1/2} = 231(1.77 \times 10^{-4})^{1/2} = 3.07$$

The heat transfer coefficient for boiling at zero quality is

$$h_{nb} = F_{nb} h_{LO} = 3.07(432.55) = 1327.93 \text{ W/m}^2 \cdot \text{K}$$

where $h_{LO} = 432.55$ W/m² · K (see Example 8.5).
Calculation of F is as follows:

$$Co_H > 1.0, \quad F = F_{cb} \text{ or } F_{vb}, \text{ whichever is greater}$$
$$0.02 > Co_H > 1.0, \quad F = F_{cb} \text{ or } F_{nb}, \text{ whichever is greater}$$

At $x = 0.05$, $Co_H = 0.8878$ and $F_{cb} = 1.9798$ (see Example 8.5):

$$h_{nb} = 0.00122 \frac{(0.112)^{0.79}(1122)^{0.45}(1360)^{0.49}(0.71)(4.5)^{0.24}(38500)^{0.75}}{(0.0155)^{0.5}(0.000282)^{0.29}(221900 \times 9.64)^{0.24}}$$

$$= 1297 \text{ W/m}^2 \cdot \text{K}$$

Another way to calculate h_{nb} is by considering

$$h_{nb} = f_{nbi} S \theta_B$$

where f_{nbi} is calculated by the equation given in theory as 405.0 W/m² · K.
The combined convection and nucleate boiling coefficient is

$$h_{cnb} = h_{cb} + h_{nb}$$
$$q'' = h_{cb} \theta_B + f_{nbi} S_B^2$$
$$= (842.7)(4.5) + (405)(0.71)(4.5)^2 = 9629 \text{ W/m}^2$$

where $\theta_B = T_w - T_s$.
The temperature difference across the boiling fluid, $(T_H - T_s)$ must be determined by trial and error. The value q'' must satisfy

$$T_H - T_s = q''/h_H + \theta_B = 9629/5400 + 4.5 = 6.28° \text{ C}$$

where H stands for the hot fluid. But from the statement of the problem, $T_H - T_s = 11°C$ therefore $\theta_B = 4.5$ is too small.
The first assumption for θ_B was much too small; a new value of $\theta_B = T_w - T_s = 7.2$ can be assumed, and by repeating the calculations, it is proven acceptable. The heat flux calculated for combined nucleate boiling and convective boiling for $x = 0.05$ is much higher than the heat flux for convection boiling only (2.5 times higher).

Nomenclature

A	area, m^2
Bo	boiling number $= q''/\Delta h_v \dot{m}$
C	constant
Co	convection number $= [(1-x)/x]^{0.8} (\rho_v/\rho_l)^{0.5}$
c_p	constant pressure specific heat, J/kg · K
d	tube diameter, m
E	enhancement factor
F	correction factor
Fr	Froude number $= G^2/g\, D_i\, \rho_h$
G	mass velocity, kg/m^2 · s
g	gravitational acceleration, m/s^2
h	heat transfer coefficient, W/m^2 · K
i_{lg}	latent heat of vaporization $(h_g - h_f)$, J/kg
i	enthalpy, J/kg
j_g^*	dimensionless mass velocity
K	correction factor
k	thermal conductivity
L	length, m
m	constant
$\dot{m}$	mass flow rate, kg/s
N	number of tubes in a vertical column
n	constant
Nu	Nusselt number $= hd/k$
p	pressure, N/m^2
Pr	Prandtl number $= c_p\, \mu/k$
Q	cumulative heat release rate, W
q'	heat input per unit tube length, W/m
q''	heat flux, W/m^2
r	radius of curvature, bubble radius, m
Re	Reynolds number $= \rho DU/\mu,\ \rho u d/\mu$
Re$_F$	film Reynolds number $= 4F/\mu_L$
$\tilde{Re}$	two-phase Reynolds number $= u_g\, D/v_l$
S	suppression factor
T	temperature, °C, K
t	time, s
ΔT	temperature difference, K
ΔT_s	wall superheat, °C, K
ΔT_{sub}	subcooling, C, K
U	overall heat transfer coefficient, W/m^2 · K
u	velocity, m/s
v	specific volume, m^3/kg
w	mean flow width, m

X_{tt}	Lockhart–Martinelli parameter, defined by Eq. (8.53)
x	vapor quality
Z	coefficient
z	distance from heater inlet, m

Greek Symbols

α	void fraction of vapor
Γ	film flow rate per unit length, kg/(m · s)
δ	film thickness, m
μ	dynamic viscosity, Pa · s
υ	kinematic viscosity = μ/ρ, m^2/s
ρ	density, kg/m^3
σ	surface tension, N/m
τ	shear stress, N/m^2
θ_B	wall superheat, °C, K

Subscripts

b	bulk
c	cold, critical point
cb	convective boiling
cnb	combined convection and nucleate boiling
f	liquid
g	gas phase
H	heating fluid
LO	liquid alone
l	liquid phase
m	mean
N	corresponding to N tubes
nb, n	nucleate boiling
ONB	onset of nucleate boiling
p	pool boiling
r	reduced
S	saturated fluid
sat	saturation
sh	vapor shear
TP	two-phase
v	vapor phase, saturated vapor
w	wall

PROBLEMS

8.1. Repeat Example 8.3 for refrigerant-22 and compare with R-134A.
8.2. Repeat Example 8.4 for refrigerant-22 and compare with R-134A.
8.3. Under the conditions given in Example 8.6, calculate the local heat transfer coefficient for the 20th row of tubes using the method given by Butteworth.

8.4. Calculate the average heat transfer coefficient for film type condensation of water at pressure of 10 kPa for
   a. outside surface of a 19 mm OD horizontal tube 2 m long
   b. a 12-tube vertical bank of 19 mm horizontal tube 2 m long.

   It is assumed that the vapor velocity is negligible and the surface temperatures are constant at 10°C below saturation temperature.

8.5. A horizontal 2 cm OD tube is maintained at a temperature of 27°C on its outer surface. Calculate the average heat transfer coefficient if saturated steam at 6.22 kPa is condensing on this tube.

8.6. In a shell-and-tube type steam condenser, assume that there are 81 tubes arranged in a square pitch with 9 tubes per column. The tubes are made of copper with an outside diameter of 1 in. The length of the condenser is 1.5 m. The shell-side condensation occurs at saturation pressure. Water flows inside the tubes with a mass flow rate of 4 kg/s. The tube outside wall temperature is 90°C. With reasonable assumptions, perform thermal and hydraulic analysis of the condenser.

8.7. A water-cooled, shell-and-tube freon condenser with in-tube condensation will be designed to satisfy the following specifications:

Cooling load of the condenser:	125 kW
Refrigerant:	R-22
Condensing temperature:	37°C
Coolant water:	City water
	Inlet temperature: 18°C
	Outlet temperature: 26°C
	Mean pressure: 0.4 Mpa
Heat transfer matrix:	3/4 in. OD, 20 BWG
	Brass tubes

It is proposed that the following heat exchanger parameters are fixed: one tube pass with a shell diameter of 15.25 in. and pitch size is 1 in. with baffle spacing of 35 cm. The number of tubes is 137.

   a. Calculate the shell-and-tube size heat transfer coefficients.
   b. Assuming proper fouling factors, calculate the length of the condenser.
   c. Space available is 6 m. Is the design acceptable?

8.8. R-134a flows in a horizontal 8 mm diameter circular tube. The mass flux is 400 kg/m²s, the entering quality is 0.0, and the existing quality is 0.8. The tube length is 3 m. Assuming a constant saturation temperature of 20°C, plot the variation of h_{TP} as a function of x.

References

1. Marto, P. J., Heat transfer in condensation, in *Boilers, Evaporators and Condensers*, Kakaç S., Ed., John Wiley & Sons, New York, 1991.
2. Marto, P. J., Fundamentals of condensation, in *Two-Phase Flow Heat Exchangers: Thermal-Hydraulic Fundamentals and Design*, Kakaç, S., Bergles, A. E., and Fernandes, E. O., Eds., Kluwer, The Netherlands, 1988, 221.
3. Nusselt, W., The condensation of steam on cooled surfaces, *Z.d. Ver. Deut. Ing.*, 60, 541, 1916.
4. Shekriladze, I. G. and Gomelauri, V. I., Theoretical study of laminar film condensation of flowing vapor, *Int. J. Heat Mass Transfer*, 9, 581, 1966.
5. Rose, J. W., Fundamentals of condensation heat transfer: laminar film condensation, *JSME Int. J.*, 31, 357, 1988.
6. Kern, D. Q., Mathematical development of loading in horizontal condensers, *AIChE J.*, 4, 157, 1958.
7. Eissenberg, D. M., An Investigation of the Variables Affecting Steam Condensation on the Outside of a Horizontal Tube Bundle, Ph.D. Thesis, University of Tennessee, Knoxville, 1972.
8. Butterworth, D., Developments in the design of shell and tube condensers, ASME Winter Annual Meeting, Atlanta, ASME Preprint 77-WA/HT-24, 1977.
9. Shklover, G. G. and Buevich, A. V., Investigation of steam condensation in an inclined bundle of tubes, *Thermal Eng.*, 25(6), 49, 1978.
10. Kutateladze, S. S., Gogonin, N. I., Dorokhov, A. R., and Sosunov, V. I., Film condensation of flowing vapor on a bundle of plain horizontal tubes, *Thermal Eng.*, 26, 270, 1979.
11. Fujii, T., Uehara, H., Hirata, K., and Oda, K., Heat transfer and flow resistance in condensation of low pressure steam flowing through tube banks, *Int. J. Heat Mass Transfer*, 15, 247, 1972.
12. Cavallini, A., Frizzerin, S., and Rossetto, L., Condensation of R-11 vapor flowing downward outside a horizontal tube bundle, *Proc. 8th Int. Heat Transfer Conf.*, 4, 1707, 1986.
13. McNaught, J. M., Two-phase forced convection heat transfer during condensation on horizontal tube bundles, *Proc. 7th Int. Heat Transfer Conf.*, 5, 125, 1982.
14. Marto, P. J., Heat transfer and two-phase flow during shell-side condensation, *Heat Transfer Eng.*, 5(1-2), 31, 1984.
15. Barness, E. J., Calculation of the performance of surface condensers by digital computer, ASME Paper No. 63-PWR-2, National Power Conference, Cincinnati, Ohio, September, 1963.
16. Emerson, W. H., The application of a digital computer to the design of surface condenser, *Chem. Eng.*, 228(5), 178, 1969.
17. Wilson, J. L., The design of condensers by digital computers, *I. Chem. E. Symp. Series*, Pergamon Press, London, No. 35, 21, 1972.
18. Hafford, J. A., ORCONI: A Fortran Code for the Calculation of a Steam Condenser of Circular Cross Section, ORNL-TM-4248, Oak Ridge National Laboratory, Oak Ridge, Tennessee, 1973.

19. Hopkins, H. L., Loughhead, J., and Monks, C. J., A computerized analysis of power condenser performance based upon an investigation of condensation, in *Condensers: Theory and Practice, I. Chem. E. Symp. Series*, Pergamon Press, London, No. 75, 152, 1983.
20. Shida, H., Kuragaska, M., and Adachi, T., On the numerical analysis method of flow and heat transfer in condensers, *Proc. 7th Int. Heat Transfer Conf.*, 6, 347, 1982.
21. Al-Sanea, S., Rhodes, N., Tatchell, D. G., and Wilkinson, T. S., A computer model for detailed calculation of the flow in power station condensers, in *Condensers: Theory and Practice, I. Chem. E. Symp. Series*, Pergamon Press, London, No. 75, 70, 1983.
22. Caremoli, C., Numerical computation of steam flow in power plant condensers, in *Condensers: Theory and Practice, I. Chem. E. Symp. Series*, Pergamon Press, London, No. 75, 89, 1983.
23. Beckett, G., Davidson, B. J., and Ferrison, J. A., The use of computer programs to improve condenser performance, in *Condensers: Theory and Practice, I. Chem. E. Symp. Series*, Pergamon Press, London, No. 75, 97, 1983.
24. Zinemanas, D., Hasson, D., and Kehat, E., Simulation of heat exchangers with change of phase, *Comput. Chem. Eng.*, 8, 367, 1984.
25. Kakaç, S., Ed., *Boilers, Evaporators and Condensers*, John Wiley & Sons, New York, 1991.
26. Soliman, H. M. and Azer, N. Z., Visual studies of flow patterns during condensation inside horizontal tubes, *Proc. 5th Int. Heat Transfer Conf.*, 3, 241, 1974.
27. Rahman, M. M., Fathi, A. M., and Soliman, H. M., Flow pattern boundaries during condensation: new experimental data, *Can. J. Chem. Eng.*, 63, 547, 1985.
28. Soliman, H. M., Flow pattern transitions during horizontal in-tube condensation, in *Encyclopedia of Fluid Mechanics*, Gulf Publishing Co., Houston, TX, 1986, ch. 12.
29. Breber, G., Palen, J. W., and Taborek, J., Prediction of horizontal tubeside condensation of pure components using flow regime criteria, *J. Heat Transfer*, 102, 471, 1980.
30. Tandon, T. N., Varma, H. K., and Gupta, C. P., A new flow regime map for condensation inside horizontal tubes, *J. Heat Transfer*, 104, 763, 1982.
31. Butterworth, D., Film condensation of pure vapor, in *Heat Exchanger Design Handbook*, Vol. 2, Schlünder, E. U., Ed., Hemisphere, Washington, DC, 1983.
32. Jaster, H. and Kosky, P. G., Condensation heat transfer in a mixed flow regime, *Int. J. Heat Mass Transfer*, 19, 95, 1976.
33. Traviss, D. P., Rohsenow, W. M., and Baron, A. B., Forced convection condensation inside tubes: a heat transfer equation for condenser design, *ASHRAE Trans.*, 79, 157, 1972.
34. Cavallini, A. and Zecchin, R., A dimensionless correlation for heat transfer in forced convection condensation, *Proc. 5th Int. Heat Transfer Conf.*, 309–313, Sept. 5–7, 1974.
35. Shah, M. M., A general correlation for heat transfer during film condensation inside pipes, *Int. J. Heat Mass Transfer*, 22, 547, 1979.
36. Schlager, L. M., Pate, M. B., and Bergles, A. E., Performance predictions of refrigerant–oil mixtures in smooth and internally finned tubes, II: design equations, *ASHRAE Trans.*, 96(1), 1990.

37. Pate, M. B., Evaporators and condensers for refrigeration and air conditioning, in *Boilers, Evaporators and Condensers*, Kakaç, S., Ed., John Wiley & Sons, New York, 635–712, 1991.
38. Wallis, G. B., Flooding Velocities for Air and Water in Vertical Tubes, UKAEA report AEEW-R123, 1961.
39. Butterworth, D. and Shock, R. A., Flow boiling, *Proc. 7th Int. Heat Transfer Conf.*, Hemisphere, Washington, DC, 1982.
40. Davis, E. J. and Anderson, G. H., The incipience of nucleate boiling in convective flow, *AIChE J.*, 12, 774, 1966.
41. Bergles, A. E. and Rohsenow, W. M., The determination of forced-convection surface-boiling heat transfer, *J. Heat Transfer*, 86, 356, 1964.
42. Kitto, J. B. and Albrecht, M. J., Elements of two-phase flow in fossil boilers, in *Two-Phase Flow Heat Exchangers: Thermal-Hydraulic Fundamentals and Design*, Kakaç, S., Bergles, A. E., and Fernandes, E. O., Eds., Kluwer, The Netherlands, 1988, 221.
43. Collier, J. G., *Convective Boiling and Condensation*, 2nd ed., McGraw-Hill, London, 1981.
44. Chen, J. C., A correlation for boiling heat transfer to saturated fluids in convective flow, *Industrial Eng. Chem. Process Design Dev.*, 5, 322, 1966.
45. Shah, M. M., A new correlation for heat transfer during boiling flow through pipes, *ASHRAE Trans.*, 82, 66, 1976.
46. Shah, M. M., Chart correlation for saturated boiling heat transfer: equations and further study, *ASHRAE Trans.*, 88, 185, 1982.
47. Güngör, K. E. and Winteron, R. H. S., A general correlation for flow boiling in tubes and annuli, *Int. J. Heat Mass Transfer*, 19(3), 351, 1986.
48. Kandlikar, S. G., A model for correlating flow boiling heat transfer in augmented tubes and compact evaporators, *J. Heat Transfer*, 113, 966, 1991.
49. Kandlikar, S. S., A general correlation for saturated two-phase flow boiling heat transfer inside horizontal and vertical tubes, *J. Heat Transfer*, 112, 219, 1990.
50. Brown, W. T., Jr., A Study of Flow Surface Boiling, Ph.D. Thesis, Massachusetts Institute of Technology, Cambridge, MA, 1967.
51. Jens, W. H. and Lottes, P. A., An Analysis of Heat Transfer, Burnout, Pressure Drop, and Density Data for High Pressure Water, Argonne National Laboratory Report ANL-4627, 1951.
52. Guglielmini, G., Nannei, E., and Pisoni, C., *Warme und Stoffübertragung*, 13, 177, 1980.
53. Pokhvalov, Y. E., Kronin, G. H., and Kurganova, I. V., *Teploenergetika*, 13, 63, 1966.
54. Thom, J. R. S., Walker, V. M., Fallon, J. S., and Reising, G. F. S., *Proc. Inst. Mech. Eng.*, 3C/80, 226, 1965/66.
55. Rassokhin, N. G., Shvetsvov, N. K., and Kuzmin, A. V., *Thermal Eng.*, 9, 86, 1970.
56. Forster, H. K. and Zuber, N., Dynamics of vapor bubbles and boiling heat transfer, *AIChE J.*, 1(4), 531, 1955.
57. Mayinger, F. and Ahrens, K. H., Boiling heat transfer in the transition region from bubble flow to annular flow, International Seminar on Momentum Heat Mass Transfer in Two-Phase Energy and Chemical Systems, Dubrovnik, Yugoslavia, 1978.
58. Smith, R. A., *Vaporisers: Selection Design and Operation*, Longman Scientific and Technical and John Wiley & Sons, New York, 1976.

59. Cooper, M. G., Saturation nucleate pool boiling: a simple correlation, *1st U.K. Natl. Conf. Heat Transfer, I. Chem. Eng. Symp. Series*, Pergamon Press, London, Vol. 2, No. 86, 785, 1984.
60. Mumm, J. F., Heat Transfer Boiling to Water Forced Through a Uniformly Heated Tube, ANL-5276, 1954.
61. Wright, R. M., Downflow Forced Convection Boiling of Water in Uniformly Heated Tube, UCRL Report, University of California at Berkley, 1961.
62. Jallouk, P. A., Two Phase Flow Pressure Drop and Heat Transfer Characteristics of Refrigerants in Vertical Tubes, Ph.D. Dissertation, University of Tennessee, 1974.
63. Chawla, J. M., Warmeubergang und Druckabfall in Waagrechten Rohren Bei Der Stromung Von Verdampfenden Kaltemitteln, VDI-Forschungsheft 523, 1967.
64. Steiner, D. and Schlünder, E. U., Heat transfer and pressure drop for boiling nitrogen flowing in a horizontal tube, in *Heat Transfer in Boiling*, Hahne, E. and Grigull, U., Eds., Hemisphere, Washington, DC, 1977, 263.
65. Mohr, V. and Runge, R., Forced convection boiling of neon in horizontal tubes, in *Heat Transfer in Boiling*, Hahne, E. and Grigull, U., Eds., Hemisphere, Washington, DC, 1977, 307.
66. Bandel, J. and Schlünder, E. U., Frictional pressure drop and convective heat transfer of gas liquid flow in horizontal tubes, *Proc. 5th Int. Heat Transfer Conf.*, 1974, 190.
67. Kandlikar, S. G., A General Correlation for Saturated Two-Phase Flow Boiling Heat Transfer Inside Horizontal and Vertical Tubes, Mechanical Engineering Dept., Rochester Institute of Technology, Rochester, NY, 1988.

9
Shell-and-Tube Heat Exchangers

9.1 Introduction

Shell-and-tube heat exchangers are the most versatile type of heat exchangers. They are used in the process industries, in conventional and nuclear power stations as condensers, steam generators in pressurized water reactor power plants, and feed water heaters, and they are proposed for many alternative energy applications, including ocean, thermal, and geothermal. They are also used in some air conditioning and refrigeration systems.

Shell-and-tube heat exchangers provide relatively large ratios of heat transfer area to volume and weight and they can be easily cleaned. They offer great flexibility to meet almost any service requirement. Reliable design methods and shop facilities are available for their successful design and construction. Shell-and-tube heat exchangers can be designed for high pressures relative to the environment and high-pressure differences between the fluid streams.

9.2 Basic Components

Shell-and-tube heat exchangers are built of round tubes mounted in a cylindrical shell with the tubes parallel to the shell. One fluid flows inside the tubes, while the other fluid flows across and along the axis of the exchanger. The major components of this exchanger are tubes (tube bundle), shell, front-end head, rear-end head, baffles, and tube sheets. Typical parts and connections, for illustrative purposes only, are shown in Figure 9.1.[1]

9.2.1 Shell Types

Various front and rear head types and shell types have been standardized by TEMA (Tubular Exchanger Manufacturers Association). They are identified by an alphabetic character, as shown in Figure 9.2.[1]

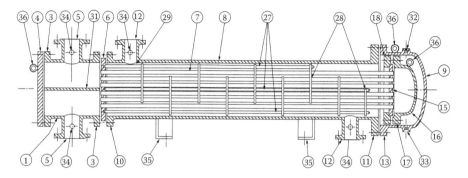

FIGURE 9.1
Constructional parts and connections: (1) stationary head—channel; (2) stationary head—bonnet; (3) stationary head flange—channel or bonnet; (4) channel cover; (5) stationary head nozzle; (6) stationary tube sheet; (7) tubes; (8) shell; (9) shell cover; (10) shell flange—rear head end; (11) shell flange—rear head end; (12) shell nozzle; (13) shell cover flange; (14) expansion joint; (15) floating tube sheet; (16) floating head cover; (17) floating head backing device; (18) floating head backing device; (19) split shear ring; (20) slip-on backing flange; (21) floating head cover—external; (22) floating tubesheet skirt; (23) packing box; (24) packing; (25) packing gland; (26) lantern ring; (27) tierods and spacers; (28) transverse baffle or support plates; (29) impingement plate; (30) longitudinal baffle; (31) pass partition; (32) vent connection; (33) drain connection; (34) instrument connection; (35) support saddle; (36) lifting lug; (37) support bracket; (38) weir; (39) liquid level connection. (Courtesy of the Tubular Exchanger Manufacturers Association.)

Figure 9.3 shows the most common shell types as condensers (v symbolizes the location of the vent).[2] The E-shell is the most common due to its low cost and simplicity. In this shell, the shell fluid enters at one end of the shell and leaves at the other end, i.e., there is one pass on the shell side. The tubes may have a single or multiple passes and are supported by transverse baffles. This shell is the most common for single-phase shell fluid applications. With a single tube pass, a nominal counterflow can be obtained.

To increase the effective temperature differences and, hence, exchanger effectiveness, a pure counterflow arrangement is desirable for a two tube-pass exchanger. This is achieved by the use of an F-shell with a longitudinal baffle and resulting in two shell passes. It is used when units in series are required, with each shell pass representing one unit. The pressure drop is much higher than the pressure drop of a comparable E-shell.

Other important shells are the J- and X-shell. In the divided flow J-shell, fluid entry is centrally located and split into two parts. The single nozzle is at the midpoint of the tubes, and two nozzles are near the tube ends. This shell is used for low pressure drop design applications such as a condenser in a vacuum, since the J-shell has approximately 1/8 the pressure drop of a comparable E-shell. When it is used for a condensing shell fluid, it will have two inlets for the vapor phase and one central outlet for the condensate.

The X-shell has a centrally located fluid entry and outlet, usually with a distributor dome. The two fluids are over the entire length of the tubes and are in crossflow arrangement. No baffles are used in this shell type.

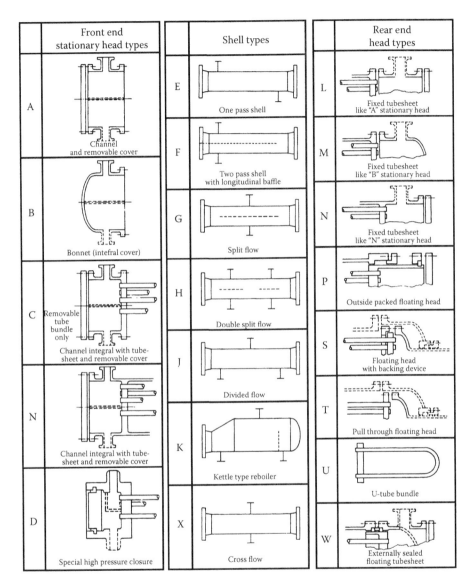

FIGURE 9.2
Standard shell types and front end and rear end head types. (Courtesy of the Tubular Exchanger Manufacturers Association.)

Consequently, the pressure drop is extremely low. It is used for vacuum condensers and low-pressure gases.

The split flow shells such as the G- and H-shell are used for specific applications. The split flow G-shell has horizontal baffles with the ends removed; the shell nozzles are 180° apart at the midpoint of tubes. The G-shell has the same pressure drop as that for the E-shell, but the LMTD factor F and, hence,

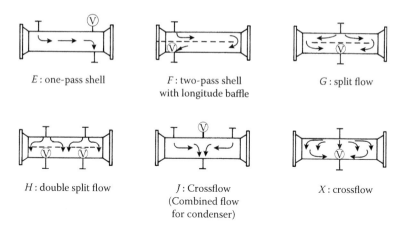

FIGURE 9.3
Schematic sketches of most common TEMA shell types. (From Butterworth, D., *Two-Phase Flow Heat Exchangers: Thermal-Hydraulic Fundamentals and Design*, Kluwer, The Netherlands, 1988 With permission.)

the exchanger effectiveness are higher for the same surface area and flow rates. The G-shell can be used for single-phase flows but is very often used as a horizontal thermosiphon reboiler. In this case, the longitudinal baffle serves to prevent flashing out of the lighter components of the shell fluids and provides increased mixing. The double split flow H-shell is similar to the G-shell but with two outlet nozzles and two horizontal baffles.

The K-shell is a kettle reboiler with the tube bundle in the bottom of the shell covering about 60% of the shell diameter. The liquid covers the tube bundle, and the vapor occupies the upper space without tubes. This shell is used when a portion of a stream needs to be vaporized, typically to a distillation column. The feed liquid enters the shell at the nozzle near the tube sheet, the nearly dry vapor exits out the top nozzle, and nonvaporized liquid overflows the end weir and exits through the right-hand nozzle. The tube bundle is commonly a U-tube configuration.

9.2.2 Tube Bundle Types

The most representative tube bundle types are shown in Figures 9.4–9.6.[3] The main design objectives here are to accommodate thermal expansion, to furnish ease of cleaning, or to provide the least expensive construction if other features are of no importance.

One design variation that allows independent expansion of tubes and shell is the U-tube configuration (Figure 9.4). Therefore, thermal expansion is unlimited. The U-tube is the least expensive construction because only one tube sheet is needed. The tube side cannot be cleaned by mechanical means because of the U-bend. Only an even number of tube passes can be accommodated. Individual tubes cannot be replaced except in the outer row.

Shell-and-Tube Heat Exchangers

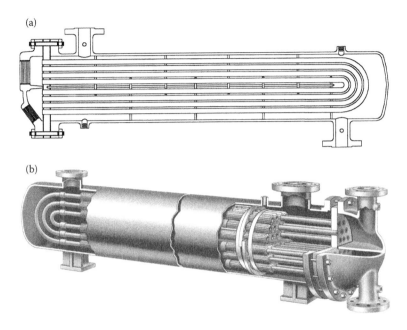

FIGURE 9.4
(a) A bare U-tube, baffled single pass shell, shell-and-tube heat exchanger (Courtesy of the Patterson-Kelley Co.); and (b) finned U-tube shell-and-tube heat exchanger. (Courtesy of Brown Fintube.)

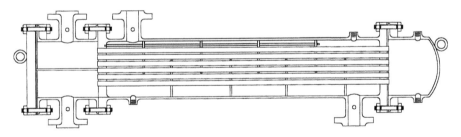

FIGURE 9.5
A two pass tube, baffled single pass shell, shell-and-tube heat exchanger designed for mechanical cleaning of the inside of the tubes. (Courtesy of the Patterson-Kelley Co.)

A fixed tube sheet configuration is shown in Figure 9.5. The shell is welded to the tube sheets and there is no access to the outside of the tube bundle for cleaning. This low-cost option has only limited thermal expansion, which can be somewhat increased by expansion bellows. Individual tubes are replaceable. Cleaning of the tubes is mechanically easy.

Several designs have been developed that permit the tube sheet to float, that is, to move with thermal expansion. A classic type of pull-through floating head design is shown in Figure 9.6. The bundle can be removed with minimum disassembly, which is important for heavy fouling units.

FIGURE 9.6
A heat exchanger similar to that of Figure 9.5 except that this one is designed with floating head to accommodate differential thermal expansion between the tubes and the shell. (Courtesy of the Patterson-Kelley Co.)

9.2.3 Tubes and Tube Passes

Only an E-shell with one tube pass and an F-shell with two tube passes result in nominal counterflow. All other multiple tube passes require a temperature profile correction (factor F), or, in some cases, simply cannot deliver the desired temperatures because of temperature cross. The next resort is to use multiple units in series.

Generally, a large number of tube passes are used to increase tube-side fluid velocity and the heat transfer coefficient (within the available pressure drop) and to minimize fouling. If, for architectural reasons, the tube-side fluid must enter and exit on the same side, an even number of tube passes is mandatory.

Tube metal is usually low carbon steel, low alloy steel, stainless steel, copper, admiralty, cupronickel, inconel, aluminum (in the form of alloys), or titanium. Other materials can also be selected for specific applications. The wall thickness of heat exchanger tubes is standardized in terms of the Birmingham Wire Gage (BWG) of the tube. Tables 9.1 and 9.2 give data on heat exchanger tubes.

Small tube diameters (8–15 mm) are preferred for greater area/volume density but are limited, for purposes of in-tube cleaning, to 20 mm (3/4 in.). Larger tube diameters are often required for condensers and boilers.

The tubes may be either bare or have low fins on the outside. Low fin tubes are used when the fluid on the outside of the tubes has a substantially lower heat transfer coefficient than the fluid on the inside of the tubes.

Tube length affects the cost and operation of heat exchangers. Basically, the longer the tube (for any given total surface), the fewer tubes are needed, fewer holes are drilled, and the shell diameter decreases, resulting in lower cost. There are, of course, several limits to this general rule, best expressed by the shell-diameter-to-tube-length ratio, which should be within limits of about 1/5 to 1/15. Maximum tube length is sometimes dictated by architectural layouts and, ultimately, by transportation to about 30 m.

TABLE 9.1

Dimensional Data for Commercial Tubing

OD of Tubing (in.)	BWG Gauge	Thickness (in.)	Internal Flow Area (in.^{2})	Sq. Ft. External Surface per Ft. Length	Sq. Ft. Internal Surface per Ft. Length	Weight per Ft. Length, Steel (lb.)	ID Tubing (in.)	OD/ID
1/4	22	0.028	0.0295	0.0655	0.0508	0.066	0.194	1.289
1/4	24	0.022	0.0333	0.0655	0.0539	0.054	0.206	1.214
1/4	26	0.018	0.0360	0.0655	0.0560	0.045	0.214	1.168
3/8	18	0.049	0.0603	0.0982	0.0725	0.171	0.277	1.354
3/8	20	0.035	0.0731	0.0982	0.0798	0.127	0.305	1.233
3/8	22	0.028	0.0799	0.0982	0.0835	0.104	0.319	1.176
3/8	24	0.022	0.0860	0.0982	0.0867	0.083	0.331	1.133
1/2	16	0.065	0.1075	0.1309	0.0969	0.302	0.370	1.351
1/2	18	0.049	0.1269	0.1309	0.1052	0.236	0.402	1.244
1/2	20	0.035	0.1452	0.1309	0.1126	0.174	0.430	1.163
1/2	22	0.028	0.1548	0.1309	0.1162	0.141	0.444	1.126
5/8	12	0.109	0.1301	0.1636	0.1066	0.602	0.407	1.536
5/8	13	0.095	0.1486	0.1636	0.1139	0.537	0.435	1.437
5/8	14	0.083	0.1655	0.1636	0.1202	0.479	0.459	1.362
5/8	15	0.072	0.1817	0.1636	0.1259	0.425	0.481	1.299
5/8	16	0.065	0.1924	0.1636	0.1296	0.388	0.49s	1.263
5/8	17	0.058	0.2035	0.1636	0.1333	0.350	0.509	1.228
5/8	18	0.049	0.2181	0.1636	0.1380	0.303	0.527	1.186
5/8	19	0.042	0.2298	0.1636	0.1416	0.262	0.541	1.155
5/8	20	0.035	0.2419	0.1636	0.1453	0.221	0.555	1.136
3/4	10	0.134	0.1825	0.1963	0.1262	0.884	0.482	1.556
3/4	11	0.120	0.2043	0.1963	0.1335	0.809	0.510	1.471
3/4	12	0.109	0.2223	0.1963	0.1393	0.748	0.532	1.410
3/4	13	0.095	0.2463	0.1963	0.1466	0.666	0.560	1.339
3/4	14	0.083	0.2679	0.1963	0.1529	0.592	0.584	1.284
3/4	15	0.072	0.2884	0.1963	0.1587	0.520	0.606	1.238
3/4	16	0.065	0.3019	0.1963	0.1623	0.476	0.620	1.210
3/4	17	0.058	0.3157	0.1963	0.1660	0.428	0.634	1.183
3/4	18	0.049	0.3339	0.1963	0.1707	0.367	0.652	1.150
3/4	20	0.035	0.3632	0.1963	0.1780	0.269	0.680	1.103
7/8	10	0.134	0.2892	0.2291	0.1589	1.061	0.607	1.441
7/8	11	0.120	0.3166	0.2291	0.1662	0.969	0.635	1.378
7/8	12	0.109	0.3390	0.2291	0.1720	0.891	0.657	1.332
7/8	13	0.095	0.3685	0.2291	0.1793	0.792	0.685	1.277
7/8	14	0.083	0.3948	0.2291	0.1856	0.704	0.709	1.234
7/8	16	0.065	0.4359	0.2291	0.1950	0.561	0.745	1.174
7/8	18	0.049	0.4742	0.2291	0.2034	0.432	0.777	1.126

(Continued)

TABLE 9.1 (CONTINUED)

Dimensional Data for Commercial Tubing

OD of Tubing (in.)	BWG Gauge	Thickness (in.)	Internal Flow Area (in.²)	Sq. Ft. External Surface per Ft. Length	Sq. Ft. Internal Surface per Ft. Length	Weight per Ft. Length, Steel (lb.)	ID Tubing (in.)	OD/ID
7/8	20	0.035	0.5090	0.2291	0.2107	0.313	0.805	1.087
1	8	0.165	0.3526	0.2618	0.1754	1.462	0.670	1.493
1	10	0.134	0.4208	0.2618	0.1916	1.237	0.732	1.366
1	11	0.120	0.4536	0.2618	0.1990	1.129	0.760	1.316
1	12	0.109	0.4803	0.2618	0.2047	1.037	0.782	1.279
1	13	0.095	0.5153	0.2618	0.2121	0.918	0.810	1.235
1	14	0.083	0.5463	0.2618	0.2183	0.813	0.834	1.199
1	15	0.072	0.5755	0.2618	0.2241	0.714	0.856	1.167
1	16	0.065	0.5945	0.2618	0.2278	0.649	0.870	1.119
1	18	0.049	0.6390	0.2618	0.2361	0.496	0.902	1.109
1	20	0.035	0.6793	0.2618	0.2435	0.360	0.930	1.075
1-1/4	7	0.180	0.6221	0.3272	0.2330	2.057	0.890	1.404
1-1/4	8	0.165	0.6648	0.3272	0.2409	1.921	0.920	1.359
1-1/4	10	0.134	0.7574	0.3272	0.2571	1.598	0.982	1.273
1-1/4	11	0.120	0.8012	0.3272	0.2644	1.448	1.010	1.238
1-1/4	12	0.109	0.8365	0.3272	0.2702	1.329	1.032	1.211
1-1/4	12	0.095	0.8825	0.3272	0.2773	1.173	1.060	1.179
1-1/4	14	0.083	0.9229	0.3272	0.2838	1.033	1.084	1.153
1-1/4	16	0.065	0.9852	0.3272	0.2932	0.823	1.120	1.116
1-1/4	18	0.049	1.042	0.3272	0.3016	0.629	1.152	1.085
1-1/4	20	0.035	1.094	0.3272	0.3089	0.456	1.180	1.059
1-1/2	10	0.134	1.192	0.3927	0.3225	1.955	1.232	1.218
1-1/2	12	0.109	1.291	0.3927	0.3356	1.618	1.282	1.170
1-1/2	14	0.083	1.398	0.3927	0.3492	1.258	1.334	1.124
1-1/2	16	0.065	1.474	0.3927	0.3587	0.996	1.370	1.095
2	11	0.120	2.433	0.5236	0.4608	2.410	1.760	1.136
2	13	0.095	2.573	0.5236	0.4739	1.934	1.810	1.105
2-1/2	9	0.148	3.815	0.6540	0.5770	3.719	2.204	1.134

Source: Courtesy of the Tubular Exchanger Manufacturers Association.

9.2.4 Tube Layout

Tube layout is characterized by the included angle between tubes, as shown in Figure 9.7. A layout of 30° results in the greatest tube density and is therefore used, unless other requirements dictate otherwise. For example, clear lanes (1/4 in. or 7 mm) are required because of external cleaning using a square 90° or 45° layout. Tube pitch, P_T, is usually chosen so that the pitch ratio

TABLE 9.2

Heat Exchanger and Condenser Tube Data

Nominal Pipe Size (in.)	Outside Diameter (in.)	Schedule Number or Weight	Wall Thickness (in.)	Inside Diameter (in.)	Surface Area Outside (ft.²/ft.)	Surface Area Inside (ft.²/ft.)	Cross-Sectional Area Metal Area (in.²)	Cross-Sectional Area Flow Area (in.²)
		40	0.113	0.824	0.275	0.216	0.333	0.533
3/4	1.05	80	0.154	0.742	0.275	0.194	0.434	0.432
		40	0.133	1.049	0.344	0.275	0.494	0.864
1	1.315	80	0.179	0.957	0.344	0.250	0.639	0.719
		40	0.140	1.38	0.434	0.361	0.668	1.496
1-1/4	1.660	80	0.191	1.278	0.434	0.334	0.881	1.283
		40	0.145	1.61	0.497	0.421	0.799	2.036
1-1/2	1.900	80	0.200	1.50	0.497	0.393	1.068	1.767
		40	0.154	2.067	0.622	0.541	1.074	3.356
2	2.375	80	0.218	1.939	0.622	0.508	1.477	2.953
		40	0.203	2.469	0.753	0.646	1.704	4.79
2-1/2	2.875	80	0.276	2.323	0.753	0.608	2.254	4.24
		40	0.216	3.068	0.916	0.803	2.228	7.30
3	3.5	80	0.300	2.900	0.916	0.759	3.106	6.60
		40	0.226	3.548	1.047	0.929	2.680	9.89
3-1/2	4.0	80	0.318	3.364	1.047	0.881	3.678	8.89
		40	0.237	4.026	1.178	1.054	3.17	12.73
4	4.5	80	0.337	3.826	1.178	1.002	4.41	11.50
		10 S	0.134	5.295	1.456	1.386	2.29	22.02
5	5.563	40	0.258	5.047	1.456	1.321	4.30	20.01
		80	0.375	4.813	1.456	1.260	6.11	18.19
		10 S	0.134	6.357	1.734	1.664	2.73	31.7
6	6.625	40	0.280	6.065	1.734	1.588	5.58	28.9
		80	0.432	5.761	1.734	1.508	8.40	26.1
		10 S	0.148	8.329	2.258	2.180	3.94	54.5
8	8.625	30	0.277	8.071	2.258	2.113	7.26	51.2
		80	0.500	7.625	2.258	1.996	12.76	45.7
		10 S	0.165	10.420	2.81	2.73	5.49	85.3
10	10.75	30	0.279	10.192	2.81	2.67	9.18	81.6
		Extra heavy	0.500	9.750	2.81	2.55	16.10	74.7
		10 S	0.180	12.390	3.34	3.24	7.11	120.6
	12.75	30	0.330	12.09	3.34	3.17	12.88	114.8
		Extra heavy	0.500	11.75	3.34	3.08	19.24	108.4
		10	0.250	13.5	3.67	3.53	10.80	143.1

(Continued)

TABLE 9.2 (CONTINUED)

Heat Exchanger and Condenser Tube Data

Nominal Pipe Size (in.)	Outside Diameter (in.)	Schedule Number or Weight	Wall Thickness (in.)	Inside Diameter (in.)	Surface Area		Cross-Sectional Area	
					Outside (ft.²/ft.)	Inside (ft.²/ft.)	Metal Area (in.²)	Flow Area (in.²)
14	14.0	Standard	0.375	13.25	3.67	3.47	16.05	137.9
		Extra heavy	0.500	13.00	3.67	3.40	21.21	132.7
		10	0.250	15.50	4.19	4.06	12.37	188.7
16	16.0	Standard	0.375	15.25	4.19	3.99	18.41	182.7
		Extra heavy	0.500	15.00	4.19	3.93	24.35	176.7
		10 S	0.188	17.624	4.71	4.61	10.52	243.9
18	18.0	Standard	0.375	17.25	4.71	4.52	20.76	233.7
		Extra heavy	0.500	17.00	4.71	4.45	27.49	227.0

Source: Courtesy of the Tubular Exchanger Manufacturers Association.

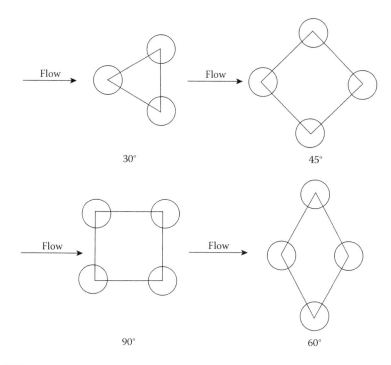

FIGURE 9.7
Tube layout angles.

P_T/d_o, is between 1.25 and 1.5. When the tubes are too close, the tube sheet becomes structurally weak. The tube layout and the tube locations have been standardized. The number of tubes (tube counts) that can be placed within a shell depends on tube layout, tube outside diameter, pitch size, number of passes, and the shell diameter. Tube counts are given in Table 9.3, which actually gives the maximum number of tubes that can be accommodated under the conditions specified.

For example, consider 1 in. OD tubes laid out on a 1.25 in. square pitch with a shell diameter of 31 in. If the tube arrangement is that for one pass (1-P), then Table 9.3 gives the maximum tubes as 406. If the heat exchanger is that for two passes (2-P), then the maximum number of tubes is 398.

9.2.5 Baffle Type and Geometry

Baffles serve two functions: most importantly to support the tubes for structural rigidity, preventing tube vibration and sagging, and second, to divert the flow across the bundle to obtain a higher heat transfer coefficient. Baffles may be classified as transverse and longitudinal types (for example, the F-shell has a longitudinal baffle). The transverse baffles may be classified as plate baffles and rod baffles. The most commonly used plate baffle types are shown in Figure 9.8 and are briefly described below.[4]

The single and double segmental baffles are most frequently used. They divert the flow most effectively across the tubes. The baffle spacing, however, must be chosen very carefully. Optimum baffle spacing is somewhere between 0.4 and 0.6 of the shell diameter and a baffle cut of 25% to 35% is usually recommended. The triple and no-tubes-in-window segmental baffles are used for low pressure drop applications, which are approximately 0.5 and 0.3 of the segmental value. No-tubes-in-window construction eliminates the tubes that are otherwise supported only by every second baffle, thus minimizing tube vibration.

Disc-and-ring (doughnut) baffles are composed of alternating outer rings and inner discs, which direct the flow radially across the tube field. The potential bundle-to-shell bypass stream is thus eliminated; there are some indications that the baffle type is very effective in pressure drop to heat transfer conversion. At present, these baffles are rarely used in the United States but are popular in Europe. Another type of plate baffle is the orifice baffle, in which shell-side fluid flows through the clearance between tube outside diameter and baffle-hole diameter.

Rod or grid baffles are formed by a grid of rod or strip supports (Figure 9.9). The flow is essentially longitudinal, resulting in very low pressure drops. Because of the close baffle spacing, the tube vibration danger is virtually eliminated. This construction can be used effectively for vertical condensers and reboilers.

TABLE 9.3

Tube-Shell Layouts (Tube Counts)

Shell ID (in.)	1-P	2-P	4-P	6-P	8-P
3/4-in. OD tubes on 1-in. triangular pitch					
8	37	30	24	24	
10	61	52	40	36	
12	92	82	76	74	70
13 ¼	109	106	86	82	74
15 ¼	151	138	122	118	110
17 ¼	203	196	178	172	166
19 ¼	262	250	226	216	210
21 ¼	316	302	278	272	260
23 ¼	384	376	352	342	328
25	470	452	422	394	382
27	559	534	488	474	464
29	630	604	556	538	508
31	745	728	678	666	640
33	856	830	774	760	732
35	970	938	882	864	848
37	1074	1044	1012	986	870
39	1206	1176	1128	1100	1078
1-in. OD tubes on 1 1/4-in. triangular pitch					
8	21	16	16	14	
10	32	32	26	24	
12	55	52	48	46	44
13 ¼	68	66	58	54	50
15 ¼	91	86	80	74	72
17 ¼	131	118	106	104	94
19 ¼	163	152	140	136	128
21 ¼	199	188	170	164	160
23 ¼	241	232	212	212	202
25	294	282	256	252	242
27	349	334	302	296	286
29	397	376	338	334	316
31	472	454	430	424	400
33	538	522	486	470	454
35	608	592	562	546	532
37	674	664	632	614	598
39	766	736	700	688	672

TABLE 9.3 (CONTINUED)
Tube-Shell Layouts (Tube Counts)

Shell ID (in.)	1-P	2-P	4-P	6-P	8-P
3/4-in. OD tubes on 1-in. square pitch					
8	32	26	20	20	
10	52	52	40	36	
12	81	76	68	68	60
13 ¼	97	90	82	76	70
15 ¼	137	124	116	108	108
17 ¼	177	166	158	150	142
19 ¼	224	220	204	192	188
3/4-in. OD tubes on 1-in. square pitch					
21 ¼	277	270	246	240	234
23 ¼	341	324	308	302	292
25	413	394	370	356	346
27	481	460	432	420	408
29	553	526	480	468	456
31	657	640	600	580	560
33	749	718	688	676	648
35	845	824	780	766	748
37	934	914	886	866	838
39	1049	1024	982	968	948
1-in. OD tubes on 1 1/4-in. square pitch					
8	21	16	14		
10	32	32	26	24	
12	48	45	40	38	36
13 ¼	61	56	52	48	44
15 ¼	81	76	68	68	64
17 ¼	112	112	96	90	82
19 ¼	138	132	128	122	116
21 ¼	177	166	158	152	148
23 ¼	213	208	192	184	184
25	260	252	238	226	222
27	300	288	278	268	260
29	341	326	300	294	286
31	406	398	380	368	358
33	465	460	432	420	414
35	522	518	488	484	472
37	596	574	562	544	532
39	665	644	624	612	600

(*Continued*)

TABLE 9.3 (CONTINUED)
Tube-Shell Layouts (Tube Counts)

Shell ID (in.)	1-P	2-P	4-P	6-P	8-P
3/4-in. OD tubes on 15/16-in. triangular pitch					
8	36	32	26	24	18
10	62	56	47	42	36
12	109	98	86	82	78
13 ¼	127	114	96	90	86
15 ¼	170	160	140	136	128
17 ¼	239	224	194	188	178
19 ¼	301	282	252	244	234
21 ¼	361	342	314	306	290
23 ¼	442	420	386	378	364
25	532	506	468	446	434
27	637	602	550	536	524
29	721	692	640	620	594
31	847	822	766	722	720
33	974	938	878	852	826
35	1102	1068	1004	988	958
37	1240	1200	1144	1104	1072
39	1377	1330	1258	1248	1212
1 1/4-in. OD tubes on 1 9/16-in. square pitch					
10	16	12	10		
12	30	24	22	16	16
13 ¼	32	30	30	22	22
15 ¼	44	40	37	35	31
17 ¼	56	53	51	48	44
19 ¼	78	73	71	64	56
21 ¼	96	90	86	82	78
23 ¼	127	112	106	102	96
25	140	135	127	123	115
27	166	160	151	146	140
29	193	188	178	174	166
31	226	220	209	202	193
33	258	252	244	238	226
35	293	287	275	268	258
37	334	322	311	304	293
39	370	362	348	342	336

Shell-and-Tube Heat Exchangers

TABLE 9.3 (CONTINUED)
Tube-Shell Layouts (Tube Counts)

Shell ID (in.)	1-P	2-P	4-P	6-P	8-P
1 1/2-in. OD tubes on 1 7/8-in. square pitch					
12	16	16	12	12	
13 ¼	22	22	16	16	
15 ¼	29	29	24	24	22
17 ¼	29	39	34	32	29
19 ¼	50	48	45	43	39
21 ¼	62	60	57	54	50
23 ¼	78	74	70	66	62
25	94	90	86	84	78
27	112	108	102	98	94
29	131	127	120	116	112
31	151	146	141	138	131
33	176	170	164	160	151
35	202	196	188	182	176
37	224	220	217	210	202
39	252	246	237	230	224
1 1/2-in. OD tubes on 1 7/8-in. triangular pitch					
12	18	14	14	12	12
13 ¼	27	22	18	16	14
15 ¼	26	34	32	30	27
17 ¼	48	44	42	38	36
19 ¼	61	58	55	51	48
21 ¼	76	78	70	66	61
23 ¼	95	91	86	80	76
25	115	110	105	98	95
27	136	131	125	118	115
29	160	154	147	141	136
31	184	177	172	165	160
33	215	206	200	190	184
35	246	238	230	220	215
37	275	268	260	252	246
39	307	299	290	284	275
1 1/4-in. OD tubes on 9/16-in. triangular pitch					
10					
10	20	18	14		
12 ¼	32	30	26	22	20
13 ¼	38	36	32	28	26

(Continued)

TABLE 9.3 (CONTINUED)

Tube-Shell Layouts (Tube Counts)

Shell ID (in.)	1-P	2-P	4-P	6-P	8-P
15 ¼	54	51	45	42	38
17 ¼	69	66	62	58	54
19 ¼	95	91	86	78	69
21 ¼	117	112	105	101	95
23 ¼	140	136	130	123	117
25	170	164	155	150	140
27	202	196	185	179	170
29	235	228	217	212	202
31	275	270	255	245	235
33	315	305	297	288	275
35	357	348	335	327	315
37	407	390	380	374	357
39	449	436	425	419	407

Source: From Kern, D. Q., *Process Heat Transfer*, McGraw-Hill, New York, 1950. With permission.

9.2.6 Allocation of Streams

A decision must be made as to which fluid will flow through the tubes and which will flow through the shell. In general, the following considerations apply:

- The more seriously fouling fluid flows through the tube, since the tube side is easier to clean, especially if mechanical cleaning is required.
- The high pressure fluid flows through the tubes. Because of their small diameter, normal thickness tubes are available to withstand higher pressures and only the tube-side channels and other connections need to be designed to withstand high pressure.
- The corrosive fluid must flow through the tubes; otherwise, both the shell and tubes will be corroded. Special alloys are used to resist corrosion, and it is much less expensive to provide special alloy tubes than to provide both special tubes and a special alloy shell.
- The stream with the lower heat transfer coefficient flows on the shell side since it is easy to design outside finned tubes. In general, it is better to put the stream with the lower mass flow rate on the shell side. Turbulent flow is obtained at lower Reynolds numbers on the shell side.

Problems arise when the above requirements are in conflict. Then, the designer must estimate trade-offs and find the most economical choices.

Shell-and-Tube Heat Exchangers

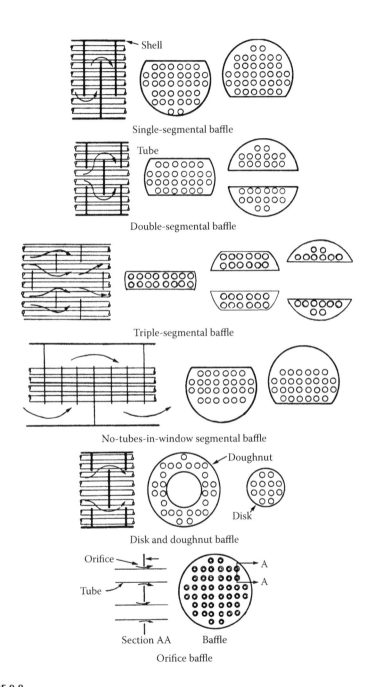

FIGURE 9.8
Plate baffle types. (Adapted from Kakaç, S., Bergles, A. E., and Mayinger, F., Eds., *Heat Exchangers: Thermal-Hydraulic Fundamentals and Design*, Taylor and Francis, Washington, D.C., 1981.)

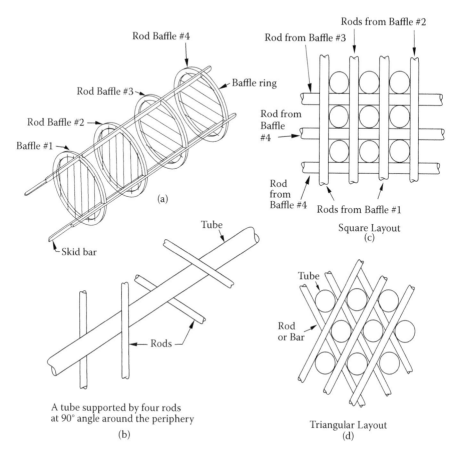

FIGURE 9.9
(a) Four rod baffles held by skid bars (no tube shown), (b) a tube supported by four rods, (c) a square layout of tubes with rods, and (d) a triangle layout of tubes with rods. (Adapted from Kakaç, S., Bergles, A. E., and Mayinger, F., Eds., *Heat Exchangers: Thermal-Hydraulic Fundamentals and Design*, Taylor and Francis, Washington, D.C., 1981.)

9.3 Basic Design Procedure of a Heat Exchanger

A selected shell-and-tube heat exchanger must satisfy the process requirements with the allowable pressure drops until the next scheduled cleaning of the plant. The basic logical structure of the process heat exchanger design procedure is shown in Figure 9.10.[5]

First, the problem must be identified as completely as possible. Not only matters like flow rates and compositions (condensation or boiling), inlet and outlet temperatures, and pressures of both streams but also the exact

Shell-and-Tube Heat Exchangers

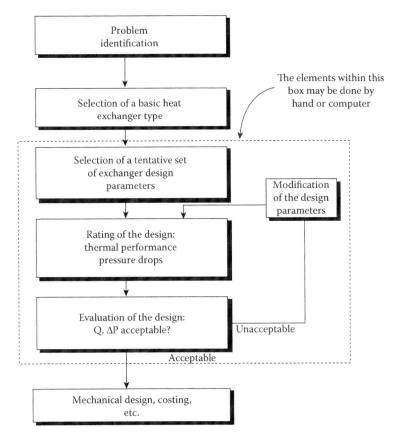

FIGURE 9.10
Basic logic structure for process heat exchanger design. (Based on Bell, K. J., *Heat Exchangers: Thermal-Hydraulic Fundamentals and Design*, Taylor and Francis, Washington, D.C., 1981.)

requirements of the process engineer and the additional information needed for the design engineer must be discussed in detail. The main duty of the process engineer is to supply all the information to the heat exchanger designer.

At this point in the design process, the basic configuration of the heat exchanger must be tentatively selected, i.e., whether it is to be U-tube, baffled single-pass shell, a tube pass, baffled single-pass shell with fixed tubes, or a shell-and-tube heat exchanger with a floating head to accommodate differential thermal expansion between the tube and the shell, if one is not unconditionally desired. The next step is to select a tentative set of exchanger design parameters. A preliminary estimate of the heat exchanger size can be made, as outlined in Section 9.3.1 below. Then, the initial design will be rated; that is, the thermal performance and the pressure drops for both streams will be calculated for this design.

9.3.1 Preliminary Estimation of Unit Size

The size of the heat exchanger can be obtained from Equation 2.36:

$$A_o = \frac{Q}{U_o \Delta T_m} = \frac{Q}{U_o F \Delta T_{lm,cf}} \tag{9.1}$$

where A_o is the outside heat transfer surface area based on the outside diameter of the tube, and Q is the heat duty of the exchanger.

First, we estimate the individual heat transfer coefficients with fouling factors. Tables, such as Tables 9.4 and 9.5, for the estimation of individual heat transfer coefficients or overall heat transfer coefficients are available in various handbooks. The estimation of heat transfer coefficients is preferable for estimating the overall heat transfer coefficient; then the designer can get a feel for the relative magnitude of the resistances.[6,7]

The overall heat transfer coefficient, U_o, based on the outside diameter of tubes, can be estimated from the estimated values of individual heat transfer coefficients, wall and fouling resistance, and the overall surface efficiency using Equation 2.17.

$$\frac{1}{U_o} = \frac{A_o}{A_i}\left(\frac{1}{\eta_i h_i} + \frac{R_{fi}}{\eta_i}\right) + A_o R_w + \frac{R_{fo}}{\eta_o} + \frac{1}{\eta_o h_o} \tag{9.2}$$

At this stage, it is useful to determine the distribution of the thermal resistances under clean and fouled conditions.

For the single tube pass, purely countercurrent heat exchanger, $F = 1.00$. For a preliminary design shell with any even number of tube side passes, F may be estimated as 0.9. Heat load can be estimated from the heat balance as

$$Q = (\dot{m} c_p)_c (T_{c_2} - T_{c_1}) = (\dot{m} c_p)_h (T_{h_1} - T_{h_2}) \tag{9.3}$$

If one stream changes phase,

$$Q = \dot{m} h_{fg} \tag{9.4}$$

where $\dot{m}$ is the mass of the stream changing phase per unit time and h_{fg} is the latent heat of the phase change.

We need to calculate the LMTD for countercurrent flow from the four given inlet/outlet temperatures. If three temperatures are known, the fourth one can be found from the heat balance:

$$\Delta T_{lm,cf} = \frac{(T_{h_1} - T_{c_2}) - (T_{h_2} - T_{c_1})}{\ln \frac{T_{h_1} - T_{c_2}}{T_{h_2} - T_{c_1}}} \tag{9.5}$$

TABLE 9.4

Typical Film Heat Transfer Coefficients for Shell-and-Tube Heat Exchangers

	Fluid Condition	W/(m² · K)
Sensible heat transfer		
Water	Liquid	5,000–7,500
Ammonia	Liquid	6,000–8,000
Light organics	Liquid	1,500–2,000
Medium organics	Liquid	750–1,500
Heavy organics	Liquid	
	Heating	250–750
	Cooling	150–400
Very heavy organics	Liquid	
	Heating	100–300
	Cooling	60–150
Gas	1–2 bar abs	80–125
Gas	10 bar abs	250–400
Gas	100 bar abs	500–800
Condensing heat transfer		
Steam, ammonia	No noncondensable	8,000–12,000
Light organics	Pure component, 0.1 bar abs, no noncondensable	2,000–5,000
Light organics	0.1 bar, 4% noncondensable	750–1,000
Medium organics	Pure or narrow condensing range, 1 bar abs	1,500–4,000
Heavy organics	Narrow condensing range, 1 bar abs	600–2,000
Light multicomponent mixture, all condensable	Medium condensing range, 1 bar abs	1,000–2,500
Medium multicomponent mixture, all condensable	Medium condensing range, 1 bar abs	600–1,500
Heavy multicomponent mixture, all condensable	Medium condensing range, 1 bar abs	300–600
Vaporizing heat transfer		
Water	Pressure < 5 bar abs, $\Delta T = 25$ K	5,000–10,000
Water	Pressure 5–100 bar abs, $\Delta T = 20$ K	4,000–15,000
Ammonia	Pressure < 30 bar abs, $\Delta T = 20$ K	3,000–5,000
Light organics	Pure component, pressure < 30 bar abs, $\Delta T = 20$ K	2,000–4,000
Light organics	Narrow boiling range, pressure 20–150 bar abs, $\Delta T = 15$–20 K	750–3,000
Medium organics	Narrow boiling range, pressure < 20 bar abs, $\Delta T_{max} = 15$ K	600–2,500
Heavy organics	Narrow boiling range, pressure < 20 bar abs, $\Delta T_{max} = 15$ K	400–1,500

TABLE 9.5

Approximate Overall Heat Transfer Coefficients for Preliminary Analysis

Fluids	U (W/m² · K)
Water to water	1300–2500
Ammonia to water	1000–2500
Gases to water	10–250
Water to compressed air	50–170
Water to lubricating oil	110–340
Light organics ($\mu < 5 \times 10^{-4}$ Ns/m²) to water	370–750
Medium organics ($5 \times 10^{-4} < \mu < 10 \times 10^{-4}$ Ns/m²) to water	240–650
Heavy organics ($\mu > 10 \times 10^{-4}$ Ns/m²) to lubricating oil	25–400
Steam to water	2200–3500
Steam to ammonia	1000–3400
Water to condensing ammonia	850–1500
Water to boiling Freon-12	280–1000
Steam to gases	25–240
Steam to light organics	490–1000
Steam to medium organics	250–500
Steam to heavy organics	30–300
Light organics to light organics	200–350
Medium organics to medium organics	100–300
Heavy organics to heavy organics	50–200
Light organics to heavy organics	50–200
Heavy organics to light organics	150–300
Crude oil to gas oil	130–320
Plate heat exchangers: water to water	3000–4000
Evaporators: steam/water	1500–6000
Evaporators: steam/other fluids	300–2000
Evaporators of refrigeration	300–1000
Condensers: steam/water	1000–4000
Condensers: steam/other fluids	300–1000
Gas boiler	10–50
Oil bath for heating	30–550

The problem now is to convert the area calculated from Equation 9.1 into reasonable dimensions of the first trial. The objective is to find the right number of tubes of diameter d_o and the shell diameter, D_s, to accommodate the number of tubes, N_t, with given tube length, L:

$$A_o = \pi d_o N_t L \tag{9.6}$$

One can find the shell diameter, D_s, which would contain the right number of tubes, N_t, of diameter d_o.

Shell-and-Tube Heat Exchangers

The total number of the tubes, N_t, can be predicted in fair approximation as a function of the shell diameter by taking the shell circle and dividing it by the projected area of the tube layout (Figure 9.7) pertaining to a single tube A_1:

$$N_t = (CTP)\frac{\pi D_s^2}{4A_1} \tag{9.7}$$

where CTP is the tube count calculation constant which accounts for the incomplete coverage of the shell diameter by the tubes due to necessary clearances between the shell and the outer tube circle and tube omissions due to tubes' pass lanes for multitube pass design.

Based on a fixed tube sheet, the following values are suggested:

one tube pass: $CTP = 0.93$
two tube passes: $CTP = 0.90$
three tube passes: $CTP = 0.85$

$$A_1 = (CL)P_T^2 \tag{9.8}$$

where CL is the tube layout constant:

$CL = 1.0$ for $90°$ and $45°$
$CL = 0.87$ for $30°$ and $60°$

Equation 9.7 can be written as

$$N_t = 0.785\left(\frac{CTP}{CL}\right)\frac{D_s^2}{(PR)^2 d_o^2} \tag{9.9}$$

where PR is the tube pitch ratio (P_T/d_o).

Substituting N_t from Equation 9.6 into Equation 9.9, an expression for the shell diameter in terms of main constructional diameters can be obtained as[7]

$$D_s = 0.637\sqrt{\frac{CL}{CTP}}\left[\frac{A_o(PR)^2 d_o}{L}\right]^{1/2} \tag{9.10}$$

Example 9.1

A heat exchanger is to be designed to heat raw water by the use of condensed water at 67°C and 0.2 bar, which will flow in the shell side with a mass flow rate of 50,000 kg/hr. The heat will be transferred to 30,000 kg/hr of city water coming from a supply at 17°C ($c_p = 4184$ J/kg · K). A single shell and a single tube pass is preferable. A fouling resistance of

0.000176 m² · K/W is suggested and the surface over design should not be over 35%. A maximum coolant velocity of 1.5 m/s is suggested to prevent erosion. A maximum tube length of 5 m is required because of space limitations. The tube material is carbon steel ($k = 60$ W/m · K). Raw water will flow inside of 3/4in. straight tubes (19 mm OD with 16 mm ID). Tubes are laid out on a square pitch with a pitch ratio of 1.25. The baffle spacing is approximated by 0.6 of shell diameter, and the baffle cut is set to 25%. The permissible maximum pressure drop on the shell side is 5.0 psi. The water outlet temperature should not be less than 40°C. Perform the preliminary analysis.

Solution

Preliminary Analysis—The cold water outlet temperature of at least 40°C determines the exchanger configuration to be considered. Heat duty can be calculated from the fully specified cold stream:

$$Q = (\dot{m}c_p)_c (T_{c_2} - T_{c_1})$$

$$Q = \frac{30{,}000}{3600} \times 4179(40 - 17) = 801 \text{ kW}$$

The hot water outlet temperature becomes

$$T_{h_2} = T_{h_1} - \frac{Q}{(\dot{m}c_p)_h} = 67 - \frac{801 \times 10^3}{\frac{50{,}000}{3600} \times 4184} = 53.2°C$$

First, we have to estimate the individual heat transfer coefficients from Table 9.4. We can assume the shell-side heat transfer coefficient and the tube-side heat transfer coefficient as 5000 W/m² · K and 4000 W/m² · K, respectively. Assuming bare tubes, one can estimate the overall heat transfer coefficient from Equation 9.2 as

$$\frac{1}{U_f} = \frac{1}{h_o} + \frac{r_o}{r_i}\frac{1}{h_i} + R_{ft} + r_o \frac{\ln(r_o/r_i)}{k}$$

$$U_f = \left[\frac{1}{5{,}000} + \frac{19}{16}\frac{1}{4{,}000} + 0.000176 + \frac{0.019}{2}\frac{\ln(19/16)}{60} \right]^{-1} = 1428.4 \text{ W/m}^2 \cdot \text{K}$$

and

$$\frac{1}{U_c} = \frac{1}{h_o} + \frac{r_o}{r_i}\frac{1}{h_i} + r_o\frac{\ln(r_o/r_i)}{k}$$

$$U_c = \left[\frac{1}{5{,}000} + \frac{19}{16}\frac{1}{4{,}000} + \frac{0.019}{2}\frac{\ln(19/16)}{60} \right]^{-1} = 1908.09 \text{ W/m}^2 \cdot \text{K}$$

Shell-and-Tube Heat Exchangers

We need to calculate ΔT_m from the four inlet and outlet temperatures.

$$\Delta T_{lm,cf} = \frac{\Delta T_1 - \Delta T_2}{\ln(\Delta T_1/\Delta T_2)} = \frac{27 - 36.2}{\ln\left(\dfrac{27}{36.2}\right)} = 31.4°C$$

Assuming $F = 0.90$, then

$$\Delta T_m = 0.90 \, \Delta T_{lm,cf} = 0.90 \times 31.4 \approx 28°C$$

Next we can estimate the required areas A_f and A_c:

$$A_f = \frac{Q}{U_f \Delta T_m} = \frac{801.93 \times 10^3}{1428.40 \times 28} = 20.05 \text{ m}^2$$

$$A_c = \frac{Q}{U_c \Delta T_m} = \frac{801.93 \times 10^3}{1908.09 \times 28} = 15.01 \text{ m}^2$$

The surface over design is $A_f/A_c = 1.34$ (34%), which is acceptable.

Shell diameter can be calculated from Equation 9.10, where $d_o = 0.019$ m, $PR = 1.25$, $CTP = 0.93$, $CL = 1.0$, and let us assume $L = 3$ m:

$$D_s = 0.637 \sqrt{\frac{CL}{CTP}} \left[\frac{A_o (PR)^2 d_o}{L} \right]^{1/2}$$

$$= 0.637 \sqrt{\frac{1.0}{0.93}} \left[\frac{20.05 \times (1.25)^2 \times 0.019}{3} \right]^{1/2}$$

$$= 0.294 \text{ m (round off to 0.30 m)}$$

The number of tubes can be calculated from Equation 9.9 as

$$N_t = 0.785 \left(\frac{CTP}{CL} \right) \frac{D_s^2}{(PR)^2 d_o^2}$$

$$N_t = \frac{0.785 \times 0.93 \times (0.3)^2}{1.0 \times (1.25)^2 \times (0.019)^2} = 116.48 \approx 117$$

Baffle spacing can be taken as 0.4 to 0.6 of the shell diameter, so let assume $0.6\,D_s$. This will give us $B = 0.18$ m, which can be rounded off to 0.2 m. Therefore, the preliminary estimation of the unit size is

Shell diameter	$D_S = 0.3$ m
Tube length	$L = 3$ m
Tube diameter	OD = 19 mm, ID = 16 mm
Baffle spacing	$B = 0.20$ m, baffle cut 25%
Pitch ratio	$P_T/d_o = 1.25$, square pitch.

Then, a rating analysis must be performed, which is presented in the following sections.

9.3.2 Rating of the Preliminary Design

After determining the tentative selected and calculated constructional design parameters, that is, after a heat exchanger is available with process specifications, then this data can be used as inputs into a computer rating program or for manual calculations. The rating program is shown schematically in Figure 9.11.[5]

In some cases, a heat exchanger may be available, and the performance analysis for this available heat exchanger needs to be done. In that case, a preliminary design analysis is not needed. If the calculation shows that the required amount of heat cannot be transferred to satisfy specific outlet temperatures or if one or both allowable pressure drops are exceeded, it is necessary to select a different heat exchanger and rerate it (see Example 9.2).

For the rating process, all the preliminary geometrical calculations must be carried out as the input into the heat transfer and the pressure drop correlations. When the heat exchanger is available, then all the geometrical parameters are also known. In the rating process, the other two basic calculations are the calculations of heat transfer coefficients and the pressure drops for each stream specified. If the length of the heat exchanger is fixed, then the rating program calculates the outlet temperatures of both streams. If the heat duty (heat load) is fixed, then the result from the rating program

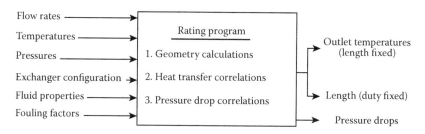

FIGURE 9.11
The rating program. (Based on Bell, K. J., *Heat Exchangers: Thermal-Hydraulic Fundamentals and Design*, Taylor and Francis, Washington, D.C., 1981.)

is the length of the heat exchanger required to satisfy the fixed heat duty of the exchanger. In both cases, the pressure drops for both streams in the heat exchanger are calculated.

The correlations for heat transfer and pressure drop are needed in quantitative forms, which may be available from theoretical analyses or from experimental studies. The correlations for tube-side heat transfer coefficient and pressure drop calculations for single-phase flow are given in Chapters 3 and 4, respectively. The correlations for two-phase flow heat transfer are discussed in Chapter 8. The most involved correlations are those for the heat transfer and the pressure drop on the shell-side stream, which will be discussed in the following sections.

If the output of the rating analysis is not acceptable, a new geometrical modification must be made. If, for example, the heat exchanger cannot deliver the amount of heat that should be transferred, then one should find a way to increase the heat transfer coefficient or to increase the area of the exchanger. To increase the tube-side heat transfer coefficient, one can increase the tube-side velocity, so one should increase the number of tube passes. One can decrease baffle spacing or decrease the baffle cut to increase the shell-side heat transfer coefficient. To increase the area, one can increase the length of the heat exchanger or increase the shell diameter; one can also go to multiple shells in series.

If the pressure drop on the tube side is greater than the allowable pressure drop, then the number of tube passes can be decreased or the tube diameter can be increased which can decrease the tube length and increase the shell diameter and the number of tubes.

If the shell-side pressure drop is greater than the allowable pressure drop, then baffle spacing, tube pitch, and baffle cut can be increased, or one can change the type of baffles.

9.4 Shell-Side Heat Transfer and Pressure Drop

Predicting the overall heat transfer coefficient requires calculation of the tube-side and shell-side heat transfer coefficients from available correlations. For tubes in a shell-and-tube exchanger, the correlations given in Chapters 3 and 8 or from the available literature can be applied depending on the flow conditions, as was done for the double-pipe exchanger. The shell-side analysis described below is called the Kern method.[8]

9.4.1 Shell-Side Heat Transfer Coefficient

The heat transfer coefficient outside the tube bundles is referred to as the shell-side heat transfer coefficient. When the tube bundle employs baffles, the heat transfer coefficient is higher than the coefficient for undisturbed flow conditions along the axis of tubes without baffles. If there are no baffles, the flow will be along the heat exchanger inside the shell. Then, the heat transfer coefficient can be based on the equivalent diameter, D_e, as is done in a double-pipe heat exchanger, and the correlations of Chapter 3 are applicable. For baffled heat exchangers, the higher heat transfer coefficients result from the increased turbulence. In a baffled shell-and-tube heat exchanger, the velocity of fluid fluctuates because of the constricted area between adjacent tubes across the bundle. The correlations obtained for flow in tubes are not applicable for flow over tube bundles with segmental baffles.

McAdams[3] suggested the following correlations for the shell-side heat transfer coefficient:

$$\frac{h_o D_e}{k} = 0.36 \left(\frac{D_e G_s}{\mu}\right)^{0.55} \left(\frac{c_p \mu}{k}\right)^{1/3} \left(\frac{\mu_b}{\mu_w}\right)^{0.14}$$

$$\text{for } 2 \times 10^3 < Re_s = \frac{G_s D_e}{\mu} < 1 \times 10^6 \qquad (9.11)$$

where h_o is the shell-side heat transfer coefficient, D_e is the equivalent diameter on the shell side, and G_s is the shell-side mass velocity.

The properties are evaluated at the average fluid temperature in the shell. In the above correlation, the equivalent diameter, D_e, is calculated along (instead of across) the long axes of the shell. The equivalent diameter of the shell is taken as four times the net flow area as layout on the tube sheet (for any pitch layout) divided by the wetted perimeter:

$$D_e = \frac{4 \times \text{free-flow area}}{\text{wetted perimeter}} \qquad (9.12)$$

For example, Figure 9.12 shows both a square and triangular-pitch layout. For each pitch layout, Equation 9.12 applies. For the square pitch, the perimeter is the circumference of a circle and the area is a square of pitch size (P_T^2) minus the area of a circle (the hatched section).

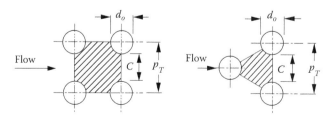

FIGURE 9.12
Square and triangular pitch-tube layouts.

Thus, one can write the following for the square pitch:

$$D_e = \frac{4\left(P_T^2 - \pi d_o^2/4\right)}{\pi d_o} \qquad (9.13)$$

and for the triangular pitch:

$$D_e = \frac{4\left(\frac{P_T^2 \sqrt{3}}{4} - \frac{\pi d_o^2}{8}\right)}{\pi d_o/2} \qquad (9.14)$$

where d_o is the tube outside diameter.

There is no free-flow area on the shell side by which the shell-side mass velocity, G_s, can be calculated. For this reason, fictitious values of G_s can be defined based on the bundle crossflow area at the hypothetical tube row possessing the maximum flow area corresponding to the center of the shell.

The variables that affect the velocity are the shell diameter, D_s, the clearance, C, between adjacent tubes, the pitch size, P_T, and the baffle spacing, B. The width of the flow area at the tubes located at center of the shell is $(D_s/P_T)\,C$ and the length of the flow area is taken as the baffle spacing, B. Therefore, the bundle crossflow area, A_s, at the center of the shell is

$$A_s = \frac{D_s C B}{P_T} \qquad (9.15)$$

where D_s is the inside diameter of the shell. Then, the shell-side mass velocity is found with

$$G_s = \frac{\dot{m}}{A_s} \qquad (9.16)$$

9.4.2 Shell-Side Pressure Drop

The shell-side pressure drop depends on the number of tubes the fluid passes through the tube bundle between the baffles as well as the length of each crossing. If the length of a bundle is divided by four baffles, for example, all the fluid travels across the bundle five times.

A correlation has been obtained using the product of distance across the bundle, taken as the inside diameter of the shell, D_s, and the number of times the bundle is crossed. The equivalent diameter used for calculating the pressure drop is the same as for heat transfer. The pressure drop on the shell side is calculated by the following expression.[8]

$$\Delta p_s = \frac{f G_s^2 (N_b + 1) \cdot D_s}{2\rho D_e \phi_s} \qquad (9.17)$$

where $\phi_s = (\mu_b/\mu_w)^{0.14}$, $N_b = L/B - 1$ is the number of baffles, and $(N_b + 1)$ is the number of times the shell fluid passes the tube bundle. The friction factor, f, for the shell is calculated from

$$f = \exp(0.576 - 0.19 \ln Re_s) \tag{9.18}$$

where

$$400 < Re_s = \frac{G_s D_e}{\mu} \leq 1 \times 10^6$$

The correlation has been tested based on data obtained on actual exchangers. The friction coefficient also takes entrance and exit losses into account.

9.4.3 Tube-Side Pressure Drop

The tube-side pressure drop can be calculated by knowing the number of tube passes, N_p, and the length, L, of the heat exchanger. The pressure drop for the tube-side fluid is given by Equation 4.17:

$$\Delta p_t = 4f \frac{LN_p}{d_i} \rho \frac{u_m^2}{2} \tag{9.19}$$

or

$$\Delta p_t = 4f \frac{LN_p}{d_i} \frac{G_t^2}{2\rho} \tag{9.20}$$

The change of direction in the passes introduces an additional pressure drop, Δp_r, due to sudden expansions and contractions that the tube fluid experiences during a return, which is accounted for allowing four velocity heads per pass[8]:

$$\Delta p_r = 4N_p \frac{\rho u_m^2}{2} \tag{9.21}$$

The total pressure drop of the tube side becomes

$$\Delta p_{total} = \left(4f \frac{LN_p}{d_i} + 4N_p\right) \frac{\rho u_m^2}{2} \tag{9.22}$$

Example 9.2

Example 9.1 involved estimating the size of the unit. Using the estimated D_s and N_t, a selection must be made from Table 9.3. The selection

Shell-and-Tube Heat Exchangers

depends on the closest number of tubes in the table that exceed N_t of the preliminary analysis. By selecting a shell diameter of 15.25 in., according to TEMA standards from Table 9.3 with 124 tubes for a 2-P shell-and-tube heat exchanger, rerate this heat exchanger for the given process specifications by using the Kern method. Note that heat duty is fixed, so the heat exchanger length and pressure drops for both streams are to be calculated.

Solution

The selected shell-and-tube heat exchanger for this purpose has the following geometrical parameters:

Shell internal diameter	$D_S = 15\ 1/4$ in. $(=0.39\ \text{m})$
Number of tubes	$N_t = 124$
Tube diameter	OD = 19 mm, ID = 16 mm
Tube material	$k = 60\ \text{W/m}^2 \cdot \text{K}$
Baffle spacing	$B = 0.2$ m, baffle cut 25%
Pitch size	$P_T = 0.0254\ \text{m}$
Number of tube passes	$N_p = 2$

Heat duty is fixed with the assumed outlet temperature of 40°C. The properties of the shell-side fluid can be taken at

$$T_b = \frac{67 + 53.2}{2} = 60° \text{C} \ (=333\ \text{K})$$

from Appendix B (Table B.2):

$$\rho = 983.2\ \text{kg/m}^3$$
$$c_p = 4184\ \text{J/kg} \cdot \text{K}$$
$$\mu = 4.67 \times 10^{-4}\ \text{N} \cdot \text{s/m}^2$$
$$k = 0.652\ \text{W/m} \cdot \text{K}$$
$$Pr = 3.00$$

Properties of the tube-side water at 28.5°C (≈ 300 K) from Appendix B (Table B.2), are

$$\rho = 996.8\ \text{kg/m}^3$$
$$c_p = 4179\ \text{J/kg} \cdot \text{K}$$
$$\mu = 8.2 \times 10^{-4}\ \text{N} \cdot \text{s/m}^2$$
$$k = 0.610\ \text{W/m} \cdot \text{K}$$
$$Pr = 5.65$$

The specifications are
Maximum tube length $L_{max} = 5$ m.
Maximum pressure drop on the shell side, $\Delta P_s = 5$ psi:

$$\frac{h_o D_e}{k} = 0.36 \left(\frac{D_e G_s}{\mu}\right)^{0.55} \left(\frac{c_p \mu}{k}\right)^{1/3} \left(\frac{\mu_b}{\mu_w}\right)^{0.14}$$

for $2 \times 10^3 < Re_s < 10^6$.
For a square-pitch tube layout,

$$D_e = \frac{4(P_T^2 - \pi d_o^2/4)}{\pi d_o} = \frac{4\left[(0.0254)^2 - \pi(0.019^2/4)\right]}{\pi(0.019)} = 0.0242 \, \text{m}$$

$$C = P_T - d_o = 0.0254 - 0.019 = 0.0064 \, \text{m}$$

$$A_s = \frac{D_s C B}{P_T} = \frac{(0.39 \, \text{m})(0.0064 \, \text{m})(0.2 \, \text{m})}{0.0254 \, \text{m}} = 0.0197 \, \text{m}^2$$

$$G_s = \frac{\dot{m}}{A_s} = \frac{50,000 \, \text{kg/hr}}{0.0197 \, \text{m}^2}\left(\frac{1 \, \text{hr}}{3600 \, \text{s}}\right) = 705 \, \text{kg/s m}^2$$

$$Re_s = \frac{G_s D_e}{\mu} = \frac{(705 \, \text{kg/s} \cdot \text{m}^2)(0.0242 \, \text{m})}{4.67 \times 10^{-4} \, \text{N} \cdot \text{s/m}^2} = 36,534$$

$$T_w = \frac{1}{2}\left(\frac{T_{c_1} + T_{c_2}}{2} + \frac{T_{h_1} + T_{h_2}}{2}\right) = \frac{1}{2}\left(\frac{17 + 40}{2} + \frac{67 + 53}{2}\right) = 44.25°\text{C}$$

and $T_{c1} = 17°\text{C}$, $T_{h1} = 67°\text{C}$, $T_{c2} = 40°\text{C}$, and $T_{h2} = 53°\text{C}$.
From Table (B.2) in Appendix B, at the approximate wall temperature of 317 K,

$$\mu_w = 6.04 \times 10^{-4} \, \text{N} \cdot \text{s/m}^2$$

$$\frac{h_o D_e}{k} = 0.36\left(\frac{(0.0242 \, \text{m})(705 \, \text{kg/s} \cdot \text{m}^2)}{4.67 \times 10^{-4} \, \text{N} \cdot \text{s}/\text{m}^2}\right)^{0.55} \left(\frac{(4184 \, \text{J/kg} \cdot \text{K})(4.67 \times 10^{-4})}{0.652 \, \text{W/m} \cdot \text{K}}\right)^{1/3}$$

$$\left(\frac{4.67 \times 10^{-4}}{6.04 \times 10^{-4}}\right)^{0.14}$$

$$= 161.88$$

$$h_o = \frac{(161.88)(0.652)}{0.0242} = 4361.3 \, \text{W/m}^2 \cdot \text{K}$$

For the tube-side heat transfer coefficient,

$$A_{tp} = \frac{\pi d_i^2}{4} \cdot \frac{N_t}{2} = \frac{\pi (0.016 \text{ m})^2}{4} \times \frac{124}{2} = 1.246 \times 10^{-2} \text{ m}^2$$

$$u_m = \frac{\dot{m}_t}{\rho_t A_{tp}} = \frac{30{,}000 \text{ kg/hr}}{(996.8 \text{ kg/m}^3)(1.246 \times 10^{-2} \text{ m}^2)} \left(\frac{1 \text{ hr}}{3600 \text{ s}}\right) = 0.67 \text{ m/s}$$

$$Re = \frac{\rho u_m d_i}{\mu} = \frac{(996.8 \text{ kg/m}^3)(0.67 \text{ m/s})(0.016 \text{ m})}{8.2 \times 10^{-4} \text{ N} \cdot \text{s/m}^2} = 13{,}049.9$$

Since $Re > 10^4$, the flow is turbulent.
Using Gnielinski's correlation,

$$Nu_b = \frac{(f/2)(Re - 1000) Pr}{1 + 12.7 (f/2)^{1/2} (Pr^{2/3} - 1)}$$

$$f = (1.58 \ln Re - 3.28)^{-2} = [1.58 \ln(13{,}049.9) - 3.28]^{-2}$$

$$f = 0.00731$$

$$Nu_b = \frac{(0.0037)(13049.9 - 1000)(5.65)}{1 + 12.7(0.0037)^{1/2}(5.65^{2/3} - 1)} = 94.06$$

$$h_i = \frac{Nu_b k}{d_i} = \frac{(94.06)(0.61 \text{ W/m} \cdot \text{K})}{0.016 \text{ m}} = 3586.1 \text{ W/m}^2 \cdot \text{K}$$

To calculate the overall heat transfer coefficient,

$$U_f = \frac{1}{\dfrac{d_o}{d_i h_i} + \dfrac{d_o R_{fi}}{d_i} + \dfrac{d_o \ln(d_o/d_i)}{2k} + R_{fo} + \dfrac{1}{h_o}}$$

$$= \frac{1}{\dfrac{0.019}{(0.016)(3586.1)} + \dfrac{(0.019)(0.000176)}{0.016} + \dfrac{(0.019)\ln(0.019/0.016)}{2(60)} + 0.000176 + \dfrac{1}{4361.3}}$$

$$= 1028.2 \text{ W/m}^2 \text{ K}$$

$$U_c = \frac{1}{\dfrac{d_o}{d_i h_i} + \dfrac{d_o \ln(d_o/d_i)}{2k} + \dfrac{1}{h_o}}$$

$$= \frac{1}{\dfrac{0.019}{(0.016)(3586.1)} + \dfrac{(0.019)\ln(0.019/0.016)}{2(60)} + \dfrac{1}{4361.3}}$$

$$= 1701.7 \text{ W/m}^2 \text{ K}$$

To determine the shell-side pressure drop,

$$\Delta p_s = \frac{f G_s^2 (N_b + 1) D_s}{2 \rho D_e \phi_s}$$

$$f = \exp(0.576 - 0.19 \ln Re_s)$$
$$= \exp[0.576 - 0.19 \ln(36,534)] = 0.242$$

$$\phi_s = \left(\frac{\mu_b}{\mu_w}\right)^{0.14} = \left(\frac{4.67 \times 10^{-4}}{6.04 \times 10^{-4}}\right)^{0.14} = 0.9646$$

$$N_b = \frac{L}{B} - 1 = \frac{5}{0.2} - 1 = 24$$

$$\Delta p_s = \frac{(0.242)(705^2)(24+1)(0.39)}{2(983.2)(0.0242)(0.9646)} = 25548 \text{ Pa} = 3.7 \text{ psi}$$

Since 3.7 < 5.0, the shell-side pressure drop is acceptable.
For the tube length,

$$Q = (\dot{m} c_p)_c (T_{c_2} - T_{c_1}) = (8.33 \text{ kg/s})(4184 \text{ J/kg K})(40 - 17) \text{K} = 801.6 \text{ kW}$$

$$A_{of} = \frac{Q}{U_{of} \Delta T_m}$$

$$\Delta T_{lm,cf} = \frac{\Delta T_1 - \Delta T_2}{\ln(\Delta T_1 / \Delta T_2)} = \frac{(67-40) - (53-17)}{\ln(67-40)/(53-17)} = 31.3 \text{ K}$$

The F factor can be estimated as 0.95 from Figure 2.7; so,

$$\Delta T_m = F \Delta T_{lm,cf} = (0.95)(31.3) = 29.8 \text{ K}$$

$$A_{of} = \frac{801,600 \text{ W}}{(1028.2 \text{ W/m}^2\text{K})(29.8 \text{ K})} = 26.2 \text{ m}^2$$

$$A_o = \pi d_o L N_t$$

$$L = \frac{A_o}{\pi d_o N_t} = \frac{26.2 \text{ m}^2}{\pi (0.019 \text{ m})(124)} = 3.54 \text{ m}$$

This is rounded-off to 4 m. Since 4 m < 5 m, the length of the heat exchanger is acceptable.
To calculate the tube-side pressure drop,

$$\Delta p_t = \left(4 f \frac{L N_p}{d_i} + 4 N_p\right) \frac{\rho u_m^2}{2}$$

$$\Delta p_t = \left(4 \times 0.000731 \times \frac{4 \times 2}{0.016} + 4 \times 2\right) \times \frac{996.8 \times (0.67)^2}{2} = 2116.95 \text{ Pa} = 0.307 \text{ psi}$$

The heat exchanger satisfies pressure drop requirement; however, the OS for this design is 66%, which is unacceptable. The design can be improved/optimized with several iterations such as by choosing less fouling liquids or by improving heat transfer coefficient.

9.4.4 Bell–Delaware Method

The calculation given in Section 9.4 for shell-side heat transfer and pressure drop analysis (Kern method) is a simplified method. The shell-side analysis is not as straightforward as the tube-side analysis because the shell flow is complex, combining crossflow and baffle window flow as well as baffle-shell and bundle-shell bypass streams and complex flow patterns, as shown in Figures 9.13 and 9.14.[5,6,9–11]

As indicated in Figure 9.13, five different streams are identified. The A-stream is leaking through the clearance between the tubes and the baffle. The B-stream is the main stream across the bundle; this is the stream desired on the shell-side of the exchanger. The C-stream is the bundle bypass stream flowing around the tube bundle between the outermost tubes in the bundle and inside the shell. The E-stream is the baffle-to-shell leakage stream flowing through the clearance between the baffles and the shell

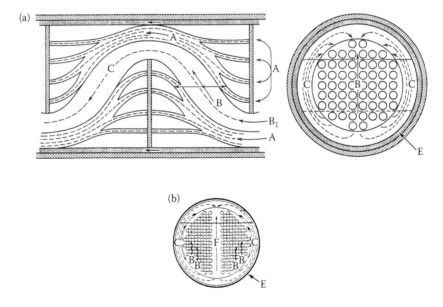

FIGURE 9.13
(a) Diagram indicating leaking paths for flow bypassing the tube matrix, through both the baffle clearances between the tube matrix and shell. (b) F-stream for a two tube pass exchanger.[2,8] (Adapted from Butterworth, D., *Two-Phase Flow Heat Exchangers: ThermalHydraulic Fundamentals and Design*, Kluwer, The Netherlands, 1988. With permission; Kern, D. Q., *Process Heat Transfer*, McGraw-Hill, New York, 1950.)

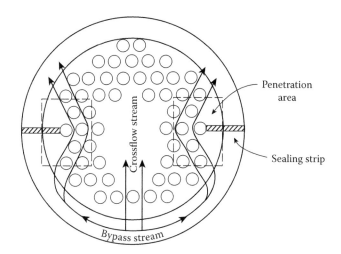

FIGURE 9.14
Radial baffles designed to reduce the amount of bypass flow through the gap between the side of the tube matrix and the shell.[2,8] (Adapted from Spalding, D. B. and Taborek, J., *Heat Exchanger Design Handbook*, Section 3.3, Hemisphere, Washington, D.C., 1983. With permission.)

inside diameter. Finally, the F-stream flows through any channel within the tube bundle caused by the provision of pass dividers in the exchanger header for multiple tube passes. Figure 9.13 is an idealized representation of the streams. The streams shown can mix and interact with one another, and a more complete mathematical analysis of the shell-side flow would take this into account.[9]

The Bell–Delaware method takes into account the effects of various leakage and bypass streams on the shell-side heat transfer coefficient and pressure drop. The Bell–Delaware method is the most reliable method at present for the shell-side analysis. In the Bell–Delaware method, the B-stream is the main essential stream. The other streams reduce the B-stream and alter the shell-side temperature profile, resulting in a decrease in heat transfer coefficient.

A brief discussion of the Bell–Delaware method for the shell-side heat transfer coefficient and the pressure drop analysis is given in this section.

9.4.4.1 Shell-Side Heat Transfer Coefficient

The basic equation for calculating the average shell-side heat transfer coefficient is given by[11,12]

$$h_o = h_{id} J_c J_l J_b J_s J_r \tag{9.23}$$

Shell-and-Tube Heat Exchangers

where h_{id} is the ideal heat transfer coefficient for pure crossflow in an ideal tube bank and is calculated from

$$h_{id} = j_i c_{ps} \left(\frac{\dot{m}_s}{A_s} \right) \left(\frac{k_s}{c_{ps} \mu_s} \right)^{2/3} \left(\frac{\mu_s}{\mu_{s,w}} \right)^{0.14} \quad (9.24)$$

where j_i is the Colburn j-factor for an ideal tube bank, s stands for shell, and A_s is the crossflow area at the centerline of the shell for one crossflow between two baffles.

Graphs are available for j_i as a function of shell-side Reynolds number, $Re_s = d_o \dot{m}_s / \mu_s A_s$, tube layout, and pitch size. Such graphs are shown in Figures 9.15–9.17.[6] A_s is given by Equation 9.15, i.e., the Reynolds number is based on the outside tube diameter and on the minimum cross-section flow area at the shell diameter. On the same graphs, friction coefficients for ideal tube banks are also given for the pressure drop calculations. Therefore, depending on the shell-side constructional parameters, the correction factors must be calculated.[6,10]

Although the ideal values of j_i and f_i are available in graphical forms, for computer analysis, a set of curve-fit correlations are obtained in the following forms:[6,11]

$$j_i = a_1 \left(\frac{1.33}{P_T/d_o} \right)^a (Re_s)^{a_2} \quad (9.25)$$

where

$$a = \frac{a_3}{1 + 0.14(Re_s)^{a_4}}$$

and

$$f_i = b_1 \left(\frac{1.33}{P_T/d_o} \right)^b (Re_s)^{b_2} \quad (9.26)$$

where

$$b = \frac{b_3}{1 + 0.14(Re_s)^{b_4}}$$

Table 9.6 gives the coefficients of Equations 9.25 and 9.26.[6]

J_c is the correction factor for baffle cut and spacing. This factor takes into account the heat transfer in the window and calculates the overall average heat transfer coefficient for the entire heat exchanger. It depends on the shell diameter and the baffle cut distance from the baffle tip to the shell inside diameter. For a large baffle cut, this value may decrease to a value of 0.53,

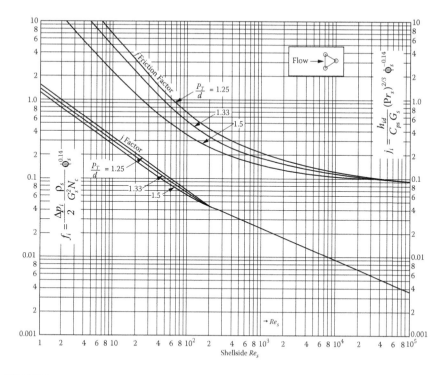

FIGURE 9.15
Ideal tube bank j_i and f_i factors for 30°C staggered layout. (From Spalding, D. B. and Taborek, J., *Heat Exchanger Design Handbook*, Section 3.3, Hemisphere, Washington, D.C., 1983. With permission.)

and it is equal to 1.0 for a heat exchanger with no tubes in the window. It may increase to a value as high as 1.15 for small windows with a high window velocity.

J_l is the correlation factor for baffle leakage effects including tube-to-baffle and shell-to-baffle leakage (A- and E-streams). If the baffles are put too close together, then the fraction of the flow in the leakage streams increases compared with the crossflow. J_l is a function of the ratio of total leakage area per baffle to the crossflow area between adjacent baffles and also of the ratio of the shell-to-baffle leakage area to the tube-to-baffle leakage area. A typical value of J_l is in the range of 0.7 and 0.8.

J_b is the correction factor for bundle bypassing effects due to the clearance between the outermost tubes and the shell and pass dividers (C- and F-streams). For a relatively small clearance between the outermost tubes and the shell for fixed tube sheet construction, $J_b \approx 0.90$. For a pull-through floating head, larger clearance is required and $J_b \approx 0.7$. The sealing strips (see Figure 9.14) can increase the value of J_b.

J_s is the correction factor for variable baffle spacing at the inlet and outlet. Because of the nozzle spacing at the inlet and outlet and the changes in local

Shell-and-Tube Heat Exchangers

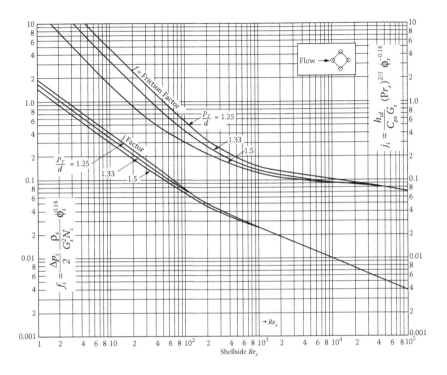

FIGURE 9.16
Ideal tube bank j_i and f_i factors for 45°C staggered layout. (From Spalding, D. B. and Taborek, J., *Heat Exchanger Design Handbook*, Section 3.3, Hemisphere, Washington, D.C., 1983. With permission.)

velocities, the average heat transfer coefficient on the shell side will change. The J_s value will usually be between 0.85 and 1.00. J_r applies if the shell-side Reynolds number, Re_s, is less than 100. If $Re_s < 20$, it is fully effective. This factor is equal to 1.00 if $Re_s > 100$. The combined effects of all these correction factors for a reasonable well-designed shell-and-tube heat exchanger is of the order of 0.60.[5,12]

Example 9.3

Distilled water with a flow rate of 50 kg/s enters a baffled shell-and-tube heat exchanger at 32°C and leaves at 25°C. Heat will be transferred to 150 kg/s of raw water coming from a supply at 20°C. Design a heat exchanger for this purpose. A single shell and single tube pass is preferable. The tube diameter is 3/4" (19 mm OD with 16 mm ID) and tubes are laid out on a 1 in. square pitch. Use the baffle spacing of 0.5 m. A maximum length of the heat exchanger of 8 m is required because of space limitations. The tube material is 0.5 Cr-alloy ($k = 42.3$ W/m · K). Assume a total fouling resistance of 0.000176 m² · K/W. Note that surface over

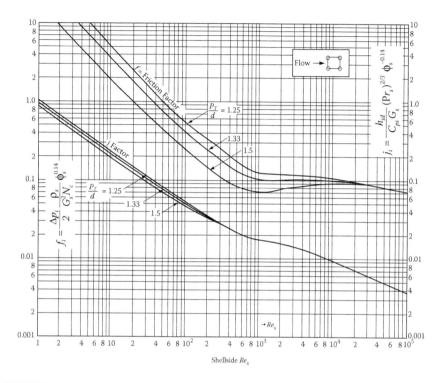

FIGURE 9.17
Ideal tube bank j_i and f_i factors for 90°C inline layout. (From Spalding, D. B. and Taborek, J., *Heat Exchanger Design Handbook*, Section 3.3, Hemisphere, Washington, D.C., 1983. With permission.)

design should not exceed 30%. The maximum flow velocity through the tube is also suggested to be 2 m/s to prevent erosion. Perform a thermal analysis of the heat exchanger.

Solution

The properties of the tube-side fluid at 20°C, from Appendix B, are

$$c_p = 4182 \text{ J/kg} \cdot \text{K}$$
$$k = 0.598 \text{ W/m}^2 \cdot \text{K}$$
$$Pr = 7.01$$
$$\rho = 998.2 \text{ kg/m}^3$$
$$\mu = 10.02 \times 10^{-4} \text{ N} \cdot \text{s/m}^2$$

The properties of the shell-side fluid at average temperature, from the appendix, are

Shell-and-Tube Heat Exchangers

TABLE 9.6
Correlation Coefficients for j_i and f_i, Equations 9.25 and 9.26

Layout Angle	Reynolds Number	a_1	a_2	a_3	a_4	b_1	b_2	b_3	b_4
30°	10^5–10^4	0.321	−0.388	1.450	0.519	0.372	−0.123	7.00	0.500
	10^4–10^3	0.321	−0.388			0.486	−0.152		
	10^3–10^2	0.593	−0.477			4.570	−0.476		
	10^2–10	1.360	−0.657			45.100	−0.973		
	<10	1.400	−0.667			48.000	−1.000		
45°	10^5–10^4	0.370	−0.396	1.930	0.500	0.303	−0.126	6.59	0.520
	10^4–10^3	0.370	−0.396			0.333	−0.136		
	10^3–10^2	0.730	−0.500			3.500	−0.476		
	10^2–10	0.498	−0.656			26.200	−0.913		
	<10	1.550	−0.667			32.00	−1.000		
90°	10^5–10^4	0.370	−0.395	1.187	0.370	0.391	−0.148	6.30	0.378
	10^4–10^3	0.107	−0.266			0.0815	+0.022		
	10^3–10^2	0.408	−0.460			6.0900	−0.602		
	10^2–10	0.900	−0.631			32.1000	−0.963		
	<10	0.970	−0.667			35.0000	−1.000		

$$c_p = 4179 \text{ J/kg} \cdot \text{K}$$
$$k = 0.612 \text{ W/m}^2 \cdot \text{K}$$
$$Pr = 5.75$$
$$\rho = 995.9 \text{ kg/m}^3$$
$$\mu = 8.15 \times 10^{-4} \text{ N} \cdot \text{s/m}^2$$

To solve the problem, first estimate the number of tubes:

$$\dot{m}_t = \rho u_m A_c N_t$$

$$N_t = \frac{\dot{m}}{\rho u_m A_c} = \frac{4\dot{m}_t}{\rho \pi d_i^2 u_m} = \frac{4 \times 150}{998.2 \times \pi \times (0.016)^2 \times 2} = 373.88$$

$$N_t \approx 374$$

The flow area through the tubes is

$$A_t = \frac{\pi d_i^2}{4} N_t = \frac{\pi (0.016)^2}{4} \times 374 = 0.075 \text{ m}^2$$

Now estimate the shell diameter from Equation 9.9:

$$D_s^2 = \frac{N_t(CL)(PR)^2 d_o^2}{0.785(CTP)}$$

$CTP = 0.93$

$CL = 1.0$

$$D_s = \left[\frac{374 \times 1 \times \left(\frac{2.54 \times 10^{-2}}{1.9 \times 10^{-2}}\right)^2 (0.019)^2}{0.785 \times 0.93}\right]^{1/2} = 575 \text{ mm}$$

which is rounded off as 580 mm.

The selected and estimated constructional parameters can be summarized as

Shell inside diameter	$D_S = 0.58$ m
Number of tubes	$N_t = 374$
Tube outside diameter	$d_o = 19$ mm
Tube inside diameter	$d_i = 16$ mm
Square tube pitch	$P_T = 0.0254$ m
Baffle spacing	$B = 0.5$ m (25% cut)

Shell-Side heat Transfer Coefficient
(a) Kern Method: Estimate the crossflow area at the shell diameter:

$$A_s = (D_s - N_{TC}d_o)B$$

where

$$N_{TC} = \frac{D_s}{P_T} = \frac{580}{25.4} = 22.83$$

$$A_s = (0.58 - 22.83 \times 0.019) \times 0.5 = 0.073 \text{ m}^2$$

The equivalent diameter can be calculated from Equation 9.13:

$$D_e = \frac{4\left(P_T^2 - \frac{\pi d_o^2}{4}\right)}{\pi d_o} = \frac{4\left[(2.54 \times 10^{-2})^2 - \frac{\pi(0.019)^2}{4}\right]}{\pi \times 0.019} = 0.024 \text{ m}$$

Calculate the Reynolds number:

$$Re = \left(\frac{\dot{m}_s}{A_s}\right)\frac{D_e}{\mu} = \frac{50}{0.073} \times \frac{0.024}{8.15 \times 10^{-4}} = 20169.76$$

Shell-and-Tube Heat Exchangers

Assuming constant properties, the heat transfer coefficient can be estimated from Equation 9.11:

$$h_o = \frac{0.36k}{D_e}Re^{0.55}Pr^{\frac{1}{3}} = \frac{0.36 \times 0.612}{0.024}(20169.76)^{0.55}(5.57)^{\frac{1}{3}} = 3793.6 \text{ W/m}^2 \cdot \text{K}$$

(b) Taborek Method: Consider the following. Taborek gives the following correlation for the shell-side heat transfer coefficient for turbulent flow[7]:

$$Nu = 0.2\, Re_s^{0.6} Pr_s^{0.4}$$

where the Reynolds number is based on the tube outside diameter and the velocity on the crossflow area at the diameter of the shell:

$$Re_s = \frac{\rho u_s d_o}{\mu} = \frac{\dot{m}_s d_o}{A_s\, \mu} = \frac{50}{0.073}\frac{0.019}{8.15 \times 10^{-4}} = 15967.8$$

$$Nu = 0.2(15967.8)^{0.6}(5.57)^{0.4} = 132.2$$

$$h_s = \frac{Nu \cdot k}{d_o} = \frac{132.2 \times 0.612}{0.019} = 4259.09 \text{ W/m}^2 \cdot \text{K}$$

(c) Bell–Delaware Method: The shell-side heat transfer coefficient is given by Equation 9.24:

$$h_{id} = j_i c_{ps} \left(\frac{\dot{m}_s}{A_s}\right)\left(\frac{k_s}{c_{ps}\mu_s}\right)^{2/3}\left(\frac{\mu_s}{\mu_{s,w}}\right)^{0.14}$$

j_i can be obtained from Figure 9.17 or from the correlation of Equation 9.25 with Table 9.6:

$$j_i = 0.37\, Re_s^{-0.395}$$

where

$$Re_s = \frac{d_o \dot{m}}{\mu_s A_s} = 15967.8$$

$$j_i = 0.37(15967.8)^{-0.395} = 0.0081$$

$$h_{id} = 0.0081 \times 4179 \left(\frac{50}{0.073}\right)\left(\frac{0.612}{4179 \times 8.15 \times 10^{-4}}\right)^{2/3}$$

$$h_{id} = 7382.8 \text{ W/m}^2 \cdot \text{K}$$

It is assumed that the properties are constant. In the Bell–Delaware method, the correction factors due to bypass and leakage stream are provided in graphical forms depending on the constructional features of the heat exchanger. Let us assume that the combined effect of all these correction factors is 60%:

$$h_o = 0.60 \times 7382.8 = 4429.7 \text{ W/m}^2 \cdot \text{K}$$

Therefore, the three above methods have demonstrated comparable results to estimate the shell-side heat transfer coefficient.

$$Re_t = \frac{\rho u_m d_i}{\mu} = \frac{998.2 \times 2 \times 0.016}{10.02 \times 10^{-4}} = 31878.6$$

The tube-side heat transfer coefficient, h_i, can be calculated from the Petukhov–Kirillov correlation as given by Equation 3.29:

$$Nu_b = \frac{(f/2) Re_b Pr_b}{1.07 + 12.7(f/2)^{1/2} (Pr_b^{2/3} - 1)}$$

$$f = (1.58 \ln Re_b - 3.28)^{-2} = [1.58 \ln(31878.6 - 3.28)]^{-2} = 0.0058$$

$$f/2 = 0.0029$$

$$Nu_b = \frac{(0.0029)(31878.6)(7.01)}{1.07 + 12.7(0.0029)^{1/2}(7.01^{2/3} - 1)} = 224.16$$

$$h_i = Nu_b \cdot \frac{k}{d_i} = (224.16) \frac{0.598}{0.016} = 8377.98 \text{ W/m}^2 \cdot \text{K}$$

The overall heat transfer coefficient for the clean surface is

$$\frac{1}{U_c} = \frac{1}{h_o} + \frac{1}{h_i} \frac{d_o}{d_i} + \frac{r_o \ln(r_o/r_i)}{k}$$

$$= \frac{1}{4375} + \frac{1}{8378} \frac{0.019}{0.016} + \frac{9.5 \times 10^{-3} \ln(19/16)}{42.3}$$

$$U_c = 2445.5 \text{ W/m}^2 \cdot \text{K}$$

and for the fouled surface is

$$\frac{1}{U_f} = \frac{1}{U_c} + Re_{ft} = \frac{1}{2445.5} + 0.000176$$

$$U_f = 1709.7 \text{ W/m}^2 \cdot \text{K}$$

$$T_{c_2} = \frac{(\dot{m}c_p)_h (T_{h_1} - T_{h_2})}{(\dot{m}c_p)_c} + T_{c_1}$$

$$T_{c_2} = \frac{50 \times 4179 \times (32-25)}{150 \times 4182} + 20 = 22.33° \text{ C}$$

$$\text{LMTD} = \frac{(32-22.3)-(25-20)}{\ln\left(\frac{32-22.3}{25-20}\right)} = 7.09° \text{ C}$$

$$Q = (\dot{m}c_p)_h (T_{h_1} - T_{h_2}) = 50 \times 4179 \times (32-25)$$

$$Q = 1,462,650 \text{ W}$$

$$Q = U_f A_f \Delta T_m$$

$$A_f = \frac{1,462,650}{1709.7 \times 7.09} = 120.66 \text{ m}^2$$

$$A_c = \frac{1,462,650}{2445.5 \times 7.09} = 84.36 \text{ m}^2$$

The over surface design is

$$OS = \frac{A_f}{A_c} = \frac{U_c}{U_f} = \frac{2445.5}{1709.7} = 1.43(43\%)$$

which is the clean vs. fouling safety factor. The over surface design should not be more than about 30%. Assuming 20% surface over design, the cleaning scheduling must be arranged accordingly:

$$\frac{U_c}{U_f} = 1.20$$

$$U_f = \frac{U_c}{1.20} = \frac{2445.5}{1.20} \cong 2037.9 \text{ W/m}^2 \cdot \text{K}$$

The corresponding total resistance can be calculate from Equation 6.2:

$$\frac{1}{U_f} = \frac{1}{U_c} + R_{ft}$$

$$R_{ft} = 0.0000817 \text{ m}^2 \cdot \text{K/W}$$

For 20% over surface design, the surface area of the exchanger becomes

$$A_f = 1.20 A_c$$

$$A_f = 1.20 \times 84.36 = 101.2 \text{ m}^2$$

The length of the heat exchanger is calculated as follows:

$$L = \frac{A_f}{N_t \pi d_o} = \frac{101.2}{374 \times \pi \times 0.019} = 4.54 \text{ m}$$

which is rounded off to $L = 5$ m.

The shell diameter can be recalculated from Equation 9.10:

$$D_s = 0.637 \sqrt{\frac{CL}{CTP}} \left[\frac{A_f (PR)^2 d_o}{L} \right]^{1/2}$$

$$D_s = 0.637 \sqrt{\frac{1}{0.93}} \left[\frac{101.2 \times (2.54/1.9)^2 \times 19 \times 10^{-3}}{5} \right]^{1/2} = 0.548 \text{ m}$$

which is rounded off to $D_s = 0.60$ m.

Further analysis is suggested. Now we have a new heat exchanger to be rerated. After calculating D_s, compare with Table 9.3 for tube counts and shell diameter. Tabulate new constructional parameters:

$$D_s = 0.60 \text{ m}$$
$$N_t = 374$$
$$L = 5 \text{ m}$$
$$d_o = 19 \text{ mm}$$
$$d_i = 16 \text{ mm}$$
$$P_T = 0.0254 \text{ m}$$
$$B = 0.50 \text{ m}$$

Calculate the pressure drops on the shell side and tube side by the use of Equations 9.17 and 9.22 and compare with the results of Example 9.4.

9.4.4.2 Shell-Side Pressure Drop

For a shell-and-tube type heat exchanger with bypass and leakage streams, the total nozzle-to-nozzle pressure drop, determined by the Bell–Delaware method, is calculated as the sum of the following three components (Figure 9.18a–c).

1. The pressure drop in the interior crossflow section (baffle tip to baffle tip); the combined pressure drop of the entire interior cross-flow section is[6,11,12]

$$\Delta p_c = \Delta p_{bi}(N_b - 1) R_l R_b \tag{9.27}$$

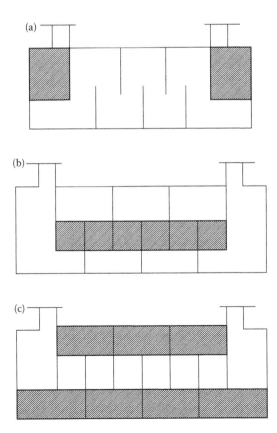

FIGURE 9.18
(a) Entrance, (b) internal, and (c) window.

where Δp_{bi} is the pressure drop in an equivalent ideal tube bank in one baffle compartment of central baffle spacing and R_l is the correction factor for baffle leakage effects (A- and E-streams). Typically, $R_l = 0.4$ to 0.5. R_b is the correction factor for bypass flow (C- and F-streams). Typically, $R_b = 0.5$ to 0.8, depending on the construction type and number of sealing strips. N_b is the number of baffles.

2. The pressure drop in the window is affected by leakage but not bypass. The combined pressure drop in all of the windows is calculated from

$$\Delta p_w = \Delta p_{wi} N_b R_l \tag{9.28}$$

where Δp_{wi} is the pressure drop in an equivalent ideal tube bank in the window section.

3. The pressure drop in the entrance and exit sections is affected by bypass but not by leakage. Additionally, there is an effect due to variable baffle spacing. The combined pressure drop for entrance and exit section is given by

$$\Delta p_e = 2\Delta p_{bi} \frac{N_c + N_{cw}}{N_c} R_b R_s \tag{9.29}$$

where N_c is the number of tube rows crossed during flow through one crossflow in the exchanger, and N_{cw} is the numbers of tube rows crossed in each baffle window. R_s is the correction factor for the entrance and exit section having different baffle spacing than the internal sections due to the existence of the inlet and outlet nozzles. The correction factors are available in graphical forms in the literature.[7,10,12,13]

The total pressure drop over the shell side of the heat exchanger is

$$\Delta p_s = \Delta p_c + \Delta p_w + \Delta p_e \tag{9.30}$$

$$\Delta p_s = \left[(N_b - 1)\Delta p_{bi} R_b + N_b \Delta p_{wi}\right] R_l + 2\Delta p_{bi} \left(1 + \frac{N_{cw}}{N_c}\right) R_b R_s \tag{9.31}$$

The pressure drops in the nozzles must be calculated separately and added to the total pressure drop.

In Equation 9.31, Δp_{bi} is calculated from

$$\Delta p_{bi} = 4 f_i \frac{G_s^2}{2\rho_s} \left(\frac{\mu_{s,w}}{\mu_s}\right)^{0.14} \tag{9.32}$$

Friction coefficients are given in Figures 9.15–9.17 and by Equation 9.26. For an ideal baffle window section, Δp_{bi} is calculated from

$$\Delta p_{wi} = \frac{\dot{m}_s^2 (2 + 0.6 N_{cw})}{2 \rho_s A_s A_w} \tag{9.33}$$

if $Re_s \geq 100$, and from

$$\Delta p_{wi} = 26 \frac{\mu_s \dot{m}_s}{\sqrt{A_s A_w} \rho} \left(\frac{N_{cw}}{p - d_o} + \frac{B}{D_w^2} \right) + \frac{\dot{m}_s}{A_s A_w \rho_s} \tag{9.34}$$

if $Re_s \leq 100$.

Calculation of the equivalent diameter of the window, D_w, the area for flow through window, A_w, and the correction factors are given in the literature.[7,11,12]

The number of tube rows crossed in one crossflow section, N_c, can be estimated from

$$N_c = \frac{d_i \left(1 - 2 \frac{L_c}{D_s} \right)}{P_p} \tag{9.35}$$

P_p is defined in Figure 9.19 and can be obtained from Table 9.7, and L_c is the baffle cut distance from baffle tip to the inside of the shell.

The number of effective crossflow rows in each window, N_{cw}, can be estimated from

$$N_{cw} = \frac{0.8 L_c}{P_p} \tag{9.36}$$

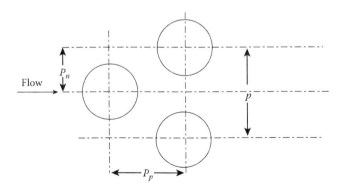

FIGURE 9.19
Tube pitches parallel and normal to flow (equilateral triangular arrangement shown).

TABLE 9.7
Tube Pitches Parallel and Normal to Flow

Tube OD (d_o, in.)	Tube Pitch (p, in.)	Layout	P_p (in.)	P_n (in.)
5/8 = 0.625	13/16 = 0.812	→◁	0.704	0.406
3/4 = 0.750	15/16 = 0.938	→◁	0.814	0.469
3/4 = 0.750	1.000	→□	1.000	1.000
3/4 = 0.750	1.000	→◊	0.707	0.707
3/4 = 0.750	1.000	→◁	0.866	0.500
1	1 1/4 = 1.250	→□	1.250	1.250
1	1 1/4 = 1.250	→◊	0.884	0.884
1	1 1/4 = 1.250	→◁	1.082	0.625

Source: From Bell, K. J., *Heat Exchangers — Thermal–Hydraulic Fundamentals and Design*, 1981. With permission.

The number of baffles, N_b, can be calculated from

$$N_b = \frac{L - B_i - B_o}{B} + 1 \tag{9.37}$$

If $B_i = B_o = B$, Equation 9.37 reduces to

$$N_b = \frac{L}{B} - 1 \tag{9.38}$$

The total shell-side pressure drop of a typical shell-and-tube exchanger is on the order of 20 to 30% of the pressure drop that would be calculated without taking into account baffle leakages and tube bundle bypass effects.[11]

Example 9.4

Assume that a preliminary analysis of a heat exchanger is performed using the Kern method as in Example 9.3 (or assume that such a heat exchanger is available), and the design parameters are as given below. $D_s = 0.58$ m, number of tubes $N_t = 374$, the length of the heat exchanger $L = 5.0$, the tube diameter is 3/4 in (19 mm OD with 16 mm ID), and tubes are laid out on a 1 in. square pitch. The baffle spacing $B = 0.5$ m, and the baffle cut is 25% of the shell inside diameter D_s. Inlet and outlet baffle spacing and central baffle spacing are equal. Flow specifications on the shell side and the tubes are specified in Example 9.3. The shell-side mass flow rate is 50 kg/s. Allowable pressure drop on the shell side is 15 kPa. The flow area through the window is calculated to be $A_w = 0.076$ m2. Calculate the shell-side pressure drop using the Bell–Delaware method and state if this heat exchanger is suitable.

Solution

Now we have to rate the heat exchanger using the Bell–Delaware method and assuming $D_s = 0.58$ m, $N_t = 374$, $L = 5$ m, $d_o = 19$ mm, $d_i = 16$ m, $B = 0.50$ m, a 25% baffle cut, $P_T = 0.0254$ m, $L_c = 0.25\ D_s$, the tubes are laid out in square pitch, and the area of the baffle windows, $A_w = 0.076$ m².

The estimated crossflow area at the shell diameter from Example 9.3 is

$$A_s = 0.073 \text{ m}^2$$

The number of the rows crossed in one crossflow section, N_c, can be calculated from Equation 9.35:

$$\frac{L_c}{D_s} = \frac{0.25\,D_s}{D_s} = 0.25$$

From Table 9.7,

$$P_p = 1 \text{ in.} = 0.0254 \text{ m}$$

$$N_c = \frac{D_s\left[1 - 2\dfrac{L_c}{D_s}\right]}{P_p}$$

$$N_c = 0.58 \times [1 - 2 \times (0.25)]/0.0254 \cong 12$$

The Reynolds number is based on the tube outside diameter and the velocity on the crossflow area at the diameter of the shell. Note that distilled water circulates on the shell side:

$$Re_s = \frac{\rho u_s d_o}{\mu} = \frac{\dot{m}}{A_s}\frac{d_o}{\mu}$$

which was calculated in Example 9.3 as

$$Re_s = 15{,}968$$

Now, calculate the Fanning friction coefficient, which is given by Equation 9.26:

$$f_i = b_1 \left(\frac{1.33}{P_T/d_o}\right)^b (Re_s)^{b_2}$$

Since $P_T/d_o \approx 1.33$, this equation can be simplified to

$$f_i = b_1 (Re_s)^{b_2}$$

From Table 9.6, $b_1 = 0.391$, $b_2 = -0.148$; therefore, the friction coefficient

$$f_i = 0.391(15{,}968)^{-0.148} = 0.093$$

If there were no leakage or bypass, the pressure drop in one crossflow section can be calculated from Equation 9.32:

$$\Delta p_{bi} = 4 f_i \frac{G_s^2}{2\rho_s} \left(\frac{\mu_{s,w}}{\mu_s} \right)^{0.14} N_c$$

$$G_s = \frac{\dot{m}_s}{A_s} = \frac{50}{0.073} = 684.9 \text{ kg/m}^2\text{s}$$

$$\Delta p_{bi} = 4 \times 0.0933 \times \frac{684.9^2}{2 \times 995.9} \times 12 = 1055 \text{ Pa}$$

Assuming $R_b = 0.60$ and $R_l = 0.4$, the combined pressure drop of the entire interior crossflow section can be calculated from Equation 9.27:

$$\Delta p_c = \Delta p_{bi} (N_b - 1) R_l R_b$$

$$\Delta p_c = 1137(9-1)0.60 \times 0.4 = 2.18 \text{ kPa}$$

where N_b is the number of baffles:

$$N_b = \frac{L}{B} - 1$$

For an ideal baffle window section, Δp_{wi} is calculated from Equation 9.33. The number of effective crossflow rows in each window, N_{cw}, can be estimated from Equation 9.36:

$$N_{cw} = \frac{0.8 L_c}{P_p}$$

where P_p is given in Table 9.7 as 0.0254 m. L_c is the baffle cut distance from the baffle tip to the shell inside diameter:

$$L_c = 0.25 D_s = 0.25 \times 0.580 = 0.145 \text{ m}$$

$$N_{cw} = \frac{0.8 \times 0.145}{0.0254} = 4.6 \cong 5$$

The area of flow through the baffle windows is

$$A_w = A_{wg} - A_{wt}$$

where A_{wg} is the gross window area and A_{wt} is the window area occupied by tubes. The expressions to calculate A_{wg} and A_{wt} are given by Taborek[6] and Bell.[11,12] Here it is given as $A_w = 0.076$ m², then, from Equation 9.33,

$$\Delta p_{wi} = \frac{\dot{m}_s^2 (2 + 0.6 N_{cw})}{2\rho A_s A_w}$$

$$\Delta p_{wi} = \frac{50^2 (2 + 0.6 \times 5)}{2 \times 995.9 \times 0.073 \times 0.076} = 1131 \text{ Pa}$$

The total pressure drop in all the windows is

$$\Delta p_w = \Delta p_{wi} N_b R_l$$

$$\Delta p_w = 1131 \times 9 \times 0.4 = 4072 \text{ Pa}$$

The total pressure drop over the heat exchanger on the shell side can be calculated from Equation 9.31:

$$\Delta p_s = \left[(N_b - 1) \Delta p_{bi} R_b + N_b \Delta p_{wi} \right] R_l + 2\Delta p_{bi} \left(1 + \frac{N_{cw}}{N_c} \right) R_b R_s$$

Baffle spacing in the inlet, exit, and central regions are equal, so $R_s = 1$:

$$\Delta p_s = \left[(9-1) \times 1105 \times 0.60 + 9 \times 1131 \right] 0.4 + 2 \times 1105 \times \left(1 + \frac{5}{12} \right)$$

$$\times 0.60 = 8.07 \text{ kPa}$$

which is less than the allowable pressure drop, so the heat exchanger is suitable. The shell-side pressure drop could be overestimated if it were calculated without baffle leakage and without tube bundle bypass effects.

Now calculate the shell-side pressure drop by the use of the Kern method, which does not take into account the baffle leakage and bypass effects. The shell-side pressure drop can be calculated from Equation 9.17:

$$\Delta p_s = \frac{f G_s^2 (N_b + 1) D_s}{2\rho D_e \phi_s}$$

where D_e is the equivalent diameter, which is calculated from Equation 9.13 and is given in Example 9.3 as

$$D_e = 0.024 \text{ m}$$

The friction coefficient is calculated from Equation 9.18, where

$$Re_s = \left(\frac{\dot{m}_s}{A_s} \right) \frac{D_e}{A_s} = G_s \frac{D_e}{\mu}$$

which is given in Example 9.3 as

$$Re_s \approx 20170$$

Then,

$$f = \exp(0.576 - 0.19 \ln Re_s)$$
$$f = 0.271$$

Assuming that $\phi_s = 1$ and inserting the calculated values into Equation 9.17, Δp_s becomes

$$\Delta p_s = f G_s^2 \frac{(N_b + 1) D_s}{2 \rho D_e \phi_s} = 0.271 \left(\frac{50}{0.073}\right)^2 \frac{(9+1) \times 0.58}{2 \times 995.9 \times 0.024 \times 1}$$

$$= 15400 \text{ Pa} = 15.4 \text{ kPa}$$

The shell-side pressure drop obtained by Bell-Delaware method is about 48% lower than that obtained by the Kern method.

Example 9.5

Distilled water with a mass flow rate of 80,000 kg/hr enters the shell-side of an exchanger at 35°C and leaves at 25°C. The heat will be transferred to 140,000 kg/hr of raw water coming from a supply at 20°C. The baffles will be spaced 12 in. apart. A single shell and single tube pass is preferable. The tubes are 18 BWG tubes with a 1 in. outside diameter (OD = 0.0254 m, ID = 0.0229 m) and they are laid out in square pitch. Shell diameter is 15.25 in. A pitch size of 1.25 in. and a clearance of 0.25 in. are selected. Calculate the length of the heat exchanger and the pressure drop for each stream. If the shell-side allowable maximum pressure drop is 200 kPa, will this heat exchanger be suitable?

Solution

The tube-side specifications are

Outer diameter	$d_o = 1$ in. $= 0.0254$ m
Inner diameter	$d_i = 0.902$ in. $= 0.0229108$ m
Flow area	$A_c = 0.639$ in.$^2 = 0.00041226$ m^2
Wall thickness	$t_w = 0.049$ in. $= 0.0012446$ m

Calculate the mass flow rate (from problem statement) by

$$\dot{m}_t = \frac{14,000 \text{ kg/hr}}{3600 \text{ s/hr}} = 38.89 \text{ kg/s}$$

Shell-and-Tube Heat Exchangers

The shell-side specifications are

Pitch size	$P_T = 1.25$ in. $= 0.03175$ m
Clearance	$C = 0.25$ in. $= 0.00635$ m
Baffle spacing	$B = 12$ in. $= 0.3048$ m
Shell diameter	$D_s = 15.25$ in. $= 0.38735$ m

Calculate the mass flow rate (from problem statement) by

$$\dot{m}_s = \frac{80,000 \text{ kg/hr}}{3600 \text{ s/hr}} = 22.22 \text{ kg/s}$$

For a single pass shell-and-tube heat exchanger with a diameter of 15.25 in., from Table 9.3, the number of tubes, N_t, in a 1.25 in. square pitch with an outer tube diameter of 1 in. is 81 tubes.

The shell-side fluid properties at a 30°C average temperature from Appendix B (Table B.2) are

$$c_p = 4.1785 \text{ kJ/kg} \cdot \text{K}$$
$$\mu_b = 0.000797 \text{ kg/m} \cdot \text{s}$$
$$k = 0.614 \text{ W/m} \cdot \text{K}$$
$$\rho = 995.7 \text{ kg/m}^3$$
$$Pr = 5.43$$

The properties of water at 22.5°C are

$$c_p = 4.179 \text{ kJ/kg} \cdot \text{K}$$
$$\mu_b = 0.00095 \text{ kg/m} \cdot \text{s}$$
$$k = 0.6065 \text{ W/m} \cdot \text{K}$$
$$\rho = 997 \text{ kg/m}^3$$
$$Pr = 6.55$$

The mean temperature difference is determined as

$$\Delta T_{lm,cf} = \frac{\Delta T_1 - \Delta T_2}{\ln \frac{\Delta T_1}{\Delta T_2}} = \frac{(35-25)-(25-20)}{\ln\left(\frac{35-25}{25-20}\right)} = 7.21°C$$

$$R = \frac{T_{h_1} - T_{h_2}}{T_{c_2} - T_{c_1}} = \frac{(35-25)}{(25-20)} = 2$$

$$P = \frac{T_{c_1} - T_{c_2}}{T_{h_1} - T_{c_1}} = \frac{25-20}{35-20} = 0.333$$

The correction factor is assumed to be $F \approx 1$.
Calculate the shell-side heat transfer coefficient by

$$A_s = \frac{(D_s CB)}{P_t} = \frac{(0.38735 \times 0.00635 \times 0.3048)}{0.03175} = 0.02361 \text{ m}^2$$

$$G_s = \frac{\dot{m}_s}{A_s} = \frac{22.22 \text{ kg/s}}{0.02361 \text{ m}^2} = 941.107 \text{ kg/m}^2 \cdot \text{s}$$

$$D_e = \frac{4(P_t^2 - \pi d_o^2/4)}{\pi d_o} = \frac{4\left[(0.03175)^2 - \pi \cdot (0.0254)^2/4\right]}{\pi \cdot 0.0254} = 0.02513 \text{ m}$$

$$Re_s = \frac{D_e G_s}{\mu} = \frac{0.02513 \cdot 941.107}{0.000797} = 29673.8$$

Therefore, the flow of the fluid on the shell side is turbulent. Using McAdam's correlation, Equation 9.11, we get the Nusselt number:

$$Nu = 0.36 \left(\frac{D_e G_s}{\mu_b}\right)^{0.55} \left(\frac{c_p \mu_b}{k}\right)^{0.33} \left(\frac{\mu_b}{\mu_w}\right)^{0.14}$$

$$= 0.36 \left(\frac{0.02513 \times 941.107}{0.000797}\right)^{0.55} \left(\frac{4178.5 \times 0.000797}{0.614}\right)^{0.33} \left(\frac{0.000797}{0.00086}\right)^{0.14}$$

$$= 179.39$$

It is assumed that the tube wall temperature is 26°C and $\mu_w = 0.00086$ kg/m·s.

The shell-sidve heat transfer coefficient, h_o, is then calculated as

$$h_o = \frac{Nu \cdot k}{D_e} = \frac{179.30 \times 0.614}{0.02513} = 4,383.09 \text{ W/m}^2 \cdot \text{K}$$

Calculate the tube side heat transfer coefficient by

$$A_t = \frac{\pi d_i^2}{4} = \frac{\pi \times (0.02291)^2}{4} = 0.0004122 \text{ m}^2$$

$$A_{tp} = \frac{N_t A_t}{\text{Number of passes}} = \frac{81 \times 0.0004122}{1} = 0.03339 \text{ m}^2$$

$$G_t = \frac{\dot{m}_t}{A_{tp}} = \frac{38.889}{0.03339} = 1164.6 \text{ kg/m}^2 \cdot \text{s}$$

$$u_t = \frac{G_t}{\rho} = \frac{1164.6}{997} = 1.1682 \text{ m/s}$$

$$Re_t = \frac{u_t \rho d_i}{\mu} = \frac{1.1682 \times 997 \times 0.02291}{0.00095} = 28,087.5$$

Therefore, the flow of the fluid on the tube side is turbulent. Using the Petukhov–Kirillov correlation,

$$Nu = \frac{(f/2)Re\,Pr}{1.07 + 12.7(f/2)^{1/2}(Pr^{2/3} - 1)}$$

where $f = (1.58 \ln Re - 3.28)^{-2} = [1.58 \times \ln(28{,}087.5) - 3.28]^{-2} = 0.0060$

$$Nu = \frac{(0.006/2) \times 28087.5 \times 6.55}{1.07 + 12.7 \times (0.006/2)^{1/2} \times (6.55^{2/3} - 1)} = 196.45$$

The tube-side heat transfer coefficient, h_i, is then found as

$$h_i = \frac{Nu \cdot k}{d_i} = \frac{196.45 \times 0.6065}{0.02291} = 5200.5 \text{ W/m}^2 \cdot \text{K}$$

The overall heat transfer coefficient, U_o, is determined by the following equation:

$$U_o = \frac{1}{\dfrac{d_o}{d_i h_i} + \dfrac{d_o \ln(d_o/d_i)}{2k} + \dfrac{1}{h_o}}$$

$$= \frac{1}{\dfrac{0.0254}{0.02291 \times 5200.5} + \dfrac{0.0254 \times \ln(0.0254/0.02291)}{2 \times 54} + \dfrac{1}{4383.09}}$$

$$= 2147.48 \text{ W/m}^2 \cdot \text{K}$$

To find the area and, consequently, the length of heat exchanger, the required heat transfer rate must first be determined. This heat transfer rate is determined by

$$Q = \dot{m}_s c_p (T_{h_1} - T_{h_2}) = 22.22 \times 4178.5 \times (35 - 25) = 928.5 \text{ kW}$$

The heat transfer rate is also defined as

$$Q = U_o A F \Delta T_{lm,cf}$$

Therefore, the area can be determined by

$$A = \frac{Q}{U_o F \Delta T_{lm,cf}} = \frac{928.5 \times 1000}{2147.48 \times 1 \times 7.2135} = 59.93 \text{ m}^2$$

and the length is

$$L = \frac{A}{N_t \pi D_o} = \frac{59.77}{81 \times \pi \times 0.0254} = 9.28 \text{ m}$$

The shell-side pressure drop can be calculated from Equation 9.17:

$$\Delta p_s = \frac{f G_s^2 (N_b + 1) D_s}{2 \rho D_e \phi_s}$$

$$N_L = \frac{L}{B} - 1 = \frac{9.28}{0.3048} - 1 \approx 29$$

$$f = \exp(0.576 - 0.19 \ln R_s)$$

$$f = \exp(0.576 - 0.19 \ln 29{,}673.8) = 0.2514$$

$$\Delta p_s = \frac{0.2514 \times (941.107)^2 \times (29+1) \times 0.38735}{2 \times 995.7 \times 0.02513 \times \left(\frac{7.97}{8.6}\right)^{0.14}} = 52{,}257 \text{ Pa} = 52.3 \text{ kPa}$$

$$\Delta p_s = 52.3 \text{ kPa} < 200 \text{ kPa}$$

Therefore, this heat exchanger is suitable.

The tube-side pressure drop can be calculated from Equation 9.22:

$$\Delta p_t = \left(4f \frac{LN_p}{d_i} + 4N_p\right) \frac{\rho u_m^2}{2}$$

$$\Delta p_t = \left(\frac{4 \times 0.0060 \times 9.28 \times 1}{0.02291} + 4 + 1\right) \times 997 \times \frac{1.1682^2}{2}$$

$$= 9300 \text{ Pa} = 9.3 \text{ kPa}$$

The above analysis can be repeated using the Bell–Delaware method. The Bell–Delaware method assumes that in addition to flow specifications on the shell side, the shell-side geometrical data must be known or specified. With this geometrical information, all remaining geometrical parameters needed in the shell-side calculations can be calculated or estimated by methods given by Palen and Taborek[10] and Example 9.5 can be repeated with the Bell–Delaware method (a thermal design project).

Application of the complex Bell–Delaware method to calculate correction factors to the heat transfer and the pressure drops in different sections of the shell side is provided in detail by Biniciogullari[14] for given process specifications.

Nomenclature

A_c	heat transfer area without fouling, m²
A_f	heat transfer area with fouling, m²
A_i	heat transfer area based on the inside surface area of tubes, m²
A_o	heat transfer area based on the outside surface area of tubes, m²
A_s	crossflow area at or near shell centerline, m²
A_w	area for flow through baffle window, m²
B	baffle spacing, m
B_i	baffle spacing at the inlet, m
B_o	baffle spacing at the outlet, m
C	clearance between the tubes, m
C_p	specific heat at constant pressure, J/kg · K
D_s	shell inside diameter, m
D_w	equivalent diameter of baffle window, m
d_o	tube outside diameter, m
d_i	tube inside diameter, m
F	correction factor to LMTD for noncounterflow systems
f_i	friction factor for flow across an ideal tube bank
G	mass velocity, kg/m² · s
h_{fg}	latent heat of evaporation, J/kg
h_i	tube-side heat transfer coefficient, W/m² · K
h_{id}	shell-side heat transfer coefficient for ideal tube bank, W/m² · K
h_o	shell-side heat transfer coefficient for the exchanger, W/m² · K
J_b	bundle bypass correction factor for heat transfer
J_c	segmental baffle window correction factor for heat transfer
J_l	baffle leakage correction factor for heat transfer
J_r	laminar flow heat transfer correction factor
J_s	heat transfer correction factor for unequal end baffle spacing
j_i	Colbum j-factor for an ideal tube bank
k_s	thermal conductivity of shell-side fluid, W/m · K
k_w	thermal conductivity of tube wall, W/m · K
L	effective tube length of heat exchanger between tube sheets, m
L_c	baffle cut distance from baffle tip to inside of shell
$\dot{m}_s$	shell-side mass flow rate, kg/s
$\dot{m}_t$	tube-side mass flow rate, kg/s
N_b	number of baffles in the exchanger
N_c	number of tube rows crossed between baffle tips of one baffle compartment
N_{cw}	number of tube rows crossed in one baffle window

N_t	total number of tubes or total number of holes in tube sheet for U-tube bundle
Nu	Nusselt number
P_n	pitch size, m
Pr	Prandtl number
P_T	pitch size, m
Q	heat duty of heat exchanger, W
R_b	bundle bypass correction factor for pressure drop
$R_{f,i}$	tube-side fouling resistance referring to the inside tube surface, m² · K/W
$R_{f,o}$	shell-side fouling resistance referring to the outside tube surface, m² · K/W
R_{ft}	total fouling, m² · K /W
R_s	baffle end zones correction factor for pressure drop
Re_s	shell-side Reynolds number
T	temperature, °C, K
T_c	cold fluid temperature, °C, K
T_h	hot fluid temperature, °C, K
T_w	wall temperature, °C, K
U_c	overall heat transfer coefficient for clean surface based on the outside tube area, W/m² · K
U_{of}	overall heat transfer coefficient for fouled surface based on the outside tube area, W/m² · K
u_m	average velocity inside tubes, m/s
Δp_{bi}	pressure drop for one baffle compartment in crossflow, based on ideal tube bank, Pa
Δp_s	total shell side pressure drop, Pa
Δp_{wi}	pressure drop in one ideal window section of a segmentally baffled exchanger, Pa
$\Delta T_c, \Delta T_h$	cold and hot end terminal temperature differences, °C, K
ΔT_{lm}	log mean temperature differences, °C, K
ΔT_m	effective or true mean temperature difference, °C, K
μ_s	shell fluid dynamic viscosity at average temperature, mPa/s
μ_t	tube fluid dynamic viscosity at average temperature, mPa/s
ρ_s, ρ_t	shell- or tube-side fluid density, respectively, at average temperature of each fluid, kg/m³
ϕ_s	viscosity correction factor for shell-side fluids, $(\mu_w/\mu_b)^{0.14}$.

PROBLEMS

9.1. Crude oil at a flow rate of 63.77 kg/s enters the exchanger at 102°C and leaves at 65°C. The heat will be transferred to 45 kg/s of tube water (city water) coming from a supply at 21°C. The exchanger data are given below: 3/4" OD, 18 BWG tubes on a 1 in. square pitch; 2 tube passes and 4 tube passes will be considered. Tube material is carbon steel. The heat exchanger has one shell. Two different shell diameters of ID 35 and 37 inches should be studied. Baffle spacing is 275 mm. Calculate the length of the heat exchanger for clean and fouled surfaces. Also calculate:

a. tube side velocity for 1–2 and 1–4 arrangements
b. overall heat transfer coefficients for clean and fouled surfaces
c. pressure drops
d. pumping powers.

The allowable shell-side and tube-side pressure drops are 60 and 45 kPa, respectively. The following properties are given:

	Shell Side	Tube Side
Specific heat, J/kg · K	2177	4186.8
Dynamic viscosity, N · s/m²	0.00189	0.00072
Thermal conductivity, W/m · K	0.122	0.605
Density, kg/m³	786.4	995
Prandtl number	33.73	6.29
Maximum pressure loss, Pa	60,000	45,000

9.2. Water at a flow rate of 60 kg/s enters a baffled shell-and-tube heat exchanger at 35°C and leaves at 25°C. The heat will be transferred to 150 kg/s of raw water coming from a supply at 15°C. You are requested to design the heat exchanger for this purpose. A single shell and single tube pass is preferable. The tube diameter is 3/4" (19 mm OD with 16 mm ID) and tubes are laid out on a 1 in. square pitch. A maximum length of the heat exchanger of 8 m is required because of space limitations. The tube material is 0.5 Cr-alloy. Assume a total fouling resistance of 0.000176 m² · K/W. Note that surface over design should not exceed 30%. Reasonable design assumptions can be made along with the calculation if necessary. Calculate the shell diameter, number of tubes, fluid velocities, shell-side heat transfer coefficient, overall heat transfer coefficient, mean temperature difference, total area, and pressure drops. Is the final design acceptable? Discuss your findings.

9.3. Distilled water at a flow rate of 80,000 kg/hr enters an exchanger at 35°C and leaves at 25°C. The heat will be transferred to 140,000 kg/hr of raw water coming from a supply at 20°C. The baffles will be spaced 12 in. apart. Write a computer program to determine the effects of varying tube size and configuration as well as shell diameter on thermal characteristics, size, and fluid flow characteristics of the shell-and-tube type heat exchanger. 1-P, 2-P, and 4-P configurations will be considered. The exchanger data are given below:

3/4" OD, 18 BWG tubes on a 1 in. square pitch
3/4" OD, 18 BWG tubes on a 1 in. triangular pitch.

Study three different shell diameters of 15 1/4, 17 1/4, and 19 1/4 in. Calculate the length of this heat exchanger for clean and fouled surfaces. A 120,000 N/m² pressure drop may be expended on both streams. Will this heat exchanger be suitable? For the pumping power, assume a pump efficiency of 80%.

9.4. A heat exchanger is available to heat raw water by the use of condensed water at 67°C which flows in the shell side with a mass flow rate of 50,000 kg/hr. Shell side dimensions are as follows: $D_s = 19.25$ in., $P_T = 1.25$ in. (square), and baffle spacing = 0.3 m. The raw water enters the tubes at 17°C with a mass flow rate of 30,000 kg/hr. Tube dimensions are as follows: $d_o = 1$ in. = 0.0254 m (18 BWG tubes), $d_i = 0.902$ in. The length of the heat exchanger is 6 m with two passes. The permissible maximum pressure drop on the shell side is 1.5 psi. Water outlet temperature should not be less than 40°C. Calculate:
a. outlet temperatures
b. heat load of the heat exchanger
Is the heat exchanger appropriate for this purpose?

9.5. Distilled water at a flow rate of 20 kg/s enters an exchanger at 35°C and leaves at 25°C. The heat will be transferred to 40 kg/s of raw water coming from supply at 20°C. Available for this service is a 17.25" ID exchanger having 166, 3/4" OD tubes (18 BWG) and laid out on a 1 in. square pitch. The bundle is arranged for two passes and baffles are spaced 12 in. apart. Calculate the length of this heat exchanger if all the surfaces are clean. Will this heat exchanger be suitable? Assume fouling factors. A $12 \times 10^4 \, N/m^2$ pressure drop may be expended on both streams.

9.6. A 1–2 baffled shell-and-tube type heat exchanger is used as an engine oil cooler. Cooling water flows through tubes at 25°C at a rate of 8.16 kg/s and exits at 35°C. The inlet and outlet temperatures of the engine oil are 65°C and 55°C, respectively. The heat exchanger has a 15.25" ID shell and 18 BWG, 0.75" OD tubes. A total of 160 tubes are laid out on a 15/16" triangular pitch. Assuming $R_{fo} = 1.76 \times 10^{-4} \, m^2 \cdot K/W$, $A_o R_w = 1.084 \times 10^{-5} \, m^2 \cdot K/W$, $h_o = 686 \, W/m^2 \cdot K$, $A_o/A_i = 1.1476$, and $R_{fi} = 8.8 \times 10^{-5} \, m^2 \cdot K/W$, find:

   a. the heat transfer coefficient inside the tubes
   b. the total surface area of the heat exchanger.

9.7. A water-to-water system is used to test the effects of changing tube length, baffle spacing, tube pitch, pitch layout, and tube diameter. Cold water at 25°C and 100,000 kg/hr is heated by hot water at 100°C, also at 100,000 kg/hr. The exchanger has a 31 in. ID shell. Perform calculations on this 1–2 shell-and-tube heat exchanger for the following conditions, as outlined in the examples, and put together an overall comparison chart.

   a. 3/4 in. OD tubes, 12 BWG laid out on a 1 in. triangular pitch; 3 baffles per meter of tube length. Analyze the exchanger for tube lengths of 2, 3, 4, and 5 m.
   b. 3/4 in. OD tubes, 12 BWG laid out on a 1 in. triangular pitch; 2 m long. Analyze for baffle placement of 1 baffle per meter of tube length, 2 baffles per meter of tube length, 3 baffles per meter of tube length, 4 baffles per meter of tube length, and 5 baffles per meter of tube length.
   c. 3/4 in. OD tubes, 12 BWG; 2 m long; 4 baffles per meter of tube length. Analyze the exchanger for tube layouts of 15/16 in. triangular pitch, 1 in. triangular pitch, and 1 in. square pitch.
   d. 1 m OD tubes, 12 BWG; 4 m long; 9 baffles. Analyze the exchanger for tube layouts of 1.25 in. triangular pitch and 1.25 in. square pitch.
   e. 3/4 in. OD tubes, 12 BWG laid out on a 1 in. triangular pitch; 2 m long; 4 baffles per meter of tube length. Analyze the exchanger for 2, 4, 6, and 8 tube passes, and compare to the case of true counterflow (i.e., 1 tube pass).

9.8. A sugar solution ($\rho = 1080 \, kg/m^3$, $c_p = 3601 \, J/kg \cdot K$, $k_f = 0.5764 \, W/m \cdot K$, $\mu = 1.3 \times 10^{-3} \, N \cdot s/m^2$) flows at rate of 60,000 kg/hr and is to be heated from 25°C to 50°C. Water at 95°C is available at a flow rate of 75,000 kg/hr (c_p = 4004 J/kg · K). It is proposed to use a one shell pass and two tube pass shell-and-tube heat exchanger containing 3/4 in. OD, 16 BWG tubes. The velocity of the sugar solution through the tube is 1.5 m/s, and the length of the heat exchanger should not be more than 3 m because of the space limitations. Assume that the shell-side heat transfer coefficient is 700 W/m² · K. Thermal conductivity of the tube material is 52 W/m · K.

a. Calculate the number of tubes and the tube-side (sugar) heat transfer coefficient.
b. Calculate the overall heat transfer coefficient.
c. Calculate the length of this heat exchanger. Is this heat exchanger acceptable?

DESIGN PROJECT 9.1
Design of an Oil Cooler for Marine Applications

The following specifications are given:

Fluid	SAE-30 Oil	Sea Water
Inlet temperature (°C)	65	20
Outlet temperature (°C)	56	32
Pressure drop limit (kPa)	140	40
Total mass flow rate (kg/s)	20	—

A shell-and-tube heat exchanger type with geometrical parameters can be selected. The heat exchanger must be designed and rated. Different configurations of shell-and-tube types can be tested. A parametrical study is expected to develop with a suitable final design; mechanical design will be performed and the cost will be estimated; follow the chart given in Figure 9.10.

DESIGN PROJECT 9.2

The shell-and-tube type oil cooler for marine applications given in Design Project 9.1 can be repeated for a lubricating oil mass flow rate of 10 and 5 kg/s while keeping the other specifications constant.

DESIGN PROJECT 9.3

Design a shell-and-tube heat exchanger to be used as a crude oil cooler. 120 kg/s of crude oil enters the shell side of the heat exchanger at 102°C and leaves at 65°C. The coolant to be used is city water entering the tube side at 21°C with a flow rate of 65 kg/s. Shell-side pressure drop is limited to

150 kPa. Design and rating expectations are the same as outlined in Design Project 9.1.

DESIGN PROJECT 9.4

Problems 9.1, 9.2, 9.3, and 9.7 can be assigned as students' thermal design projects. Materials selection, parametric study, mechanical design, technical drawings of various components, assembly, and the cost analysis must be included.

References

1. Tubular Exchanger Manufacturers Association (TEMA), *Standards of the Tubular Exchanger Manufacturers Association*, 7th ed., Tarrytown, NY, 1988.
2. Butterworth, D., Condensers and their design, in *Two-Phase Flow Heat Exchangers: ThermalHydraulic Fundamentals and Design*, Kakaç, S., Bergles, A. E., and Fernandes, E. O., Eds., Kluwer, The Netherlands, 1988.
3. Fraas, A. P., *Heat Exchanger Design*, John Wiley & Sons, New York, 1989.
4. Kakaç, S., Bergles, A. E., and Mayinger, F., Eds., *Heat Exchangers: Thermal-Hydraulic Fundamentals and Design*, Taylor and Francis, Washington, D.C., 1981.
5. Bell, K. J., Preliminary design of shell and tube heat exchangers., in *Heat Exchangers: Thermal-Hydraulic Fundamentals and Design*, Kakaç, S., Bergles, A. E., and Mayinger, F., Eds., Taylor and Francis, Washington, D. C., 1981, 559.
6. Taborek, J., Shell-and-tube heat exchangers, in *Heat Exchanger Design Handbook*, Schlünder, E. U., Ed., Hemisphere, Washington, D.C., 1983, Section 3.3.
7. Taborek, J., Industrial heat exchanger design practices, in *Boilers, Evaporators and Condensers*, Kakaç, S., Ed., John Wiley & Sons, New York, 1991, 143.
8. Kern, D. Q., *Process Heat Transfer*, McGraw-Hill, New York, 1950.
9. Tinker, T., Shell-side characteristics of shell-and-tube heat exchangers, in *General Discussion on Heat Transfer*, Institute for Mechanical Engineering and ASME, New York and London, 1951, 97.
10. Palen, J. W. and Taborek, J., Solution of shell-side flow pressure drop and heat transfer by stream analysis method, *Heat Transfer—Philadelphia*, CEP Symp. Series No. 92, Vol. 66, 1969, 53.
11. Bell, K. J., Delaware method of shell-side design, in *Heat Transfer Equipment Design*, Shah, R. K., Sunnarao, E. C., and Mashelkar, R. A., Eds., Taylor and Francis, Washington, D.C., 1988.
12. Bell, K. J., Delaware method for shell side design, in *Heat Exchangers: Thermal-Hydraulic Fundamentals and Design*, Kakaç, S., Bergles, A. E., and Mayinger, F., Eds., Taylor and Francis, Washington D. C., 1981, 581.
13. Hewitt, G. F., Shires, G. L., and Bott, T. R., *Process Heat Transfer*, CRC Press, Boca Raton, FL, 1994.
14. Biniciogullari, D., Comparative study of shell-and-tube heat exchangers versus gasketed-plate heat exchangers, M.S. Thesis, METU, Ankara, 2001.

10
Compact Heat Exchangers

10.1 Introduction

Compact heat exchangers of plate- and tube-fin types, tube bundles with small diameters, and regenerative type are generally used for applications where gas flows. A heat exchanger having a surface area density greater than about 700 m²/m³ is quite arbitrarily referred to as a *compact heat exchanger*. The heat transfer surface area is increased by fins to increase the surface area per unit volume and there are many variations available.[1-3] A heat exchanger is referred to as a microheat exchanger if the surface area density is above 10,000 m²/m³.

Compact heat exchangers are widely used in industry especially as gas-to-gas or liquid-to-gas heat exchangers; some examples are vehicular heat exchangers, condensers and evaporators in air-conditioning and refrigeration industry, aircraft oil coolers, automotive radiators, oil coolers, unit air heaters, intercoolers of compressors, and aircraft and space applications. Compact heat exchangers are also used in cryogenics process, electronics, energy recovery, conservation and conversion, and other industries. Some examples of compact heat exchanger surfaces are shown in Figure 10.1.[1,2]

10.1.1 Heat Transfer Enhancement

The main thermal hydraulic objective of compact heat exchangers is to produce more efficient heat exchange equipment for minimizing cost, which is to reduce the physical size of a heat exchanger for a given duty.

In heat exchangers, depending on the application, plain or enhanced/augmented heat transfer surfaces are used; either the heat transfer coefficient is increased or the surface is augmented. The main goal is to obtain high heat transfer rates under the specified conditions. Heat transfer between a wall and the fluid is given by

$$Q = hA(T_w - T_f) \tag{10.1}$$

or

$$Q = (hA)_p (T_w - T_f) \tag{10.2}$$

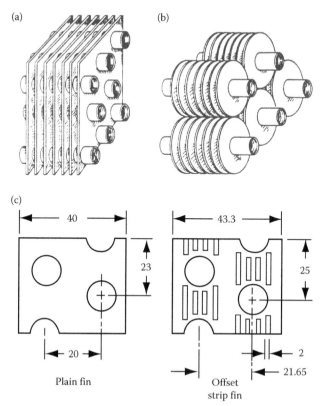

FIGURE 10.1
Finned-tube geometries used with circular tubes: (a) plate fin-and-tube used for gases; (b) individually finned tubes; (c) plainfin and offsetfin. (From Webb, R. L., *Principles of Enhanced Heat Transfer*, John Wiley & Sons, New York, 1994. With Permission.)

The ratio of the hA of an augmented surface to that of a plain surface is defined as the enhancement ratio, which is[1]

$$E = \frac{hA}{(hA)_p} \qquad (10.3)$$

There are several methods to increase the hA value:

1. The heat transfer coefficient can be increased without an appreciable increase in the surface area.
2. The surface area can be increased without appreciable changes in the heat transfer coefficient.

3. Both the heat transfer coefficient and the surface area can be increased.

Heat transfer from a finned surface (augmented) is given by Equation 2.15:

$$Q = \eta h A (T_w - T_f) \qquad (10.4)$$

or

$$Q = \eta h A \Delta T \qquad (10.5)$$

where

$$\eta = \left[1 - \frac{A_f}{A}(1 - \eta_f) \right] \qquad (10.6)$$

where A is the total surface area of the finned surface. The total heat transfer rate, Q, between the fluids can be determined from Equation 2.7:

$$Q = UA\Delta T_m \qquad (2.7)$$

The term $1/UA$ is the overall thermal resistance, which is expressed by Equation 2.16:

$$R_t = \frac{1}{UA} = \frac{1}{\eta_i h_i A_i} + \frac{R_{fi}}{\eta_i A_i} + R_w + \frac{R_{fo}}{\eta_o A_o} + \frac{1}{\eta_o h_o A_o} \qquad (2.16)$$

which includes fouling on both sides of the heat exchanger surface. The fouling characteristics of an enhanced surface must be considered.

As can be seen from Equations 2.7 and 2.16, the performance of the heat exchanger will be increased if the UA term is increased or the overall thermal resistance is reduced. The details of the principles of performance evaluation for single-phase and two-phase flows can be found in the literature.[1,3] The heat transfer enhancement increases with either or both of the hA terms, that is, decreased convective resistance on both sides. It can be seen from the above, that the increased UA term will contribute to the following:

1. Reduction of the size of the heat exchanger: if the heat duty is kept constant, under the same temperature conditions, the length of the heat exchanger may be reduced. If the length and the inlet temperatures are kept constant, then the heat exchange rate will increase.

2. If the heat duty Q and the total length L are kept constant, then ΔT_m may be reduced. This provides increased thermodynamic process efficiency (the second law efficiency) and yields lower system operational cost.
3. For fixed heat duty, Q, the pumping power can be reduced. However, this will dictate that the enhanced heat exchanger operate at a smaller velocity than the plain surface. This will require increased frontal area.

Selection of one of the above improvements will depend on the objectives of designing the heat exchanger. The most important factor, in general, is the size reduction, which results in cost reduction.

There are various techniques used for heat transfer enhancement, which are segregated in two groups: (1) the active techniques that require external power to the surface (surface vibration, acoustic, or electric field), and (2) the passive techniques, which use specific surface geometries with surface augmentation, which is the case generally applied in designing compact heat exchangers; the heat transfer coefficient may or may not be increased. The heat transfer coefficient of an enhanced surface, which is given by the Colburn modulus, J, and frictional factor of the surface, f, are usually presented as a function of the Reynolds number ($D_h G/\mu$):

$$j = \frac{h}{G c_p} Pr^{2/3} \tag{10.7}$$

The friction factor of an enhanced surface in a single-phase flow is higher than the smooth surface when operated at the same Reynolds number. Therefore, the enhanced surface would be allowed to operate at a higher pressure drop. The plain surface would give a higher heat transfer coefficient if it were also allowed to operate at a higher velocity, giving the same pressure drop as the enhanced surface. Therefore, the actual performance improvement cannot be obtained by calculating h/h_p and f/f_p at the same Reynolds number (at equal velocities).

Performance evaluation allows us to make a simple surface performance comparison between the thermal resistances of both fluid streams. The performance analysis allows the designer to quickly identify the most effective surface or other enhancement techniques of variable choices. If one of the fluids in the heat exchanger undergoes a phase change, then this exchanger is called a two-phase heat exchanger. Some of the examples are evaporators, boilers, reboilers, and condensers. In two-phase flow heat exchangers, pressure drop of phase-changing fluid may affect the mean temperature difference in the heat exchanger, and the performance evaluation must take this effect into account. In a single-phase flow, the reduced pumping power can be considered an objective function, which is not applicable to the two-phase flow situation. A detailed analysis of performance evaluation criteria for two-phase flow heat exchangers is presented by Webb.[1]

As mentioned before, compact heat exchangers are widely used in industry. Some examples are vehicular heat exchangers, condensers and evaporators in air conditioning and refrigeration units, oil coolers in aircrafts, radiators in automobiles, etc.

Most common compact heat exchangers are plate- and tube-fin heat exchangers. Adding fins increases the heat transfer surface area of these types of heat exchangers. Due to the small passage dimensions, passage-to-passage nonuniformities, and geometrical variations, accurate characterization of the heat transfer surface is difficult. In addition, due to the ever-decreasing flow passages and the ever-increasing compactness, some of the assumptions applicable to normal heat exchangers may not be valid for compact heat exchangers, especially for microheat exchangers, which find wider and wider applications. In this chapter, we are going to introduce the design of only tube- and plate-fin heat exchangers.

10.1.2 Plate-Fin Heat Exchangers

In this type, each channel is defined by two parallel plates separated by fins or spacers. Fins or spacers are sandwiched between parallel plates or formed tubes. Examples of compact plate heat exchangers are shown in Figure 10.2.[4] Fins are attached to the plates by brazing, soldering, adhesive bonding, welding, mechanical fit, or extrusion. Alternate fluid passages are connected in parallel by end heads to form two sides of a heat exchanger. Fins are employed on both sides in gas-to-gas heat exchangers.

In gas-to-liquid heat exchanger applications, fins are usually employed only on the gas side where the heat transfer coefficient is lower; if fins are employed on the liquid side, the fins provided a structural strength. The fins used in a plate heat exchanger may be plain and straight fins, plain but wavy fins, or interrupted fins such as strip, louver, and perforated fins.

10.1.3 Tube-Fin Heat Exchangers

In a tube-fin exchanger, round, rectangular, and elliptical tubes are used and fins are employed either on the outside, the inside, or both the outside and the inside of the tubes, depending upon the application. In a gas-to-liquid heat exchanger, the gas-side heat transfer coefficient is very low compared with the liquid-side heat transfer coefficient; therefore, no fins are needed on the liquid side. In some applications, fins are also used inside the tubes. The liquids flow in the tube side, which can accommodate high pressures. Tube-fin heat exchangers are less compact than plate-fin heat exchangers. Examples of tube-fin heat exchangers are shown in Figure 10.3.[4]

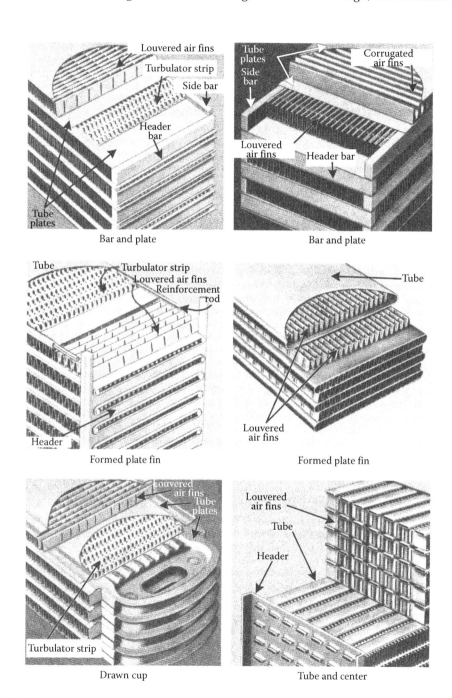

FIGURE 10.2
Plate-fin heat exchangers. (Courtesy of Harrison Division, General Motors Corp., Lockport, NY.)

Compact Heat Exchangers

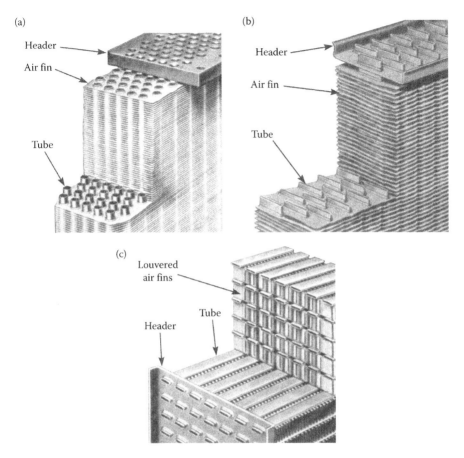

FIGURE 10.3
Tube-fin heat exchangers. (Courtesy of Harrison Division, General Motors Corp., Lockport, NY.)

Fins on the outside tubes may be categorized as (1) flat or continuous (plain, wavy, or interrupted) external fins on an array of tubes, (2) normal fins on individual tubes, or (3) longitudinal fins on individual tubes.

10.2 Heat Transfer and Pressure Drop

10.2.1 Heat Transfer

As outlined earlier, compact heat exchangers are available in a wide variety of configurations of the heat transfer matrix. Their heat transfer and pressure-drop characteristics have been studied by Kays and London.[3]

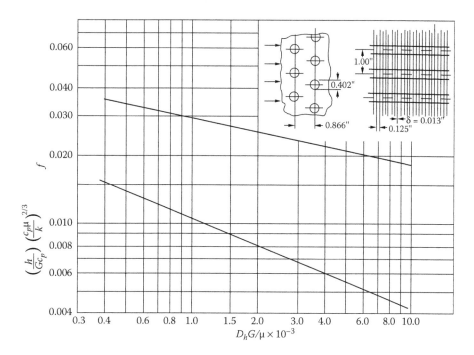

FIGURE 10.4
Heat transfer and friction factor for a circular tube continuous fin heat exchanger. Surface 8.0-3/8 T: tube OD = 1.02 cm; fin pitch = 3.15/cm; fin thickness = 0.033 cm; fin area/total area = 0.839; air-passage hydraulic diameter = 0.3633 cm; free-flow area/frontal area, σ = 0.534; heat transfer area/total volume = 587 m²/m³. (From Kays, W. M. and London, A. L., *Compact Heat Exchangers*, 3rd ed., McGraw-Hill, New York, 1984. With permission.)

The heat transfer and pressure-drop characteristics of various configurations for use as compact heat exchangers are determined experimentally. Figures 10.4–10.7 and 10.9 show typical heat transfer and friction factor data for different configurations of compact heat exchangers.[3,5] For fin-tube type configuration, five different surface geometries are given in Table 10.1 and shown in Figure 10.8. Also, data for the heat transfer and friction coefficients for four different plain plate-fins types are shown in Figure 10.9, and geometrical data are given in Table 10.1.

Note that there are three dimensionless groups governing these correlations, which are the Stanton, Prandtl, and Reynolds numbers:

$$St = \frac{h}{Gc_p}; \quad Pr = \frac{c_p \mu}{k}; \quad Re = \frac{GD_h}{\mu} \tag{10.8}$$

Compact Heat Exchangers

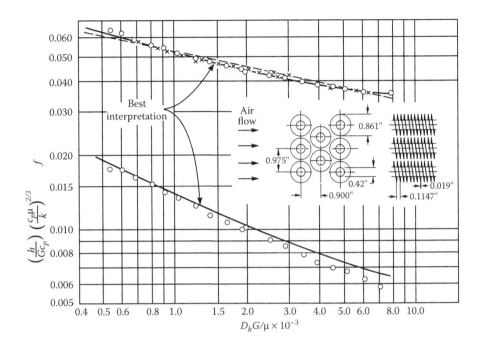

FIGURE 10.5
Heat transfer and friction factor for flow across circular finned-tube matrix. Surface CF-8.72(c): tube OD = 1.07 cm; fin pitch = 3.43/cm; fin thickness = 0.048 cm; fin area/total area = 0.876; air-passage hydraulic diameter, d_h = 0.443 cm; free-flow area/frontal area, σ = 0.494; heat transfer area/total volume = 446 m²/m³. (From Kays, W. M. and London, A. L., *Compact Heat Exchangers*, 3rd ed., McGraw-Hill, New York, 1984. With permission.)

where G is the mass velocity or mass flux, defined as

$$G = \rho u_{max} = \frac{\dot{m}}{A_{min}} \qquad (10.9)$$

$\dot{m}$ is the total mass flow rate of fluid, and A_{min} is the minimum free-flow cross-sectional area, regardless of where this minimum occurs.

The value of the hydraulic diameter for each configuration is specified in Figures 10.4–10.6 and in Table 10.1.

The hydraulic diameter is defined as four times the flow passage volume divided by the total heat transfer area:

$$D_h = 4\frac{LA_{min}}{A} \qquad (10.10)$$

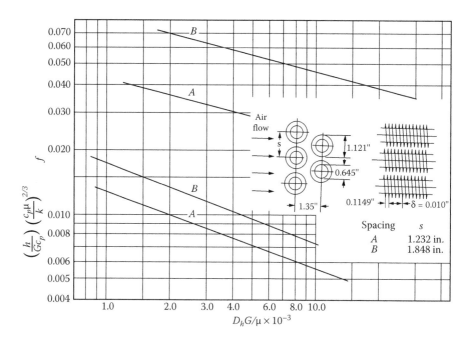

FIGURE 10.6
Heat transfer and friction factor for flow across finned-tube matrix. Surface CF-8.7-5/8 J: tube OD = 1.638 cm; fin pitch = 3.43/cm; fin thickness = 0.0254 cm; fin area/total area = 0.862; air-passage hydraulic diameter, D_h = 0.5477 cm (a), 1.1673 cm (b); free-flow area/frontal area, σ = 0.443 (a), 0.628 (b); heat transfer area/total volume = 323.8 m²/m³ (a), 215.6 m²/m³ (b). (From Kays, W. M. and London, A. L., *Compact Heat Exchangers*, 3rd ed., McGraw-Hill, New York, 1984. With permission.)

where L is the flow length of the exchanger matrix, LA_{min} is the minimum free flow passage volume, and A is the total heat transfer area. Once the Reynolds number for flow is known, the heat transfer coefficient (Colburn modules, $J_H = St\, Pr^{2/3}$) and the friction factor, f, for flow across the matrix can be evaluated.

The overall heat transfer coefficient based on the gas-side surface area in a gas–liquid heat exchanger neglecting fouling effects can be written from Equation 2.17 as

$$\frac{1}{U_o} = \frac{A_t}{A_i}\frac{1}{h_i} + A_t R_w + \frac{1}{\eta_o h_o} \qquad (10.11)$$

where η_o is the outside overall surface efficiency, defined as

$$\eta_o = 1 - \frac{A_f}{A_t}(1-\eta_f)$$

Compact Heat Exchangers

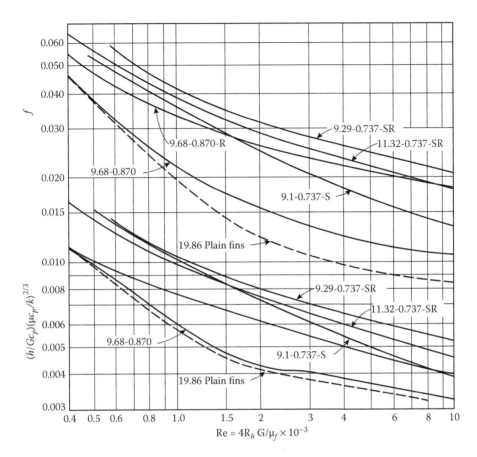

FIGURE 10.7
Heat transfer and friction factor for flow across finned flat-tube matrix for the surfaces shown in Figure 10.8 and Table 10.1. (From Kays, W. M. and London, A. L., *Compact Heat Exchangers*, 3rd ed., McGraw-Hill, New York, 1984. With permission.)

and A_t ($=A_u + A_f$) is the total external air-side surface area.

The overall heat transfer coefficient based on the internal surface area becomes

$$\frac{1}{U_i} = \frac{1}{h_i} + A_i R_w + \frac{A_i}{A_t} \frac{1}{\eta_o h_o} \qquad (10.12)$$

If both sides are finned, Equation 10.12 becomes

$$\frac{1}{U_i} = \frac{1}{\eta_i h_i} + A_i R_w + \frac{A_i}{A_t} \frac{1}{\eta_o h_o} \qquad (10.13)$$

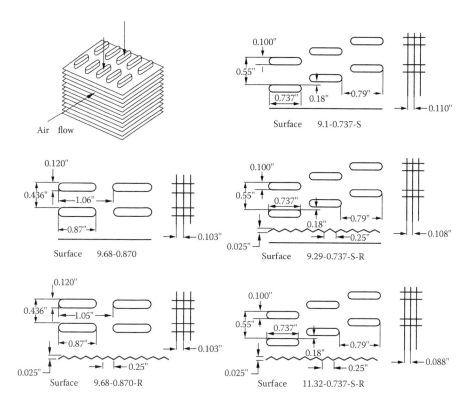

FIGURE 10.8
Various flattened tube-platefin compact surfaces for which test data are presented in Figure 10.7. (From Kays, W. M. and London, A. L., *Compact Heat Exchangers*, 3rd ed., McGraw-Hill, New York, 1984. With permission.)

The air-side conductance is

$$K = \frac{\eta_o h_o A_t}{A_i} \tag{10.14}$$

Generally, an increase in the number of fins per centimeter will increase the conductance by increasing the ratio A_t/A_i. Also, the use of more closely spaced fins will increase the heat transfer coefficient, h_o, because of a smaller hydraulic diameter. Alternatively, the use of a special fin configuration, such as a wavy fin, will produce a higher heat transfer coefficient. The outside surface efficiency, η_o, is influenced by the fin thickness, thermal conductivity, and fin length. The fin efficiency, η_f, may be calculated from appropriate graphs or equations given in most heat transfer textbooks.

Compact Heat Exchangers

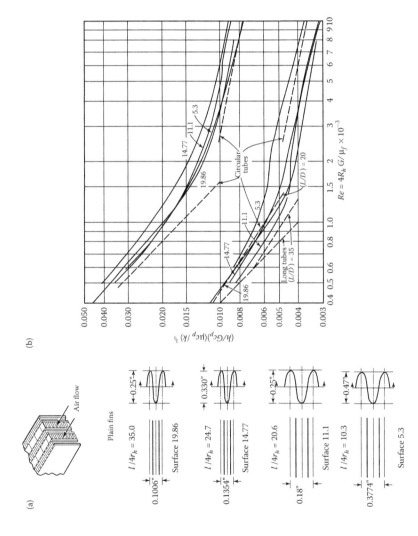

FIGURE 10.9
Heat transfer and friction factor for four plain plate-fin heat transfer matrixes of Table 10.1. (From Kays, W. M. and London, A. L., *Compact Heat Exchangers*, 3rd ed., McGraw-Hill, New York, 1984. With permission.)

TABLE 10.1

Heat Transfer Matrix Geometries for Plate Plain-Fin and Fin Flat-Tube Types for Which Test Data are Presented in Figures 10.7 and 10.9

Surface Designation	Fins (per cm)	Hydraulic Diameter (D_h, cm)	Plate Spacing (b, cm)	Tube or Fin Thickness (cm)	Extended Area Total Area	Area Volume Between Plates (β, m²/m³)	Area Core Volume (β, m²/m³)	Free Flow Area Frontal Area (σ)
Plate plain-fin type								
5.30	13.46	0.051	1.194	0.0152	0.719	511.8		
11.10	28.19	0.257	0.635	0.0152	0.730	1095.8		
14.77	37.52	0.215	0.838	0.0152	0.831	1210.6		
19.86	50.44	0.152	0.635	0.0152	0.833	1493.0		
Fin flat-tube type								
9.68-0.870	24.587	0.2997		0.0102	0.795		751.3	0.697
9.68-0.870-R	24.587	0.2997		0.0102	0.795		751.3	0.697
9.1-0.737-S	23.114	0.3565		0.0102	0.813		734.9	0.788
9.29-0.737-S-R	28.753	0.3510		0.0102	0.845		885.8	0.788
11.32-0.737-S-R	23.596	0.3434		0.0102	0.814		748.0	0.780

Source: From Kays, W. M. and London, A. L., *Compact Heat Exchangers*, 3rd ed., McGraw-Hill, New York, 1984. With permission.

10.2.2 Pressure Drop for Finned-Tube Exchangers

For flow normal to finned-tube banks such as illustrated in Figures 10.4–10.7, the total pressure drop, namely, the difference between the pressures at the inlet and outlet, is given by Kays and London[3:]

$$\Delta p = \frac{G^2}{2\rho_i}\left[f\frac{A_t}{A_{min}}\frac{\rho_i}{\rho}+(1+\sigma^2)\left(\frac{\rho_i}{\rho_o}-1\right)\right] \quad (10.15)$$

where

$$\sigma = \frac{A_{min}}{A_{fr}} = \frac{\text{minimum free-flow area}}{\text{frontal area}} \quad (10.16)$$

$$\frac{A_t}{A_{min}} = \frac{4L}{D_h} = \frac{\text{total heat transfer area}}{\text{minimum flow area}} \quad (10.17)$$

$$G = \frac{\rho u_\infty A_{fr}}{A_{min}} = \frac{\rho u_\infty}{\sigma} \quad (10.18)$$

In this equation, ρ is the average density evaluated at the average temperature between the inlet and outlet, or it can also be estimated by averaging the fluid-specific volume between the inlet and outlet as

$$\frac{1}{\rho} = \frac{1}{2}\left(\frac{1}{\rho_i}+\frac{1}{\rho_o}\right) \quad (10.19)$$

The friction factor, f, for some tube-plate fin heat exchangers has been found experimentally and plotted in Figures 10.4–10.7 (Refs. [3,5]) and accounts for fluid friction against solid walls and for the entrance and exit losses. The second term on the right-hand-side of Equation 10.15 accounts for the acceleration or deceleration of flow. This term is negligible for liquids for which the density is essentially constant.

10.2.3 Pressure Drop for Plate-Fin Exchangers

Consider plate-fin exchanger surfaces as shown in Figure 10.8. The total pressure drop for flow across the heat exchanger matrix is given as[3,6]

$$\Delta p = \frac{G^2}{2\rho_i}\left[(k_c+1-\sigma^2)+2\left(\frac{\rho_i}{\rho_o}-1\right)+f\frac{A}{A_{min}}\frac{\rho_i}{\rho}-(1-k_e-\sigma^2)\frac{\rho_i}{\rho_o}\right] \quad (10.20)$$

The first term inside the bracket shows the entrance effect, the second term shows the flow acceleration effect, the third term shows the core friction, and the last term represents the exit effect.

In Equation 10.20, the frictional pressure drop generally dominates and accounts for about 90% or more of the total pressure drop across the core. The friction factor, f, has been found experimentally and plotted in Figure 10.9 for the surfaces shown in Figure 10.8. The entrance and exit losses become important for short cores, small values of σ, high values of the Reynolds number, and gases. They are negligible for liquids. Typical values of the contraction loss coefficient, k_c, and the enlargement loss coefficient, k_e, are given by Kays and London.[3]

Example 10.1

Air at 1 atm and 400 K and with a velocity of $U_\infty = 10$ m/s flows across a compact heat exchanger matrix having the configuration shown in Figure 10.4. Calculate the heat transfer coefficient, h, and frictional pressure drop for the air side. The length of the matrix is 0.6 m.

Solution

At 400 K and 1 atm, the properties of air from Appendix B are

$$\rho = 0.8825 \text{ kg/m}^3$$
$$\mu = 2.29 \times 10^{-5} \text{ kg/m} \cdot \text{s}$$
$$c_p = 1013 \text{ Jk/g} \cdot \text{K}$$
$$Pr = 0.719$$

Figure 10.4 demonstrates

$$\frac{A_{min}}{A_{fr}} = \sigma = 0.534 \text{ and } D_h = 0.3633 \text{ cm}$$

Then,

$$G = \frac{\rho u_\infty A_{fr}}{A_{min}} = \frac{\rho u_\infty}{\sigma} = \frac{0.8825 \times 10}{0.534} = 16.53 \text{ kg/}(\text{m}^2 \cdot \text{s})$$

$$Re = \frac{GD_h}{\mu} = \frac{16.53 \times 0.3633 \times 10^{-2}}{2.29 \times 10^{-5}} = 2622$$

Compact Heat Exchangers

From Figure 10.4, for $Re = 2,622$, we can obtain

$$\frac{h}{Gc_p} Pr^{2/3} = 0.0071$$

$$h = 0.0071 \times \frac{Gc_p}{Pr^{2/3}} = 0.0071 \times \frac{16.53 \times 1013}{(0.719)^{2/3}}$$

$$h = 148.1 \text{ W/m}^2 \cdot \text{K}$$

For $Re = 2,622$, from Figure 10.4, $f = 0.025$ and so

$$\Delta p_f = f \frac{G^2}{2\rho_a} \frac{A_t}{A_{min}}$$

$$\frac{A_t}{A_{min}} = \frac{4 \times L}{D_h} = \frac{4 \times 0.6}{0.3633 \times 10^{-2}} = 660.6$$

Then,

$$\Delta p_f = 0.025 \times \frac{16.53^2}{2 \times 0.8825} \times 660.6 = 2556 \text{ Pa}$$

Example 10.2

Air at 2 atm and 500 K with a velocity of $U_\infty = 20$ m/s flows across a compact heat exchanger matrix having the configuration shown in Figure 10.8 (surface 11.32-0737-S-R). Calculate the heat transfer coefficient and the frictional pressure drop. The length of the matrix is 0.8 m.

Solution

For air at 500 K and 2 atm, from Appendix B,

$$\rho = 1.41 \text{ kg/m}^3$$
$$c_p = 1030 \text{ J/kg} \cdot \text{K}$$
$$\mu = 2.69 \times 10^{-5} \text{ kg/m} \cdot \text{s}$$
$$Pr = 0.718$$

The mass flux G is

$$G = \frac{\dot{m}}{A_{min}} = \frac{\rho u_\infty A_{fr}}{A_{min}} = \frac{\rho u_\infty}{\sigma} = \frac{1.41 \times 20}{0.78} = 36.15 \text{ kg/m}^2 \cdot \text{s}$$

and the Reynolds number becomes

$$Re = \frac{GD_h}{\mu} = \frac{36.15 \times 0.3434 \times 10^{-2}}{2.69 \times 10^{-5}} = 4615$$

From Figure 10.7, for $Re = 4{,}615$, we get

$$\frac{h}{Gc_p} \cdot Pr^{2/3} = 0.0058$$

$$h = 0.0058 \frac{Gc_p}{Pr^{2/3}}$$

$$h = 0.0058 \times \frac{36.15 \times 1030}{(0.718)^{2/3}} = 278.7 \text{ W/m}^2 \cdot \text{K}$$

$$\Delta p_f = f \frac{A_t}{A_{min}} \frac{\rho_i}{\rho} \frac{G^2}{2\rho_i}$$

$$\frac{A_t}{A_{min}} = \frac{4 \times L}{D_h} = \frac{4 \times 0.8}{0.3434 \times 10^{-2}} = 932$$

For $Re = 4{,}615$, from Figure 10.7, $f = 0.023$ and $\rho_i/\rho_o \approx 1$:

$$\Delta p_f = 0.023 \times 932 \times \frac{(36.15)^2}{2 \times 1.41} = 9934 \text{ N/m}^2$$

Example 10.3

Air enters the core of a finned-tube heat exchanger of the type shown in Figure 10.5 at 1 atm and 30°C. The air flows at a rate of 1,500°kg/h perpendicular to the tubes and exits with a mean temperature of 100°C. The core is 0.5 m long with a 0.25 m² frontal area. Calculate the total pressure drop between the air inlet and outlet and the average heat transfer coefficient on the air side.

Solution

The air densities at the inlet and the outlet are, from Appendix B,

$$\rho_i = 1.177 \text{ kg/m}^3$$
$$\rho_o = 0.954 \text{ kg/m}^3$$

The properties at the bulk temperature of $(100 + 30)/2 = 65°C$, from Appendix B, are

$$\rho = 1.038 \text{ kg/m}^3$$
$$Pr = 0.719$$
$$\mu = 2.04 \times 10^{-5} \text{ kg/m} \cdot \text{s}$$
$$c_p = 1.007 \text{ kJ/kg} \cdot \text{K}$$

One can also use information from the caption of Figure 10.5, $D_h = 0.443$ cm and $\sigma = 0.494$, or the pressure drop is calculated from Equation 10.15:

$$\frac{A_t}{A_{min}} = \frac{\beta A_{fr} L}{A_{fr}\sigma} = \frac{\beta L}{\sigma}$$

where

$$\frac{A_t}{A_{min}} = \frac{4L}{D_h} = \frac{4 \times 0.5}{0.00443} = 451.5$$

$$A_{min} = \sigma A_f = 0.494 \times 0.25 = 0.124 \text{ m}^2$$

$$G = \frac{\dot{m}}{A_{min}} = \frac{1500}{3600} \times \frac{1}{0.124} = 3.36 \text{ kg/m}^2 \cdot \text{s}$$

$$Re = \frac{GD_h}{\mu} = \frac{3.36 \times 0.00443}{2.04 \times 10^{-5}} = 729$$

(Note the deletion of kg/m s in the second line of this group of equations.)
From Figure 10.5, for $Re = 729$ and $f = 0.057$,

$$\Delta p = (3.36)^2 \frac{1}{2 \times 1.177}\left[0.057 \times 451.5 \times \frac{1.177}{1.038} + (1+0.494^2)\left(\frac{1.177}{0.954} - 1\right)\right]$$

$$= 141.3 \text{ N/m}^2$$

For the heat transfer coefficient, the Colburn modulus $(h/Gc_p) Pr^{2/3}$ can be read from Figure 10.5 for $Re = 729$ as 0.0165.

$$\frac{h}{Gc_p} Pr^{2/3} = 0.0165$$

$$h = 0.0165 \times 3.36 \times 1.007 \times 10^3 \times (0.719)^{-2/3}$$

$$h \approx 70 \text{ W/m}^2 \cdot \text{K}$$

Nomenclature

A	area, m^2
A_f	finned area, m^2
A_{fr}	frontal area, m^2
A_{min}	minimum free-flow area, m^2
A_t	total heat transfer area on the outside, m^2
A_u	unfinned area, m^2
d	diameter of the tube, m
D_h	hydraulic diameter, m
f	friction factor
G	mass velocity, kg/m$^2 \cdot$ s
k	thermal conductivity, W/m $\cdot$ K
h	heat transfer coefficient, W/m$^2 \cdot$ K
U_∞	free stream velocity, m/s
V	volume of the core, m^3
β	heat transfer area density, m^2/m^3
Δp	pressure drop, N/m^2
ΔT	temperature difference, K, °C
η_f	fin efficiency
η_o	overall surface efficiency
μ	viscosity, kg/m $\cdot$ s
ρ	density, kg/m^3

PROBLEMS

10.1. Air at 1 atm and 400 K and with a velocity of 10 m/s flows across the compact heat exchanger shown in Figure 10.6a and exits with a mean temperature of 300 K. The core is 0.6 m long. Calculate the total frictional pressure drop between the air inlet and outlet and the average heat transfer coefficient on the air side.

10.2. Air enters the core of a finned tube heat exchanger of the type shown in Figure 10.4 at 2 atm and 150°C. The air mass flow rate is 10 kg/s and flows perpendicular to the tubes. The core is 0.5 m long with a 0.30 m^2 frontal area. The height of the core is 0.5 m. Water at 15°C and at a flow rate of 50 kg/s flows inside the tubes. Airside data are given on Figure 10.4. For water side data, assume that $\sigma_w = 0.129$, $D_h = 0.373$ cm, and waterside heat transfer area/total volume = 138 m^2/m^3. This is a rating problem. Calculate:
   a. the airside and waterside heat transfer coefficients
   b. overall heat transfer coefficient based on the outer (airside) surface area
   c. total heat transfer rate and outlet temperatures of air and water.

10.3. Hot air at 2 atm and 500 K at a rate of 8 kg/s flows across a circular finned-tube matrix configuration shown in Figure 10.6. The frontal area of the heat exchanger is 0.8 m × 0.5 m and the core is 0.5 m long. Geometrical configurations are shown in Figure 10.6. Calculate:
   a. the heat transfer coefficient
   b. the total frictional pressure drop between the air inlet and outlet.

10.4. Repeat Problem 10.3 for a finned-tube matrix shown in Figure 10.5 and discuss the results.

10.5. Repeat Problem 10.1 for the heat exchanger matrix configuration in Figure 10.8 for the matrix 9.1-0.737-s as given in Table 10.1.

10.6. Repeat Problem 10.2 for the heat exchanger matrix configuration shown in Figure 10.8 for the surface 11.32-0.737-S-R (see Table 10.1).

10.7. Air at 2 atm and 500 K and at a rate of 12 kg/s flows across a plain plate-fin matrix of the configuration shown in Figure 10.8 and in Table 10.1 for the surface 9.68–0.870. The frontal area is 0.8 m × 0.6 m and the length of the matrix is 0.6 m. Calculate:
   a. the heat transfer coefficient
   b. the friction coefficient.

10.8. Repeat Problem 10.5 for a finned-tube matrix of the configuration shown in Figure 10.8 for the surface 9.68-0.870-R (see Figure 10.7) geometrical configuration given in Table 10.1.

10.9. An air-to-water compact heat exchanger is to be designed to serve as an intercooler in a gas turbine plant. Geometrical details of the proposed surface (surface 9.29-0.737-S-R) for the air side are given in Figure 10.8 and Table 10.1. Hot air at 2 atm and 400 K with a flow rate of 20 kg/s flows across the matrix. The outlet temperature of air is 300 K and the allowable pressure drop is 0.3 bar. Water at 17°C and a flow rate of 50 kg/s flows inside the flat tubes. Water velocity is 1.5 m/s. Waterside geometrical details are

$$D_h = 0.373 \text{ cm}$$

$$\sigma_w = 0.129$$

$$\frac{\text{water} - \text{side heat transfer area}}{\text{total volume}} = 138 \text{ m}^2/\text{m}^3$$

10.10. Repeat Problem 10.7 for air at a flow rate of 15 kg/s while everything else remains the same.

10.11. An aircooled refrigerant condenser is to be designed. A flattened tube with corrugated fins will be used. The surface selected for the matrix is similar to Figure 10.4. The cooling load (heat duty) is 125 kW. The refrigerant 134A condenses inside the tubes at 310 K. Air enters the condenser at 18°C and leaves at 26°C. The mean pressure is 2 atm. Calculate:

a. the air side heat transfer coefficient

b. the tube side condensation heat transfer coefficient.

The airside geometrical configurations for the surface 7.75-5/8 T are³

Tube OD = 1.717 m

Tube arrangement = staggered

Fins/inch = 7.75

Fin type = plain

Fin thickness = 0.41×10^{-3} m

Minimum free-flow area/frontal area, $\sigma = 0.481$

Hydraulic diameter, $D_h = 3.48 \times 10^{-3}$ m

Heat transfer area/total volume, $\alpha = 554$ m²/m³

10.12. An air-to-water compact heat exchanger is to be designed to serve as an intercooler for a 4,000 hp gas turbine plant. The heat exchanger is to meet the following heat transfer and pressure-drop performance specifications:

Air-side operating conditions:

Flow rate = 25 kg/s

Inlet temperature = 500°C

Outlet temperature = 350°C

Inlet pressure = 2×10^5 N/m²

Pressure drop rate = 8%

Water side operating conditions:

Flow rate = 50 kg/s

Inlet temperature = 290 K

10.13. Design a heat exchanger asking the surface given in Figure 10.8 (surface 9.29-0.737-S-R). Fins are continuous aluminum. The geometrical data for the air side are given in Table 10.1. On the water side, the flatted tube is 0.2 cm × 1.6 cm. The inside diameter of the tube before it was flattened was 1.23 cm, with a wall thickness of 0.025 cm. Water velocity inside is 1.5 m/s. The design should specify the core size and the core pressure drop.

DESIGN PROJECT 10.1

Design of the Cooling System and Radiator of a Truck

Some of the design specifications as applicable to a typical truck are given as

Heat load:	100 kW
Water inlet temperature:	80°C
Water outlet temperature:	70°C
Air inlet temperature:	35°C
Air outlet temperature:	46°C
Air pressure:	100 kPa
Air pressure drop:	0.3 kPa
Core matrix of the radiator:	to be selected

Different compact surfaces must be studied and compared for the thermal and hydraulic analysis of the radiator. A parametric study is expected to develop an acceptable final design which includes the selection of a pump, materials selection, mechanical design, drawings, and cost estimation.

DESIGN PROJECT 10.2

Design of an Air-Cooled Refrigeration Condenser

The process specifications and the geometrical parameters of the condenser are given in Design Project 10.1. Two different compact surfaces given in this chapter, including 7.75-5/8,[3] will be analyzed and compared. The final design will include materials selection, mechanical design, technical drawings, and cost estimation.

References

1. Webb, R. L., *Principles of Enhanced Heat Transfer*, John Wiley & Sons, New York, 1994.
2. Kakaç, S., Shah, R. K., and Mayinger, F., Eds., *Low Reynolds Number Flow Heat Exchangers*, Hemisphere, Washington, DC, 1982.
3. Kays, W. M. and London, A. L., *Compact Heat Exchangers*, 3rd ed., McGraw-Hill, New York, 1984.
4. Kakaç, S., Bergles, A. E., and Mayinger, F., Eds., *Heat Exchangers: Thermal-Hydraulic Fundamentals and Design*, Hemisphere, Washington, DC, 1981.
5. Fraas, A. P., *Heat Exchanger Design*, 2nd ed., John Wiley & Sons, New York, 1989.
6. Özisik, M. N., *Heat Transfer — A Basic Approach*, McGraw-Hill, New York, 1985.
7. Bejan, A., *Heat Transfer*, John Wiley & Sons, New York, 1993.

11
Gasketed-Plate Heat Exchangers

11.1 Introduction

Gasketed-plate heat exchangers (the plate and frame) were introduced in the 1930s mainly for the food industries because of their ease of cleaning, and their design reached maturity in the 1960s with the development of more effective plate geometries, assemblies, and improved gasket materials. The range of possible applications has widened considerably and, at present, under specific and appropriate conditions, overlaps and successfully competes in areas historically considered to be the domain of tubular heat exchangers. They are capable of meeting an extremely wide range of duties in many industries. Therefore, they can be used as an alternative to tube-and-shell type heat exchangers for low- and medium-pressure liquid-to-liquid heat transfer applications.

The design of plate heat exchangers is highly specialized in nature considering the variety of designs available for the plates and arrangements that possibly suit various duties. Unlike tubular heat exchangers for which design data and methods are easily available, plate heat exchanger design continues to be proprietary in nature. Manufacturers have developed their own computerized design procedures applicable to the exchangers they market.

11.2 Mechanical Features

A typical plate and frame heat exchanger is shown in an exploded view in Figures 11.1 and 1.14.[1] The elements of the frame are a fixed plate, compression plate, pressing equipment, and connecting ports. The heat transfer surface is composed of a series of plates with parts for fluid entry and exit in the four corners (Figure 11.2). The flow pattern through a gasketed-plate heat exchanger is illustrated in Figure 11.3.[1]

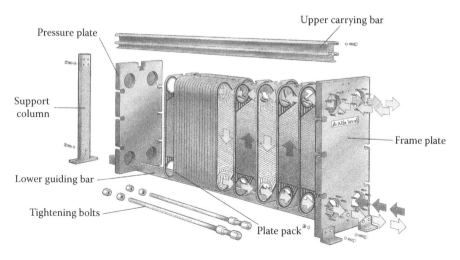

FIGURE 11.1
Gasketed-plate heat exchangers. (Courtesy of Alfa-Laval Thermal AB.)

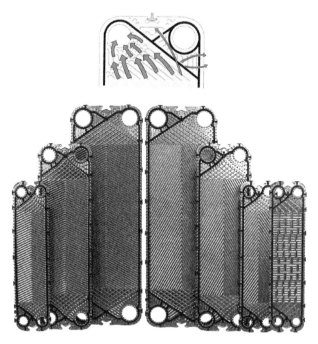

FIGURE 11.2
A chevron-type heat exchanger plate. (Courtesy of Alfa-Laval Thermal AB.)

Gasketed-Plate Heat Exchangers

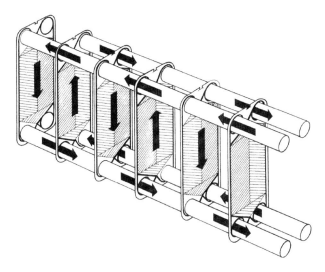

FIGURE 11.3
Flow diagram in a single-pass counterflow arrangement. (Courtesy of Alfa-Laval Thermal AB.)

11.2.1 Plate Pack and the Frame

When a package of plates are pressed together, the holes at the corners of the plates form continuous tunnels or manifolds, leading the media from the inlets into the plate package, where they are distributed into the narrow channels between the plates. The plate pack is tightened by means of either a mechanical or hydraulic tightening device that permits control of tightening pressure to the desired level. These passages formed between the plates and corner ports are arranged so that the two heat transfer media can flow through alternate channels, always in countercurrent flow. During the passage through the apparatus, the warmer medium will give some of its heat energy through the thin plate wall to the colder medium on the other side. Finally, the medium are led into similar hole-tunnels as in the inlets at the other end of the plate package and are then discharged from the heat exchanger. The plates can be stacked up to several hundred in a frame and are held together by the bolts that hold the stack in compression (Figure 11.4).[1] Figure 11.5 shows gasket arrangements.

The plate pack is compressed between the head plates using bolts. Plates and removable parts of the frame are suspended from the upper carrier bar and supported at their lower ends by a guide bar. The carrier and guide bars are bolted to the fixed part of the frame and, on all but smaller types, are attached to the end support.

The plate pack corresponds to the tube bundle in shell-and-tube exchangers with the important difference being that the two sides of a plate heat exchanger are normally of identical hydrodynamic characteristics. The basic element of the plate pack is the plate, a sheet of metal precision-pressed into

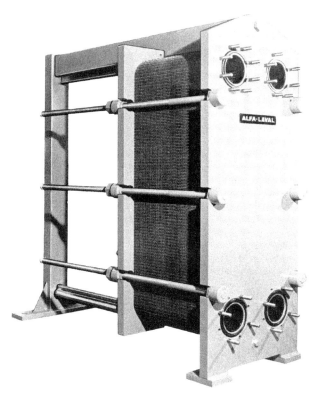

FIGURE 11.4
Gasketed-plate heat exchanger assembly. (Courtesy of Alfa-Laval Thermal AB.)

a corrugated pattern, as shown in Figure 11.2.[1] The largest single plate is of the order of 4.3 m high × 1.1 m wide. The heat transfer area for a single plate lies in the range 0.03–2.2 m². The fluid should be equally distributed over the full width of the plate. To avoid poor distribution of the fluid across the plate width, the minimum length/width ratio is of the order of 1.8. Plate thickness ranges between 0.5 and 1.2 mm, and plates are spaced with nominal gaps of 1.5–5.0 mm, yielding hydraulic diameters for the flow channels of 4–10 mm.

Leakage from the channels between the plates to the surrounding atmosphere is prevented by the gasketing around the exterior of the plate. Plates can be made from all malleable materials. The most common materials are (Table 11.1) stainless steel, titanium, titanium–palladium, incoloy 825, hastelloy C-276, diabon F100, Monel 400, aluminum, and aluminum brass.[2,3]

The number and size of the plates are determined by the flow rate, physical properties of the fluids, pressure drop, and temperature requirements.

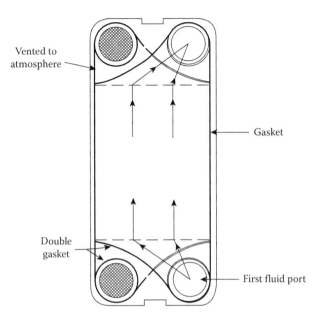

FIGURE 11.5
Gasket arrangements.

TABLE 11.1

Plate Materials

Material	Thermal Conductivity (W/m² · K)
Stainless steel (316)	16.5
Titanium	20
Inconel 600	16
Incolay 825	12
Hastelloy C-276	10.6
Monel 400	66
Nickel 200	66
9/10 Cupro-nickel	52
70/30 Cupro-nickel	35

Source: From Raju, K. S. N. and Jagdish, C. B., *Low Reynolds Number Flow Heat Exchangers*, Hemisphere, Washington, DC, 1983. With permission.

11.2.2 Plate Types

A wide variety of corrugation types are available in practical applications (Figure 11.2). Although most modern plate heat exchangers are the chevron type, a number of commercial plates have a surface corrugated pattern called washboard.[4] Because of major variations in individual corrugation patterns of this plate type, the prediction methods rely on experimental data particular to a specific pattern. In the washboard type, turbulence is promoted by a continuously changing flow direction and velocity of the fluids. In the chevron type, adjacent plates are assembled such that the flow channel provides swirling motion to the fluids; the corrugated pattern has an angle β, which is referred to as the chevron angle (Figure 11.6a). The chevron angle is reversed on adjacent plates so that when plates are clamped together, the corrugations provide numerous contact points (Figure 11.7). Because of the many support points of contact, the plates can be made from very thin material, usually 0.6 mm. The chevron angle varies between the extremes of about 65° and 25° and determines the pressure drop and heat transfer characteristics of the plate.[4]

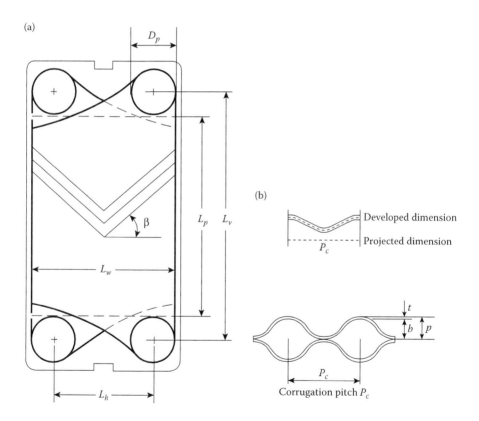

FIGURE 11.6
(a) Main dimensions of a chevron plate; (b) developed and projected dimensions of a chevron plate cross-section normal to the direction of troughs.

Gasketed-Plate Heat Exchangers

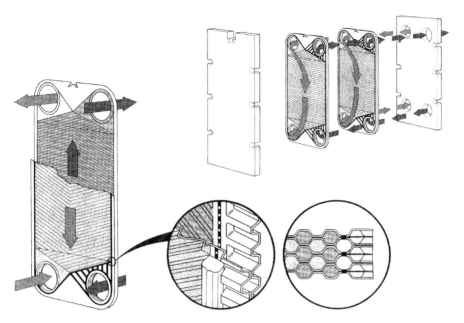

FIGURE 11.7
The chevron angle is reversed on adjacent plates. (Courtesy of Alfa-Laval Thermal AB.)

11.3 Operational Characteristics

Gasketed-plate heat exchangers can be opened for inspection, cleaning, maintenance, or rebuilding within the length of the frame (Figure 11.8).

11.3.1 Main Advantages

The gasket design minimizes the risk of internal leakage. Any failure in the gasket results in leakage to the atmosphere, which is easily detectable on the exterior of the unit. The additional main advantages and benefits offered by the gasketed-plate heat exchangers are[1,3]

- Flexibility of design through a variety of plate sizes and pass arrangements
- Easily accessible heat transfer area, permitting changes in configuration to suit changes in processes requirements through changes in the number of plates

FIGURE 11.8
Cleaning the plates. (Courtesy of Alfa-Laval Thermal AB.)

- Efficient heat transfer; high heat transfer coefficients for both fluids because of turbulence and a small hydraulic diameter
- Very compact (large heat transfer area/volume ratio), and low in weight; in spite of their compactness, 1,500 m² of surface area is available in a single unit
- Only the plate edges are exposed to the atmosphere; the heat losses are negligible and no insulation is required
- Intermixing of the two fluids cannot occur under gasket failure
- Plate units exhibit low fouling characteristics due to high turbulence and low residence time
- More than two fluids may be processed in a single unit.

The transition to turbulence flow occurs at low Reynolds numbers of 10–400. The high turbulence in a gasketed-plate heat exchanger leads to very high transfer coefficients, low fouling rates, and reduced size. The thin plates cut down the metal wall resistance to the minimum. These units can utilize up to about 82% of the theoretical LMTD, while shell-and-tube units are capable of utilizing only 50%, mainly because of absence of cross-flow in a plate unit. Over 90% of heat recovery is possible with a plate unit with very close temperature approaches.

The same frame can accommodate heat transfer duties for more than two liquids with the use of connecting plates. Costwise, a plate unit competes favorably with a tubular one if the tubes are to be of a costly material like stainless steel.

The plate unit is exceptionally compact and requires considerably less floor space than a tubular unit for the same duty. Low holdup volume, less weight, and lower costs for handling, transportation, and foundations are the other advantages involved in a plate unit.

11.3.2 Performance Limits

Capabilities of the gasketed-plate heat exchangers are limited due to the plates and the gaskets. The gaskets impose restrictions on operating temperatures (150°C–260°C), pressures (minimum operating pressure 25–30 bar), and the nature of fluids that can be handled.[5,6] Complex channel geometries result in the plate heat exchangers having high-shear characteristics. Friction factors are rather high for fully developed turbulent flow, but channel lengths are rather short and velocities are low so that pressure drops can be maintained within allowable limits for single-phase flow operations.

The main hindrance to the application of gasketed-plate heat exchangers in process industries is the upper limit on their size, which has been limited by the presses available to stamp out the plates from the sheet metal. Exchangers with sizes larger than 1,500 m^2 are not normally available. Since the flow passages between the plates are thin, high liquid rates will involve excessive pressure drops, thus limiting the capacity. For specific cases, it is possible to have a maximum design pressure of up to 2.5 MPa; normally, the design pressure is around one MPa. Operating temperatures are limited by the availability of suitable gasket materials.

Gasketed-plate heat exchangers are not suitable as air coolers. They are also not quite suitable for air-to-air or gas-to-gas applications. Fluids with very high viscosity present some problems due to flow distribution effects, particularly when cooling is taking place. Flow velocities less than 0.1 m/s give low heat transfer coefficients and low heat exchanger efficiency. As such, velocities lower than 0.1 m/s are not used in plate heat exchangers.

Gasketed-plate heat exchangers are less suitable for condensing duties. This applies particularly to vapors under a vacuum because the narrow plate gaps and induced turbulence result in appreciable pressure drops on the vapor side. Specially designed gasketed-plate heat exchangers are now available for duties involving evaporation and condensation systems.

11.4 Passes and Flow Arrangements

The term gasketed-plate heat exchanger "pass" refers to a group of channels in which the flow is in the same direction. Figure 11.9 shows a single-pass arrangement of the so-called U- and Z-arrangement types. All four ports in U-arrangement will be on the fixed-head plate, thus permitting disassembly of the exchanger for cleaning or repair without disturbing any external piping. In this arrangement, the flow distribution is less uniform than the Z-arrangement.[7]

A multipass arrangement consists of passes connected in series, and Figure 11.10 shows an arrangement of a two-pass configuration (2/2 configuration) with 3 channels and 4 channels which is often abbreviated as $2 \times 3/2 \times 3$ and $2 \times 4/2 \times 4$. The system is in counterflow, except in the central plate where parallel flow prevails.

Figure 11.11 shows a two-pass/one-pass flow system (2/1 configuration) where one fluid flows in a single pass, while the other fluid flows in two passes in series. In this configuration, one-half of the exchanger is in counterflow and the other half is in parallel flow (asymmetrical systems). This configuration is used when one of the fluids has a much greater mass flow rate or smaller allowable pressure drop.

Multipass arrangements always require that the ports are located on both the fixed and movable head plates. Usually, the number of passes and number of channels per pass are identical for the two fluids (symmetrically).

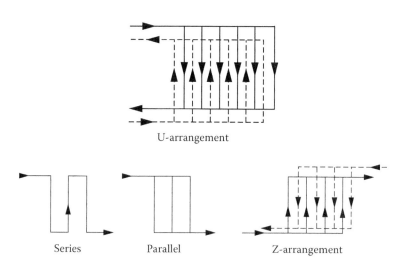

FIGURE 11.9
Flow pattern: (a) schematic of a U-type arrangement—counterflow, single-pass flow ($1 \times 6/1 \times 6$) (b) Z-arrangement ($1 \times 4/1 \times 4$ configuration).

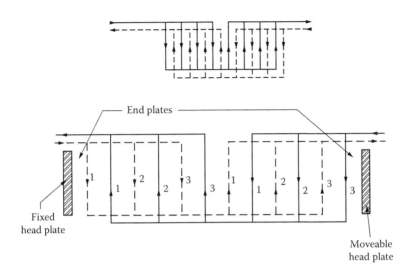

FIGURE 11.10
Schematic arrangement of a 2 pass/2 pass flow system (2 × 3/2 × 3 configuration).

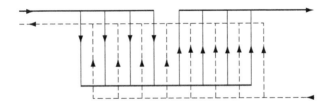

FIGURE 11.11
Schematic arrangement of a two-pass/one-pass flow system (2 × 4/1 × 8 configuration).

Fluid maldistribution is a potential problem in any system of interconnected channels, including gasketed-plate heat exchangers, and must be considered carefully in the design of gasketed-plate heat exchangers.

11.5 Applications

Gasketed-plate heat exchangers are extensively used in chemical, pharmaceutical, hygiene products, biochemical processing, food, and dairy industries, to name a few, because of the ease of disassembling the heat exchanger for cleaning and sterilization to meet health and sanitation requirements; they are used as conventional process heaters and coolers as well as condensers.[3,8]

Typical applications are mainly liquid-to-liquid turbulent flow situations. One example of very great importance, especially in at-sea or coastal

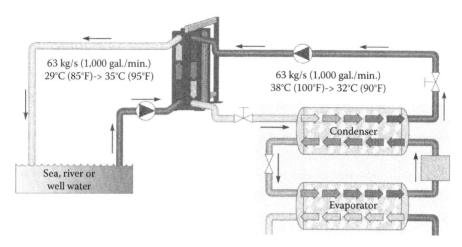

FIGURE 11.12
Closed-circuit cooling system. (Courtesy of Alfa-Laval Thermal AB.)

locations in central cooling systems using sea, river, or well water, is a heat sink, for example, in compressor intercooler systems. In central cooling, a closed-circuit of high quality water which is used in the intercooler is then passed through the gasketed-plate heat exchanger where heat is transferred from closed-circuit water to water of lower quality, such as sea or river water. Two closed-circuit systems using the gasketed-plate heat exchanger as an intermediate exchanger are shown in Figure 11.12. Corrosion and fouling problems are transferred from the intercooler to the plate heat exchanger system. Equipment in the closed system (process plant equipment) can be fabricated out of low carbon steel, while the plate exchangers can be made out of titanium to withstand sea water corrosion. Fouling problems can be ultimately solved by the ability to disassemble the plate heat exchanger system for cleaning whenever it is necessary. Therefore, a gasketed-plate heat exchanger is ideal for offshore and on-ship applications. Some of the useful data on gasketed-plate heat exchangers are presented in Table 11.2.

11.5.1 Corrosion

When corrosion-resistant materials are essential, gasketed-plate heat exchangers become more attractive than conventional exchangers even though the plates are made out of expensive materials. The range of construction materials used has already been mentioned in Table 11.1. Since the plates are very thin compared to the tube thickness in shell-and-tube units, corrosion allowance normally recommended for tubular units becomes meaningless for gasketed-plate units. The corrosion allowance for gasketed-plate heat exchangers is much smaller than for tubular units.

TABLE 11.2
Some Useful Data on Plate Heat Exchangers

Unit		
Largest size	1,540 m^2	
Number of plates	Up to 700	
Port size	Up to 39 cm	
Plates		
Thickness	0.5–1.2 mm	
Size	0.03–2.2 m^2	
Spacing	1.5–5.0 mm	
Contact points	For every 1.5–20 cm^2	Depends on plate size and type of corrugations
Operation		
Pressure	0.1–1.5 MPa	Up to 2.5 MPa in special cases
Temperature	−25°C to 150°C −40°C to 260°C	With rubber gaskets
Port velocities	5 m/s	With compressed asbestos fiber gaskets
Channel flow rates	0.05–12.5 m^3/h	
Maximum flow rates	2,500 m^3/h	
Performance		
Temperature approach	As low as 1°C	
Heat recovery	As high as 90%	
Heat transfer coefficients	3,000–7,000 W/m^2 °C	Water-to-water duties with normal fouling resistance
Number of transfer units	0.4–4.0	
Optimum pressure drops	30 kPa per NTU	

Source: From Raju, K. S. N. and Jagdish, C. B., *Low Reynolds Number Flow Heat Exchangers*, Hemisphere, Washington, DC, 1983. With permission.

As a consequence of high turbulence over the entire plate surface, erosion problems can be more significant in gasketed-plate heat exchangers (Figure 11.13). This requires the use of superior plate materials such as Monel or titanium (Table 11.3).

The increased concentration of a corrosive component in a scale deposit combined with the effect of a high metal wall temperature under the deposit will enhance localized corrosion of the heat transfer surface leading to eventual failure. In plate heat exchangers, this type of concentration corrosion is least because of low fouling tendencies. The following example illustrates this point.

In a particular case, a gasketed-plate heat exchanger with stainless steel 316 plates operated for 5 years, whereas a heating coil with a costlier material like incolay 825 failed after 6 weeks of operation due to concentration corrosion due to the heavy scale formed.[3]

FIGURE 11.13
Flow regime between plates. (Courtesy of Alfa-Laval Thermal AB.)

TABLE 11.3

Materials Selection

Application	Material
Natural cooling water, cooling tower water, or demineralized water	Stainless steel 316
Sea or brackish water	Titanium
Dilute sulphuric and nitric acids up to 10% concentration, and for temperatures up to 70°C	Titanium, titanium–palladium alloy, Incoloy 825, Hastelloy
Chloride solution	
Chloride content < 200 ppm	Stainless steel
Chloride content > 200 ppm	Titanium
Caustic solutions (50–70%)	Nickel
Wet chloride, chlorinated brines, hypochlorite solutions	Titanium
Copper sulphate solution in electrolyte refining	Stainless steel
Cooling hydrogen gas saturated with water vapor and mercury carryover in electrolysis plants	Incoloy

Source: From Raju, K. S. N. and Jagdish, C. B., *Low Reynolds Number Flow Heat Exchangers*, Hemisphere, Washington, DC, 1983. With permission.

As introduced in Chapter 6, the phenomenon of fouling is complex and very little information is available on the subject. It is often necessary to carry out prolonged experimental studies under actual operating conditions to incorporate the influence of fouling on the heat transfer rates. A careful choice of fouling factors while designing exchangers is very necessary in order to minimize the chances of over- or under-design. Besides adding to the costs, overdesign of a plate heat exchanger often reduces the flow velocities, resulting in increased fouling rates.

Fouling is much less in a gasketed-plate heat exchanger than in a tubular unit for the following reasons:

- High turbulence maintains solids in suspension
- Velocity profiles across a plate are uniform with zones of low velocities absent

- The plate surfaces are generally smooth and can be further electropolished
- Deposits of corrosion products, to which fouling can adhere, are absent because of low corrosion rates
- In cooling duties, the high film coefficients maintain a moderately low metal wall temperature; this helps prevent crystallization growth of the inverse solubility compounds
- Because of the ease with which plate units can be cleaned in place, deposits that grow with time can be kept to a minimum by frequent cleaning.

Table 11.4 gives recommended fouling factors for gasketed-plate heat exchangers assuming operation at the economic pressure drop of about 30 kPa per NTU.[3,9] Manufacturer's recommendations based on test data should be a better guide while choosing fouling resistances. The general recommendations in this respect are 5% excess NTU for low fouling duties, 10% for moderate fouling, and 15–20% for high fouling.

11.5.2 Maintenance

A gasketed-plate heat exchanger can be easily opened for inspection, mechanical cleaning, gasket replacement, extension or reduction to the number of

TABLE 11.4

Recommended Fouling Factors for Plate Heat Exchangers

Service	Fouling Factor, $m^2 \cdot K/W$
Water	
Demineralized or distilled	0.0000017
Soft	0.0000034
Hard	0.0000086
Cooling tower (treated)	0.0000069
Sea (coastal) or estuary	0.000086
Sea (ocean)	0.000052
River, canal, tube well, etc.	0.000086
Engine jacket	0.0000103
Steam	0.0000017
Lubricating oils	0.0000034–0.0000086
Vegetable oils	0.0000017–0.0000052
Organic solvents	0.0000017–0.0000103
General process fluids	0.0000017–0.00000103

Source: From Raju, K. S. N. and Jagdish, C. B., *Low Reynolds Number Flow Heat Exchangers*, Hemisphere, Washington, DC, 1983; Cooper, A. et al., *Heat Transfer Eng.*, Vol. 1, No. 3, 50–55, 1980. With permission.

plates, or other modifications of the duties (Figure 11.8). In-place cleaning by circulating suitable cleaning chemicals is relatively easy. Because the plate units are built from standard components that are interchangeable and replaceable, stocking of spare parts is minimal. Replacing the gasket does not pose much of a problem. In a single-pass unit, it is possible to locate all pipe connections in the fixed part of the frame. This is of great advantage as the unit can be opened for access to the heat transfer surface without opening any connections.

11.6 Heat Transfer and Pressure Drop Calculations

The design of gasketed-plate heat exchangers is highly specialized in nature considering the variety of designs available for the plates and arrangements that may possibly suit various duties. Unlike tubular heat exchangers for which design data and methods are easily available, a gasketed-plate heat exchanger design continues to be proprietary in nature. Manufacturers have developed their own computerized design procedures applicable to the exchangers that they market.

Attempts have been made to develop heat transfer and pressure-drop correlations for use with plate heat exchangers, but most of the correlations cannot be generalized to give a high degree of prediction ability.[10–24] In these exchangers, the fluids are much closer to countercurrent flow than in shell-and-tube heat exchangers. In recent years, some design methods have been reported. These methods are mostly approximate in nature to suit preliminary sizing of the plate units for a given duty. No published information is available on the rigorous design of plate heat exchangers.

11.6.1 Heat Transfer Area

The corrugations increase the surface area of the plate as compared to the original flat area. To express the increase of the developed length in relation to projected length, a surface enlargement factor, φ, is then defined as the ratio of the developed length to the flat or projected length (Figure 11.6):

$$\phi = \frac{\text{developed length}}{\text{projected length}} \tag{11.1}$$

The value of ϕ is a function of the corrugation pitch and the corrugation depth or plate pitch (Figure 11.6). The enlargement factor varies between 1.15 and 1.25. The value of 1.17 can be assumed to be a typical average.[2,11] The value of ϕ as given by Equation 11.1 is the ratio of the actual effective area as specified by the manufacturer, A_1, to the projected plate area A_{1p}:

$$\phi = \frac{A_1}{A_{1p}} \tag{11.2}$$

where A_{1p} can be approximated from Figure 11.10 as

$$A_{1p} = L_p \cdot L_w \tag{11.3}$$

and L_p and L_w can be estimated from the port distance L_v and L_h and port diameter D_p as

$$L_p \approx L_v - D_p \tag{11.4}$$

$$L_w \approx L_h + D_p \tag{11.5}$$

The value of φ is used to calculate the effective flow path.

11.6.2 Mean Flow Channel Gap

Flow channel is the conduit formed by two adjacent plates between the gaskets. The cross section of a corrugated surface being very complex, the mean channel spacing, b, is defined as shown in Figure 11.6b.

$$b = p - t \tag{11.6}$$

where p is the plate pitch or the outside depth of the corrugated plate, t is the plate thickness, and b is also the thickness of a fully compressed gasket, as the plate corrugations are in metallic contact. Plate pitch is not to be confused by corrugation pitch. Channel spacing b is required for calculation of the mass velocity and Reynolds number and is, therefore, a very important value that is usually not specified by the manufacturer. If not known, the plate pitch p can be determined from the compressed plate pact length (between the head plates) L_c, which is usually specified on drawings. Then, p is determined as

$$p = \frac{L_c}{N_t} \tag{11.7}$$

where N_t is the total number of plates.

11.6.3 Channel Hydraulic Diameter

The hydraulic diameter of the channel, D_h, is defined as

$$D_h = \frac{4 \times \text{channel flow area}}{\text{wetted perimeter}} = \frac{4A_c}{P_w} \tag{11.8}$$

$$D_h = \frac{4(b)(L_w)}{2(b + L_w \phi)} \approx \frac{2b}{\phi} \tag{11.9}$$

with the approximation that $b \ll L_w$.

11.6.4 Heat Transfer Coefficient

With gasketed-plate heat exchangers, heat transfer is enhanced. In this section, enhanced thermal-hydraulic characteristics of chevron plates are presented. The heat transfer enhancement will strongly depend on the chevron inclination angle β, relative to flow direction. Heat transfer and the friction factor increase with β. On the other hand, the performance of a chevron plate will also depend upon the surface enlargement factor φ, corrugation profile, channel aspect ratio, $2b/P_c$, gab b, the temperature dependent physical properties, and especially, the variable viscosity effects. In spite of extensive research on plate heat exchangers, generalized correlations for heat transfer and friction factor are not available.

Any attempt to estimate the film coefficient of heat transfer in gasketed-plate heat exchangers involves extension of correlations that are available for heat transfer between flat flow passages. The conventional approach for such passages employs correlations applicable for tubes by defining an equivalent diameter for the noncircular passage, which is substituted for diameter, d.

For gasketed-plate heat exchangers with chevron plates, some selected correlations for the friction factor f, and the Nusselt number Nu, are listed in Table 11.5.[14] In these correlations, the Nusselt and Reynolds numbers are based on the equivalent diameter ($D_e = 2b$) of the chevron plate.

Martin[32], Wang and Sunden[33], and Rao[34] also studied the performance of plate heat exchanger surfaces with chevron plates to determine heat transfer and flow friction coefficients and also provided correlations for Nusselt numbers and friction coefficients under different conditions which are alos listed in Table 11.5.

As can be seen from Table 11.5, except for the correlation given by Savostin and Tikhonov[15] and Tovazhnyanski et al.,[16] all the correlations give separate

TABLE 11.5

Fanning Friction Factor f and Nusselt Number $Nu(2hb/k)^a$ Correlations for Gasketed-Plate Heat Exchangers with Chevron Plates

1. Savostin and Tikhonov[15]: Air flows; $57° \leq \beta \leq 90°$; $\psi = p - 2\beta$; β in radian

    $200 \leq Re/\phi \leq 600$
    $f = 6.25(1+0.95\psi^{1.72})\phi^{1.84} Re^{-0.84}$
    $Nu = 1.26[0.62+0.38\cos(2.3\psi)]\phi^{1-a_1} Pr^{1/3} Re^{a_1}$
    $a_1 = 0.22[1+1.1\psi^{1.5}]$
    $600 \leq Re/\phi \leq 4000$
    $f = 0.925[0.62+0.38\cos(2.6\psi)]\phi^{1+a_2} Re^{-a_2}$
    $a_2 = 0.53[0.58+0.42\cos(1.87\psi)]$
    $Nu = 0.072 e^{0.5\psi+0.17\psi^2} \phi^{0.33} Pr^{1/3} Re^{0.67}$

2. Tovazhnyanski et al.[16]: $\beta = 30°, 45°,$ and $60°$; $2{,}000 < Re < 25{,}000$; $\psi = p - 2\beta$; β in radian

    $f = 0.085 \exp[1.52\tan\psi] Re^{-(0.25-0.06\tan\beta)}$
    $Nu = 0.051 e^{[0.64\tan\psi]} Re^{0.73} Pr^{0.43} (Pr/Pr_w)^{0.25}$

3. Chisholm and Wanniarachchi[17]: $30° \leq \beta \leq 80°$; β in degree

    $f = 0.8 Re^{-0.25} \phi^{1.25} ((90-\beta)/30)^{3.6}$ $10^3 < Re < 4\times 10^3$
    $Nu = 0.72 Re^{0.59} Pr^{0.4} \phi^{0.41} ((90-\beta)/30)^{0.66}$ $10^3 < Re < 4\times 10^3$

4. Heavner et al.[18]: Mixed chevron plate arrangements; $400 < Re \, \phi < 10{,}000$; $3.3 < Pr < 5.9$

    $f = \begin{cases} 1.458 \phi^{1.0838} & Re^{-0.0838} & \beta = 45°/0° \\ 1.441 \phi^{1.1353} & Re^{-0.1353} & \beta = 67°/0° \\ 0.687 \phi^{1.1405} & Re^{-0.1405} & \beta = 45°/45° \\ 0.545 \phi^{1.1555} & Re^{-0.1555} & \beta = 67°/45° \\ 0.490 \phi^{1.1814} & Re^{-0.1814} & \beta = 67°/67° \end{cases}$

    $\dfrac{Nu}{Pr^{1/2}} \left(\dfrac{\mu}{\mu_w}\right)^{-0.17} = \begin{cases} 0.278 \phi^{0.317} & Re^{0.683} & \beta = 45°/0° \\ 0.308 \phi^{0.333} & Re^{0.667} & \beta = 67°/0° \\ 0.195 \phi^{0.308} & Re^{0.692} & \beta = 45°/45° \\ 0.118 \phi^{0.280} & Re^{0.720} & \beta = 67°/45° \\ 0.089 \phi^{0.282} & Re^{0.718} & \beta = 67°/67° \end{cases}$

5. Talik et al.[19]: $\beta = 30°$ chevron plates; $\phi = 1.22$; $D_h = 4.65$ mm; $L_w = 0.346$ m; $t = 0.61$ mm

    $f = \begin{cases} 12.065 Re^{-0.74} & 10 < Re < 80; \text{ water/glycol} \\ 0.3323 Re^{-0.042} & 1450 < Re < 11460; \text{ water} \end{cases}$

    $Nu = \begin{cases} 0.2 Re^{0.75} Pr^{0.4} & 10 < Re < 720; 70 < Pr < 450; \text{ water/glycol} \\ 0.248 Re^{0.75} Pr^{0.4} & 1450 < Re < 11460; 2.5 < Pr < 5.0; \text{ water} \end{cases}$

(Continued)

TABLE 11.5 (CONTINUED)
Fanning Friction Factor f and Nusselt Number $Nu(2hb/k)^a$ Correlations for Gasketed-Plate Heat Exchangers with Chevron Plates

6. Muley and Manglik[20]: chevron plates with $\beta = 30°$

$$f = \begin{cases} 51.5/Re & Re < 16 \\ 17.0\,Re^{-0.6} & 16 \le Re \le 100 \\ 2.48\,Re^{-0.2} & Re \ge 800 \end{cases}$$

$$Nu = \begin{cases} 0.572\,Re^{0.5}\,Pr^{1/3}(\mu/\mu_w)^{0.14} & 20 \le Re \le 210 \\ 0.1096\,Re^{0.78}\,Pr^{1/3}(\mu/\mu_w)^{0.14} & Re \ge 800 \end{cases}$$

7. Muley and Manglik[21]: Mixed chevron plates with $\beta = 30°$ and $60°$ ($\beta_{avg} = 45°$)

$$f = \begin{cases} \left[(40.32/Re)^5 + (8.12\,Re^{-0.5})^5\right]^{0.2} & 2 \le Re \le 200 \\ 1.274\,Re^{-0.15} & Re \ge 1000 \end{cases}$$

$$Nu = \begin{cases} 0.471\,Re^{0.5}\,Pr^{1/3}(\mu/\mu_w)^{0.14} & 20 \le Re \le 400 \\ 0.10\,Re^{0.76}\,Pr^{1/3}(\mu/\mu_w)^{0.14} & Re \ge 1000 \end{cases}$$

8. Martin[32]; β in radian

$Re_h = \phi Re$

$Nu_h = \phi Nu$

$$Nu_h = 0.122\,Pr^{1/3}\left(\frac{\eta_m}{\eta_w}\right)^{1/6}\left(f\,Re^2\,\sin 2\left(\frac{\pi}{2}-\beta\right)\right)^{0.374}$$

$$\frac{1}{\sqrt{f}} = \frac{\cos\left(\frac{\pi}{2}-\beta\right)}{\left(0.18\tan\left(\frac{\pi}{2}-\beta\right)+0.36\sin\left(\frac{\pi}{2}-\beta\right)+f_0/\cos\left(\frac{\pi}{2}-\beta\right)\right)^{1/2}} + \frac{1-\cos\left(\frac{\pi}{2}-\beta\right)}{\sqrt{3.8\,f_1}}$$

$Re_h < 2000 \Rightarrow \begin{cases} f_0 = \dfrac{64}{Re_h} \\ f_1 = \dfrac{597}{Re_h} + 3.85 \end{cases}$

$Re_h \ge 2000 \Rightarrow \begin{cases} f_0 = (1.8\log_{10} Re_h - 1.5)^{-2} \\ f_1 = \dfrac{39}{Re_h^{0.289}} \end{cases}$

TABLE 11.5 (CONTINUED)
Fanning Friction Factor f and Nusselt Number $Nu(2hb/k)^a$ Correlations for Gasketed-Plate Heat Exchangers with Chevron Plates

9. Wang and Sunden[33]: $10° \leq \beta \leq 80°$; β in degree

$$Nu = 0.205 Pr^{1/3} \left(\frac{\mu}{\mu_w}\right)^{1/6} \left(f Re^2 \sin(180-2\beta)\right)^{0.374}$$

$$\frac{1}{\sqrt{f}} = \frac{\cos(90-\beta)}{(0.045\tan(90-\beta)+0.09\sin(90-\beta)+f_0/\cos(90-\beta)} + \frac{1-\cos(90-\beta)}{\sqrt{3.8 f_1}}$$

$Re < 2000 \Rightarrow \begin{cases} f_0 = 16/Re \\ f_1 = (149/Re)+0.9625 \end{cases}$

$Re \geq 2000 \Rightarrow \begin{cases} f_0 = (1.56 \ln(Re)-3.0)^{-2} \\ f_1 = 9.75/Re^{0.289} \end{cases}$

10. Rao[34]: $600 \leq Re \leq 6{,}000$, $3.5 \leq Pr \leq 5.7$, $\beta = 30°$

$Nu = 0.218 Re^{0.65} Pr^{1/3}$ $500 < Re$

$f = 21.41 Re^{-0.301}$ $1000 < Re < 7000$

Note: The chevron angle β in correlations 1 to 10 is defined as the angle between the corrugated channels and the vertical edge of the plate.

[a] Properties are evaluated at average temperature.

equations for different values of β and do not specifically take into account the effects of the different parameters of the corrugated passage.

The channel flow geometry in chevron plate packs is quite complex, which is why most of the correlations are generally presented for fixed value of β in symmetric ($\beta = 30°/30°$ or $\beta = 60°/60°$) plate arrangements and mixed ($\beta = 30°/60°$) plate arrangements. The various correlations are compared by Manglik[14] and discrepancies have been found. These discrepancies originated from the differences of plate surface geometries, which include the surface enlargement factor φ, the metal-to-metal contact pitch p, the wavelength P_c, the amplitude b, and the profile or shape of the surface corrugation (Figure 11.6) as well as other factors such as port orientation, flow distribution channels, and plate width and length. It should be noted that in some correlations, variable viscosity effects have not been taken into account.

As mentioned above, the relatively high heat transfer coefficients obtained in plate heat exchangers depend on the geometrical parameters of the plate-surface corrugations; the heat transfer enhancement mechanisms may include swirl or vortex flow generation (Figure 11.13), surface-area enlargement, distraction and reattachment of the boundary layers, and small hydraulic-diameter flow channels.

TABLE 11.6

Constants for Single-Phase Heat Transfer and Pressure Loss Calculation in Gasketed-Plate Heat Exchangers[2,10]

Chevron Angle (degree)	Heat Transfer			Pressure Loss		
	Reynolds Number	C_h	n	Reynolds Number	K_p	m
≤30	≤10	0.718	0.349	<10	50.000	1.000
	>10	0.348	0.663	10–100	19.400	0.589
				>100	2.990	0.183
45	<10	0.718	0.349	<15	47.000	1.000
	10–100	0.400	0.598	15–300	18.290	0.652
	>100	0.300	0.663	>300	1.441	0.206
50	<20	0.630	0.333	<20	34.000	1.000
	20–300	0.291	0.591	20–300	11.250	0.631
	>300	0.130	0.732	>300	0.772	0.161
60	<20	0.562	0.326	<40	24.000	1.000
	20–400	0.306	0.529	40–400	3.240	0.457
	>400	0.108	0.703	>400	0.760	0.215
≥65	<20	0.562	0.326	50	24.000	1.000
	20–500	0.331	0.503	50–500	2.800	0.451
	>500	0.087	0.718	>500	0.639	0.213

Source: From Saunders, E. A. D., *Heat Exchangers—Selection, Design, and Construction*, John Wiley & Sons, New York, 1988; Kumar, H., *Inst. Chem. Symp. Series*, No. 86, 1275–1286, 1984.

As can be seen from Table 11.6, both heat transfer coefficient and friction factor increase with β. From the experimental data base, Muley et al.[22] and Muley and Manglik[23,24] proposed the following correlation for various values of β:

For $Re \leq 400$, $30° \leq \beta \leq 60°$

$$Nu = \frac{2hb}{k} = 0.44 \left(\frac{\beta}{30}\right)^{0.38} Re^{0.5} Pr^{1/3} \left(\frac{\mu_b}{\mu_w}\right)^{0.14} \quad (11.10)$$

$$f = \left(\frac{\beta}{30}\right)^{0.83} \left[\left(\frac{30.2}{Re}\right)^5 + \left(\frac{6.28}{Re^{0.5}}\right)^5\right]^{0.2} \quad (11.11)$$

For $Re \geq 1{,}000$, $30° \leq \beta \leq 60°$, $1 \leq \phi \leq 1.5$

$$Nu = \left[0.2668 - 0.006967\,\beta + 7.244 \times 10^{-5}\beta^2\right]$$
$$\times \left[20.78 - 50.94\phi + 41.1\phi^2 - 10.51\phi^3\right] \quad (11.12)$$
$$\times Re^{[0.728 + 0.0543\sin\{(\pi\beta/45)+3.7\}]} Pr^{1/3} \left(\mu_b/\mu_w\right)^{0.14}$$

$$f = [2.917 - 0.1277\beta + 2.016 \times 10^{-3} \beta^2]$$
$$\times [5.474 - 19.02\phi + 18.93\phi^2 - 5.341\phi^3] \quad (11.13)$$
$$\times Re^{-[0.2+0.0577\sin\{(\pi\beta/45)+2.1\}]}$$

The heat transfer coefficient and the Reynolds number are based on the equivalent diameter De (= $2b$). To evaluate the enhanced performance of chevron plates, predictions from the following flat-plate channel equations[25] are compared with the results of the chevron plates for $\varphi = 1.29$ and $\gamma = 0.59$:

$$Nu = \begin{cases} 1.849(L/D_e)^{-1/3}(RePr)^{1/3}(\mu_b/\mu_w)^{0.14} & Re \leq 2000 \\ 0.023 Re^{0.8} Pr^{1/3}(\mu_b/\mu_w)^{0.14} & Re > 4000 \end{cases} \quad (11.14)$$

$$f = \begin{cases} 24/Re & Re \leq 2000 \\ 0.1268 Re^{-0.3} & Re > 2000 \end{cases} \quad (11.15)$$

Depending on β and the Reynolds number, chevron plates produce up to five times higher Nusselt numbers than those in flat-plate channels. The corresponding pressure-drop penalty, however, is considerably higher: depending on the Reynolds number, from 1.3 to 44 times higher friction factors than those in equivalent flat-plate channels.[23]

A correlation in the form of Equation 11.16 has also been proposed by Kumar,[7-10] and the values of constants C_h and n are given in Table 11.6.[2,10]

$$\frac{hD_h}{k} = C_h \left(\frac{D_h G_c}{\mu}\right)^n \left(\frac{c_p \mu}{k}\right)^{1/3} \left(\frac{\mu}{\mu_w}\right)^{0.17} \quad (11.16)$$

where D_h is the hydraulic diameter defined by Equation 11.8. Values of C_h and n depend on flow characteristics and chevron angles. The transition to turbulence occurs at low Reynolds numbers and, as a result, the gasketed-plate heat exchangers give high heat transfer coefficients (Figure 11.13).

A fully developed documented method of estimating the heat transfer and pressure drop of gasketed-plate heat exchangers based on a wide range of commercially available chevron-pattern plates is possible and will be outlined here (Figure 11.2).

The Reynolds number, Re, based on channel mass velocity and the hydraulic diameter, D_h, of the channel is defined as

$$Re = \frac{G_c D_h}{\mu} \quad (11.17)$$

The channel mass velocity is given by

$$G_c = \frac{\dot{m}}{N_{cp} b L_w} \quad (11.18)$$

where N_{cp} is the number of channels per pass and is obtained from

$$N_{cp} = \frac{N_t - 1}{2N_p} \tag{11.19}$$

where N_t is the total number of plates and N_p is the number of passes.

11.6.5 Channel Pressure Drop

The total pressure drop is composed of the frictional channel pressure drop, Δp_c, and the port pressure drop, Δp_p. The friction factor, f, is defined by the following equation for the frictional pressure drop Δp_c (Refs. 2,10,11):

$$\Delta p_c = 4f \frac{L_{eff} N_p}{D_h} \frac{G_c^2}{2\rho} \left(\frac{\mu_b}{\mu_w}\right)^{-0.17} \tag{11.20}$$

where L_{eff} is the effective length of the fluid flow path between inlet and outlet ports and must take into account the corrugation enlargement factor ϕ; this effect is included in the definition of friction factor. Therefore, $L_{eff} = L_v$, which is the vertical part distance as indicated in Figure 11.6.

The friction factor in Equation 11.14 is given by

$$f = \frac{K_p}{Re^m} \tag{11.21}$$

Values of K_p and m are given in Table 11.6 (Refs. 2,10) as functions of Reynolds number for various values of chevron angles. For various plate-surface configurations, friction coefficient versus. Reynolds number must be provided by the manufacturer. In the following analysis, Correlations 11.16 and 11.21 will be used for the heat transfer coefficient and the Fanning friction factors.

11.6.6 Port Pressure Drop

The pressure drop in the port ducts, Δp_p, can be roughly estimated as 1.4 velocity head:

$$\Delta p_p = 1.4 N_p \frac{G_p^2}{2\rho} \tag{11.22}$$

where

$$G_p = \frac{\dot{m}}{\frac{\pi D_p^2}{4}} \tag{11.23}$$

where $\dot{m}$ is the total flow rate in the port opening and D_p is the port diameter. The total pressure drop is then

$$\Delta p_t = \Delta p_c + \Delta p_p \qquad (11.24)$$

11.6.7 Overall Heat Transfer Coefficient

The overall heat transfer coefficient for a clean surface is

$$\frac{1}{U_c} = \frac{1}{h_h} + \frac{1}{h_c} + \frac{t}{k_w} \qquad (11.25)$$

and under fouling conditions (fouled or service overall heat transfer coefficient) is

$$\frac{1}{U_f} = \frac{1}{h_h} + \frac{1}{h_c} + \frac{t}{k_w} + R_{fh} + R_{fc} \qquad (11.26)$$

where h and c stand for hot and cold streams, respectively.

The relationship between U_c, fouled U_f, and the cleanliness factor, CF, can be written as

$$U_f = U_c(CF) = \frac{1}{\dfrac{1}{U_c} + R_{fh} + R_{fc}} \qquad (11.27)$$

11.6.8 Heat Transfer Surface Area

The heat balance relations in gasketed-plate heat exchangers are the same as for tubular heat exchangers. The required heat duty, Q_r, for cold and hot streams is

$$Q_r = (\dot{m}c_p)_c (T_{c_2} - T_{c_1}) = (\dot{m}c_p)_h (T_{h_1} - T_{h_2}) \qquad (11.28)$$

On the other hand, the actually obtained heat duty, Q_f, for fouled conditions is defined as

$$Q_f = U_f A_e \Delta T_m \qquad (11.29)$$

where A_e is the total developed area of all thermally effective plates, that is $N_t - 2$ (see Figure 11.9) which accounts for the two plates adjoining the head plates. In multipass arrangements, the separating plate is in parallel flow, which will have a negligible effect on the mean temperature difference. If an entire pass is in parallel flow, the correction factor, F, must be applied to obtain the true mean temperature difference, ΔT_m.[8,11,26–28] Otherwise, the true mean temperature difference, ΔT_m, for the counterflow arrangement is

$$\Delta T_{lm,cf} = \frac{\Delta T_1 - \Delta T_2}{\ln \frac{\Delta T_1}{\Delta T_2}} \tag{11.30}$$

where ΔT_1 and ΔT_2 are the terminal temperature differences at the inlet and outlet. A comparison between Q_r and Q_f defines the safety factor, C_s, of the design

$$C_s = \frac{Q_f}{Q_r} \tag{11.31}$$

where Q_r is the rated heat duty and Q_f is the heat duty under fouled condition.

11.6.9 Performance Analysis

Assume that for a specific process, a manufacturer proposes a gasketed-plate heat exchanger. A stepwise calculation as documented in the preceding section will be used for rating (performance analysis of the proposed gasketed-plate heat exchanger as an existing heat exchanger).

Example 11.1

Cold water will be heated by a wastewater stream. The cold water with a flow rate of 140 kg/s enters the gasketed-plate heat exchanger at 22°C and will be heated to 42°C. The wastewater has the same flow rate entering at 65°C and leaving at 45°C. The maximum permissible pressure drop for each stream is 50 psi.

The process specifications are as follows:

Items	Hot Fluid	Cold Fluid
Fluids	Wastewater	Cooling water
Flow rates (kg/s)	140	140
Temperature in (°C)	65	22
Temperature out (°C)	45	42
Maximum permissible pressure drop (psi)	50	50
Total fouling resistance ($m^2 \cdot K/W$)	0.00005	0
Specific heat ($J/kg \cdot K$)	4,183	4,178
Viscosity ($N \cdot s/m^2$)	5.09×10^{-4}	7.66×10^{-4}
Thermal conductivity ($W/m \cdot K$)	0.645	0.617
Density (kg/m^3)	985	995
Prandtl number	3.31	5.19

Gasketed-Plate Heat Exchangers

The constructional data for the proposed plate heat exchanger are

Plate Material	SS304
Plate thickness (mm)	0.6
Chevron angle (degrees)	45
Total number of plates	105
Enlargement factor, ϕ	1.25
Number of passes	One pass/one pass
Overall heat transfer coefficient (clean/fouled) W/m² · K	8,000/4,500
Total effective area (m²)	110
All port diameters (mm)	200
Compressed plate pack length, L_c, (m)	0.38
Vertical port distance, L_v, (m)	1.55
Horizontal port distance, L_h, (m)	0.43
Effective channel width, L_w, (m)	0.63
Thermal conductivity of the plate material (SS304), W/m · K	17.5

Solution

Stepwise Performance Analysis—The required heat load can be calculated from the heat balance as

$$Q_{rh} = 140 \times 4,183 \times (65 - 45) = 11,712.4 \text{ kW}$$

$$Q_{rc} = 140 \times 4,178 \times (42 - 22) = 11,698.4 \text{ kW}$$

The mean temperature difference is

$$\Delta T_1 = \Delta T_2 = \Delta T_m = 23° \text{ C}$$

Therefore,

$$\Delta T_{lm,cf} = \Delta T_{lm} = 23° \text{ C}$$

The effective number of plates is

$$N_e = N_t - 2 = 103$$

The effective flow length between the vertical ports is

$$L_{eff} \approx L_v = 1.55 \text{ m}$$

The plate pitch can be determined from Equation 11.7:

$$p = \frac{L_c}{N_t} = \frac{0.38}{105} = 0.0036 \text{ m}$$

From Equation 11.6, the mean channel flow gap is

$$b = 0.0036 - 0.0006 = 0.0030 \text{ m}$$

The one channel flow area is

$$A_{ch} = b \times L_w = 0.0030 \times 0.63 = 0.00189 \text{ m}^2$$

and the single-plate heat transfer area is

$$A_1 = \frac{A_e}{N_e} = \frac{110}{103} = 1.067 \text{ m}^2$$

The projected plate area A_{1p} from Equation 11.3 is

$$A_{1p} = L_p \cdot L_w = (1.55 - 0.2) \times 0.63 = 0.85 \text{ m}^2$$

The enlargement factor has been specified by the manufacturer, but it can be verified from Equation 11.2:

$$\phi = \frac{1.067}{0.85} = 1.255$$

The channel hydraulic/equivalent diameter from Equation 11.3 is

$$D_h = \frac{2b}{\phi} = \frac{2 \times 0.0030}{1.255} = 0.00478 \text{ m}$$

The number of channels per pass, N_{cp}, from Equation 11.19 is

$$N_{cp} = \frac{N_t - 1}{2N_p} = \frac{105 - 1}{2 \times 1} = 52$$

Heat Transfer Analysis—Calculate the mass flow rate per channel:

$$\dot{m}_{ch} = \frac{140}{52} = 2.69 \text{ kg/s}$$

and the mass velocity, G_{ch}:

$$G_{ch} = \frac{2.69}{0.00189} = 1423.8 \text{ kg/m}^2 \cdot \text{s}$$

The hot fluid Reynolds number can be calculated as

$$Re_h = \frac{G_h D_h}{\mu_h} = \frac{1423.28 \times 0.00478}{5.09 \times 10^{-4}} = 13{,}366$$

and the cold fluid Reynolds number can be calculated as

$$Re_c = \frac{G_c D_h}{\mu_c} = \frac{1423.28 \times 0.00478}{7.66 \times 10^{-4}} = 8881.6$$

The hot fluid heat transfer coefficient, h_h, can be obtained by using Equation 11.16. From Table 11.6, $C_h = 0.3$ and $n = 0.663$, so the heat transfer coefficient from Equation 11.16:

$$Nu_h = \frac{h_h D_h}{k} = 0.3(Re)^{0.663}(Pr)^{1/3}\left(\frac{\mu_b}{\mu_w}\right)^{0.17}$$

Assuming $\mu_b \approx \mu_w$,

$$Nu_h = 0.3(13366)^{0.663}(3.31)^{1/3} = 243.4$$

$$h_h = \frac{Nu_h k}{D_h} = \frac{0.645 \times 243.4}{0.00478} = 32844 \text{ W/m}^2 \cdot \text{K}$$

To determine the cold fluid heat transfer coefficient, h_c, using Table 11.6 and Equation 11.16, we have

$$Nu_c = \frac{h_c D_h}{k} = 0.3(8881.6)^{0.663}(5.19)^{1/3} = 215.46$$

$$h_c = \frac{Nu_c k}{D_h} = \frac{0.617 \times 215.46}{0.00478} = 27{,}811 \text{ W/m}^2 \cdot \text{K}$$

For the overall heat transfer coefficient, the clean overall heat transfer coefficient, from Equation 11.25, is

$$\frac{1}{U_c} = \frac{1}{h_c} + \frac{1}{h_h} + \frac{0.0006}{17.5}$$

$$= \frac{1}{27{,}811} + \frac{1}{32{,}844} + \frac{0.0006}{17.5}$$

$$U_c = 9931.5 \text{ W/m}^2 \cdot \text{K}$$

The fouled (or service) overall heat transfer coefficient is calculated from Equation 11.26:

$$\frac{1}{U_f} = \frac{1}{U_c} + 0.00005$$

$$U_f = 6636 \text{ W/m}^2 \cdot \text{K}$$

The corresponding cleanliness factor is

$$CF = \frac{U_f}{U_c} = \frac{6636}{9931.5} = 0.67$$

which is rather low because of the high fouling factor.

The actual heat duties for clean and fouled surfaces are

$$Q_c = U_c A_e \Delta T_m = 9931.5 \times 110 \times 23 = 25,127 \text{ kW}$$
$$Q_f = U_f A_e \Delta T_m = 6636 \times 110 \times 23 = 16,789 \text{ kW}$$

The safety factor is

$$C_s = \frac{Q_f}{Q_r} = \frac{16,789}{11,705.4} \approx 1.43$$

The percent over surface design, from Equation 6.20, is

$$OS = 100 U_c R_{ft} = 49.7\%$$

This is a rather large heat exchanger, which could be reduced to satisfy the process specifications with more frequent cleaning. It may be preferable to use a 30% over surface design to decrease the investment cost, and then the cleaning scheduling could be arranged accordingly. Therefore, the proposed design may be modified and rerated.

Pressure Drop Analysis—The hot fluid friction coefficient from Equation 11.21 and Table 11.6 is

$$f_h = \frac{1.441}{(Re_h)^{0.206}} = \frac{1.441}{(13366)^{0.206}} = 0.204$$

$$f_c = \frac{1.441}{(Re_c)^{0.206}} = \frac{1.441}{(8881.6)^{0.206}} = 0.221$$

The frictional pressure drop from hot and cold streams is calculated from Equation 11.20.

For hot fluid,

$$(\Delta p_c)_h = 4 \times 0.204 \times \frac{1.55 \times 1}{0.00478} \times \frac{(1423.28)^2}{2 \times 985} = 272,087.4 \text{ Pa}$$

$$(\Delta p_c)_h = 39.5 \text{ psi}$$

and for cold fluid,

$$(\Delta p_c)_c = 4 \times 0.221 \times \frac{1.55 \times 1}{0.00478} \times \frac{(1043.28)^2}{2 \times 995} = 291{,}799 \text{ Pa}$$

$$(\Delta p_c)_c = 42.3 \text{ psi}$$

The pressure drop in the port ducts is calculated from Equation 11.22. To determine the port mass velocity,

$$(G_p)_h = (G_p)_c = \frac{\dot{m}}{\pi \left(\frac{D_p^2}{4}\right)} = \frac{140}{\pi \left(\frac{0.2^2}{4}\right)} = 4458.6 \text{ kg/m}^2 \cdot \text{s}$$

For hot fluid,

$$(\Delta p_p)_h = 1.4 \times 1 \times \frac{G_p^2}{2 \times 985} = 14{,}127.29 \text{ Pa}$$

$$(\Delta p_p)_h = 2.05 \text{ psi}$$

$$(\Delta p_p)_c = 1.4 \times 1 \times \frac{G_p^2}{2 \times 995} = 13{,}985.31 \text{ Pa}$$

$$(\Delta p_p)_c = 2.03 \text{ psi}$$

The total pressure drop is calculated from Equation 11.24 for hot fluid:

$$(\Delta p_t)_h = (\Delta p_c)_h + (\Delta p_p)_h = 39.5 + 2.05 = 41.55 \text{ psi}$$

and for cold fluid:

$$(\Delta p_t)_c = (\Delta p_c)_c + (\Delta p_p)_c = 42.3 + 2.03 = 44.33 \text{ psi}$$

The above calculation shows that the proposed unit satisfies the process required and pressure drop constraint, but it could be smaller unless it is used under heavy fouling conditions. Interesting design examples and performance analysis are also given by Hewitt et al.[29]

11.7 Thermal Performance

In order to quantitatively assess the thermal performance of plates, it would be necessary to define the thermal duty of the plates. The plates are of two types, short duty or soft plates and long duty or hard plates. The soft plates involve low pressure drop and low heat transfer coefficients. Plates having a high chevron angle provide this feature. A short and wide plate is of this type. The hard plates involve high pressure drop and give high heat transfer coefficients. A plate having a low chevron angle provides high heat transfer combined with high pressure drop. Long and narrow plates belong to this category. The type of embossing, apart from the overall dimensions, has a significant effect on the thermal performance.

Manufacturers produce both hard and soft plates, so that a wide range of duties can be handled in the most efficient manner. In deciding the type of plate that will be most suitable for a given duty, the most convenient term to use is the number of transfer units (NTU), defined as

$$\text{NTU}_c = \frac{UA}{(\dot{m}c_p)_c} = \frac{T_{c_2} - T_{c_1}}{\Delta T_m} \tag{11.32}$$

$$\text{NTU}_h = \frac{UA}{(\dot{m}c_p)_h} = \frac{T_{h_1} - T_{h_2}}{\Delta T_m} \tag{11.33}$$

Manufacturers specify the plates that have low values of chevron angle as high-θ plates and plates with high values of chevron angle as low-θ plates. A low chevron angle is around 25°–30°, while a high chevron angle is around 60°–65°.

The ε-NTU method is described in Chapter 2; the total heat transfer rate, from Equation 2.43, is

$$Q = \varepsilon (\dot{m}c_p)_{min} (T_{h_1} - T_{h_2}) \tag{11.34}$$

The heat capacity rate ratio is given by Equation 2.35 as

$$R = \frac{T_{h_1} - T_{h_2}}{T_{c_2} - T_{c_1}} \tag{11.35}$$

When $R < 1$,

$$(\dot{m}c_p)_c = (\dot{m}c_p)_{min} = C_{min} \tag{11.36}$$

$$\text{NTU} = \frac{UA}{C_{min}} = \frac{UA}{(\dot{m}c_p)_c} \tag{11.37}$$

and when $R > 1$,

$$(\dot{m}c_p)_h = (\dot{m}c_p)_{min} = C_{min} \tag{11.38}$$

$$NTU = \frac{UA}{C_{min}} = \frac{UA}{(\dot{m}c_p)_h} \tag{11.39}$$

In calculating the value of NTU for each stream, the total mass flow rates of each stream must be used.

Graphs and tables of heat exchanger effectiveness, ε, and correction factor, F, to ΔT_m as functions of R and NTU for various plate heat exchangers are available in the literature.[5,29,30]

For pure counterflow and parallel flow, the heat exchanger effectiveness is given by Equations 2.46 and 2.47, respectively. Heat exchanger effectiveness, ε, and $(NTU)_{min}$ for counterflow can be expressed as

$$\varepsilon = \frac{\exp\left[(1 - C_{min}/C_{max})NTU_{min}\right] - 1}{\exp\left[(1 - C_{min}/C_{max})NTU_{min}\right] - C_{min}/C_{max}} \tag{11.40}$$

and

$$NTU_{min} = \frac{\ln\left[(1 - C_{min}/C_{max})/(1 - \varepsilon)\right]}{(1 - C_{min}/C_{max})} \tag{11.41}$$

which are useful in rating analysis when the outlet temperatures of both streams are not known.

One of the problems associated with plate units concerns the exact matching of the thermal duties. It is very difficult to achieve the required thermal duty and at the same time fully utilize the available pressure drop. One solution to this problem is to install more than one kind of channel in the same plate pack with differing NTU values.[13] The types could be mixed to produce any desired value of NTU between the highest and lowest. The high NTU plate is corrugated with a large included chevron angle, giving a comparatively larger pressure drop, while the low NTU plate has chevron corrugations with a much smaller included angle leading to a relatively low pressure drop.

A comparative study of the shell-and-tube vs. gasketed-plate heat exchangers is given by Biniciogullari,[31] in which a sample calculation is given for a set of specifications of a process for water to water heat exchangers to show how to make thermal and hydraulic analyses of these two types of heat exchangers. The experimental studies are performed on plate package heat exchangers, and further correlations for heat transfer and pressure drop analysis are provided.[34-36]

Nomenclature

A_t	total effective plate heat transfer area, m^2
A_1	single plate effective area, m^2
A_{1p}	single plate projected area, m^2
b	mean mass channel gap
c_p	specific heat of fluid, J/kg · K
CF	cleanliness factor
C_h	constant in Equation 11.10
C_s	Safety factor defined by Equation 11.31
D_e	channel equivalent diameter ($D_e = 2b$)
D_h	channel hydraulic defined by Equation 11.9, m
D_p	port diameter, m
f	Fanning friction factor
K_p	constant in Equation 11.21
k	thermal conductivity, W/m · K
G_c	mass velocity through a channel, kg/m^2 · s
G_p	heat transfer coefficient, W/m · K
L_e	length of compressed plate pack, m
L_h	horizontal port distance, m
L_p	projected plate length, m
L_v	vertical port distance (flow length in one pass), m
L_w	plate width inside gasket, m
L_{eff}	effective flow length between inlet and outlet port ($\approx L_v$), m
$\dot{m}$	mass flow rate, kg/s
NTU	number of transfer units defined by Equation 11.32
P	plate pitch, m
P_c	longitudinal corrugation pitch, wavelength of surface corrugation, m
Pr	Prandtl number, $c_p\mu/k$
P_w	wetted perimeter, m
Q	duty, W
Q_c	heat load under clean conditions, W
Q_f	heat load under fouled conditions, W
Q_r	heat transfer required, W
R_{fc}	fouling factor for cold fluid, m^2 · K/W
R_{fh}	fouling factor for hot fluid, m^2 · K/W
Re	Reynolds number, $G_c D_e/\mu$, $G_c D_h/\mu$, or $G_p D_p/\mu$
t	plate thickness, m
U_c	clean overall heat transfer coefficient, W/m^2 · K
U_f	fouled (service) overall heat transfer coefficient, W/m^2 · K

Gasketed-Plate Heat Exchangers

Greek Symbols

β	Chevron angle, deg
γ	channel aspect ratios, $2b/P_c$
Δp_c	channel flow pressure drop, kPa
Δp_p	port pressure drop, kPa
ΔT_m	mean temperature difference, °C, K
ρ	fluid density, kg/m³
ϕ	ratio of effective to projected area of corrugated plate
μ	dynamic viscosity at average inlet temperature, $N \cdot s/m^2$ or $Pa \cdot s$

PROBLEMS

11.1. The following constructional information is available for a gasketed-plate heat exchanger:

Chevron angle	50°
Enlargement factor	1.17
All port diameters	15 cm
Plate thickness	0.006 m
Vertical port distance	1.50 m
Horizontal port distance	0.50 m
Plate pitch	0.0035 m

Calculate:
a. Mean channel flow gap
b. One channel flow area
c. Channel equivalent diameter
d. Projected plate area
e. Effective surface area per plate.

11.2. A gasketed-plate heat exchanger will be used for heating city water ($R_{fc} = 0.00006$ m² · K/W) using the waste water available at 90°C. The vertical distance between the ports of the plate is 1.60 m and the width of the plate is 0.50 m with a gap between the plates of 6 mm. The enlargement factor is given by the manufacturer as 1.17 and the chevron angle is 50°. The plates are made of titanium ($k = 20$ W/m · K) with a thickness of 0.0006 m. The port diameter is 0.15 m. The cold water enters the plate heat exchanger at 15°C and leaves at 45°C at a rate of 6 kg/s and it will be heated by the hot water available at 90°C, flowing at a rate of 12 kg/s. Considering single-pass arrangements for both streams, calculate:

a. The effective surface area and the number of plates of this heat exchanger
b. The pressure drop for both streams.

11.3. Solve problem 11.2 with the use of the correlations proposed by Kumar, and Mulay and Manglik. List the resutls in a Table and compare the results.

11.4. Repeat Example 11.1 for 30% or less over surface design.

11.5. Solve Problem 11.2 for a two pass/two pass arrangement.

11.6. A heat exchanger is required to heat treated cooling water with a flow rate of 60 kg/s from 10°C to 50°C using the waste heat from water, cooling from 60°C to 20°C with the same flow rate as the cold water. The maximum allowable pressure drop for both streams is 120 kPa. A gasketed-plate heat exchanger with 301 plates having a channel width of 50 cm, a port diameter of 0.15 m and a vertical distance between ports of 1.5 m is proposed and the plate pitch is 0.0035 m with an enlargement factor of 1.25. The spacing between the plates is 0.6 mm. Plates are made of stainless steel (316) (k = 16.5 W/m · K). For a two-pass/two-pass arrangement, analyze the problem to see if the proposed design is feasible. Could this heat exchanger be smaller or larger?

11.7. Analyze problem 11.6 with the use of the correlations proposed by Kumar and correlation proposed by Chisholm and Wanniarachchi to see if the proposed design is feasible. Then, compare the results of both correlations. Could this heat exchanger be smaller or larger?

11.8 Repeat Example 11.1 for different water conditions:

a. Coastal ocean (R_{fc} = 0.0005 K · m²/W), demineralized closed loop (0.000001 K · m²/W)

b. Wastewater (0.00001 K · m²/W)/cooling tower (R_{fc} = 0.000069 m² · K/W)

Explain your conclusions.

11.9. A one-pass counter-current flow heat exchanger has 201 plates. The exchanger has a vertical port distance of 2 m and is 0.6 m wide, with a gap between the plates of 6 mm. This heat exchanger will be used for the following process: cold water from the city supply with an inlet temperature of 10°C is fed to the heat exchanger at a rate of 15 kg/s and will be heated to 75°C with wastewater entering at an inlet temperature of 90°C. The flow rate of hot water is 30 kg/s, which is distilled water. The other construction parameters are as given in Problem 11.2. There is no limitation on the pressure drop. Is this heat exchanger suitable for this purpose (larger or smaller)?

11.10. In Example 11.1, a new mass flow rate is provided as 90 kg/s. Assume that the inlet temperatures of the hot and cold fluid are changed to 120°C and 20°C respectively. If the flow arrangement is single pass for both streams, use the ε – NTU method to calculate the outlet temperatures for the specific heat exchanger. There may be a constraint on the outlet temperature of the cold stream, therefore this type of analysis is important to see if the required outlet temperature of the cold stream is satisfied.

DESIGN PROJECT 11.1
A Comparative Study of Gasketed-Plate Heat Exchangers vs. Double-Pipe (Hairpin) Heat Exchangers

The main considerations involved in the choice of gasketed-plate heat exchangers for a specific process in comparison with double-pipe heat exchanger/shell-and-tube heat exchanger units may be made on the basis of weight and space limitations, temperature approach, operating temperatures, pressure drop limitations, maintenance requirements, and capital and operating costs which must be considered in a comparative study. For given specifications, make a comparative study of a gasketed-plate heat exchanger with a double-pipe heat exchanger. Use the specifications given in design Project 7.1 or in Problem 11.5 and the constructional parameters given in Example 11.1, or select alternative parameters.

DESIGN PROJECT 11.2
A Comparative Study of Gasketed-Plate Heat Exchangers vs. Shell-and-Tube Heat Exchangers

The specifications given in Example 11.1 or in Problem 11.5 can be used for such a comparative study.

DESIGN PROJECT 11.3
Design a Heat Exchanger for an Open Heart Operation—Blood Heat Exchanger

The objective of a blood heat exchanger is to shorten the time normally required to cool a patient's blood prior to open-heart surgery. Since blood behaves in a non-Newtonian manner, necessary assumptions are made to complete the calculations. The material in contact needs to have the smoothest possible surface, therefore a stainless steel with a specific non-wetting silicon resin shell-and-tube heat exchanger is chosen for the design. For a sufficient design, the pressure drop must be minimal due to the fragile nature of blood, and the heat transfer rate must be high due to the required rapid temperature change. This design project will take into account thermal, dynamic heat transfer and a parametric analysis for a blood heat exchanger. Assume that the patient's blood will be cooled from 37°C to 27°C by cooling water available at 15°C. The mass flow rate of blood is 0.03 kg/s and the properties at 32°C are: $\rho = 1055$ kg/m^3, $\mu = 0.00045$ kg/m · s, $k = 0.426$ W/m · K, $Pr = 3.52$, and $c_p = 3330$ J/kg · K. Cooling water may be available at 0.20 kg/s. One can directly select a gasketed-plate heat exchanger, or a comparative study as in design Projects 11.1 and 11.2 can be carried out.

References

1. *Thermal Handbook*, Alfa-Laval Thermal AB, Lund, Sweden.
2. Saunders, E. A. D., *Heat Exchangers — Selection, Design, and Construction*, John Wiley & Sons, New York, 1988.
3. Raju, K. S. N. and Bansal, J. C., Plate heat exchangers and their performance, in *Low Reynolds Number Flow Heat Exchangers*, Kakaç, S., Shah, R. K., and Bergles, A. E., Eds., Hemisphere, Washington, DC, 899–912, 1983.
4. Cooper, A. and Usher, J. D., Plate heat exchangers, in *Heat Exchanger Design Handbook*, Schlünder, E. U., Ed., Hemisphere, Washington, DC, 1988, ch. 3.
5. Raju, K. S. N. and Bansal, J. C., Design of plate heat exchangers, in *Low Reynolds Number Flow Heat Exchangers*, Kakaç, S., Shah, R. K., and Bergles, A. E., Eds., Hemisphere, Washington, DC, 913–932, 1983.
6. Edwards, M. F., Heat transfer in plate heat exchangers at low Reynolds numbers, in *Low Reynolds Number Flow Heat Exchangers*, Kakaç, S., Shah, R. K., and Bergles, A. E., Eds., Hemisphere, Washington, DC, 933–947, 1983.
7. Wilkinson, W. L., Flow distribution in plate exchangers, *Chem. Eng.*, No. 285, 289, 1974.
8. Marriott, J., Where and how to use plate heat exchangers, *Chem. Eng.*, 78(8), 127, 1971.
9. Cooper, A., Suitor, J. W., and Usher, J. D., Cooling water fouling in plate heat exchangers, *Heat Transfer Eng.*, 1(3), 50, 1980.
10. Kumar, H., The plate heat exchanger: construction and design, 1st UK National Conference on Heat Transfer, University of Leeds, 3-5 July, *Inst. Chem. Symp. Series*, No. 86, 1275–1286, 1984.
11. Edwards, M. F., Changal, A. A., and Parrott, D. L., Heat transfer and pressure drop characteristics of a plate heat exchanger using Newtonian and non-Newtonian liquids, *Chem. Eng.*, 286–293, May 1974.
12. McKillop, A. A. and Dunkley, W. L., Heat transfer, *Indus. Eng. Chem.*, 52(9), 740, 1960.
13. Clark, D. F., Plate heat exchanger design and recent development, *Chem. Eng.*, 285, 275, 1974.
14. Manglik, R. M., Plate heat exchangers for process industry applications: enhanced thermal-hydraulic characteristics of chevron plates, in *Process, Enhanced and Multiphase Heat Transfer*, Manglik, R. M. and Kraus, A. D., Eds., Begel House, New York, 1996.

15. Savostin, A. F. and Tikhonov, A. M., Investigation of the characteristics of plate-type heating surfaces, *Thermal Eng.*, 17(9), 113, 1976.
16. Tovazhnyanski, L. L., Kapustenko, P. A., and Tsibulnik, V. A., Heat transfer and hydraulic resistance in channels of plate heat exchangers, *Energetika*, 9, 123, 1980.
17. Chisholm, D. and Wanniarachchi, A. S., Maldistribution in single-pass mixed-channel plate heat exchangers, in *Compact Heat Exchangers for Power and Process Industries*, HTD-Vol. 201, ASME, New York, 1992, 95.
18. Heavner, R. L., Kumar, H., and Wanniarachchi, A. S., Performance of an industrial plate heat exchanger: effect of chevron angle, AICHE Symposium Series No. 295, Vol. 89, AICHE, New York, 262, 1993.
19. Talik, A. C. and Swanson, L. W., Heat transfer and pressure drop characteristics of a plate heat exchanger using a propylene-glycol/water mixture as the working fluid, *Proc. 30th Natl. Heat Transfer Conf.*, 12, 83, 1995.
20. Muley, A. and Manglik, R. M., Experimental investigation of heat transfer enhancement in a PHE with $\beta = 60°$ chevron plates, in *Heat and Mass Transfer*, Tata McGraw-Hill, New Delhi, India, 1995, 737.
21. Muley, A. and Manglik, R. M., Enhanced heat transfer characteristics of single-phase flows in a plate heat exchanger with mixed chevron plates, *J. Enhanced Heat Transfer*, 4(3), 187, 1997.
22. Muley, A., Manglik, R. M., and Metwally, H. M., Enhanced heat transfer characteristics of viscous liquid flows in a chevron plate heat exchanger, *J. Heat Transfer*, 121(4), 1011, 1999.
23. Muley, A. and Manglik, R. M., Enhanced thermal-hydraulic performance of chevron plate heat exchangers, *Int. J. Heat Exchangers*, 1, 3, 2000.
24. Muley, A. and Manglik, R. M., Experimental study of turbulent flow heat transfer and pressure drop in a plate heat exchanger with chevron plates, *J. Heat Transfer*, 121(1), 110, 1999.
25. Kakaç, S., Shah, R. K., and Aung, W., *Handbook of Single-Phase Convective Heat Transfer*, John Wiley & Sons, New York, 1987.
26. Buonopane, R. A., Troupe, R. A., and Morgan, J. C., Heat transfer design method for plate heat exchangers, *Chem. Eng. Prog.*, 59, 57, 1963.
27. Foote, M. R., Effective mean temperature difference in multi-pass plate heat exchangers, Nat. Eng. Lab., U.K., Report No. 303, 1967.
28. Bell, J. K., Plate heat exchanger, in *Heat Exchangers — Thermal Hydraulic Fundamentals and Design*, Kakaç, S., Bergles, A. E., and Mayinger, F., Eds., Hemisphere, Washington, DC, 165–176, 1981.
29. Hewitt, G. F., Shirs, G. L., and Bott, T. R., *Process Heat Transfer*, CRC Press, Boca Raton, FL, 1994.
30. Shah, R. K. and Focke, W. W., Plate heat exchangers and their design theory, in *Heat Transfer Equipment Design*, Shah, R. K., Subbarao, E. C., and Mashelkar, R. A., Eds., Hemisphere, Washington, DC, pp. 227–254, 1988.
31. Biniciogullari, D., Comparative study of shell-and-tube heat exchangers versus gasketed-plate heat exchangers, M.S. Thesis, METU, Ankara, 2001.
32. Martin, H., A theoretical approach to predict the performance of chevron-type plate heat exchangers, *Chem. Eng. Process.*, 35, 301, 1996.

33. Wang, L., and Sundén, B., Optimal design of plate heat exchangers with and without pressure drop specifications, *Appl. Thermal Eng.*, 23, 295, 2003.
34. Rao, B. P., Sunden, B. and Das, S., An experimental and theoretical investigation of the effect of flow maldistribution on the thermal performance of plate heat exchangers. *J. Heat Transfer*, 127, 332, 2005.
35. Warnakulasuriya, F. S. K. and Worek, W. M., Heat transfer and pressure drop properties of high viscous solutions in plate heat exchangers, *Int. J. Heat Mass Transfer*, 51, 52, 2008.
36. Gulben, G., Development of a computer program for designing a gasketed-plate heat exchanger for various working conditions and verification of the computer program with experimental data, M.S. Thesis, TOBB-ETU, Ankara, 2011.

12

Condensers and Evaporators

12.1 Introduction

A condenser is a two-phase flow heat exchanger in which heat is generated from the conversion of vapor into liquid (condensation) and the heat generated is removed from the system by a coolant. Condensers may be classified into two main types: those in which the coolant and condensate stream are separated by a solid surface, usually a tube wall, and those in which the coolant and condensing vapor are brought into direct contact.

The direct contact type of condenser may consist of a vapor which is bubbled into a pool of liquid, a liquid which is sprayed into a vapor, or a packed-column in which the liquid flows downward as a film over a packing material against the upward flow of vapor. Condensers in which the streams are separated may be subdivided into three main types: air-cooled, shell-and-tube, and plate. In the air-cooled type, condensation occurs inside tubes with cooling provided by air blown or sucked across the tubes. Fins with large surface areas are usually provided on the air side to compensate for the low air-side heat transfer coefficients. In shell-and-tube condensers, the condensation may occur inside or outside the tubes. The orientation of the unit may be vertical or horizontal. In the refrigeration and air-conditioning industry, various types of two-phase flow heat exchangers are used. They are classified according to whether they are coils or shell-and-tube heat exchangers. Evaporator and condenser coils are used when the second fluid is air because of the low heat transfer coefficient on the air side.

In the following sections, the basic types of condensers, air-conditioning evaporators, condenser coils, and their rating and sizing problems are briefly discussed. Further information can be obtained from Butterworth.[1,2]

12.2 Shell and Tube Condensers

12.2.1 Horizontal Shell-Side Condensers

The shell-and-tube condensers for process plants are covered by TEMA; Figure 9.1 shows various TEMA shell types as described in Chapter 9. The most common condenser types are shown in Figure 9.3; these condensers may be used as shell-side condensers with the exception of the F-shell type, which is unusual. The E-type is the simplest form of these types of condensers. In shell-and-tube condensers, condensation may occur inside or outside the tubes, depending on the design requirements (Figures 12.1 and 12.2).[3] A very important feature of a condenser as compared with any other type of heat exchanger is that it must have a vent for the removal of noncondensable

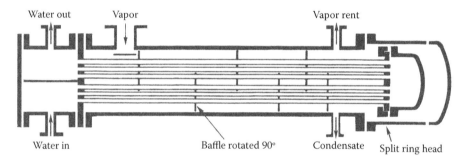

FIGURE 12.1
Horizontal shell-side condenser. (From Mueller, A. C., *Heat Exchanger Design Handbook*, Hemisphere, Washington, DC, 1983.)

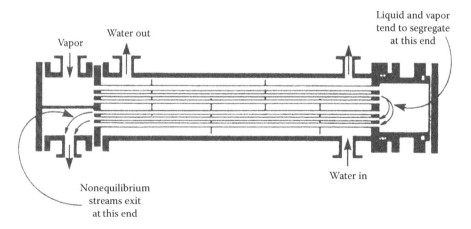

FIGURE 12.2
Horizontal in-tube condenser. (From Mueller, A. C., *Heat Exchanger Design Handbook*, Hemisphere, Washington, DC, 1983; Breber, G., *Heat Transfer Equipment Design*, Hemisphere, Washington, DC, 477, 1988.)

gas. Thus, an E-type condenser will have two outlet nozzles: one for the vent and the other for the condensate outlet. Noncondensable has the effect of depressing the condensation temperature and, thus, reducing the temperature difference between the streams. Therefore, it is clear that there is no chance of noncondensable accumulation during condenser operation. Hence, the vent is provided.

There are some advantages and disadvantages of each of the different shell-type condensers, which are briefly discussed below.

The J-shell type has a great advantage over the E-shell type in that it can be arranged with two nozzles, one at either end for the vapor inlet, and one small nozzle in the middle for the condensate outlet. One would normally have a small nozzle in the middle at the top in order to vent noncondensable gas. By having these two inlet nozzles, a large vapor volume coming into the condenser can be accommodated more easily. Also, by splitting the vapor flow into two and by halving the path length for vapor flow, the pressure drop may be reduced substantially over that for a similar size E-shell. It is a good practice with the J-shell to make sure that the heat loads in both halves of the exchanger are the same to prevent the possibility that noncondensed vapor coming from one end of the exchanger meets subcooled liquid from the other. This could give rise to periodic violent vapor collapse and possible exchanger damage. This problem usually means that the J-shell condenser should not be designed with a single tube-side pass if there is a large temperature variation in the tube-side fluid as it flows from one end of the exchanger to the other. J-shells would normally have baffles similar to those found in E-shells, except that a full-circle tube support plate may be placed in the center of the exchanger (Figure 12.3).

The F-, G-, and H-shell condensers can also have transverse baffles in addition to the longitudinal baffle (Figures 9.1 and 12.4). Full-circle tube support plates may be placed in line with the inlet nozzles and, for H-shells, additional full-circle tube support plates can be placed halfway along the shell. An H-shell would, therefore, have three tube support plates along the lengths of the tubes, so it may be possible to avoid having any further

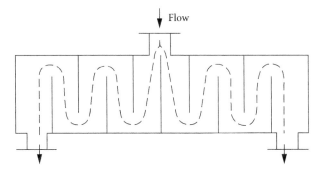

FIGURE 12.3
Schematic sketch of TEMA J-shell.

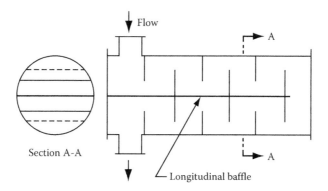

FIGURE 12.4
Schematic sketch of TEMA J-shell.

segmented baffles supporting the tubes. In such circumstances, an H-shell gives a fairly low pressure drop. The vent nozzles in G- and H-shell types have to be placed in the side of the shell above the condensate outlet nozzles but, of course, below the longitudinal baffles. If there are multiple tube-side passes in G- and H-shells, these should be arranged so that the coldest pass is at the bottom and the warmest at the top so that there is some degree of countercurrent flow.[1]

The crossflow, or X-type exchanger, is a very useful unit for vacuum operation. In such operations, large volumes of vapor must be handled, and it is possible to avoid the chance of tube vibration. The large flow area combined with the short flow path also means that pressure drops can be kept low. It is important, particularly, to keep pressure drops low in vacuum operation so as to avoid reducing the saturation temperature and, therefore, losing temperature difference. Figure 12.5 shows a typical crossflow unit. This particular unit has three inlet nozzles to avoid having a very large single inlet nozzle which may lead to difficulties in mechanical construction. A large space above the top of the bundle is necessary to give good vapor distribution along the exchanger length, and this may be assisted by the introduction of a perforated distributor inserted as is necessary to give sufficient tube support to prevent vibration. Noncondensable gases must be vented from as low as possible in the exchanger, as shown in Figure 12.5. Variations on the tube bundle layout are possible in crossflow condensers.[1]

In shell-and-tube condensers, a variety of alternative baffle arrangements like single-segmental, double-segmental, triple-segmental, and no-tube-in-the-window are used. Baffles in condensers are usually arranged with the cut vertical so that the vapor flows from side to side. The baffles are notched on the bottom to allow drainage of the condensate from one compartment to the next and, finally, to the condensate exit.

With most of the baffle types, having the baffles closer together increases the shell-side fluid velocities. This would be undesirable if vibration or high

Condensers and Evaporators

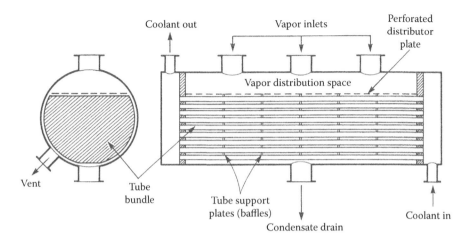

FIGURE 12.5
Main features of a crossflow condenser (TEMA X-type). (From Butterworth, D., *Boilers, Evaporators and Condensers*, John Wiley & Sons, New York, 571, 1991; Butterworth, D., *Two-Phase Flow Heat Exchangers*, Kluwer, The Netherlands, 1988. With permission.)

pressure drop is a problem. The design with no tubes in the window, however, allows for additional intermediate baffles which support the tubes without having any significant effect on the flow. However, such a design is expensive because there is a large volume of empty shell.

It is almost a universal practice to use an impingement plate under the vapor inlet nozzle on a shell-side condenser to prevent tube erosion from the high velocity of the incoming vapor (Figure 12.6). Otherwise, the outer tubes of the first row of the bundle could be subject to serious damage.[1]

When flow-induced vibration is a problem, an extra tube support plate may be inserted near the inlet nozzle, as shown in Figure 12.7.[1]

12.2.2 Vertical Shell-Side Condensers

A vertical E-shell condenser is shown in Figure 12.8. The vent is shown near the condensate exit at the cold end of the exchanger. Vertical shell-side condensation has definite thermodynamic advantages when a wide condensing range mixture is to be condensed. There will be very good mixing between the condensate and vapor. In this arrangement, it may not be easy to clean the inside of the tubes. Thus, this arrangement should not be used with a fouling coolant.

12.2.3 Vertical Tube-Side Condensers

These condensers are often designed with down-flows for both the vapor and the condensate. Vapor enters at the top and flows down through the tubes

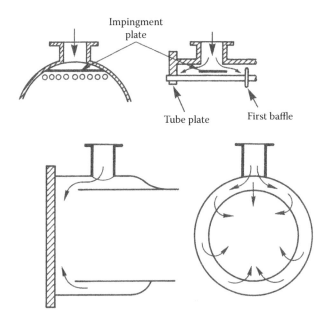

FIGURE 12.6
Impingement plate and vapor belt. (From Butterworth, D., *Boilers, Evaporators and Condensers*, John Wiley & Sons, New York, 571, 1991; Butterworth, D., *Two-Phase Flow Heat Exchangers*, Kluwer, The Netherlands, 1988. With permission.)

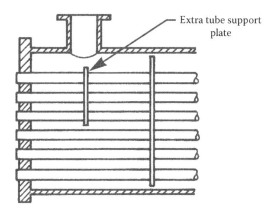

FIGURE 12.7
Extra tube support plate to help prevent vibration of tubes near the inlet nozzle.

with the condensate draining from the tubes by gravity and vapor shear (Figure 12.9). If the shell-side cleaning can be done chemically, a fixed tube sheet construction can be designed. Two tube-side passes, using U-tubes, would be possible where there is upward flow in the first pass and downward flow in the second. Tube-side condensers must also have adequate

Condensers and Evaporators 497

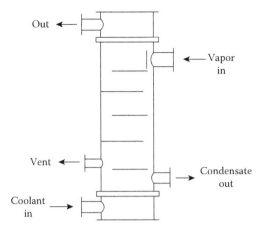

FIGURE 12.8
Vertical E-shell condenser. (From Butterworth, D., *Boilers, Evaporators and Condensers*, John Wiley & Sons, New York, 571, 1991; Butterworth, D., *Two-Phase Flow Heat Exchangers*, Kluwer, The Netherlands, 1988. With permission.)

venting. The vent line in vertical units with down-flow should be placed in the oversized lower header above any possible pool of condensate formed there.

Vertical tube-side condensers may also be designed to operate in the reflux mode, with an upward flow of vapor but a downward counterflow of any condensate formed on the tube walls. Clearly, such units can only operate provided the flooding phenomenon is avoided (Figure 12.10). Capacity is limited due to flooding of the tubes. The vapor condenses on the cold tube walls, and the condensate film drains downward by gravity. Any noncondensable gases pass up the tube and can be evacuated through a vent.

The tubes extend beyond the bottom tube sheet and are cut at a certain angle to provide drip points for the condensate. The shell should be vented through the upper tube sheet. Reflux condenser tubes are usually short, only 2–5 m long, with large diameters. In this type of condenser, it must be ensured that the vapor velocity entering the bottom of the tubes is low so that the condensate can drain freely from the tubes.[5]

12.2.4 Horizontal in-Tube Condensers

Horizontal tube-side condensation is most frequently used in air-cooled condensers and in kettle or horizontal thermosiphon reboilers. Horizontal in-tube condensers can be in single-pass, multipass, or U-tube arrangements. More than two passes is unusual. In U-tube bundles, the different lengths between the inner and outer U-tubes will result in different condensing than the same area in a straight-tube, single-pass bundle.[3,4]

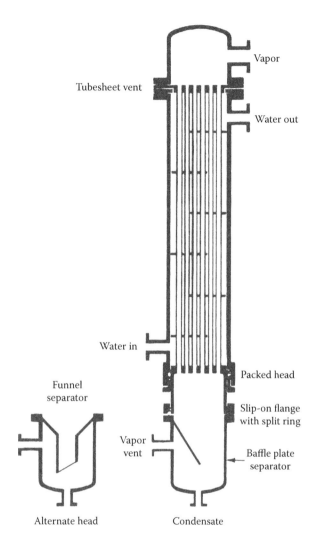

FIGURE 12.9
Vertical in-tube downflow condenser. (From Mueller, A. C., *Heat Exchanger Design Handbook*, Hemisphere, Washington, DC, 1983. With permission.)

Figure 12.2 shows a typical two-pass arrangement for horizontal tube-side condensation. The vapor coming in is partially condensed in the first pass in the top of the bundle. When the vapor–condensate mixture exits from the first pass into the turnaround header, it is difficult, using existing methods, to estimate the distribution of the liquid and the vapor among the tubes of the second pass. The condensate dropout between passes will cause misdistribution, and its effect on the condensation processes must be considered. The misdistribution of the condensate and

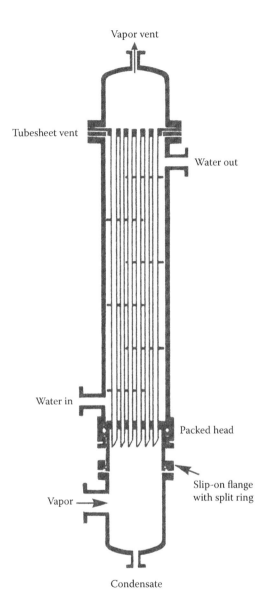

FIGURE 12.10
Reflux condenser. (From Mueller, A. C., *Heat Exchanger Design Handbook*, Hemisphere, Washington, DC, 1983; Breber, G., *Heat Transfer Equipment Design*, Hemisphere, Washington, DC, 477, 1988. With permission.)

vapor into certain tubes alters the heat transfer and fluid flow characteristics of these tubes, resulting in a net reduction of the average effective heat transfer coefficient for the entire pass. If a multicomponent mixture is being condensed, the separation of the liquid and vapor will result in a partial fractionation of the mixture being condensed. It is possible in that case to have substantial quantities of noncondensed vapor exiting the condenser.

A vent must be located where the noncondensables are finally concentrated. Tube layouts depend on the coolant requirements. Because of the blanketing effect of condensate accumulation in the bottom portion of the tubes, this configuration is not very effective in gravity-controlled flow.

It is desirable to subcool any condensate leaving a condenser to prevent flashing in the piping circuit and equipment downstream of the condenser. For condensation inside vertical tubes, subcooling can be achieved by having a liquid level control and running the tubes full of liquid up to a certain height. However, this is clearly only possible if there is a noncondensable present, otherwise there is no convenient way to vent such noncondensable.

12.3 Steam Turbine Exhaust Condensers

For historical reasons, steam turbine exhaust condensers are often referred to as surface condensers. In principle, they are no different from the shell-side condensers described in Section 12.2.1, in particular, the X-type (Figure 12.5). In practice, there are certain severe demands placed on these units that have been overcome by special design features. These special demands arise from the large heat duties that they must perform, and from the necessity to maintain a low condensing temperature to achieve the highest possible power station efficiency.[1]

The aim is to operate with the condensing temperature only a few degrees above the cooling water temperature. Typically, the cooling water is about 20°C, with condensation taking place at around 30°C. Saturation pressure of water at this temperature is 0.042 bar absolute, which is a typical operating pressure for these condensers. Clearly, there is little pressure available for pressure drop through the unit. There is also small temperature difference to spare in order to overcome the effects of noncondensable gases. Hence, the design of surface condensers is governed by the need for good venting and low pressure drop.

These very large condensers often have box-shaped shells. The smaller ones, with surface areas less than about 5,000 m², may have cylindrical shells.[1] Figure 12.5 illustrates some of the main features of a surface condenser which are common to most designs.

As with any other shell-and-tube unit, tubes in surface condensers must be supported at regular intervals along their length with tube support plates. Such support plates also have the advantage of deliberately preventing any axial flow of vapor, thus making it easier for designers to ensure that vapor flow paths through the bundles are relatively straightforward, giving rise to no recirculation pockets where noncondensable can accumulate.

It is desirable to subcool any condensate leaving a condenser to prevent flashing in the piping circuit and equipment downstream of the condenser. One solution is to design a separate unit as a subcooling exchanger. One can also overdesign the condenser; a subcooling section can be added as a second pass. This second section often has closer baffling to increase the heat transfer coefficient in this region. For example, the unit may be designed with all the required condensation achieved in the first pass. The vent line would then be placed in the header at the end of this first pass. The liquid level would be maintained in this header and the second pass would be running full of liquid. With this particular method of achieving subcooling, however, a much smaller number of tubes would normally be required in the second pass to give high velocities to ensure good heat transfer.

There is such a variety of different surface condenser designs that it is impossible to illustrate them all here, but many examples of modern condensers are described by Sebald.[6]

12.4 Plate Condensers

There are three main types of plate exchangers—plate-and-frame, spiral plate, and plate-fin exchangers—which have already been introduced in Chapters 1 and 10 (see Figures 1.13, 1.15, and 11.1).

Plate-and-frame (gasketed) heat exchangers are usually limited to fluid streams with pressures below about 25 bar and temperatures below about 250°C. Detailed information is given by Alta Laval.[7] Gasketed heat exchangers are mainly developed for single-phase flows. They are not well suited for condensers because of the size of the parts in the plates which are small for handling large volume flows of vapor. However, they are frequently used with service steam on one side to heat some process stream.

A spiral heat exchanger as a condenser is shown in Figure 12.11. These can operate at pressures up to about 20 bar and temperatures up to about 400°C.

Plate-fin heat exchangers are discussed in Chapter 1. These heat exchangers are very compact units having heat transfer surface densities around 2,000 m^2/m^3.[1] They are often used in low temperature (cryogenic) plants where the temperature difference between the streams is small (1–5°C). The flow channels in plate-fin exchangers are small and often contain many irregularities. This means that the pressure drop is high with high flow velocities. They are not mechanically cleanable; therefore, their use is restricted to clean fluids.

Plate-fin exchangers can be arranged as parallel and counterflow or the streams may be arranged in crossflow (Figures 1.17 and 1.25c). The corrugated sheets which are sandwiched between the plates serve both to give

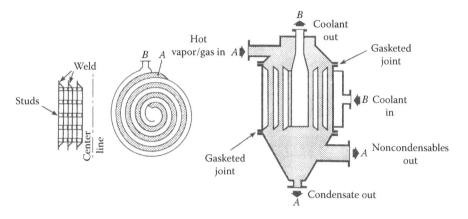

FIGURE 12.11
Spiral heat exchanger for condenser applications. (Courtesy of APV International Ltd. From Butterworth, D., *Boilers, Evaporators and Condensers*, John Wiley & Sons, New York, 571, 1991; Butterworth, D., *Two-Phase Flow Heat Exchangers*, Kluwer, The Netherlands, 1988. With permission.)

extra heat transfer and to provide structural support to the flat plate. The most common corrugated sheets are shown in Figure 1.18.

Uniform flow distribution is very essential for good performance of plate-fin exchangers as condensers. When the nominal flow rate of coolant in the adjacent channels is small, partial condensation can occur which decreases the unit efficiently and can lead to several operating problems.

12.5 Air-Cooled Condensers

Air-cooled condensers may be favored where water is in short supply. Condensation occurs inside tubes which are finned with transverse fins on the air side for a low air-side heat transfer coefficient. Figure 1.23 illustrates a typical air-cooled heat exchanger. The unit shown is a forced draught unit since the air is blown across. The alternative design includes a fan at the top which is called induced draught. Air-cooled condensers can also become economical if condensation takes place at temperatures that are about 20°C higher than the ambient temperature.[4]

Disadvantages of air-cooled condensers are their requirement of a relatively large ground area and space around the unit, noise from the fans, and problems with freezing of condensate in climates with low temperatures.

Air-cooled heat exchangers as condensers normally have only a few tube rows, and the process stream may take one or more passes through the unit. With multipass condensers, a problem arises with redistributing the two-phase mixture on entry to the next pass. This can be overcome in some cases by using U-tubes or by having separate tube rows only for the condensate (Figure 1.36). There is only one arrangement practically possible: each successive pass must be located below the previous one to enable the condensate to continue downward.

12.6 Direct Contact Condensers

Direct contact condensers are inexpensive and simple to design but have limited application because the condensate and coolant are mixed. The main advantage of these condensers, besides their low cost, is that they cannot be fouled and they have very high heat transfer rates per unit volume. There are generally three types of direct contact condensers: pool, spray, and tray condensers.

In pool condensers, vapor is injected into a pool of liquid. The coolant may be a process fluid to be heated. There may be some operational problems with this type of condenser. The first is that the condensation front may move back into the vapor inlet line, causing the liquid to be periodically ejected, often with some violence. The second is that a very large vapor bubble may form in the liquid pool and may collapse suddenly, causing damage to the vessel. These problems may be avoided by injecting the vapor through a large number of small holes or by using special ejectors which mix the incoming vapor with liquid in a special mixing tube.[1,4]

The most common type of direct-contact condenser is one in which subcooled liquid is sprayed into the vapor in a large vessel. This arrangement is illustrated in Figure 1.7a. Very often, these units are used for condensing steam using water as a coolant. In these cases, the mixing of water with condensate presents no major problems. When condensing a vapor whose condensate is immiscible with the spray liquid, however, a separator is usually required after the condenser in order to recover the product. Alternatively, the condensate product may be cooled in a single-phase exchanger, and some may be recycled as coolant spray. At first sight, there then seems to be little benefit to using a combination of a direct-contact condenser and a conventional single-phase exchanger instead of using only one shell-and-tube condenser. The advantage appears, however, when the condenser is operating under vacuum conditions. As has been seen already, tubular condensers for vacuum operation are large and complex. It can, therefore, sometimes be economical to replace such a condenser by a simple spray condenser and a compact single-phase cooler.[1]

Spray condensers cannot be used with dirty coolants since the spray nozzles may be blocked. In those circumstances, a tray condenser may be used, as illustrated in Figure 1.7b. The trays may be sloped slightly to prevent dirt accumulation on them. The tray arrangement can have a slight thermodynamic advantage over a spray unit because some degree of countercurrent flow may be achieved between the falling liquid and upward flowing vapor–gas mixture.

12.7 Thermal Design of Shell-and-Tube Condensers

Design and rating methods of a heat exchanger have already been introduced in Chapter 9. The rating problem is a performance analysis for a known exchanger geometry with given specifications. The condenser dimensions of the known condenser or tentative estimates of condenser dimensions are used as input data to the rating program. The thermal duty and outlet conditions of the condensate are calculated if the length is fixed. If the heat duty is fixed, the length of the condenser can be calculated (Figure 9.10).

Besides these parameters, the rating program will also calculate pressure drops for the condensing vapor side and the coolant side and the outlet temperature of the coolant; an advanced program will calculate many other useful parameters. Some advanced programs provide design modifications. They take the output from the rating subroutines and modify the condenser parameters in such a way that the new condenser configuration will better satisfy the required process conditions.

Two approaches are used in designing heat exchangers, especially two-phase flow heat exchangers such as condensers. In the first approach, the heat exchanger is taken as a single control volume with an average overall heat transfer coefficient and two inlets and two outlets (lumped analysis). In the second analysis, the heat exchanger is divided into segments or multiple control volumes, with the outlet of one control volume being the inlet to an adjacent control volume. This is a local analysis. The heat transfer rate (heat duty) for the heat exchanger is obtained by integrating the local values. The first approach is more common and simpler and it generally provides acceptable results for design.

Let us treat the first approach (lumped analysis). The heat transfer rate can be expressed for two control volumes defined on the condensing side (hot fluid) and coolant side, respectively, as follows:

$$Q_h = \dot{m}_h (i_i - i_o) \tag{12.1}$$

$$Q_c = (\dot{m} c_p)_c (T_{c_2} - T_{c_1}) \tag{12.2}$$

The basic design equation for a heat exchanger in terms of mean quantities that are related by Equation 2.7 is

$$Q = U_m A_o \Delta T_{lm} \tag{12.3}$$

Condensers and Evaporators

where U_m is the mean overall heat transfer coefficient and ΔT_{lm} is the mean temperature difference, which is the well-known LMTD given by Equation 2.28:

$$\Delta T_{lm} = \frac{\Delta T_1 - \Delta T_2}{\ln\left(\frac{\Delta T_1}{\Delta T_2}\right)} \tag{12.4}$$

The overall heat transfer coefficient, U, is built up of a number of components, including the individual heat transfer coefficients and wall and fouling resistances and is given by Equation 2.17 for a bare tube:

$$\frac{1}{U_m} = \frac{1}{h_o} + R_{fo} + A_o R_w + \left(R_{fi} + \frac{1}{h_i}\right)\frac{A_o}{A_i} \tag{12.5}$$

where U_m is based on the outside heat transfer surface area.

A simplified form of Equation 12.5 can also be used:

$$\frac{1}{U_m} = \frac{1}{h_o} + R_t + \frac{1}{h_i} \tag{12.6}$$

where R_t is the combined thermal resistance of the tube wall and fouling, and U_m is defined as

$$U_m = \frac{1}{A_o} \int_{A_o} U dA \tag{12.7}$$

where U is the local overall heat transfer coefficient and may vary considerable along the heat exchanger.

For variable U, the local heat transfer from a differential surface area, dA, is given by

$$\delta Q = U dA dT \tag{12.8}$$

The total heat transfer can be calculated by the procedure outlined in Chapter 2, Section 2.10.

If U varies linearly with A, Equation 12.7 can be integrated between the overall heat transfer coefficients at the inlet and outlet of the condenser.

$$U_m = \frac{1}{2}(U_1 + U_2) \tag{12.9}$$

For the overall heat transfer coefficient U and ΔT varying linearly with Q, Equation 2.60 can be used. If both $1/U$ and ΔT vary linearly with Q, Equation 12.7 can be integrated with the aid of Equation 12.8 to give[1]

$$\frac{1}{U_m} = \frac{1}{U_1} \frac{\Delta T_{lm} - \Delta T_2}{\Delta T_1 - \Delta T_2} + \frac{1}{U_2} \frac{\Delta T_1 - \Delta T_{lm}}{\Delta T_1 - \Delta T_2} \qquad (12.10)$$

Equation 12.9 is simple to use, but it is not valid if there are large differences between U_1 and U_2. It is difficult to decide which of the above equations for U_m is valid for given circumstances.[1]

Example 12.1

A shell-and-tube type condenser is to be designed for a coal fired power station of 200 MWe. Steam enters the turbine at 5 MPa and 400°C ($i_i = 3195.7$ kJ/kg). The condenser pressure is 10 kPa (0.1 bar). The thermodynamic efficiency of the turbine $\eta_t = 0.85$. The actual enthalpy of steam entering the condenser at 0.1 bar is calculated to be $i_e = 2268.4$ kJ/kg with 80% quality. The condenser is to be designed without subcooling. A single tube pass is used and the cooling water velocity is assumed to be 2 m/s. Cooling water is available at 20°C and can exit the condenser at 30°C. Allowable total pressure drop on tube side is 35 kPa. The tube wall thermal conductivity $k = 111$ W/m · K

Solution

The mass flow rate of wet steam entering the condenser can be found from the first law of thermodynamics:

$$\dot{m}_s = \frac{W_T}{\Delta i} = \frac{200 \times 10^3}{3195.7 - 2268.4} = 215.68 \text{ kg/s}$$

Cooling water properties at the mean temperature of 25°C are, from the appendix:

$$c_{ps} = 4{,}180 \text{ J/kg·K}$$
$$\mu_c = 0.00098 \text{ kg/m·s}$$
$$k_c = 0.602 \text{ W/m·K}$$
$$\rho_c = 997 \text{ kg/m}^3$$
$$Pr_c = 6.96$$

Saturated liquid properties of the condensed water at 10 kPa, $T_s = 45.8°C$, from the appendix, are

$$i_{lg} = 2392 \text{ kJ/kg}$$
$$\mu_l = 0.000588 \text{ N·s/m}^2$$
$$k_l = 0.635 \text{ W/m·K}$$

$\rho_l = 990 \text{ kg/m}^3$

$i_1 = 191.8 \text{ kJ/kg}$

Outside diameter of the tubes, $d_o = 0.0254$ m

Inside diameter of the tubes, $d_i = 0.02291$ m

Tube wall thermal conductivity, $k = 111$ W/m·K

The cooling water velocity $u_c = 2$ m/s, which is chosen as a compromise between fouling, which increase at low velocities, and erosion, which occurs at high velocities. Fouling resistances chosen from TEMA Tables in Chapter 6 are

$$R_{fi} = 0.00018 \text{ m}^2 \cdot \text{K/W}$$

$$R_{fo} = 0.00009 \text{ m}^2 \cdot \text{K/W}$$

Condenser heat load, Q, is

$$Q = \dot{m}_s (i_{in} - i_l)$$

$$= 215.68 \, (2268.4 - 1191.8) = 447.9 \text{ MW}$$

The cooling water mass flow rate, $\dot{m}_c$, is

$$\dot{m}_c = \frac{Q}{(T_{c_2} - T_{c_1}) c_{pc}}$$

$$= \frac{(4.479 \times 10^5) \times 10^3}{(30 - 20) \times 4180}$$

$$= 10{,}717.4 \text{ kg/s}$$

The number of tubes, N_T, is determined from the fixed cooling water velocity, U_c, as follows (one tube pass):

$$\dot{m}_c = u_c \rho_c \left(\frac{\pi d_i^2}{4} \right) N_T$$

or

$$N_T = \frac{4 \dot{m}_c}{u_c \rho_c \pi d_i^2}$$

$$= \frac{4 \times 1.07174 \times 10^4}{997 \times 2 \times \pi \times (0.02291)^2}$$

$$= 13{,}038 \text{ tubes}$$

In order to calculate the condensing side heat transfer coefficient, an estimate of the average number of tube rows in a vertical column, N_T, is needed. For typical condenser tube layouts, this was estimated as 70.

The coolant heat transfer coefficient, h_i, can be estimated from single-phase, in-tube heat transfer correlations, as given in Chapter 2.

The tube-side Reynolds number is calculated first:

$$Re = \frac{u_c \rho_c d_i}{\mu_c}$$

$$= \frac{0.02291 \times 2 \times 997}{0.00098}$$

$$= 46,614.8$$

Therefore, the flow inside the tubes is turbulent. The Petukhov–Kirillov correlations, for example, can be used to calculate the heat transfer coefficient:

$$Nu = \frac{(f/2)\, Re\, Pr}{1.07 + 12.7(f/2)^{1/2}(Pr^{2/3} - 1)}$$

where

$$f = (1.58 \ln Re - 3.28)^{-2}$$
$$= [11.58 \ln(46614.8) - 3.28]^{-2}$$
$$= 0.00532$$

So,

$$f/2 = 0.00266$$
$$Nu = \frac{0.00266 \times 46614.8 \times 6.96}{11.07 + 12.7(0.00266)^{1/2}\left[(6.96)^{2/3} - 1\right]} = 304.4$$

Hence,

$$h_i = \frac{Nu \cdot k_c}{d_1}$$
$$h_i = \frac{304.4 \times 0.602}{0.02291}$$
$$h_i = 8027\ W/m^2 \cdot K$$

The condenser is to be designed without subcooling, so

$$\Delta T_{in} = (45.8 - 20) = 25.8°\ C$$

and

$$\Delta T_{out} = (45.8 - 30) = 15.8° \text{ C}$$

Hence, the log mean temperature difference is

$$\Delta T_{lm} = \frac{\Delta T_{in} - \Delta T_{out}}{\ln(\Delta T_{in}/\Delta T_{out})} = \frac{25.8 - 15.8}{\ln(25.8/15.8)} = 20.4° \text{ C}$$

Next, the shell-side heat transfer coefficient is calculated in order to determine the overall heat transfer coefficient. This coefficient depends on the local heat flux and, hence, an iteration is necessary. The equations required in this iteration are developed first. The overall heat transfer coefficient, U, based on the tube outside diameter, is given by

$$\frac{1}{U} = R_t + \frac{1}{h_o}$$

where h_o is the coefficient outside the tubes (shell side) and R_t is the sum of all the other thermal resistance, given by Equation 12.5:

$$R_t = R_{fo} + \left[\frac{1}{h_i} + R_{fi}\right]\frac{d_o}{d_i} + \frac{t_w}{k_w}\frac{d_o}{D_m}$$

where D_m is approximated as

$$D_m = \frac{d_o - d_i}{\ln(d_o/d_i)} \approx \frac{1}{2}(d_o + d_i)$$

and t_w is the wall thickness.
Hence,

$$R_t = 0.00009 + \left[\frac{1}{8027} + 0.00018\right]\frac{0.0254}{0.0229} + \frac{0.0013}{111} \cdot \frac{0.0254}{0.0242}$$

$$= 4.39 \times 10^{-4}$$

and

$$\frac{1}{U} = 4.39 \times 10^{-4} + \frac{1}{h_o} \tag{1}$$

The condensing side heat transfer coefficient, h_o, may be calculated by the Nusselt method with the Kern correction for condensate inundation (Equations 8.2 and 8.10). Hence,

$$h_o = 0.728 \left[\frac{\rho_l^2 g i_{lg} k_l^3}{\mu_l \Delta T_w d_o} \right]^{1/4} \frac{1}{N^{1/6}}$$

where ΔT_w is the difference between the saturation temperature and the temperature at the surface of the fouling. Since $\rho_l \gg \rho_g$, Equation 8.2 has been simplified. Inserting the values, we get

$$h_o = 0.728 \left[\frac{(990)^2 (9.81)(2392 \times 10^3)(0.635)^3}{(5.66 \times 10^{-4}) \Delta T_w (0.0254)} \right]^{1/4} \frac{1}{70^{1/6}} \quad (2)$$

$$h_o = 8990 / \Delta T_w^{1/4}$$

The temperature difference, ΔT_w, is given by

$$\Delta T_w = \Delta T - R_t q''$$

where ΔT is the local temperature difference between the streams, R_t is the sum of all other resistances, and q'' is the local heat flux, which is given by

$$q'' = U \Delta T$$

Hence,

$$\Delta T_w = \Delta T (1 - R_t U)$$
$$\Delta T_w = \Delta T (1 - 4.39 \times 10^{-4} U) \quad (3)$$

A suggested iteration at the inlet and outlet is, therefore,

1. Guess ΔT_w
2. Calculate h_o from Equation 2 above
3. Calculate U from Equation 1 above
4. Recalculate ΔT_w from Equation 3 above
5. Repeat the calculations from step 2 and continue the iteration until U converges.

Tables 12.1 and 12.2 summarize the results of this iteration for the inlet and outlet of the condenser when $\Delta T_w = 25.8°C$ and $\Delta T_w = 15.8°C$. The initial guess of ΔT_w is 10°C.

The mean overall heat transfer coefficient can then be determined by taking the average of the inlet and outlet overall heat transfer coefficients:

$$U_m = \frac{1575 + 1632}{2} = 1603 \text{ W/m}^2 \cdot \text{K}$$

TABLE 12.1
Iteration for Overall Coefficient at the Inlet of the Power Condenser

T_w, °C (Equation 3)	h_o, W/m² · K (Equation 2)	U, W/m² · K (Equation 1)
10.00	5,049	1,569
9.47	5,118	1,575
9.45	5,121	1,575
9.44	5,122	1,575

TABLE 12.2
Iteration for Overall Coefficient at the Outlet of the Power Condenser

T_w, °C (Equation 3)	h_o, W/m² · K (Equation 2)	U, W/m² · K (Equation 1)
6.00	5,737	1,629
5.89	5,764	1,631
5.86	5,771	1,632
5.84	5,776	1,632

The required surface area is therefore calculated from Equation 12.1:

$$Q = U_{MAo} \Delta T_{lm}$$

$$A_o = \frac{Q}{U_m \Delta T_{lm}}$$

$$A_o = \frac{447.9 \times 10^6}{1603 \times 20.4} = 1.370 \times 10^4 \text{ m}^2$$

The required length is found by the following equation:

$$A_o = N_T \pi d_o L$$

or

$$L = \frac{A_o}{N_T \pi d_o} = \frac{1.370 \times 10^5}{15,340 \times \pi \times 0.0254} = 13.2 \text{ m}$$

Now, the thermal resistances of the heat exchanger will be analyzed. This is useful when cleaning practices must be considered for the exchanger. Table 12.3 shows this for the inlet of the condenser.

As can be seen from Table 12.3, a considerable amount of fouling is due to the inside tubing. Therefore, it is necessary to concentrate on the tube side when cleaning the exchanger. Several possible methods of cleaning

TABLE 12.3

Comparison of Thermal Resistances at the Inlet of the Condenser

Item	Given by	Value (m² · K/kW)	%
Tube-side fluid	$\dfrac{d_o}{h_i d_i}$	0.138	25
Tube-side fouling	$\dfrac{R_{fi} d_o}{d_i}$	0.200	36
Tube wall	$\dfrac{t_w d_o}{k_w D_m}$	0.012	2
Shell-side fouling	R_{fo}	0.090	16
Shell-side fluid	$\dfrac{1}{h_o}$	0.195	35

can be used. One method of cleaning will include the cooling water chemistry. The second method involves passing balls down the tubes.

Now that the length of the heat power condenser has been determined, the shell size of the condenser, the tube-side pressure drop, and the tube-side pumping power can all be calculated.

The expression of the shell diameter as a function of heat transfer area, A_o, tube length, L, and tube layout dimensions, P_T, PR, and d_o as parameters can be estimated from Equation 9.10:

$$D_s = 0.637 \sqrt{\dfrac{CL}{CPT}} \left[\dfrac{A(PR)^2 d_o}{L} \right]^{1/2} \tag{4}$$

where CL is the tube layout constant, $CL = 1.00$ for 90° and 45°, and $CL = 0.87$ for 30° and 60°. CPT accounts for the incomplete coverage of the shell diameter by the tubes:

$CPT = 0.93$ for 1 tube pass
$CPT = 0.9$ for 2 tube passes
$CPT = 0.85$ for 3 tube passes
P_T is the tube pitch, which equals 0.0381 m
PR is the tube pitch ratio = P_T/d_o = 0.0381/0.0254 = 1.501.

Then, using the values given for the above parameters in Equation 4, the shell diameter is determined as

$$D_s = 0.637 \sqrt{\dfrac{1}{0.93}} \left[\dfrac{1.37 \times 10^4 \times 1.501^2 \times 0.0254}{13.2} \right]^{1/2}$$

$$= 5.1 \text{ m}$$

Condensers and Evaporators

The pressure drop on the tube side is calculated as follows:

$$\Delta p_{tot} = \Delta p_t + \Delta p_r$$

The pressure drop through the tubes can be calculated from Equation 9.19:

$$\Delta p_t = 4f \frac{LN_p}{D_e} \frac{G^2}{2\rho}$$

where

$$f = 0.046\, Re^{-0.2}$$
$$= 0.046 \times (46.6 \times 10^3)^{-0.2} = 0.00535$$

Also,

N_p = the number of passes
$D_e = d_i$
$G = u_c \rho$

Therefore,

$$\Delta p_t = 4 \times 0.00535 \times \frac{13.2 \times 1}{0.02291} \times \frac{(2 \times 997)^2}{2 \times 997}$$
$$= 24,559.3\ \text{Pa}$$

The pressure drop due to the return is given by Equation 9.21:

$$\Delta p_r = 4N_p \frac{\rho u_c^2}{2}$$
$$= 4 \times 1 \times \frac{997 \times 2^2}{2}$$
$$= 7976\ \text{Pa}$$

Therefore, the total pressure drop on the tube side is determined by

$$\Delta p_{tot} = \Delta p_t + \Delta p_r$$
$$= 24,559.3 + 7976$$
$$= 32,535.3\ \text{Pa}$$

The pumping power is proportional to the pressure drop across the condenser:

$$W_p = \frac{\dot{m}\Delta p_{tot}}{\rho \eta_p}$$

where η_p is the efficiency of the pump, assumed to be 85%.

$$W_p = \frac{10{,}717.4 \times 56{,}602.1}{997 \times .85} = 82{,}9597\ \text{W} \approx 829.6\ \text{kW}$$

TABLE 12.4

The Effect of Coolant Velocity on Design

Parameters	Cooling Water Velocity		
	$u_c = 1.5$ m/s	$u_c = 2.0$ m/s	$u_c = 2.5$ m/s
Number of tubes, N_t	17,385	13,038	10,430.7
Reynolds no., Re (cool.)	34,961.2	46614.8	58,268.5
Heat transfer coefficient (cool.), h_c (W/m² · K)	6307.6	8026.7	9685.3
Heat transfer coefficient (shell), h_o (W/m² · K)	5441	5441	5443
Overall heat transfer coefficient, U_m (W/m² · K)	1476	1603.5	11,666.5
Heat transfer area, A_o (m²)	14,525.5	13,696.6	13,178.8
Length, L (m)	10.5	13.2	15.8
Shell diameter, D_s (m)	5.9	5.1	4.6
Pressure drop (tubes), ΔP_t (kPa)	16.1	32.5	56.6
Pumping power (tubes), $\eta_p = 85\%$), P_t (kW)	378.5	829.6	1540.4

This procedure was done for cooling water velocities of 1.5 m/s and 2.5 m/s, and the results are given in Table 12.4.

During the rating procedures, the thermal performance and the pressure drop for both streams are calculated for this design. The better the starting point estimate, the sooner the final design will be done. On a computer, however, it is usually faster to let the computer select a starting point—usually a very conservative case—and use its enormous computational speed to move towards the suitable design. If the calculation shows that the required amount of heat cannot be transferred or if one or both allowable pressure drops are exceeded, it is necessary to select new geometrical parameters and to recalculate the condenser. On the other hand, if pressure drops are much smaller than allowable, a better selection of parameters may result in a smaller and more economical condenser while utilizing more of the available pressure drop.

The best economic choice, then, is a result of the additional step which is an economic evaluation of several designs. The final design must meet several conditions:

1. The condenser must be operable in a range of conditions which include the design conditions.
2. The design must meet all specific physical limitations that may be imposed, such as weight, size, coolant process constraints, metallurgical constraints, etc.
3. The design must allow standard and/or economical fabrication.
4. Operation and maintenance expenses must not exceed economical limits.

12.8 Design and Operational Considerations

Muller[3] and Taborek[8] described a number of requirements that should be considered during selection, as well as design practices of condensers.

- Condensation Modes—Condensers are designed under filmwise condensation conditions. Although dropwise condensation gives higher heat transfer coefficients, it is not possible to sustain this condensation mode for a long period of time on individual condensers.
- Condensation Regimes—Depending on the flow characteristics of the vapor and condensate, the designer must determine the flow regime applicable along the vapor flow path. At low vapor velocities, the so-called "gravity controlled" or Nusselt flow regime exists. At high vapor velocities, the "vapor shear controlled" regime will predominate (see Section 8.4).
- Desuperheating—Some vapor stream enters the condenser superheated. If the wall temperature is below the dew point, condensation will take place and the desuperheating duty must be calculated as a single-phase process.
- Subcooling—It is sometimes desirable to subcool the condensate slightly before further processing. This can be accomplished by raising the level of the condensate so that it would be in contact with the cool tubes. For larger subcooling heat applications, it is more efficient to allocate a separate unit.
- Construction Considerations—Some guidelines and practices must be observed for the design of well-functioning condensers: (a) Vertical in-tube condensation is very effective, but the tube length is limited as it may fill up with condensate. Thus, the size of such condensers is restricted, as otherwise large shell diameters would be required. (b) Horizontal tube-side consideration is less effective and much more difficult to calculate because of the stratification of the condensate. Positive tube inclination must be used. (c) Horizontal shell-side condensation is very popular as it is well predictable, permits use of large surfaces, and the extremely low pressure drop required for vacuum operations can be obtained by proper unit selection (TEMA X-shell).

If noncondensable gases are present in vapors, the condensing coefficient will vary an order of magnitude between vapor inlet and outlet. Stepwise calculation is unconditionally required. A detailed stepwise analysis is given by Butterworth.[1] Furthermore, the unit construction must be suitable for effective removal (venting) of the gases which would otherwise accumulate and render the condenser inoperative.

Some of the main reasons for failure of condenser operation are given by Steinmeyer and Muller[9] and are briefly summarized here:

1. The tubes may be more fouled than expected—a problem not unique to condensers.
2. The condensate may not be drained properly, causing tubes to be flooded. This could mean that the condensate outlet is too small or too high.
3. Venting of noncondensable may be inadequate.
4. The condenser was designed on the basis of end temperatures without noticing that the design duty would involve a temperature cross in the middle of the range.
5. Flooding limits have been exceeded for condensers with backflow of liquid against upward vapor flow.
6. Excessive fogging may be occurring. This can be a problem when condensing high molecular weight vapors in the presence of noncondensable gas.
7. Severe misdistribution in parallel condensing paths is possible, particularly with vacuum operation. This occurs because there can be two flow rates which satisfy the imposed pressure drops. An example might be that in which the pressure drop is zero. One channel may have a very high vapor inlet flow but may achieve a zero pressure drop because the momentum pressure recovery cancels out the frictional pressure loss. The next channel may achieve the zero pressure drop by having no vapor inlet flow. To be stable in this case, the channel would be full of noncondensable. This problem may occur with parallel tubes in a tube-side condenser or with whole condensers when they are arranged in parallel.[1]

12.9 Condensers for Refrigeration and Air-Conditioning

Several types of heat exchangers are used in refrigeration and air-conditioning applications. They are two-phase flow heat exchangers—one side is the refrigerant and the other side is the air or liquid. In general, condensers in the refrigeration and air-conditioning industry are of two types: (1) a condenser coil, where the refrigerant flows through the tubes and air flows over the finned tubes, and (2) a shell-and-tube condenser where the refrigerant flows on the shell side and the liquid flows through the tubes. The common heat exchanger types used in the refrigeration and air-conditioning industry are shown in Figure 12.12. An illustration of a commercial two-stage vapor compression refrigeration cycle is given in Figure 12.13.

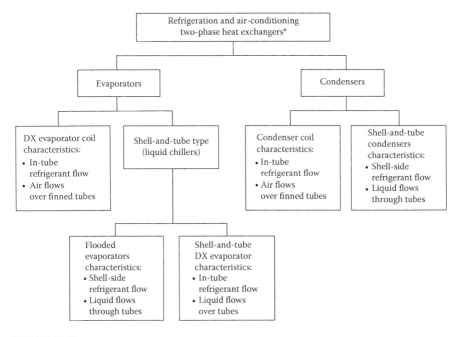

FIGURE 12.12
Common heat exchanger types used in the refrigeration and air-conditioning industry. (From Pate, M. B., *Boilers, Evaporators and Condensers*, John Wiley & Sons, New York, 1991. With permission.)

The main heat transfer components of the refrigeration system are the evaporator and condenser, which absorb and reject heat respectively. The evaporator vaporizes the refrigerant as well as heat, while the condenser rejects heat and discharges high pressure and temperature liquid refrigerant. Figure 12.13 shows a schematic of a commercial air-conditioning system cycle with all its components.[13]

The maximum efficiency in a refrigeration cycle is ultimately obtained by the combined performance of all the major components. The evaporator and condenser (along with the compressor) make-up the bulk of the system in terms of size and cost.

The most common forms of condensers may be classified on the basis of the cooling medium as

1. Water-cooled condensers
   - Horizontal shell-and-tube
   - Vertical shell-and-tube
   - Shell-and-coil
   - Double pipe
2. Air-cooled condensers
3. Evaporative condensers (air- and water-cooled).

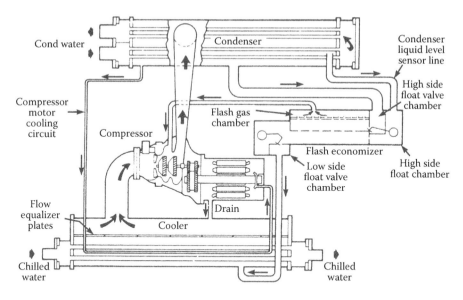

FIGURE 12.13
Illustration of a two-stage vapor compression refrigeration cycle. (Courtesy of Carrier Corp.)

12.9.1 Water-Cooled Condensers

Water-cooled condensers are shell-and-tube, horizontal and vertical, shell and U-type, shell-and-coil, and double-pipe types. An example of shell-and-tube horizontal and vertical condensers which can be used with refrigerant condensing on the shell side and cooling water circulated inside the tubes is given in Chapter 9 and in Section 12.2.

The shell-and-tube condensers with superheated refrigerant vapor entering and subcooled liquid-refrigerant exiting are the most common type of water-cooled condensers. They are usually mounted in the horizontal position to facilitate condensate drainage from the tube surface. Shell-and-tube heat exchangers are discussed in Chapter 9, and the geometrical parameters have already been outlined. In air-conditioning and refrigeration, copper tubes are usually used in shell-and-tube condensers with sizes ranging from 5/8 to 1 in. OD, with 3/4 in. OD being the most popular. Ammonia requires carbon steel tubes. Equivalent triangular pitches are normally used for tube arrangements in the shell. Water is circulated through the tubes in a single to multipass circuit.

When fixed tube sheet, straight tube construction is used, provisions and space for tube cleaning and replacement must be provided at one end of the condenser. Location of the gas and liquid outlet nozzles should be carefully considered; they should be located far enough apart to allow the entering superheated vapor to be exposed to the maximum tube surface area. As the refrigerant vapor is condensed on the outside of the tubes, it drips down onto the lower tubes and collects at the bottom of the condenser. In some cases, the condenser bottom is designed to act as a reservoir (receiver) for storing refrigerant. It must be ensured that excessive amounts of liquid refrigerant are maintained to seal the outlet nozzle from inlet gas flow.

A shell-and-tube type vertical condenser with an open water circuit is widely used in ammonia refrigeration systems. Shell-and-tube condensers with U-tubes and a single tube sheet are also manufactured. A shell-and-coil condenser has cooling water circulated through one or more continuous or assembled coils contained within the shell.

A double-pipe or hairpin condenser consists of one or more assemblies of two tubes, one within the other (see Chapter 7). The refrigerant vapor is condensed in either the annular space or the inner tube.

12.9.2 Air-Cooled Condensers

Air-cooled condensers are made of finned-tubes. The air side is finned, and the liquids flow through the tubes. The heat transfer process in air-cooled condensers has three main phases: (1) desuperheating, (2) condensing, and (3) subcooling.

Condensing takes place in approximately 85% of the condensing area at a relatively constant temperature. As a result of the frictional pressure drop, there will be a small condensing temperature drop along the surface. The geometrical parameters of air-cooled condensers are the same as compact heat exchangers, some examples of which are shown in Chapter 10.

Coils are commonly constructed of copper, aluminum, or steel tubes, ranging from 0.25 to 0.75 in. (6.35–19.05 mm) in diameter. Copper, the most expansive material, is easy to use in manufacturing and requires no protection against corrosion. Aluminum requires exact manufacturing methods, and special protection must be provided if aluminum-to-copper joints are made. Steel tubing requires weather protection. Fins are used to improve the air-side heat transfer. Most fins are made of aluminum, but copper and steel are also used.

Air-cooled condensers may be classified as remote from the compressor or part of the compressor (condensing unit). A more specific subdivision may be forced flow or free convection air flow. In this type, air is passed over the coils where in-tube condensation occurs.

12.9.3 Evaporative Condensers

An evaporative condenser is a type of water-cooled condenser that offers a means of conserving water by combining the condenser and the cooling tower into one piece of equipment. The coil containing the refrigeration gas is wet outside and then the cooling air is forced over the coils, or the air may flow freely over the coils (free convection), at which time some of the water evaporates, in turn cooling and condensing the tube-side

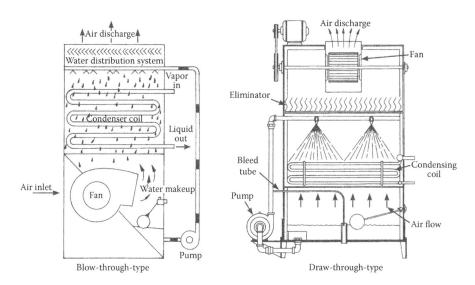

FIGURE 12.14
Operation of evaporative condenser. (From *Equipment Handbook*, ASHRAE, 1983. With permission.)

refrigerant (Figures 12.14 and 12.15). Principal components of an evaporative condenser include the condensing coil, fan, spray water pump, water distribution system, cold water pump, drift eliminators, and water makeup assembly. The heat from a condensing vapor is rejected into the environment. Evaporative condensers eliminate the need for the pumping and chemical treatment of large quantities of water associated with cooling tower/refrigerant condenser systems. They require substantially less fan power than air-cooled condensers of comparable capacity.[14] As can be seen from Figures 12.14 and 12.15, vapor to be condensed is circulated through a condensing coil that is continually water cooled on the outside by recalculating water systems. Air is simultaneously directed over the coil, causing a small portion of the recalculated water to evaporate. This evaporation results in the removal of heat from the coil, thus cooling and condensing the vapor in the coil. The high rate of energy transfer from the wetted external surface to the air eliminates the need for extended surface design. Therefore, these condensers are more easily cleaned. Coil materials are steel, copper, iron pipe, or stainless steel.

Coil water is provided by circulating water by the pump to the spray nozzles located above the coil. The water descends through the air circulated by the fan, over the coil surface, and eventually returns to the pan sump. The water distribution system is designed for complete and continuous wetting of the full coil surface. This ensures the high rate of heat transfer achieved with fully wetted tubes and prevents excessive scaling, which is more likely

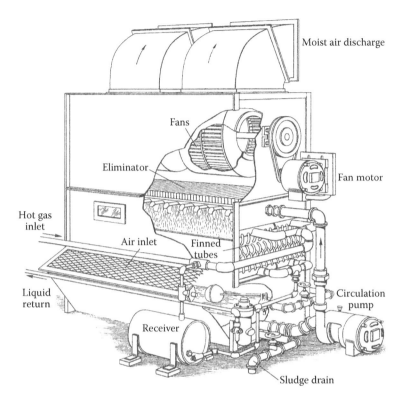

FIGURE 12.15
Evaporative condenser. (From *Trane Air Conditioning Manual*, The Trane Company, Lacrosce, Wisconsin, 1979. With permission.)

to occur on intermittently or partially wetted surfaces. Water lost through evaporation and blow down from the cold-water pump is replaced through an assembly which typically consists of a mechanical float valve or solenoid valve and float switch combination.

Most evaporative condensers employ one or more fans to either blow out or draw in air through the unit. Typically, the fans are either the centrifugal or propeller type, depending on the external pressure needs, permissible sound level, and energy usage requirements.

Drift eliminators are used to recover entrained moisture from the air stream. These eliminators strip most of the water from the discharge air system; however, a certain amount of water is discharged as drift.

Evaporative condensers require substantially less fan horsepower than air-cooled condensers of comparable capacity. Perhaps most importantly, however, systems utilizing evaporate condensers can be designed for a lower condensing temperatures and, subsequently, a lower compressor energy input than systems utilizing conventional or water-cooled condensers.

The evaporative condenser allows for a lower condensing temperature than the air-cooled condenser because the heat rejection is limited by the ambient wet-bulb temperature, which is normally 14°F to 25°F (8°C to 14°C) lower than the ambient dry bulb temperature. Though less obvious, the evaporative condenser also provides a lower condensing temperature than the cooling tower/water cooled condenser because of the reduction of heat-transfer/mass-transfer steps from two steps (the first step is between the refrigerant and the cooling water and the second step is between the water and ambient air) to one step, from refrigerant directly to ambient air.[13]

12.10 Evaporators for Refrigeration and Air-Conditioning

As can be seen from Figure 12.12, two-phase flow heat exchangers in the refrigeration and air-conditioning industry can be classified as coils when the refrigerant boils inside the tubes and air follows over finned tubes. Shell-and-tube evaporators are used when the second fluid is liquid that flows either through the tubes or shell-and-tube flooded evaporators or flows over the tubes (shell-and-tube direct expansion evaporators). Flooded evaporators and direct expansion (DX) evaporators perform similar functions. The five types of heat exchangers shown in Figure 12.12 represent the majority of heat exchanger types used in the refrigeration and air-conditioning industry. Other types of heat exchangers, such as plate-fin heat exchangers and double-pipe heat exchangers, are also used for automotive air-conditioning systems and low-tonnage liquid-cooling systems, respectively.

12.10.1 Water-Cooling Evaporators (Chillers)

As indicated above, evaporators in which the refrigerant boils inside the tubes are the most common commercial evaporators; evaporators in which the refrigerant boils in the shell, outside the tubes are an important class of liquid chilling evaporator's standard in centrifugal compressor applications. Examples of liquid-cooling flooded evaporators are shown in Figures 12.16 and 12.17.

Direct expansion evaporators are usually of the shell-and-tube type with the refrigerant inside the tubes, cooling the shell-side fluid (Figures 12.16 and 12.17). Two crucial items in the direct expansion shell-and-tube evaporators are the number of tubes (refrigerant passes) and the type of tubes.

Another type of water-cooling evaporator is the flooded shell-and-tube evaporator, which cools liquids flowing through the tubes by transferring heat to the evaporating refrigerant on the shell side; the tubes are covered (flooded) with a saturated mixture of liquid and vapor (Figure 12.18).

Condensers and Evaporators

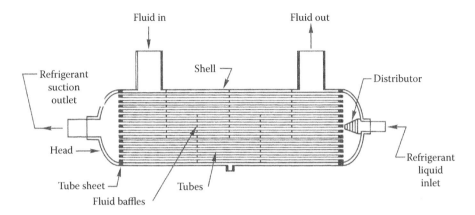

FIGURE 12.16
Direct (straight-tube type) expansion evaporator. (From *Equipment Handbook*, ASHRAE, 1988. With permission.)

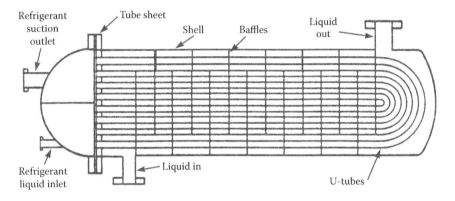

FIGURE 12.17
Direct (U-type) expansion evaporator. (From *Equipment Handbook*, ASHRAE, 1988. With permission.)

Baudelot evaporators are also used in industrial applications for cooling. Baudelot evaporators are used in industrial applications for cooling a liquid to near its freezing point. The liquid is circulated over the outside of a number of horizontal tubes positioned one above the other. The liquid is distributed uniformly along the length of the tube and flows by gravity to the tubes below. The cooled liquid is collected by a pan from which it is pumped back up to the source of heat and then again to the cooler.[13]

12.10.2 Air-Cooling Evaporators (Air Coolers)

The dry expansion coil is a type of air evaporator. It is the most popular and widely used air cooler. The refrigerant entering the coil is mostly liquid, and

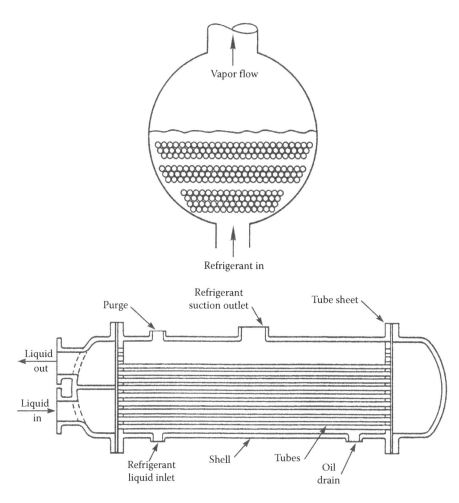

FIGURE 12.18
Flooded shell-and-tube evaporator.

as it passes through the coil, picking up heat from the air outside the tubes, it evaporates and exits mostly as vapor. The typical dry expansion coils shown in Figures 12.19 and 12.20 are multicircuited (from 2 to 22 circuits) and fed by one expansion valve. The air cooler can be forced convection, implementing the use of a fan, or may rely on natural convection.

The most common type of air-conditioning evaporator and condenser is the type in which air flows over a circular tube bank that has been finned with continuous plates, as shown in Figure 12.20, hence it is known as a plate finned-tube heat exchanger. The evaporating or condensing refrigerant flows through tubes that are mounted perpendicular to the air flow and arranged in staggered rows. The end views in Figure 12.20 show that the tubes can be connected and coiled to form any number of passes, rows, and parallel

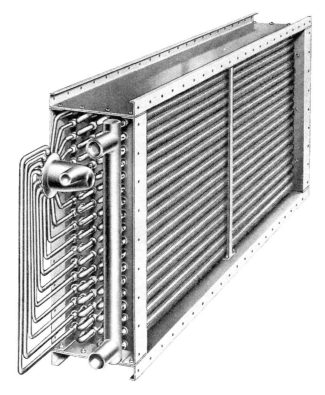

FIGURE 12.19
Dry expansion coil-finned tube bank for an air-conditioning system. (Courtesy of Aerofin Corp.)

paths, hence the names evaporator or condenser coil. Evaporator coils use round tubes for the most part. Typical tube sizes representing a wide range of applications are outside diameters of 5/16, 3/8, 1/2, 5/8, 3/4, and 1 in. Oval tubes are also used for special applications. The fins and tubes are generally made of aluminum and copper, but sometimes both components are manufactured from the same material. The fins are connected to the tubes by inserting the tubes into holes stamped in the fins and then expanding the tubes by either mechanical or hydraulic means.

12.11 Thermal Analysis

The thermal design methods of condensers and evaporators for air-conditioning and refrigeration systems are the same as single-phase applications and power and process industries, which are discussed in previous chapters and in Section 12.7. The only difference is in the selection of proper

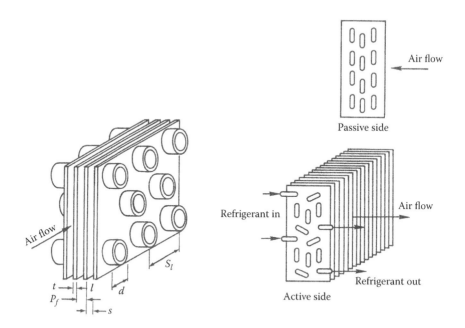

FIGURE 12.20
Dry expansion coil (air evaporator). (From Pate, M. B., *Boilers, Evaporators and Condensers*, John Wiley & Sons, New York, 1991. With permission.)

correlations to calculate heat transfer coefficients. In refrigeration and air conditioning, refrigerants like R-12, R-22, R-502, and R-134A are most commonly used. The correlations given to calculate the heat transfer coefficients for refrigerants are often different from other fluids.

Three recent correlations for in-tube flow boiling refrigerants are given by Shah,[15] Kandlikar,[16] and Güngör and Winterton.[17] All three of these correlations are discussed in Chapter 8. These correlations have been verified by experiment. For the completeness of this chapter, and in order to make the chapter comprehensible by itself, explanations of these three correlations are briefly repeated here.

12.11.1 Shah Correlation

The Shah correlation is applicable to nucleate, convection, and stratified boiling regions and it uses four dimensionless parameters given by the following equations:

$$F_0 = \frac{h_{TP}}{h_{LO}} \qquad (12.21)$$

where h_{TP} is the two-phase heat transfer coefficient and h_{LO} is the liquid only superficial convection heat transfer coefficient for liquid phase and is calculated by the Dittus–Boelter equation as

$$h_{LO} = 0.023 \left[\frac{G(1-x)d_i}{\mu_l} \right]^{0.8} Pr^{0.4} \frac{k_l}{d_i} \qquad (12.22)$$

The other three dimensionless numbers are the convection number (Co), boiling number (Bo), and Fraude number (Fr_L), which are

$$Co = \left(\frac{1}{x} - 1 \right)^{0.8} \left(\frac{\rho_g}{\rho_l} \right)^{0.5} \qquad (12.23)$$

$$Bo = \frac{q''}{G\, i_{fg}} \qquad (12.24)$$

$$Fr_L = \frac{G^2}{\rho_l^2 g d_i} \qquad (12.25)$$

When $Co > 1$, it is in the nucleate boiling regime where F_o is independent of Co and depends on the boiling number. For horizontal tubes, the surface is fully wet only if $Fr_L \geq 0.04$; for $Fr_L < 0.04$, part of the tube surface is dry and the heat transfer coefficient is lower than in the vertical tubes. We define a dimensionless parameter, N_s, as follows:

For horizontal tubes with $Fr_L < 0.04$,

$$N_s = 0.38\, Fr_L^{-0.3} Co \qquad (12.26)$$

For vertical tubes at all values of Fr_L and for horizontal tubes with $Fr_L > 0.04$,

$$N_s = Co \qquad (12.27)$$

For $N_s = Co < 1$,

$$F_{cb} = \frac{1.8}{N_s^{0.8}} \qquad (12.28)$$

$$F_{nb} = 231\, Bo^{0.5} \qquad Bo > 1.9 \times 10^{-5} \qquad (12.29)$$

$$F_{nb} = 1 + 46\, Bo^{0.5} \qquad Bo < 0.3 \times 10^{-4} \qquad (12.30)$$

F is the larger of F_{nb} and F_{cb}. Thus, if $F_{nb} > F_{cb}$, $F = F_{nb}$. If $F_{cb} > F_{nb}$, then $F = F_{cb}$. For $0.02 < N_s < 1.0$,

$$F_{bs} = \psi Bo^{0.5} \exp\left(2.74\, N_s^{-0.1}\right) \qquad (12.31)$$

F_{cb} is calculated with Equation 12.28. F equals the larger of F_{bs} and F_{cb}. For $N_s < 0.1$,

$$F_{bs} = \psi Bo^{0.5} \exp(2.47 N_s^{-0.15}) \qquad (12.32)$$

F_{cb} is calculated with Equation 12.28 and F equals the larger of F_{cb} and F_{bs}. The constant ψ in Equations 12.31 and 12.32 is as follows:

$$\psi = 14.7 \qquad Bo \geq 11 \times 10^{-4} \qquad (12.33)$$

$$\psi = 15.43 \qquad Bo < 11 \times 10^{-4} \qquad (12.34)$$

For horizontal tubes with $Fr_L < 0.04$, these equations are recommended only for $Bo \geq 11 \times 10^{-4}$. For horizontal tubes with $Fr_L \geq 0.04$ and vertical tubes without any Fr_L values, these equations are recommended for all values of Bo without restriction.[15]

As expected, at low qualities, nucleate boiling dominates, while at high qualities, convective boiling dominates.

12.11.2 Kandlikar Correlation

The two-phase boiling heat transfer coefficient, h_{TP}, was expressed as the sum of the convective and nucleate boiling terms:

$$h_{TP} = h_{cov} + h_{nucl} \qquad (12.35)$$

Kandlikar represented the nucleate boiling region by the boiling number, Bo, and the convective boiling region by the convection number, Co, as defined by Equations 12.23 and 12.24.

The final form of the proposed correlation is

$$\frac{h_{TP}}{h_{LO}} = C_1 (Co)^{C_2} (25 Fr_L)^{C_5} + C_3 Bo^{C_4} F_{fl} \qquad (12.36)$$

where h_{LO} is the single-phase liquid-only heat transfer coefficient and is calculated from Equation 12.2. F_{fl} is the fluid dependent parameter and is given in Table 12.5. The values of constants C1–C5 are given below:

For $Co < 0.65$—convective boiling region:

C1 = 1.1360

C2 = −0.90

C3 = 667.2

C4 = 0.7

C5 = 0.3

TABLE 12.5
Fluid-Dependent Parameter F_{fl} in the Proposed Correlation

Fluid	F_{fl}
Water	1.00
R-11	1.30
R-12	1.50
R-13B1	1.31
R-22	2.20
R-113	1.30
R-114	1.24
R-152a	1.10
Nitrogen	4.70
Neon	3.50

For $Co > 0.65$—nucleate boiling region:

$C1 = 0.6683$

$C2 = -0.2$

$C3 = 1058.0$

$C4 = 0.7$

$C5 = 0.3$

for vertical tubes and for horizontal tubes with $Fr_L > 0.04$. Correlation 12.36 is valid for horizontal and vertical smooth tubes.

12.11.3 Güngör and Winterton Correlation

A general correlation for forced convection boiling has been developed by Güngör and Winterton, and the basic form of the correlation is

$$h_{TP} = Eh_{LO} + Sh_{pool} \tag{12.37}$$

where the all-liquid convection heat transfer coefficient, h_{LO}, is calculated from Equation 12.22 and E and S are the enhanced and suppression factors, respectively, and are given by the following equations:

$$E = 1 + 2.4 \times 10^4 Bo^{1.16} + 1.37\left(\frac{1}{X_{tt}}\right)^{0.86} \tag{12.38}$$

and

$$S = \frac{1}{1 + 1.15 \times 10^{-6} E^2 Re_L^{1.17}} \quad (12.39)$$

where Bo is given by Equation 12.24 and X_{tt} is the well-known Martinelli parameter:

$$X_{tt} = \left(\frac{1-x}{x}\right)^{0.9} \left(\frac{\rho_g}{\rho_l}\right)^{0.5} \left(\frac{\mu_L}{\mu_g}\right)^{0.1} \quad (12.40)$$

X_{tt} is similar to the convective number, Co, used in Shah's correlation.

For the pool boiling term, h_p, the correlation given by Cooper[18] is used:

$$h_p = 55 Pr^{0.12} (-\log_{10} Pr)^{-0.55} M^{-0.5} (q'')^{0.67} \quad (12.41)$$

If the tube is horizontal and the Froude number < 0.05, E should be multiplied by

$$E_2 = Fr_L^{(0.1 - 2Fr_L)} \quad (12.42)$$

and S should be multiplied by

$$S_2 = \sqrt{Fr_L} \quad (12.43)$$

As the equations stand, it is assumed that heat flux, q'', is known, in which case it is straightforward to calculate T_w. If T_w is known, then, as with many other correlations, an iteration is required.

Boiling in annuli is treated by means of an equivalent diameter that depends on the annular gap:

$$D_e = \frac{4 \times \text{flow area}}{\text{wetted perimeter}} \quad \text{for gap} > 4 \text{ mm}$$

$$D_e = \frac{4 \times \text{flow area}}{\text{heated perimeter}} \quad \text{for gap} < 4 \text{ mm} \quad (12.44)$$

Note that, in the data, only one of the annulus walls was heated.

In subcooled boiling, the driving temperature differences for nucleate boiling and for forced convection are different, so Equation 12.37 is replaced by

$$q = h_{LO}(T_w - T_b) + Sh_p(T_w - T_s) \quad (12.45)$$

There is no enhancement factor since there is no net vapor generation, but the suppression factor remains effective (calculated according to Equations

12.38 and 12.39). It could be argued that there should still be an enhancement factor since there is still local vapor generation, but this approach gave a worse fit to the data.

12.12 Standards for Evaporators and Condensers

The designer of air-conditioning evaporators and condensers should be aware of standards that are used to rate heat exchangers once they are built according to the recommended design. In the United States, standards are published by both the Air-Conditioning and Refrigerating Institute (ARI) and the American Society of Heating, Refrigerating, and Air-Conditioning Engineers (ASHRAE). For example, the ASHRAE Standard for testing air-conditioning evaporator coils is Standard 33-78, *Methods of Testing Forced Circulation Air Cooling and Air Heating Coils*, while the standard for condenser coils is Standard 20-70, *Methods of Testing for Rating Remote Mechanical-Draft Air-Cooled Refrigerant Condensers*. Both of these standards are presently being reviewed and updated so that revised standards may be forthcoming in the near future. The ARI Standards are very similar to the ASHRAE Standards.

The standards noted above include the prescribed laboratory methods including procedures, apparatus, and instrumentation that are to be followed when determining the capacity of the heat exchanger coil. Only then can the ratings that are published by manufacturers be compared on a common basis.

Example 12.2

Consider the following data to design a water-cooled, shell-and-tube Freon condenser for the given heat duty.

Cooling load of the condenser:	125 kW
Refrigerant:	R-22
	Condensing temperature, 37°C
	In-tube condensation
Coolant:	City water
	Inlet temperature: 18°C
	Outlet temperature: 26°C
	Mean pressure: 0.4 MPa
Heat transfer matrix:	3/4" OD 20 BWG
	Brass tubes

Solution

For the preliminary design, the assumed values for the shell-and-tube condenser are

$$N = 1 \text{ tube pass}$$

$$D_s = 15.25'' = 0.387 \text{ m}$$

$$N_T = 137 \text{ tubes}$$
$$P_T = 1'' \text{ square pitch}$$
$$B = 0.35 \text{ m baffle spacing.}$$

The properties of the shell-side coolant are

Fluid: city water
$T_{c1} = 18°C$ (inlet temperature)
$T_{c2} = 26°C$ (outlet temperature)
$d_o = 0.75'' = 0.01905 \text{ m.}$

The bulk temperature is

$$T_b = (T_{c1} + T_{c2})/2$$
$$= (18+26)/2 = 22° \text{ C}$$

The properties at the bulk temperature, from the appendix, are

$v_l = 0.001002 \text{ m}^3/\text{kg}$
$c_{pl} = 4.181 \text{ kJ/kg} \cdot \text{K}$
$k_l = 0.606 \text{ W/m} \cdot \text{K}$
$\mu_l = 959 \times 10^{-6} \text{ Pa} \cdot \text{s}$
$Pr = 6.61.$

To calculate the cooling water flow rate (shell side),

$$Q = (\dot{m}c_p)_c \Delta T_c = (\dot{m}c_p)_h \Delta T_h$$
$$\dot{m}_h = \frac{(\dot{m}c_p)_c \Delta T_c}{c_p \Delta T_h} = \frac{1.384 \times 4.1794 \times 10^3 \times 15}{4.268 \times 10^3 \times 15} = 1.36 \text{ kg/s}$$

The properties of the tube side condensing fluid are

Fluid: R-22
$T_H = T_{sat} = 37°C$ (saturation temperature)
$d_i = 0.68'' = 0.01727 \text{ m}$
$A_c = 0.3632 \text{ in.}^2 = 2.34 \times 10^{-4} \text{ m}^2.$

The properties at saturation temperature, from the appendix, are

$P_{sat} = 14.17 \text{ bar}$ $c_{pl} = 1.305 \text{ kJ/kg} \cdot \text{K}$
$v_l = 8.734 \times 10^{-4} \text{ m}^3/\text{kg}$ $v_g = 0.01643 \text{ m}^3/\text{kg}$

Condensers and Evaporators

$$\mu_l = 0.000186 \text{ Pa} \cdot \text{s} \qquad \mu_g = 0.0000139 \text{ Pa} \cdot \text{s}$$
$$k_l = 0.082 \text{ W/m} \cdot \text{K} \qquad Pr = 2.96$$
$$i_{fg} = 169 \text{ kJ/kg}$$

The refrigerant flow rate (in-tube) is

$$\dot{m}_R = \frac{Q_c}{i_{fg}} = \frac{125}{169} = 0.737 \text{ kg/s}$$

The shell-side calculations are as follows. Calculate A_s, the shell-side area, by calculating the fictitious cross-sectional area:

$$A_s = \frac{D_s CB}{P_T}$$

$$A_s = \frac{(0.387)(.0254 - .01905)(0.35)}{(0.0254)} = 0.0339 \text{ m}^2$$

The mass flux is

$$G_s = \frac{\dot{m}}{A_s} = \frac{3.73}{0.0339} = 110.03 \text{ kg/m}^2 \cdot \text{s}$$

To calculate the equivalent diameter, D_e, assuming square pitch,

$$D_e = \frac{4(P_T^2 - \pi d_o^2/4)}{\pi d_o}$$

$$D_e = \frac{4\left[(0.0254)^2 - \pi(0.01905)^2/4\right]}{\pi(0.01905)} = 0.0241 \text{ m}$$

The Reynolds number for the shell side is

$$Re_s = \frac{G_s d_o}{\mu_l}$$

$$Re_s = \frac{(110.03)(0.0241)}{(959 \times 10^{-6})} = 2772$$

So, the flow is turbulent on the shell side.
To calculate the shell-side heat transfer coefficient, for $2{,}000 < Re < 10^5$,

$$\frac{h_o D_e}{k} = 0.35 \, Re_s^{0.55} \, Pr_s^{1/3} \left(\frac{\mu_b}{\mu_w}\right)^{0.14}$$

where μ_b/μ_w is assumed to be 1 because of the small temperature difference:

$$h_o = 0.35(2772)^{0.55}(6.61)^{1/3}(0.606/0.0241) = 1329\,W/m^2K$$

The tube-side calculations are as follows. The mass flux of the refrigerant is

$$G_R = \frac{\dot{m}_a}{A_R}$$

$$A_R = \frac{A_c N_T}{N_p} = \frac{(0.000234)(137)}{1} = 0.0321\,m^2$$

where A_c = cross-sectional tube area:

$$G_R = \frac{(0.737)}{0.0321} = 22.97\,kg/m^2 s$$

The Reynolds number can be calculated as

$$Re_L = \frac{G_R(1-x)d_i}{\mu_l}$$

For a quality of $x = 0.5$,

$$Re_L = \frac{(22.97)(1-0.5)(0.0173)}{186 \times 10^{-6}} = 1068$$

$$Re_v = \frac{G_R x d_i}{\mu_l}$$

$$Re_v = \frac{(22.97)(0.5)(0.0173)}{14 \times 10^{-6}} = 14,192$$

Re_L is the liquid-only Reynolds number, and Re_v is the vapor-only Reynolds number.

For the average heat transfer coefficient for condensation, h_{TP}, the tube-side h_{TP} is given by the Cavallini–Zecchin correlation, Equation 8.31:

$$h_{TP} = 0.05\,Re_{eq}^{0.8}\,Pr^{1/3}k_l/d_i$$

where the Reynolds number equivalent, Re_{eq}, is defined as

$$Re_{eq} = Re_v\left(\frac{\mu_v}{\mu_l}\right)\left(\frac{\rho_l}{\rho_v}\right)^{0.5} + Re_L$$

$$Re_{eq} = (14192)\left(\frac{14}{186}\right)\left(\frac{0.01648}{8.73 \times 10^{-4}}\right)^{0.5} + 1068 = 5709$$

$$h_{TP} = (0.05)(5709)^{0.8}(2.96)^{1/3}(0.082/0.0173) = 345\,W/m^2K$$

Condensers and Evaporators

To calculate the overall heat transfer coefficient,

$$U_o = \left[\frac{d_o}{h_i d_i} + \frac{d_o R_{fi}}{d_i} + \frac{d_o \ln(d_o/d_i)}{2k} + R_{fo} + \frac{1}{h_o}\right]^{-1}$$

where the fouling resistances are

$R_{fi} = 0.000176$ m² · K/W for refrigerant liquids

$R_{fo} = 0.000176$ m² · K/W for city water

k (brass) = 111 W/m · K

$h_i = h_{TP}$, is the inside heat transfer coefficient

$$U_o = \left[\frac{(0.0191)}{(345)(0.0173)} + \frac{(0.0191)(0.000176)}{0.0173} + \frac{(0.0191)\ln(0.0191/0.0173)}{2(111)}\right.$$

$$\left. + 0.000176 + \frac{1}{1283}\right]^{-1}$$

$U_o = 230$ W/m² · K

The log mean temperature difference (LMTD) is

$$\Delta T_{LMTD} = \frac{\Delta T_1 - \Delta T_2}{\ln(\Delta T_1 / \Delta T_2)}$$

where

$\Delta T_1 = T_H - T_{c1} = 19°C$ and $\Delta T_2 = T_H - T_{c2} = 11°C$:

$$\Delta T_{LMTD} = \frac{19-11}{\ln(19/11)} = 14.64°C$$

The mean temperature difference is

$$\Delta T_m = \Delta T_{LMTD} \cdot F = 14.64°C$$

where $F = 1$ for a condenser.
Finally, calculate the heat transfer surface area:

$$Q = U_o A_o \Delta T_m$$
$$A_o = N_t \pi d_o L$$
$$A_o = \frac{Q}{U_o \Delta T_m}$$
$$A_o = \frac{125{,}000}{(230)(14.64)} = 37.12 \text{ m}^2$$

and the length of the tube:

$$L = \frac{A_o}{N_T \pi d_o}$$

$$L = \frac{(37.12)}{(137)\pi(0.01905)} = 4.53 \,\text{m}$$

Nomenclature

A	heat transfer area, m²
Bo	boiling number, $q''/G\, i_{fg}$
Co	convection number, $[(1-x)/x]^{0.8}\,(\rho_g/\rho_l)^{0.5}$
C1–C5	constants in Equation 12.36
c_p	specific heat at constant pressure, J/kg K
d	tube diameter, m
D_e	equivalent diameter, m
E	enhancement factor
F	constant given in Equation 12.32
F_{fl}	fluid-dependent parameter in Equation 12.36
Fr_L	Froude number, $G^2/(\rho_l^2 gd)$
g	acceleration of gravity, m/s²
G	mass velocity (flux), kg/m² · s
h	heat transfer coefficient, W/m² · K
h_L	liquid heat transfer coefficient, W/m² · K
h_{TP}	local two-phase heat transfer coefficient, W/m² · K
i	specific enthalpy, J/kg
i_{lg}	latent heat of vaporization, J/kg
k	thermal conductivity, W/m² · K
L	tube length, m
M	molecular weight
$\dot{m}$	mass flow rate, kg/s
N	average number of tubes in a vertical column
N_s	parameter defined by Equations 12.26 and 12.27
N_T	number of tubes in a bundle
Nu	Nusselt number, $hD_e/k,\ hd_o/k,\ hd_i/k$
P	pressure, Pa
Pr	Prandtl number, $c_p\mu/k$
Q	cumulative heat release rate, W
q''	heat flux, W/m²
Re	Reynolds number, $GD_e/\mu,\ Gd/\mu$
Re_L	liquid Reynolds number, $Gd_i(1-x)/\mu$
S	suppression factor

t	thickness of the wall, m
T	temperature, °C, K
ΔT	temperature difference, $(T_s - T_w)$, °C, K
U	overall heat transfer coefficient, W/m² · K
x	vapor quality
X_{tt}	Lockhart–Martinelli parameter, $(1-x/x)^{0.9}(\rho_g/\rho_l)^{0.5}$
μ_g	dynamic viscosity of vapor, Ns/m²
μ_l	dynamic viscosity of liquid, Ns/m²
ρ_g	density of vapor, kg/m³
ρ_l	density of liquid, kg/m³
ψ	h_{TP}/h_L
ψ_{bs}	value of ψ in the bubble suppression regime
ψ_{cb}	value of ψ in the pure convective boiling regime
ψ_{nb}	value of ψ in the pure nucleate boiling regime

PROBLEMS

12.1. (See also Problem 8.1) Design a shell-and-tube type power condenser for a 250 MW(e) coal-fired power station. Steam enters the turbine at 500°C and 4 MPa. The thermodynamic efficiency of the turbine is 0.85. Assume that the condenser pressure is 10 kPa. Cooling water for the operation is available at 15°C. Assume a laminar film condensation on the shell side.

12.2. In a power plant, a shell-and-tube type heat exchanger is used as a condenser. This heat exchanger consists of 20,000 tubes and the fluid velocity through the tubes is 2 m/s. The tubes are made of admiralty metal and have 18 BWG, 7/8″ OD. The cooling water enters at 20°C and exits at 30°C. The average temperature of the tube walls is 55°C and the shell-side heat transfer coefficient is 4,000 W/m² · K. Fouling on both sides is neglected. Making acceptable engineering assumptions, perform the thermal design of this condenser.

12.3. A surface condenser is designed as a two-tube-pass shell-and-tube type heat exchanger at 10 kPa ($h_{fg} = 2007.5$ kJ/kg, $T_s = 45°C$). Coolant water enters the tubes at 15°C and leaves at 25°C. The coefficient is 300 W/m² · K. The tubing is thin walled, 5 cm inside diameter, made of carbon-steel, and the length of the heat exchanger is 2 m. Calculate the

  a. LMTD
  b. Steam mass flow rate
  c. Surface area of the condenser
  d. Number of tubes
  e. Effectiveness, ε.

12.4. A shell-and-tube steam condenser is to be constructed of 18 BWG admiralty tubes with 7/8" OD. Steam will condense outside these single-pass horizontal tubes at $T_s = 60°C$. Cooling water enters each tube at $T_i = 17°C$, with a flow rate of $\dot{m} = 0.7$ kg/s per tube, and leaves at $T_o = 29°C$. The heat transfer coefficient for the condensation of steam is $h_s = 4,500$ W/m2 · K. Steam-side fouling is neglected. The fouling factor for the cooling water is 1.76×10^{-4} m2 · K/W. Assume that the tube wall temperature is 50°C. The tube material is carbon steel.

12.5. (See also Problem 8.2) A shell-and-tube type condenser will be designed for a refrigerant-22 condenser for a refrigeration system that provides a capacity of 100 KW for air-conditioning. The condensing temperature is 47°C at design conditions. The tubes are copper (386 W/m · K) and are 14 mm ID and 16 mm OD. Cooling water enters the condenser tubes at 30°C with a velocity of 1.5 m/s and leaves at 35°C. The condenser will have two tube passes. Inline and triangular pitch arrangements will be considered.

12.6. Repeat Problem 12.5 for refrigerant-134A.

12.7. A shell-and-tube type condenser will be designed for a refrigerant-22 condenser for a refrigeration system that provides a capacity of 100 KW for air conditioning. The condensing temperature is 47°C at design conditions. The tubes are copper and are 14 mm ID and 16 mm OD. Cooling water enters the condenser tubes at 30°C with a velocity of 1.5 m/s and leaves at 35°C. The condenser will have two tube passes.

DESIGN PROJECT 12.1

Design of a Compact Air-Cooled Refrigerant Condenser

Specifications:

Cooling load (heat duty):	$Q = 125$ kW
Refrigerant:	R-134A condensing inside tubes at $T_s = 37°C$ (310 K)
Coolant:	Air
	Inlet temperature, $T_{c1} = 18°C$
	Outlet temperature, $T_{c2} = 26°C$
	Mean pressure $P = 2$ atm (0.2027 MPa)
Heat transfer matrix:	to be selected from Chapter 10

At least two different surfaces must be studied for thermal and hydraulic analysis and then compared; the primary attention must be given to obtaining the smallest possible heat exchanger. A parametrical study is expected to develop a suitable final design. The final design will include materials selection, mechanical design, technical drawings, and the cost estimation.

DESIGN PROJECT 12.2
Design of an Air-Conditioning Evaporator

Specifications:

Cooling load:	4 ton air-conditioning system
Evaporator temperature:	10°C
Refrigerant:	Freon - 134A
Operating temperature:	10°C
Air inlet temperature:	30°C
Outlet temperature:	20°C

Finned brass tubes will be used for the heat transfer matrix and the core matrix will be selected from one of the compact surfaces given in Chapter 10. The final report should include thermal, hydraulic analysis, material selection and mechanical design together with cost estimation and other considerations, as listed in Design Project 12.1.

DESIGN PROJECT 12.3
Water-Cooled Shell-and-Tube Type Freon Condenser

Specifications:

Cooling load:	200 kW
Refrigerant:	Freon-134A
Temperature:	27°C
Mean pressure:	0.702 MPa
Coolant:	Water
Inlet temperature:	18°C
Outlet temperature:	26°C
Mean pressure:	0.4 MPa
Heat transfer matrix:	to be selected from Chapter 10

Condensation will be on the shell side of the heat exchanger. A shell type must be selected. The size will be estimated first and then it will be rated, as described in Chapter 8. The final report should include thermal-hydraulic analysis, optimization, materials selection, mechanical design, drawings, sizing, and cost estimation.

DESIGN PROJECT 12.4
Design of a Steam Generator for a Pressured Water Reactor (PWA)—600 MWe

Specifications:

Power to be generated:	150 MWe
Reactor coolant inlet temperature:	300°C
Reactor coolant outlet temperature:	337°C
Reactor coolant pressure:	15 MPa
Secondary circuit water inlet temperature:	200°C
Secondary circuit steam outlet temperature:	285°C
Secondary fluid pressure:	6.9 MPa

One may design four steam generators (see Figure 1.27). Thermal-hydraulic analysis must be carried out upon the selection of the type of steam generator. The final report should include a parameteric study, materials selection, mechanical design, technical drawings, and the cost estimation.

References

1. Butterworth, D., Steam power plant and processes condensers, in *Boilers, Evaporators and Condensers*, Kakaç, S., Ed., John Wiley & Sons, New York, 1991, 571.
2. Butterworth, D., Condensers and their design, in *Two-Phase Flow Heat Exchangers*, Kakaç, S., Bergles, A. E., and Fernandes, E. O., Eds., Kluwer, The Netherlands, 1988.
3. Mueller, A. C., Selection of condenser types, in *Heat Exchanger Design Handbook*, Hemisphere, Washington, DC, 1983.
4. Breber, G., Condenser design with pure vapor and mixture of vapors, in *Heat Transfer Equipment Design*, Shah, R. K., Subbarao, E. C., and Mashelkar, R. A., Eds., Hemisphere, Washington, DC, 1988, 477.
5. Bell, K. J. and Mueller, A. C., *Condensation Heat Transfer Condenser Design*, AIChE Today Series, American Institute of Chemical Engineers, New York, 1971.
6. Sebald, J. F., A development history of steam surface condensers in the electrical utility industry, *Heat Transfer Eng.*, 1(3), 80, 1980.
7. Alfa Laval, *Thermal Handbook*, Lund, Sweden, 1989.
8. Taborek, J., Industrial heat transfer exchanger design practices, in *Boilers, Evaporators and Condensers*, Kakaç, S., Ed., John Wiley & Sons, New York, 1991, 143.
9. Steinmeyer, D. E. and Muller, A. C., Why condensers don't operate as they are supposed to, *Chem. Eng. Progress*, 70, 78, 1974.
10. Pate, M. B., Evaporators and condensers for refrigeration and air-conditioning systems, in *Boilers, Evaporators and Condensers*, Kakaç, S., Ed., John Wiley & Sons, New York, 1991.

11. Althouse, A. D., Turnquist, C. H., and Bracciano, A. F., *Modern Refrigeration and Air-Conditioning*, The Goodheat-Willcox Co., 1968.
12. *Trane Air Conditioning Manual*, The Trane Company, Lacrosce, Wisconsin, 1979.
13. *Equipment Handbook*, ASHRAE, Atlanta, 1988.
14. Condensers, in *Equipment Handbook*, ASHRAE, Atlanta, 1983, ch. 16.
15. Shah, M. M., Chart correlation for saturated boiling heat transfer: equation and further study, *ASHRAE Trans.*, 88, 185, 1982.
16. Kandlikar, S. G., A general correlation for saturated two-phase flow boiling heat transfer inside horizontal and vertical tubes, *J. Heat Transfer*, 11, 219, 1990.
17. Güngör, K. E. and Winterton, R. H. S., A general correlation for flow boiling in tubes and annuli, *Int. J. Heat Mass Transfer*, 29, 351, 1986.
18. Cooper, M. G., Saturated nucleate pool boiling, a simple correlation of data for isothermal two-phase two-component flow in pipes, *Chem. Eng. Progress*, 45(39), 39, 1949.

13

Polymer Heat Exchangers

13.1 Introduction

In solar energy, air conditioning, refrigeration, automotive, computer, desalination, pharmaceutical and food industries, and energy recovery applications, various types of polymer heat exchangers are used. Because of their ability to resist fouling and corrosion, the use of polymers in some applications offers substantial reduced weight, costs, water consumption, volume, space, maintenance costs, which can make these heat exchangers more competitive in some applications over metallic heat exchangers. Because of the low thermal conductivity, usually, polymers are not considered as a material to construct heat exchangers, except for specific applications; but when the material properties of polymers and the research development on the polymer matrix composites, these materials have considerable potentials to be used in the construction of heat exchangers in several applications. In addition, plastics are much simpler to shape, form, and machine; plastics have smooth surface which lower wettability resulting in dropwise condensation with higher heat transfer coefficient. ASTM and ASHRAE codes dictates minimal property requirements for materials used in this type of heat exchangers dependent on applications: thermal conductivity, specific heat capacity, maximum operating temperature, thermal expansion coefficient, density, ultimate tensile strength, and tensile modulus. A list of some commonly used polymers and their material properties are given in Tables 13.1.[1-6]

Although low thermal conductivity will result in a dominating heat transfer resistance by the walls, but using very thin polymer structures, both plate and tubular heat exchangers can be designed with their performance being comparable to conventional units at lower cost and reduced weight. Considering the advances made in composites materials, and nanoscale composites, the use of polymer materials in the constructions of plastic heat exchangers has a promising future in applications. To minimize the resistance to heat transfer across the wall, the ratio of outer diameter-to-wall thickness, termed the standard diameter ratio or SDR should be maximized. Polymers can be divided in two groups according to their properties and structures: *monolithic polymers* and *polymer matrix composites*.[1]

TABLE 13.1

Thermal and Mechanical Properties of Common Polymers

Properties	Nylon-6.6	PC	PEEK	PPS	PPSU	PP	PS	PSU	PTFE	PVDF
Density, g/cc	1.12	1.2	1.33	1.43	1.29	0.937	1.05	1.24	2.17	1.78
Water absorption, %	2.3	0.17	0.21	0.031	0.37	0.079	0.088	0.41	0.0042	0.032
Moisture absorption at equilibrium, %	2	0.27	0.46	0.03	0.85	0.1	0.089	0.32		0.018
Tensile strength, ultimate, MPa	73.1	64	110	86.7	76	36.8	44.9	72	33.6	42.8
Tensile modulus, GPa	2.1	2.3	4.5	3.6	7.2	1.9	3	2.5	0.61	1.8
Thermal conductivity, W/mK	0.26	0.2	0.25	0.3	0.35	0.11	0.14	0.22	0.27	0.19
Melting point, °C	250		340	280		160			330	160
Glass temperature, °C		150	140	88	220		90.4	190		−37.6
Specific heat capacity, J/g × °C	2.2	1.2	2			2	2.1	1.2	1.4	1.5

Source: From T'Joen, C., et al., Int. J. Refrig., 32, 763, 2009. With permission.

Monolithic polymers are structures made out of one unit, called monomers. These polymers are primarily composed of hydrogen and carbon atoms, arranged in a chain. They have two kinds: synthetic and natural. Wood and rubber can be given as example for natural monolithic polymers. Synthetic monolithic polymers have a wide range of materials. One way to classify them is by their behavior to heating and then cooling, as *thermoplastics* and *thermosets*. Thermoplastics soften when heated and harden again when cooled down. Thermosets soften when heated and can be molded.

Mechanical and thermal designs have equal importance in designing of a heat exchanger. As shown in detail in the previous chapters, important parameters in designing of a heat exchanger are; heat conductivity, specific heat capacity, maximum operating temperature, coefficient of thermal expansion, ultimate tensile strength, and density. A state-of-the art study on polymer heat exchangers is given in T'Joen et al.[1]

Advantages and disadvantages of some monolithic polymers are summarized below:

Fluoropolymers are known for their resistance to chemical corrosion.

Polyvinylidene difluoride (PVDF) is a pure thermoplastic fluoropolymer, and it is used in applications which need the highest purity, strength, and resistance to solvents, acids.

Ethylene tetrafluorethylene (ETFE) are known for their low working operating temperature intervals. They are mostly used with acidic fluids. Their working temperatures are between −1.6°C and 154°C.

The polymer known as *Polytetrafluoroethylene* or *Teflon* (PTEE) is chemically resistant to nearly every fluid and it can work up to the temperature of 204°C.

Liquid crystal polymers (LCP) have the properties of both polymer and liquid crystals. They can be used in applications that exceed 300°C. Also, they show a chemical resistance to basic and acidic solvents. They have low thermal expansion coefficients and high tensile strengths. Because of these properties, they are advantageous in heat exchanger designs.

Polypropylene (PP) does not contain toxic chemicals. They have a high resistance to corrosion. PP has the highest melting point at 165°C. It has a significant application in mechanical vapor compression units. *Polyethylene* have low densities. They can work up to 95°C for a limited time but can work constantly at 80°C.

Polycarbonate (PC) has a good chemical resistance to acids but not for alkalis and solvents. The melting point is 149°C. Their working temperatures are between −4°C and 135°C. They are resistant to mineral acids, organic acids, and oils.

Polyethylene (PE) has low density and it can be used at temperature up to 95°C for short periods. It can be used continuously for conditions less than 80°C. PE is a very useful and widely used plastic as Laboratory apparatus.

Polyphenylene sulfide (PPS) are known for their high chemical resistance to acids. They have shown a better performance than Teflon and PVDF, in experiments done in 85% sulfuric acid at 120°C, suggested that is the best performance in the acid conditions compared with Teflon and PVDF.

Polyphenylene oxide (PPO) are similar to polyphenylene ether (PPE) in chemical composition. They have a high heat resistance but a low chemical resistance. They are preferred because of their flame-retardant and low water absorption properties in various water-handling products; PPO can also be electroplated.

Ployetheretherketone (PEEK) have high working temperatures (up to 250°C) and conservation of mechanical properties at higher temperatures and high chemical resistance to acids, making them desirable.

Polysulfone (PSU) have a maximum working temperature of 190°C. They have high creep resistances and they are resistant to many solvents, oils, and acids of chemicals.

Important properties of monomers with regard to heat exchanger design are given at Table 13.1.

Significant differences can be seen from the data at the Table 13.1 and the properties of metals that are used in heat exchangers. For example, if the thermal conductivities are examined, heat conductivity coefficients of monomers are much less than thermal conductivity coefficients of metals. Depending on the big differences, it can be said that polymers are not suitable for heat exchangers. But in their studies, dated, Zaheed and Jachuck[5] compared heat exchangers with same thermal properties, one made of nickel, chrome, and molybdenum alloy and the other made of PVDF. Results showed that the heat exchanger made of PVDF is, although it was 6 times bigger than the one made of metal, cost 2.5 times less. Despite their low thermal conductivities, polymers can have overall heat transfer coefficients near to those of metals, with the use of thin-walled structures.

Another study was done by Ma et al.[7] The study showed that tube coating a heat exchanger with PTFE film can increase its heat transfer coefficient by 0.3 to 4.6 times. This increase is accomplished by the dropwise condensation properties of PTFE. The same dropwise condensation properties can be observed at PVDF too. Based on this, it can be said that, in heat exchangers where condensation takes place, the use of polymers cause an increase in overall heat transfer coefficients because of their dropwise condensation properties. But this is not valid for all the heat exchangers. For example, no increase in overall heat transfer coefficient is observed in compact heat exchangers, due to the blockage in the passage (Jacobi et al.[8]).

13.2 Polymer Matrix Composite Materials (PMC)

The properties of monomers are obtained by the help of chemical compositions and they might not have the properties that are desired. But the purpose of composite materials is to acquire materials with desired properties. Composite materials are compositions, acquired with the combining of two or more materials. Their properties are determined by their constituents, and according to the way their constituents are combined, similar or totally different materials can be acquired. Polymer matrix composite materials are composite materials in which polymers are used as constituents. The constituent used primarily in the composite material is called the matrix, and the other constituent is called reinforcement. Reinforcements used in the composite materials can be in different shapes, which are particles, fibers, flakes, and laminas, and the composite materials are called as particulate, fiber reinforced, flake, and laminated composites (Gurdal et al.[9]).

For the composite to be efficient, the matrix material should integrate fully with the reinforcement material, have good environmental degradation resistance, and be able to deform as much as the reinforcement material. Because of their advantages, studies have been done to comprehend the properties of PMC materials. Biggs[10] examined various methods developed on the thermal conductivity of PMCs. Ahmed and Jones[11] studied theoretical models developed for the tensile modulus and strength of PMCs. It was decided that there are too many parameters affecting the PMCs and no general theory that can predict the tensile strength.

Particle-Reinforced Composites. PMCs that are to be used in heat exchangers should have a high thermal conductivity; for this, aluminum, copper, silver, and other materials that have high thermal conductivity are added for reinforcement. Despite this, PMCs generally have thermal conductivities less than 4W/m · K, which makes them not suitable for many commercial uses. In his studies, Bigg[12] examined particulate composite materials made of metallic or glass spheres and he compared the results with existing models and found that Nielsen[13] model was accurate in finding the thermal conductivity of PMCs. As a result, it is found that at volume ratio of 0.6 for metallic particles, the thermal conductivity of the composite is seven times the thermal conductivity of the matrix material, and at a volume ratio of 0.3 for metallic particles, the thermal conductivity of the composite is almost two times the thermal conductivity of the matrix material which indicates strong enhancement.

Using the Nielsen method, Bigg[12] showed that the thermal conductivity enhancement depends on volume fraction of the metallic particles when the thermal conductivity of the reinforcement material is more than 100 times the thermal conductivity of the base material; the thermal conductivity of the PMC is a direct function of the volume ratio.

Mamunya[14] gave a definition of percolation theory for electrical and thermal conduction in with particle-filled PMCs. According to this theory, a sudden increase occurs while the volume ratio of the reinforcement material increases. The reason for this is the continuous conduction path composed of the reinforcement material. This behavior is valid for electrical conductivity at PMCs but not for thermal conductivity because for this to be valid the thermal conductivity of the reinforcement material should be more than 100,000 times the thermal conductivity of the matrix material.

There are considerable number of studies on the properties of PMCs, which will be summarized below:

Boudenne et al.[15] provided data for density, heat capacity, thermal conductivity, crystallinity, and thermal diffusivity of PMCs. For these, he used two aluminum spheres with diameters of 44 and 8 μm. As a result, for 58.7% volume fraction, the thermal conductivity of PMC is found to be 2.7 and 4.2 W/m · K, respectively. This increase is substantial compared to the thermal conductivity of the matrix material, which is 0.239 W/m · K.

Mamunya[14] obtained data using nickel and copper as reinforcement material for epoxy and PVC. The nickel particles had a diameter of 10 μm, and copper particles had an average size of 100 μm. The experimental results showed that, when PVC was used as matrix material, the thermal conductivity increased from 0.45 to 1.64 W/m · K at a volume ratio of 0.4. And when epoxy was used as a matrix material, the thermal conductivity increased from 0.52 to 1.6 W/m · K at a volume ratio of 0.25. When the thermal conductivities were compared with the ratio of reinforcement material taken into account, it was seen that epoxy performed better than PVC material, thermally. As a result, it was decided that the properties of the matrix material has a significant effect on the thermal performance of the PMC.

Serkan Tekce et al.[16] studied the effects of different copper reinforcement materials on the thermal conductivity of the PMC as spheres, flakes, and short fibers. With these studies, it is shown that the PMC with fiber reinforcement material with 60%, the thermal conductivity was increased from 0.21 to 11.57 W/m · K, which has the greatest thermal conductivity and after that plate and sphere, respectively.

Krupa et al.[17] examined the effects of graphite particle as reinforcement materials on different kinds of PMCs. For the volume ratio of 0.11, it is found that a sudden rise in the electrical conductivity occurred, and for the volume ratio of 0.38, the thermal conductivity increased from 0.35 to 2.4 W/m · K. Also, as a result of Krupa and Chodak,[18] they observed big differences in properties caused by different sizes of graphites, which get more significant with the increase of volume ratio.

Fiber-Filled Composites: Mostly, glass, carbon, and aramid fibers are used as fillers. Thermal conductivities and expansion coefficients are given at Table 13.2.

TABLE 13.2
Properties of Monolithic Polymers and PMCs

Reinforcement	Matrix	Density, g/cm³	Thermal Conductivity, W/m·K		Specific Thermal Conductivity[a], W/m·K		Source
			In-Plane	Thickness	In-Plane	Thickness	
—	Epoxy	1.2	1.7	1.7	1.4	1.4	Zweben[19]
E-glass fibers	Epoxy	2.1	0.16–0.26	n/a	0.1–0.2	n/a	Zweben[19]
Kevlar (aramid)	Epoxy	1.38	0.9	n/a	0.6	n/a	Zweben[20]
Milled glass fiber	Polymer	1.4–1.6	0.2–2.6	0.2–2.6	0.1–1.6	0.1–1.6	Zweben[6]
Discontinuous carbon fibers	Polymer	1.7	10–100	3–10	6–59	1.8–5.9	Zweben[6]
Continuous carbon fibers	Polymer	1.8	330	10	183	5.6	Zweben[19]
Natural graphite	Epoxy	1.94	370	6.5	190	3.4	Tzeng et al.[21]

Source: From T'Joen, C., et al., *Int. J. Refrig.*, 32, 763, 2009. With permission.
[a] Specific thermal conductivity is defined as thermal conductivity divided by specific gravity of the material. n/a, data not available.

Glass Fiber-Reinforced Polymers (GFRPs): These glass fiber-reinforced polymers are known for their high strength and elastic modulus, good corrosion resistance, and good insulating properties. Glass fiber-reinforced materials are usually composed of a laminate structure with different fiber orientations. Their volume ratios are generally between 40 and 70%.

Carbon Fiber-Reinforced Polymers (CFRP): These are electrically conductive and have really high elastic strength and corrosion- and fatigue-resistance properties. They are also advantageous because of their lower weights. Structurally, they are brittle, and so their impact resistances are low. Mostly, epoxy, polyester, and Nylon are used as matrix materials.

Nysten and Issi[22] examined carbon fiber-reinforced polymers that are already commercially available. The most advantageous composite with regard to thermal conductivity is 45% volume fraction of P-120 fibers. The thermal conductivity of this composite is found to be as high as 245 W/m·K with a density of 1.66 kg/cm^3.

Norley[23] developed a new graphite-epoxy composite material. This material has a low weight (1.9 g/cm^3), and its thermal conductivity is observed to be 370 W/m · K. This material is used as a fin material with the combination of aluminum or copper base for cooling.

The most important advancement for carbon fiber PMCs was the development of injection molding compounds. By the help of this technique, the thermal conductivities can reach 100 W/m · K by using discontinuous carbon fibers with high thermal conductivities. As a result, the biggest increase in thermal conductivity is acquired by Nylon composites with 40% volume fraction carbon fiber (from 0.3 to 16.6 W/m · K), and by polycarbonate composites that are formed from carbon black, synthetic graphite and carbon fiber, respectively, at 5% volume fraction, 30% volume fraction and 20% volume fraction (from 0.23 to 20.1 W/m · K).

Kevlar Fiber-Reinforced Polymers (KFRPs): Kevlar is a kind of aramid fiber that has a high tensile strength, low thermal expansion coefficient, and a good corrosion resistance. The disadvantages of Kevlar are its low compressive strength and the difficulty to cut it. Epoxy, vinylester, and phenolics are the most used matrix material for kevlars.[1,24]

Example 13.1

Determine the overall heat transfer coefficient, U, for liquid-to-liquid heat transfer through a 1.5-mm-thick PEEK polymer [k_{PEEK} = 0.25 W/(m · K)] for the following conditions:

$$h_i = 2500 \text{ W/m}^2 \cdot \text{K}$$

$$h_o = 1800 \text{ W/m}^2 \cdot \text{K}$$

$$R_{fi} = 0.0002 \text{ m}^2 \cdot \text{K/W}$$

Polymer Heat Exchangers

Solution

Substituting h_i, h_i, R_{fi}, t, and k_{PEEK} into Equation 2.21, we get

$$\frac{1}{U} = \frac{1}{2500} + 0.0002 + \frac{0.0015}{0.25} + \frac{1}{1800}$$
$$= 0.0004 + 0.0002 + 0.006 + 0.00056 = 0.00716 \; m^2 \cdot K/W$$
$$U \approx 139.8 \; W/m^2 \cdot K$$

In this case, no resistances are negligible.

Example 13.2

Replace one of the flowing liquids in Example 13.1 with a flowing gas, $h_o = 50 \; W/m^2 \cdot K$.

Solution

$$\frac{1}{U} = \frac{1}{2500} + 0.0002 + \frac{0.0015}{0.25} + \frac{1}{50}$$
$$= 0.0004 + 0.0002 + 0.00006 + 0.02 = 0.02066 \; m^2 \cdot K/W$$
$$U \approx 37.6 \; W/m^2 \cdot K$$

In this case, only the gas-side resistance is significant.

Example 13.3

Replace the remaining flowing liquid in Example 13.2 with another flowing gas, $h_i = 20 \; W/m^2 \cdot K$.

Solution

$$\frac{1}{U} = \frac{1}{20} + 0.0002 + \frac{0.0015}{0.25} + \frac{1}{50}$$
$$= 0.05 + 0.0002 + 0.006 + 0.02 = 0.0762 \; m^2 \cdot K/W$$
$$U \approx 13.12 \; W/m^2 \cdot K$$

Here, the wall and scale resistances are negligible.

13.3 Nanocomposites

Nanocomposites are a newly developing kind of composite. If a dimension of the particles used in the composite is smaller than 100 nm, the composite is named a nanocomposite. *Inorganic Clay Nanocomposites: For this kind of* composites, the particle sizes, the dispersion or aggregation of particles play a big role in the mechanical properties of the material. Even though they are still in development, this composites will become significantly important in long term. Jordan et al.,[25] LeBaron,[26] Ahmedi[27] and Gao[28] worked on this subject to develop a lot of new ways to manufacture nanocomposites. LeBaron and Ahmedi did research on inorganic nanocomposites. As a result of these, it was understood that the reinforcement materials should be as homogeny as possible in the composite for the nanocomposites to have the best mechanical properties they could.

Gao[28] compared clay-polymer (NC) to GFRPs in his work. The weak point of classical composites is that their estimated theoretical properties are at very low levels. For example, CFRPs reach only 3–4% of the theoretical strength of the aromatic sheets that are present in the graphite structure. In addition, in exfoliated nanocomposites physical properties can be achieved fully because the layers are connected by the polymer. But in reality, nanocomposites with low volume ratios have poorer mechanical properties than classical PMCs with high volume ratios and reach the mechanical properties of Nylon-6 with 48% glass fibers. However, for low fiber range, clay composites exhibit superior mechanical properties.

Carbon nanotube reinforced polymers (CNTRPs). There is an alternative fiber type for PMC named carbon nanotube (CNT). These materials are made of carbon chains rolled up into a cylinder in an order of nanomeasures. Carbon nanotubes are known by their high strength rates and low weights with good flexibility. Their thermal conductivity values are isotropic.

In some composite materials, polymers can be used as fillers with different matrix materials such as graphite.

13.4 Application of Polymers in Heat Exchangers

Liquid to Liquid Heat Exchangers. Morcos and Shafey[29] presented an experimental study about thermal performance of shell-and-tube heat exchangers made of PVC material for both shell- and tube-side. Heat exchanger used in the study was 1.3 m long with tubes which had outer diameters of 34 mm with 5 mm thickness (Figure 13.1). There were also five baffles placed in the heat exchanger. After measurements, it was found that, because of the thickness, overall heat transfer coefficient was limited to 90 W/m² · K. To get a greater heat transfer coefficient, some conical turbulators were used in the tubes. With this change, overall heat transfer coefficient could become 3–4 times greater with the increase in the pressure drop.

Group of solar energy at the University of Minnesota presented some studies about low cost solar water heaters. For example, Davidson et al.[30] examined the suitable materials for solar heaters. They used three constraints

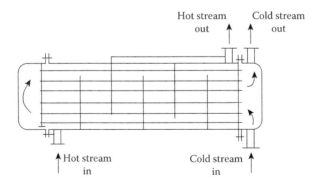

FIGURE 13.1
PVC shell-and-tube heat exchanger studied by Morcos and Shafey. (From Morcos, V. and Shafey, H., *Journal of Elastomers and Plastics*, 27, 200–213, 1995.)

to select suitable candidates. First, they selected the polymers accepted by NSF (National Sanitation Foundation) standards to use with potable water. Second constraint that they used was to select materials by mechanical properties like retention and water compatibility. For final screening, suitable polymers were sorted by their strength, stiffness, cost, and thermal conductivity. After these steps of screening, they have chosen three polymers as best candidates for solar applications, which were high temperature Nylon (HTN), cross-linked polyethylene (PEX), and polypropylene (PP). However, these constraints does not tell much about the long-term behavior of tubular heat exchangers in hot water conditions. The examination for long-term behavior was made by Freeman et al.[31] and Wu et al.[32] examined different polymers about tensile strength, creep compliance, and strain failure at 82°C. For mineral-containing water, comparisons were made and Nylon-6.6 was observed to have more scaling rate compared to HTN, PB, PP, and copper (Wang et al.[33]).

Liu et al.[34] studied on the materials can be used for solar water heating systems and designed a shell-and-tube and an immersed heat exchanger which are used frequently in these kind of applications. In the study, they fixed the parameters like number of tubes, tube arrangements, flow rate, and heat transfer rate and calculated the required length of the tubes. Shell-and-tube and immersed heat exchanger used in the study are given at Figure 13.2 to 13.3,[34] respectively. Experiments were performed for both PEX and Nylon. PEX and Nylon have thermal conductivities about 0.38 and 0.31 W/m²·K, respectively. For PEX material, outer diameter was 9.53 mm and thickness was 1.78 mm. For Nylon tube, diameter was 3.81 mm and thickness was 0.2 mm. Results showed that for PEX, wall resistance was a dominant factor limiting the overall heat transfer coefficient. However, it was not true for Nylon. Liu et al. also obtained resistances in the system, and for 5.7 L/min flow rate, they found the percentage of each resistance in the total resistance.

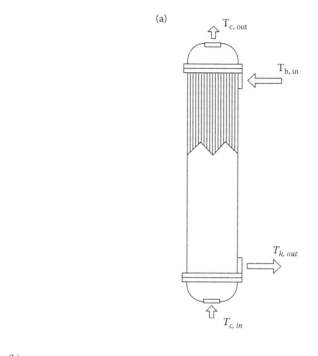

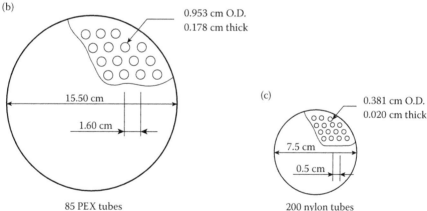

FIGURE 13.2
Polymer shell-and-tube heat exchanger studied by Liu et al. (From Liu, W., Davidson, J., and Mantell, S., *Journal of Solar Energy Engineering*, 122, 84–91, 2000. With permission.)

These percentages were 49% inside resistance, 26% outside resistance and 25% wall resistance for Nylon and 24%, 34%, 42% for PEX material. For a copper-tube material, with tubes with 6.35 mm outer diameter and 0.5 mm wall resistance, percentages were found to be 76, 24, and 0.04%. For the same heat duties, under the same mass flow rates, a Nylon heat exchanger requires

Polymer Heat Exchangers

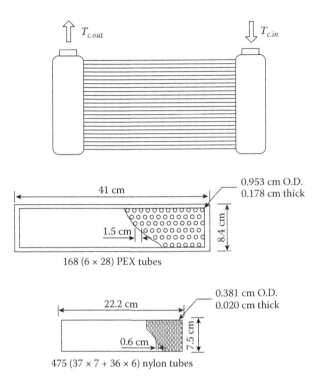

FIGURE 13.3
Immersed heat exchanger studied by Liu et al. (From Liu, W., Davidson, J., and Mantell, S., *Journal of Solar Energy Engineering*, 122, 84–91, 2000. With permission.)

similar surface areas with copper alternative. These analysis with polymer shell and tube and immersed heat exchangers showed that designing heat exchangers with thin walled is only way to compare with metal ones.

Zakardas et al.[35] came up with a novel design of polymer heat exchangers for liquid-to-liquid or condensing fluid-to-liquid heat exchanger. This design was named as hallow fiber heat exchanger, which is basically a shell-and-tube heat exchanger without baffles. However, its size is usually smaller. With constructed specimens, PP and PEEK were used. PP had thickness of 150 and 575 μm outer diameter, and PEEK one had 360 mm outer diameter and thickness of 210 μm. Shell diameter was 2.3 cm and length of tubes was 11.8 cm. Under these circumstances, hallow fiber HEX showed a surface area of 1500 m² of exterior surface. Also, overall heat transfer coefficients of 1360 W/m²·K had been observed. The study shows that these heat exchangers are excellent replacement for the metal ones since they have similar thermal performances with metal ones even under fouled conditions.

Patel and Brisson[36] studied a polymer kapton heat exchanger for cryogenic applications. In the study, heat transfer between a superfluid and the dilution was aimed. Because of the Kapitza limit, polymer heat exchangers have some advantages over metal ones under low temperature conditions like these superfluid applications.

Leigh et al.[37] aimed to reduce the first costs of absorption chillers by using alternative materials for some part of the system to reduce the cost of the chillers. On a conceptual design, a model was run and parts especially heat exchangers that could be replaced with polymer materials. Results showed that second effect generator and recuperaters were able to be produced with PEEK material. However, for some parts of chiller, it was shown that polymers could not be used because of their permeability to oxygen.

For plastic thin-film heat exchangers, a novel design was developed by Lowenstein and Sibilia.[38,39] This study showed that, for evaporators and absorbers, applications using thin plastic films heat exchangers are feasible enough to implement.

Liquid-to-Gas Heat Exchangers: Bigg[40] studied on the design of tubes for flue gas channels of gas fired boilers. He used 1.27 cm diameter PTFE, PP, PPS, and 1.9 cm diameter PSU and PEI. There were also metal tubes coated with polymer materials. After a long time of operation, tubes were examined, and it was observed that full polymer specimens had no degradation. However, there were cracks and degradation on the metal parts of coated tubes because of the cracks caused by thermal expansion difference between the coating and metal. One of the tubes which was coated with polymer including metal particles showed degradation at points where metal had contact with the gas.

El-Dessouky and Ettouney[41] studied on plate and shell-and-tube heat exchangers for single-step mechanical vapor compression unit. A developed model was used to show thermal performances of thin-walled polymer tubes and plates in the desalination applications. Results were compared with the metal alternatives. It was found that PTEE heat exchangers surface area was 3–4 times larger than that of metal alternatives using Titanium, steel and Cu-Ni alloys and plastic tubes were cheaper to be implemented despite of their sizes.

Bourouni et al.[42] presented an experiment study about evaporator and condenser used in desalination process. He used circular PP tubes with 2.5 cm diameter and 5 mm wall thickness. Results were compatible with a model that examining impact of mass flow rate and inlet temperature. However, because of economical reasons, these designs should be used with cheap energy sources.

Burns and Jachuck[43] studied a plate heat exchanger made of PEEK films (Figure 13.4). These films had 53 μm thickness and amplitude of 1 mm. Seven sheets that are sized 13.5 × 13.5 cm were placed with 90° angle. PEEK was selected because of its high chemical and fouling resistance, hydrophobic and smooth structure with high working temperature. Within the experimental setup, overall heat transfer coefficient of 60–370 $W/m^2 \cdot K$ was measured with dropwise condensation at the gas side.

Cheng and Van Der Geld[44] studied a plate-fin heat exchanger made of PVDF for air–water and air–steam fluids passing through the heat exchanger.

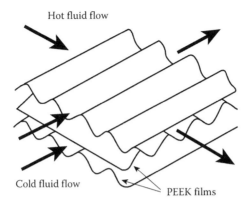

FIGURE 13.4
Corrugated PEEK films used by Zaheed and Jachuck. (From Zaheed, L. and Jachuck, R. J. J., *Applied Thermal Engineering*, 24, 2323–2358, 2004. With permission.)

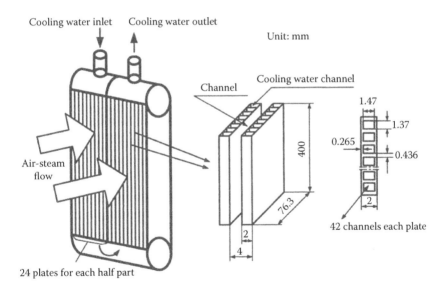

FIGURE 13.5
Heat exchanger geometry studied by Cheng and Van Der Geld. (From Cheng, L. and Van Der Geld, C., *Heat Transfer Engineering*, 26, 18–27, 2005. With permission.)

The heat exchanger is shown in Figure 13.5. Overall, transfer coefficients of 80–130 and 150–600 W/m² · K were observed with water–air and steam–air sides, respectively.

Van Der Geld[45] studied a similar experimental setup as that of Cheng and Van Der Geld[44]. He aimed to enhance the setup with changing some of its parameters. He was able to enhance the heat transfer rate by 7% by inclining the heat exchangers and changing the location of spacers. It was mainly due to improvement with drainage of the condensate.

Harris[46] studied a micro cross-flow plate heat exchanger made of PMMA and nickel. The dimensions of polymer heat exchanger were 5 × 5 mm with thickness of 1.8 mm. An analytical model was derived for the micro cross-flow heat exchanger and verified by experimental results. After taking measurements from setup, for nickel and PMMA materials, heat transfer for unit area and unit temperature difference, for unit volume and unit temperature difference, and for unit mass and unit temperature differences was calculated. It was shown that, because of low density, PMMA had a superior heat transfer rate for unit mass and unit temperature difference. Under other conditions, Nickel heat exchanger had superior heat transfer rates.

Example 13.4

In a double-pipe heat exchanger made of PTFE polymers, water will be heated from 20°C to 35°C at 40 kg/hr by a hot water at 140°C and 0.0012 kg/s. A 15°C hot water temperature drop is allowed. The tubes with an outside diameter of 0.953 cm and inside diameter of 0.597 cm can be used. The wall thickness is 0.178 cm with a thermal conductivity of 0.27 W/m · K. Heat transfer coefficients in the inner tubes and through the annulus are 4911 W/m² · K and 1345 W/m² · K, respectively. Neglecting the fouling, calculate the surface area of the heat exchanger and compare the surface area required when the tube is made of mild steel, $k = 25$ W/m × K.

Solution

At the mean temperature of 27.5°C, the properties of water are

$$\rho = 996.307 \text{ kg/m}^3$$

$$k = 0.610 \text{ W/m} \cdot \text{K}$$

$$\mu = 8.43 \times 10^{-4} \text{ Pa} \cdot \text{s}$$

$$Pr = 5.79$$

$$C_p = 4.18 \text{ kJ/kg} \cdot \text{K}$$

The hot water outlet temperature can be obtained from the heat balance:

$$Q = \left(\dot{m}c_p\right)_h \left(T_{in} - T_{out}\right)_h = \left(\dot{m}c_p\right)_c \left(T_{out} - T_{in}\right)_c$$

$$Q = \frac{4}{3600}(4180)(35-20) = 69.67 \text{ W}$$

$$T_{out,h} = T_{in,h} + \frac{Q}{\left(\dot{m}c_p\right)_h} = 140 + \frac{69.67}{0.0012 \times 4.18 \times 10^3} = 126.11°\text{ C}$$

The overall heat transfer coefficients for the clean PTFE and mild steel surfaces based on the outside surface area of the tube are

$$U_c = \frac{1}{\dfrac{d_o}{d_i}\dfrac{1}{h_i} + \dfrac{d_o \ln(d_o/d_i)}{2k} + \dfrac{1}{h_o}}$$

$$U_{c,PTFE} = \frac{1}{\dfrac{0.00953}{0.00597}\dfrac{1}{4911} + \dfrac{0.00953 \ln(0.00953/0.00597)}{2 \times 0.27} + \dfrac{1}{1345}} = 214.91$$

$$U_{c,ms} = \frac{1}{\dfrac{0.00953}{0.00597}\dfrac{1}{4911} + \dfrac{0.00953 \ln(0.00953/0.00597)}{2 \times 25} + \dfrac{1}{1345}} = 903.13$$

and the mean temperature difference is

$$\Delta T_{lm} = \frac{\Delta T_1 - \Delta T_2}{\ln \dfrac{\Delta T_1}{\Delta T_2}} = \frac{(126.11-20)-(140-35)}{\ln \dfrac{(126.11-20)}{(140-35)}} = 243.05°\ C$$

The surface areas are calculated from Equation 6.1:

$$Q = U_c A_o \Delta T_{lm}$$

$$69.67 = 214.91 \times A_{o,PTFE} \times 243.05$$

$$A_{o,PTFE} = \frac{69.67}{214.91 \times 243.05}$$

$$A_{o,PTFE} = 1.33\ m^2$$

$$69.67 = 903.13 \times A_{o,ms} \times 243.05$$

$$A_{o,ms} = \frac{69.67}{903.13 \times 243.05}$$

$$A_{o,ms} = 0.32\ m^2$$

Therefore, the size of the mild steel heat exchanger with the higher thermal conductivity of the metal is smaller than that of the PTFE heat exchanger. But it may cost less than the metallic heat exchanger.

Gas-to-Gas Heat Exchangers: Rousse[47] presented an experimental study of a shell-and-tube heat exchanger for a heat recovery unit for greenhouses. Design of this PE heat exchanger was not compact because it was designed to prove that polymer heat exchanger can be advantageous for

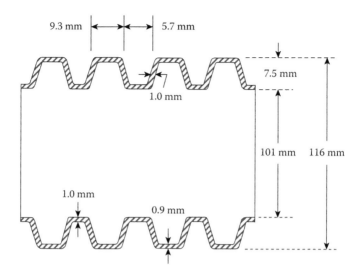

FIGURE 13.6
Corrugated PE tubes as used in a shell-and-tube heat exchanger. (From Rousse, D., Martin, D., Theriault, R., Leveillee, F., and Boily, R., *Applied Thermal Engineering*, 20, 687–706, 2000. With permission.)

greenhouses for heat recovery process. Design met the requirements like the ease of assembly, low cost repair, maintenance, operation, corrosion, and good performance under frosting conditions. For the study, five corrugated tubes were used in a single shell (Figure 13.6). Results showed that, in operation, efficiencies were defined as the ratio of the temperature differences between the inlet and outlet of the inlet air to the maximum temperature difference.

Jia et al.[48] performed experiments on a plate heat exchanger made of PTFE, which is used as a flue gas heat recovery unit. Channels had 1.5 mm thickness and 1 cm space between plates. This unit was used as a SO_2 scrubber by condensation. For this setup, measurements were taken with or without condensation and also a correlation was obtained.

Saman and Alizadeh[49,50] performed an experimental study on polymer (PE) plate heat exchangers (Figure 13.7) for dehumidification and cooling processes. To simulate this process, water was injected to air flow to achieve evaporative cooling and a liquid desiccant was injected into the other air flow for achieving dehumidification process. Experimental results were taken from setup for 0.2 mm thickness sheets of PE material which separate both streams. Results were compared to the theoretical findings, and good agreements were found. Experiment was done for different parameters like injection angle, mass flow rate, and temperature. However, PE materials did not perform well according to the summer comfort standards of some countries like Brisbane, Australia.

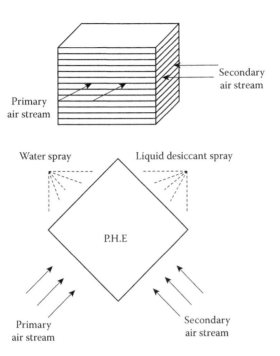

FIGURE 13.7
Plate heat exchanger for dehumidification and cooling studied by Saman and Alizadeh. (From Saman, W. and Alizadeh, S., *Solar Energy*, 70(4), 361–372, 2001. With permission.)

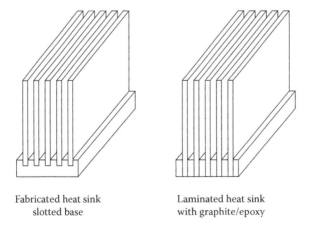

FIGURE 13.8
Comparison of hybrid and laminated graphite/epoxy heat sinks. (From Shives, G., Norley, J., Smalc, M., Chen, G., and Capp, J., *Thermomechanical Phenomena in Electronic Systems—Proceedings of the Intersociety Conference*, 1, 410–417, 2004. With permission.)

Heat Sinks: Heat sinks made of polymers can be seen as a good alternative to the classical metal heat sinks. Polymers are cheap, have low weight and electrically insulating properties. All of these factors make polymer heat sinks favorable. Portable devices that have microprocessors can be a good area of usage. However, because of the low thermal conductivity of polymer materials, using PMC materials can be a better replacement for heat sinks. Bahadur and Cohen[51] presented usage of composite materials for heat sink applications in their CFD analysis. With this work, in the existence of natural convection, fiber-filled polyphenylene sulfide showed a performance similar to the aluminum design.

Graphite-epoxy composites can be a better alternative to aluminum and copper when mass amount is 70% and 21% weight of their (aluminum and copper) weight, respectively. Norley et al.[52,53] offered a better heat sink with using a new graphite-based epoxy composite. This material has a thermal conductivity of 370 $W/m^2 \cdot K$ in-plane and 7 $W/m^2 \cdot K$ (Nysten and Issi[22]) through-thickness.

It is a big challenge to design heat sinks from anisotropically conductive material, even a bigger challenge to design for the heat sink base. Norley and Chen,[53] in their work, tried different orientations for the heat sink base with CFD analysis. Analyses showed that thermal conductivity is lowest when the lowest thermal conductivity of base material was in the through-thickness orientation.

Marotta et al.[54] designed a heat sink for usage of high performance servers. Design included graphic-base/epoxy fins and spacers bonded each other. Experimental results showed that these arrangements can achieve thermal performance comparable to the copper ones with only 21–25% of copper mass.

Yet sometimes smaller heat sink is needed, and for these situations, bonded graphite/epoxy heat sinks became inadequate. Shives et al.[55] studied on a hybrid design that bonded epoxy fins with copper base. Hybrid design was compared to pure aluminum and copper heat sinks (Figure 13.8) with CFD. Results showed that best performance is achieved by copper heat sink bonded to copper base with soldering. However, the thermal resistance of bonded graphite/epoxy was only 0.015 C/W, higher than pure copper heat sink and also graphite/epoxy specimen only weighted 60% of the copper one.

The idea of using polymers for heat transfer applications first occurred because of their chemical and corrosion resistance. But it was shown that these materials could also be used directly as tubular material for some applications. However, because of low thermal conductivity, polymer materials can be inadequate in some applications because of their high thermal resistance. To design efficient heat exchangers, it is more convenient to use PMCs instead of using polymer materials directly. PMC materials can be produced with advantages of both metal and polymer materials. It can have both high chemical and corrosion resistance and high thermal conductivity with lower amount of mass. PMCs mostly have similar thermal performance to copper ones and have advantages over metals, like their low weight, low cost, and chemical and corrosion resistances. Polymer and polymer-like materials (PMCs) have great potential for many applications.

13.5 Polymer Compact Heat Exchangers

As indicated in the previous sections, plastic can resist corrosion and fouling. Additionally, plastic compact heat exchangers can be manufactured and maintained more easily compared with the metal ones. The use of plastic heat exchangers as compact heat exchangers can also help lower such costs as transportation and installation, which consequently lowers the total investment.[56]

The excellent review of compact heat exchanger is given in Ref. 5. The types of high-performance polymers currently used in polymer compact heat exchangers are PVDF (polyvinylidene fluoride), Teflon or PTFE (polytetrafluoroethylene), PP (polypropylene), PE (polyethylene), PC (polycarbonate), PPS (polyphenylene sulfide), and (PPO) polyphenylene oxide. As discussed before, the use of polymers offers substantial weight, volume, space, and cost savings, which give them a competitive edge over exchangers manufactured from more exotic alloys.

To begin with, PVDF is a highly nonreactive and pure thermoplastic fluoropolymer. PVDF is a specialty plastic material in the fluoropolymer family; it is used generally in applications requiring the highest purity, strength, and resistance to solvents, acids, bases, and heat and low smoke generation during a fire event. Of the other fluoropolymers, it has an easier melt process because of its relatively low melting point of around 177°C.

PTFE is most well known by the DuPont brand name Teflon. PTFE is a fluorocarbon solid, as it is a high-molecular-weight compound consisting wholly of carbon and fluorine. Moreover, it is very nonreactive, partly because of the strength of carbon–fluorine bonds, and so it is often used in containers and pipework for reactive and corrosive chemicals. Where used as a lubricant, PTFE reduces friction, wear, and energy consumption of machinery. PP is a thermoplastic polymer. It is used in a wide variety of applications such as packaging, stationery, plastic parts, and reusable containers of various types, automotive components, and polymer banknotes. PP is rigidly constructed and is only prone to attack by oxidizing agents on the tertiary hydrogen. Compared to other hydrocarbon polymers, PP has the highest melting point at 165°C It is nontoxic, nonstaining and exhibits excellent corrosion resistance. It has a significant application in mechanical vapor compression (MVR) units.[5]

PE is a most useful and widely used plastic and it contains the chemical elements carbon and hydrogen. It is also used in applications whose maximum temperature does not exceed 340 K. PC has good chemical resistance to acids but poor resistance to alkalis and solvents. PC is recognized not only for its clarity, safety, security, and energy savings but also for the design freedom it provides to architects.[5]

PPS is the best performer in acidic conditions compared to Teflon and PVDF, and it is one of the most important high temperature polymers.

PPO is similar in chemical composition to PPE, and they are generally treated as equivalents. It can be said that PPO has good heat resistance but poor chemical resistance. Nevertheless, the strength, stability, and the acceptance of flame-retardants of PPO makes it desirable for machine and appliance housings.[5]

It is known that the use of polymers can cause problems during design and manufacture owing to their low strength, poor creep resistance, relatively poor thermal conductivity, and large thermal exchange surface expansion. Nevertheless, their impressive resistance to chemical attack at heat exchangers.[2] Moderate temperatures and pressures and also their lower relative cost can offset these factors. The latter is proven mathematically as follows: PVDF has a thermal conductivity of 0.17 W/m × K, which is 100–1000 times lower than that of steels and other metals. By using the same diameter, wall thickness, and fouling coefficient, a tube bundle of Ni-Cr-Mo alloy is compared with PVDF exchange surface. Considering all other conditions are equal, it is shown that the price difference is crucial enough to develop and use of the PVDF.[5]

In industry, three major categories of polymer compact heat exchangers are available. They are the polymer-gasketed plate heat exchangers, heat exchanger coils, and shell-and-tube heat exchangers. Because the plate heat exchanger is of low cost, flexibility, easy maintenance, and high thermal efficiency; it is used as the most economical and efficient type of heat exchanger on the market.

A new type of plate-type heat exchanger is designed by Dobbs[57] where the plates are constructed of ionomer membranes, such as sulfonated or carboxylated polymer membranes, which are capable of transferring a significant amount of moisture from one side of the membrane to the other side. Incorporating such ionomer membranes into a plate-type heat exchanger provides the heat exchanger with the ability to transfer a large percentage of the available latent heat in one stream to the other air streams. The ionomer membrane plates are, therefore, more efficient at transferring latent heat than plates constructed of metal or paper.

An additional plate heat exchanger worthy of interest is the Dutch LEVEL unit. This unit had a triangular flow path cross-section, not far removed from the polymer film compact heat exchanger (PFCHE) concept, but it operated in counterflow in one variant. A compactness of up to 1000 m^2/m^3 is claimed. The polymer unit could be used as a recuperator for dryers.[5,58]

A plate heat exchanger made from a mixture of polymers, PVDF and PP (Novelerg), is used intensively in heating and ventilating systems such as car radiators. It is a form of constructed prototype plate exchanger using 60 plates measuring 45 cm. The unit transfers 26.4 kW and operates successfully at 75°C with 6.5 bar differential pressure.[58]

Polymer plate heat exchangers are also used as air conditioners. Such plate heat exchangers are capable of reducing energy needs for air-conditioning, by using exhaust air to condition intake air without physical mixing. The use of plate heat exchangers allows considerable increases in ventilation rates (up to 100% fresh air) without significant increases in energy consumption and cost. A French company Favier Setrem has designed and produced plastic heat exchangers with pressures being able to exceed 60 bars.[5,59]

A new concept for a liquid–liquid plate heat exchanger has been studied by Deronzier and Bertolini[3] using thermoplastic polymers. Considering their advantages, polymers called LCP (liquid crystal polymers) have been developed in the laboratory from a base formulation of pure LCP (SPER1) to improved formulations (SPER12), using fillers such as silica powder, glass, and carbon fibers. A fully aromatic structure (SPER12FA) and a modified fully aromatic structure (SPER12IA) were also developed and synthesized. Corrosion and mechanical tests in the laboratory with a selected LCP (SPER12IA) have shown a relatively low chemical resistance and high mechanical resistance in comparison with commercial LCP Vectra.[5]

Heat exchanger coils are also one of the major categories of polymer compact heat exchangers. These exchangers are very suitable for large volume, price sensitive users who need a high quality, cost-effective part. DuPont has developed the first all plastics automotive intercooler. The unit is made entirely of plastics, using Caltrel and DuPont proprietary class of molding and extrusion materials to design fluid energy transfer systems. Caltrel offers the advantages of a component with less weight, greater corrosion resistance and durability, and the many added benefits of single-material construction. Most significantly, it possesses the design flexibility to be curved, bowed or bent to make the maximum use of the crowded space available under the hood. Typical heat exchangers are made of aluminum and copper and take the shape of a square box. By using the Caltrel exchanger, flexibility is obtained to change the shape and location of the heat exchanger. These plastic units can provide the same performance to conventional metal heat exchangers at the same or less cost.[5]

PVDF coils (Solvay) used as cooling and heating coils for aggressive fluids in tanks were built very easily. These coils resist particularly well to water-hammering.[4] Ametek has three styles of immersion heat exchangers.[30] One of them is Supercoils used for heating and cooling in essentially all plating and metal finishing solutions. Supercoils are multibraided, flexible coils that can be fitted into almost any tank configuration. Minicoils are generally suited to the electronics industry for etching, precious metal plating, and also in the deionized water rinse tanks used in the production of wafers, microcircuits, and connectors. The third type, the reactor coil, has either 2.54 or 3.12 mm OD tubes. Tubes are connected at PTFE headers. A reactor coil with 160, 2.54 mm OD, tubes has a total heat transfer surface area of 1.35 m^2.[5]

Additionally, the immersion heat exchangers by George Fischer are made of PVDF, PP, or PE tubes, connected to headers by an over-molding process.[30] A nylon car radiator, when compared to traditional metal radiators, conserves space and can be shaped in an engine compartment, which allows for greater design flexibility and significant weight savings. It can be said that these units is marketed for applications in not only automotive but also in marine engines, heating, ventilating, air conditioning units, and many industrial uses.[18]

Teflon shell-and-tube heat exchangers (DuPont) have been used for sulfuric acid cooling and another use of the Teflon shell-and-tube heat exchangers (Du Pont) was to help copper manufacturers reduce costs and improve sulfuric acid production, as experienced by Inspiration Copper Globe (Arizona) which had a water balance problem.[60]

George Fischer[61] has introduced the Calorplast compact shell-and-tube heat exchanger featuring corrosion resistant material in a compact, lightweight shell. The new heat exchanger is available in a variety of corrosion-resistant materials to meet a wide range of application requirements.[2] PP and PVDF shell-and-tube heat exchangers (Eta Process Plant Limited) were developed for external heat transfer between highly corrosive fluids. These units are particularly suitable for high-purity media such as deionized water.[62]

Morcos and Shafey[63] were taken into account the use of turbulence enhancement devices in a PVC shell-and-tube heat exchanger with water on both sides. They were able to increase overall heat transfer by a factor of 3.5 at Reynolds numbers near 1000. Softening of the material was noted at 100°C.[2]

The Process Intensification Group at Newcastle University has developed a novel concept of thin film polymer heat exchangers that provide high thermal efficiency, reduced fouling, significant reduction in weight and cost, resistance to chemically aggressive fluids, and the ability to handle both liquids and gases (single and two phase duties). Thanks to this PCHEs, a competitive edge to their metal. Although the thermal conductivity of the polymer is not as high as metals, the PFCHE made of 100 mm thick PEEK films offers negligible thermal resistance when the heat transfer coefficient is less than 4000 $W/m^2 \times K$. Some researchers are clearly show that the PFCHE has a much higher heat transfer capacity than conventional metal units and highlight the advantage of the extreme compactness of the PFCHE.

In order to further enhance the thermal performance of existing polymer compact heat exchangers, thin polymer films are adopted in a new design as opposed to the more conventional shell and tube, plate, or coil configurations.[5] In current designs, the wall thickness can only be decreased to a range of 0.5–1 mm without affecting the mechanical strength of the polymers. However, the use of polymer films of approximately 100 µm thick is possible, and this will subsequently promote a better thermal performance. In order to minimize the thermal resistance offered by polymeric materials, a PFCHE using 100 µm thick polymer films has been developed by PIIC (Process Intensification and Innovation Centre), at University of Newcastle Upon Tyne.[24] The presence of corrugations on the films aids toward heat transfer enhancement, as it encourages more mixing of the fluid flow. Thin films or sheets are chosen in preference to tubes, coils, or plates in order to deal with the low thermal conductivity in polymer compact heat exchangers.

The use of thin films may enhance the thermal performance of heat exchangers. Furthermore, the use of thin films and corrugations on the films also promote better thermal performance of the heat exchanger. The corrugated films improve the thermal performance as they break down the boundary layers of the fluid, also known as the massaging effect.[5]

13.6 Potential Applications for Polymer Film Compact Heat Exchangers

In order to increase the thermal performance of polymer compact heat exchangers, thin films are used in new designs instead of conventional shell-and-tube, plate and coil configurations. The use of polymer films of approximately 100 µm thickness is possible, which provides a better thermal performance. In addition to the use of thin films, corrugations on the films also contribute to the better thermal performance of the compact heat exchangers.

PFCHE have variety of applications; they are used in the desalination industry, for heat recovery, as evaporators, absorption chillers, in the automotive, cryogenic, computer industries, and air-conditioning and refrigeration systems, and in many others as PFCHEs. More detailed information will be given below.

There are various types of film compact heat exchangers available. First, in the desalination industry a plastic unit performing heat exchange by evaporation and condensation has been adopted.[64]

Plastic film plate heat exchangers, second of them, for heat recovery are relatively expensive and complex designs. Therefore, plastic materials are preferred to reduce the costs. Cross-corrugated polymer film exchanger as an evaporator is a gas and liquid apparatus.

Film heat exchanger as a thermocompressor evaporator, this invention is particularly suitable for use in a film evaporator or distillation apparatus, operating in the thermocompressor principle.[65]

Thin plastic film heat exchangers for absorption chillers is formed by two layers of thin plastic film material, in which flow passages are formed, by placement of seams or welds between the two layers.[66] The two layers are bonded or sealed together by thermal or ultrasonic welds, to produce a serpentine passage through the heat exchanger. Such heat exchangers would be situated in a working environment; at less than atmospheric pressure in a vacuum vessel, and during operation, these units are exposed to a continuous partial vacuum.[5]

A panel polymer heat exchanger was invented by Shuster and Cesaroni.[65] The heat exchangers are relatively economical to manufacture and may be used in a variety of end users, depending on the properties of the polymer composition, including in some instances as heat exchangers in automobiles. This invention relates to heat exchangers, particularly liquid-to-gas heat exchangers for use in vehicles. The unit is essentially a planar panel having a pair of relatively thin outer walls formed from a composition of a thermoplastic polymer, especially a polyamide. The walls, which have a thickness of less than 0.7 mm, are bonded together to form a labyrinth of fluid passages between the walls. The passages extend between the inlet and outlet entrances and occupy a substantial proportion of the area of the panel.[5]

Another variation of the panel polymer heat exchanger is formed from two sheets of thermoplastic polymer, again preferably polyamide.[4] The units may be used in a variety of end uses, including automotive oil coolers, automotive comfort heaters, refrigerators, and industrial uses. Each sheet has a thickness of 0.07–0.7 mm, and a plurality of grooves extending at least partially across the width of the sheet. The sheets are superimposed in a face-to-face relationship and bonded together in a fluid tight bond. The headers are situated at opposite ends of the panel, where the grooves form a plurality of fluid flow channels between the inlet and outlet headers. Plastic heat exchangers may be used in the automotive industry. The multipaneled heat exchanger is in the form of a radiator for an automobile.[20] These heat exchangers may be used in automotive end uses, for example, as part of the water and oil cooling systems.

A plastic solar collector that has been developed as an alternative to metal units is the advanced corrugated duct solar collector. It is known that the collector (air heater) is constructed of corrugated surfaces similar to those used for compact heat exchangers, with the air flowing normal to the corrugations.[67] Another case in point is the polymer solar energy absorber module. So, a solar energy absorber module having a molded top sheet of translucent plastic material and a molded bottom sheet of solar absorbing plastic material, bonded together by circular bond indentations in the body has been developed by Smith.[66]

An investigation on the viability of polymers at cryogenic temperatures represents that certain polymers do exhibit high brittle impact failure at low temperatures. For instance, expanded polystyrene can be used with liquid nitrogen. PEEK, on the other hand, becomes brittle at temperatures below 10°C.[10] A low thermal conductivity support has been described. It is rigid in all three directions and constructed from a taut band of Kapton, a strong plastic film.[68] The experimental performance of the first superfluid Stirling refrigerator (SSR), to use a plastic recuperator is reported by Patel and Brisson.[69]

The thin film polymer heater specified is a portable heat generating device in which fuel vapor and an oxygen supply such as air are directed through channels contained within a thin, flexible, and compliant polymer sheet.[70] The heat generation process starts, by pumping an air stream into a reservoir, which contains a fuel source such as methanol. This action saturates the air stream with fuel vapor. The fuel vapor is mixed with another stream of air to achieve a particular fuel/air ratio and directed into channels within the polymer sheet. Here, it reacts with the catalytic heat elements to produce flameless combustion. The warm exhaust gas is directed to a thermally controlled diverter valve, which senses the temperature of the liquid fuel supply and diverts some or all of the warm exhaust gas, as necessary, to heat the fuel and keep its temperature within a specified range. Exhaust by-products are passed into a miniature scrubber module adjacent to the fuel module. The scrubber absorbs any noxious components in the exhaust stream that may occur during start-up or rapid changes in the operating conditions.[5]

Plastic fin heat exchangers are used in the computer industry. In a portable notebook computer, a specially designed heat pipe-based heat exchanger assembly has been used to efficiently dissipate heat from a processor board in the computer's base housing to ambient air surrounding the computer.[71]

An improved air-to-air, heat and moisture film exchanger is designed to facilitate the simultaneous transfer of both heat and water vapor between two streams of air, while greatly inhibiting the transfer of pollutant gases. A rigid encasement supports a stack of flat thin film composite membranes that are held in a parallel spaced apart relation, thereby forming a box shaped exchanger.[72]

The PFCHE can be adopted as a chemical reactor as an attempt to diversify its functions. This idea is supported by the development of the polymer unit below. A stacked plate chemical reactor in which simple plates, each incorporating no surface features other that an opening has been developed by Georg et al.[73] These plates are stacked together and when openings in adjacent plates are properly aligned, a fluid pathway is defined between inlet ports for each chemical reactant and an outlet port for a chemical product.

The thin durable fluoropolymer film is useful as a corrosion inhibiting agent, as a heat transfer promoter and an adhesion barrier. A separation membrane formed of a polymer film, is incorporated in the dehumidifier and arranged to separate vapor from the primary air without condensation.[74] The polymer film is moisture permeable and has a hydrophobic porous nature. It contains a large number of pores of substantially the same size as free path molecules, to enable the vapor to permeate by capillary condensation and diffusion of water molecules.

Polymer films in power electronic module packaging are invented by Ozmat relates to electronic packaging technology applicable to power modules, such as ac–ac, ac–dc, and dc–dc converters and also electronic switches. Disadvantages of presently available designs include high volume, weight, and thermal resistance. Accordingly, a method to overcome such disadvantages has been developed by fabricating power circuit modules and associated connections that are compatible with the standard top layer metallization of available power devices.[5]

Heat exchangers can be made from corrosion resistant and thermally conductive compositions containing resins (particularly linear crystal polymers) and carbon (particularly impervious graphite) that are made through injection molding and like processes. The composites can be anisotropic; that is, they exhibit enhanced thermal conductivity in specific directions through the material. The resulting articles made from this composition are able to conduct heat at different rates, into, out of, and through the articles depending on the design imposed. Incorporating this anisotropic composition in the construction of heat exchangers can provide increased heat transfer between two mediums.[2] This invention relates to a system for conditioning a gaseous supply stream. The system includes a desiccant free heat and moisture exchange wheel that efficiently transfers both heat and moisture between a warmer, saturated gaseous exhaust stream, and a gaseous supply stream such that the supply stream becomes heated and substantially saturated.[75] The exchange wheel collects heat from the exhaust stream through condensation and releases heat into the supply stream through conduction. Subsequently, it also collects moisture from the exhaust stream through condensation and releases the moisture into the supply stream through evaporation.

The desiccant free heat and moisture exchange wheel comprises of a high-molecular weight synthetic polymer film in the form of strips spirally wounded around a hub. The strips consist of two layers, with at least one layer having projections providing passages between the layers of the strip. The heat exchanger matrix of the desiccant free wheel is provided with a predetermined mass and rotation speed, which allows transfer of both sensible heat and a desired amount of latent heat between the exhaust and supply stream. The units can produce large moisture transfers without employing desiccants conventionally used, hence providing the benefits of being easier and less expansive to manufacture.

All things considered, PFCHEs, hold great potential as an alternative for a wide range of industrial applications currently monopolized by metallic heat exchangers. Their energy saving benefits and resistance to fouling and corrosion are likely to be at the forefront of a new technology that will uncover new market opportunities in the various sectors of the chemical industry.[5]

The use of plastic materials and plate and frame heat exchangers are used for desalination processes. El-Dessouky and Ettouney[41] studied the performance evaluation of single-effect mechanical vapor compression desalination industry with metal (high steel alloys, Cu/Ni alloys, titanium) or plastics (PTFE). Although the analysis gives lowest heat transfer area for Cu/Ni (90/10), the lowest cost is obtained for the PTFE thin walled plastic evaporator and preheaters in the system. It should be noted that power consumption is independent of the construction material.

13.7 Thermal Design of Polymer Heat Exchangers

The most common problems in polymer heat exchangers as in metallic heat exchangers are rating and sizing. The basic design methods are given in

Polymer Heat Exchangers

Chapter 2, together with rating and design problems to meet the requirements of specified hot- and cold-fluid inlet and outlet temperatures, flow rates, and pressure drop for a given thermal process requirements. This flow chart of the heat exchanger design methodology is given in Section 2.9 (Figure 2.17), which should also be followed in the design of polymer heat exchangers. Various student design projects are given at the end of the previous chapters, which can be assigned as polymer heat exchangers design.

Thermal analysis of the compact heat exchanger is the same as the metallic heat exchanger as outlined in the previous chapters. Correlations given for the calculation of the heat transfer coefficients under single-phase laminar and turbulent flow, condensation and boiling are valid for polymer heat exchangers including friction coefficients and pressure drop calculations, which are given in Chapters 3, 4, 8, and 12. Selected problems can be assigned for polymer materials.

PROBLEMS

13.1. Water, $c_p = 4182$ J/kg · K, at a flow rate of 2 kg/hr is heated from 7°C to 25°C in a double-pipe PPS oil cooler by engine oil, $c_p = 2072$ J/kg · K, with an inlet temperature of 65°C and a flow rate of 10 kg/hr. The tubes have an outside diameter of 1 cm and the wall thickness is 0.2 cm. Take the thermal conductivity to be 0.3 W/m² · K and the heat transfer coefficient of water to be 4911 W/m · K, what are the areas required for counterflow.

13.2. Refrigerant-22 (R-22) boiling at 250 K flows through a tubular heat exchanger. Tubes are arranged horizontally with an inner diameter of 0.0172 m. The mass velocity of the R-22 is 200 kg/m². s. Heat is supplied by condensing the fluid outside at –12°C. The hot fluid-side heat transfer coefficient, h_h, is 5400 W/m² · K. The tube is made of PVDF polymer, and fouling is neglected. By the use of both Chen and Shah's methods, calculate the convection heat transfer coefficient for R-22 for various values of quality assuming no nucleate boiling and the heat transfer rate per length.

13.3. A heat exchanger is to be designed to heat raw water by the use of condensed water at 67°C and 0.2 bar, which will flow in the shell side with a mass flow rate of 50,000 kg/hr. The heat will be transferred to 30,000 kg/hr of city water coming from a supply at 17°C ($c_p = 4184$ J/kg · K). A single shell and a single tube pass without baffles are preferable. A fouling resistance of 0.000176 m² · K/W is suggested and the surface over design should not be over 35%. A maximum coolant velocity of 1.5 m/s is suggested to prevent erosion. A maximum tube length of 5 m is required because of space limitations. The tube material is PPSU polymer. Raw water will flow inside of 3/4 in. straight tubes (19 mm

OD with 16 mm ID). Tubes are laid out on a square pitch with a pitch ratio of 1.25. The permissible maximum pressure drop on the shell side is 5.0 psi. The water outlet temperature should not be less than 40°C. Perform the preliminary analysis whether the surface over design acceptable.

13.4. Air at 2 atm and 500 K with a velocity of $u_\infty = 20$ m/s flows across a PVDF compact heat exchanger matrix studied by Cheng and Van Der Geld[44] having the configuration shown in Figure 13.5. By assuming that the PVDF heat exchange surface is similar to the surface 9.68–0.870 and using the length of the matrix as 48 plates with 4 mm per plate, calculate the heat transfer coefficient and the frictional pressure drop.

13.5. Air enters the core of a finned tube PTFE heat exchanger of the type shown in Figure 10.4 at 2 atm and 150°C. The air mass flow rate is 10 kg/s and flows perpendicular to the tubes. The core is 0.5 m long with a 0.30 m² frontal area. The height of the core is 0.5 m. Water at 15°C and at a flow rate of 50 kg/s flows inside the tubes. Air-side data are given on Figure 10.4. For water-side data, assume that $\sigma_w = 0.129$, $D_h = 0.373$ cm, and water-side heat transfer area/total volume = 138 m²/m³. When consider this problem as a rating problem, calculate:

a. The airside and waterside heat transfer coefficients
b. Overall heat transfer coefficient based on the outer (airside) surface area
c. Total heat transfer rate and outlet temperatures of air and water.

DESIGN PROJECT 13.1
Design a Heat Exchanger for an Open Heart Operation—Polymer Blood Heat Exchanger

The objective of a blood heat exchanger here is to use polymer materials. The material in contact needs to have the smoothest possible surface; therefore, a PP gasketed-plate heat exchanger may be chosen for the design because the polymer does not contain toxic chemicals. For a sufficient design, the pressure drop must be minimal due to the fragile nature of blood, and the heat transfer rate must be high due to the required rapid temperature change. This design project will take into account thermal, dynamic heat transfer, and a parametric analysis for a blood heat exchanger. Assume that the patient's blood properties are the same as design Project 11.3 but using the PP properties to calculate the overall heat transfer coefficient. One can directly select a gasketed-plate heat exchanger, or a comparative study as in design Project 11.3.

Polymer Heat Exchangers 573

DESIGN PROJECT 13.2

The shell-and-tube type oil cooler for marine applications given in Design Project 9.1 can be repeated for a polymer heat exchanger by selecting a polymer material from Table 13.1. for a lubricating oil mass flow rate of 0.01 and 0.005 kg/s while keeping the other specifications constant.

References

1. T'Joen, C., Park, Y., Wang, Q., Sommers, A., Han, X., and Jacobi, A., A review on polymer heat exchangers for HVAC&R applications, *International Journal of Refrigeration*, 32, 763–779, 2009.
2. Reay, D., The use of polymers in heat exchangers, *Heat Recovery Systems & CHP 9*, 209–216, 1989.
3. Deronzier, J. C., and Bertolini, G. Plate heat exchanger in liquid crystal polymer, *Applied Thermal Engineering*, 17, 799–808, 1997.
4. Wharry, S. R., Jr., Fluoropolymer heat exchangers, *Metal Finishing*, 100, 752–762, 2002.
5. Zaheed, L. and Jachuck, R. J. J., Review of polymer compact heat exchangers, with special emphasis on a polymer film unit, *Applied Thermal Engineering*, 24, 2323–2358, 2004.
6. Zweben, C., Emerging high-volume applications for advanced thermally conductive materials, *Proceedings of the International SAMPE Symposium and Exhibition*, 49, 4046–4057, 2004a.
7. Ma, X., Chen J., Xu, D., Lin, J., Ren, C., and Long, Z., Influence of processing conditions of polymer film on dropwise condensation heat transfer, *International Journal of Heat Mass Transfer*, 45, 3405–3411, 2002.
8. Jacobi, A. M., Park, Y., Zhong, Y, Michna, G., and Xia, Y., High performance heat exchangers for air conditioning and refrigeration applications (non-circular tubes). *ARTI–CR21/605–20021–01*, Final Report, July 2005.
9. Gurdal, Z., Haftka, R. T., and Hajela, P., *Design and optimization of laminated composite materials*, Wiley, New York, 1999.
10. Bigg, D., Thermal conductivity of heterophase polymer compositions, *Advances in Polymer Science*, 119, 1–30, 1995.
11. Ahmed, S. and Jones, F., A review of particulate reinforcement theories for polymer composites, *Journal of Material Science*, 25, 4922–4942, 1990.
12. Bigg, D., Thermally conductive polymer compositions, *Polymer Composites*, 7, 125–140, 1986.
13. Nielsen, L.E., *Mechanical Properties of Polymers and Composites*, Vol. 2, Marcel Dekker, New York, 1974.
14. Mamunya, Y., Davydenko, V., Pissis, P., and Lebedev, E., Electrical and thermal conductivity of polymers filled with metal powders, *European Polymer Journal*, 38, 1887–1897, 2002.
15. Boudenne, A., Ibos, L., Fois, M., Gehin, E., and Majeste, J. C., Thermophysical properties of polypropylene-aluminum composites, *Journal of Polymer Science part B*, 42, 722–732, 2004.

16. Serkan Tekce, H., Kumlutas, D., and Tavman, I., Effect of particle shape on thermal conductivity of copper reinforced polymer composites, *Journal of Reinforced Plastics and Composites*, 26, 113–121, 2007.
17. Krupa, I., Novak, I., and Chodak, I., Electrically and thermally conductive polyethylene/graphite composites and their mechanical properties, *Synthetic Metals*, 145, 245–252, 2004.
18. Krupa, I. and Chodak, I., Physical properties of thermoplastic/graphite composites, *European Polymer Journal*, 37, 2159–2168, 2001.
19. Zweben, C., Thermal materials solve power electronics challenges, *Power Electronics Technology*, 2006.
20. Zweben, C. H., New material options for high-power diode laser packaging, *Proceedings of the SPIE, The International Society for Optical Engineering*, 5336, 166–175, 2004.
21. Tzeng, J. W., Getz, G., Fedor, B. S., and Krassowski, D. W., *Anisotropic Graphite Heat Spreaders for Electronics Thermal Management*, Graftech, 2000.
22. Nysten, B. and Issi, J. P., Composites based on thermally hyperconductive carbon fibres, *Composites*, 21, 339–343, 1990.
23. Norley, J., Tzeng, J. W., Getz, G., Klug, J., and Fedor, B., The development of a natural graphite heat-spreader, *17th Annual IEEE Semiconductor Thermal Measurement and Management Symposium*, 107–110, 2001.
24. Zaheed, L., Cross-corrugated polymer film compact heat exchanger (PFCHE), Ph.D. Thesis, School of Chemical and Advanced Materials, University of Newcastle Upon Tyne, United Kingdom, 2003.
25. Jordan, J., Jacob, K., Tannenbaum, R., Sharaf, M., and Jasiuk, I., Experimental trends in nanocomposites—A review, *Material Science and Engineering A*, 393, 1–11, 2005.
26. LeBaron, P., Wang, Z., and Pinnavaia, T., Polymer-layered silicate nanocomposites—An overview, *Applied Clay Science*, 15, 11–29, 1999.
27. Ahmadi, S., Huang, Y., and Li, W., Synthetic routes, properties and future applications of polymer-layered silicate nanocomposites, *Journal of Material Science*, 39, 1919–1925, 2004.
28. Gao, F., Clay polymer composites: the story, *Materials Today*, 7(11), 50–55, 2004.
29. Morcos, V. and Shafey, H., Performance analysis of a plastic shell and tube heat exchanger, *Journal of Elastomers and Plastics*, 27, 200–213, 1995.
30. Davidson, J., Mantell, S., Liu, W., and Raman, R., Use of polymers in liquid to liquid heat exchangers: applied to low cost solar water heating systems, Phase I, Task III report. University of Minnesota, Minnesota, 1998.
31. Freeman, A., Mantell, S., and Davidson, J., Mechanical performance of polysulfone, polybutylene and polyamide 6/6 in hot chlorinated water, *Journal of Solar Energy*, 79, 624–637, 2005.
32. Wu, C., Mantell, S., and Davidson, J., Polymers for solar domestic hot water: long term performance of PB and Nylon 6.6 Tubing in hot water, *Journal of Solar Energy Engineering*, 126, 581–586, 2004.
33. Wang, Y., Davidson, J., and Francis, L., Scaling in polymer tubes and interpretation for use in solar water heating systems, *Journal of Solar Energy Engineering*, 127, 3–14, 2005.
34. Liu, W., Davidson, J., and Mantell, S., Thermal analysis of polymer heat exchangers for solar water heating: a case study, *Journal of Solar Energy Engineering*, 122, 84–91, 2000.

35. Zakardas, D., Li, B., and Sirkar, K., Polymeric hollow fiber heat exchanger (PHFHEs): a new type of compact heat exchanger for low temperature applications, *Proceedings of the 2005 ASME Heat Transfer Conference*, HT2005-72590, 2005.
36. Patel, A. B. and Brisson, J. G., Design, construction and performance of plastic heat exchangers for sub-Kelvin use, *Cryogenics*, 40, 91–98, 2000.
37. Leigh, R. W., Isaacs, H. S., Kirley, J., and Rawe, A., Capital cost reductions in absorption chillers, *ASHRAE Transactions*, 95, 939–952, 1989.
38. Lowenstein, A. and Sibilia, M., A low-cost thin-film absorber/ evaporator for an absorption chiller, Final Report, Gas Research Inst., Illinois, May 1992- April 1993.
39. Lowenstein, A. I. and Sibilia, M. J., Thin plastic-film heat exchanger for absorption chillers, *US Patent 5992508*, 1999.
40. Bigg, D., Stickford, G., and Talbert, S., Applications of polymeric materials for condensing heat exchangers, *Polymeric Engineering and Science*, 29, 1111–1116, 1989.
41. El-Dessouky, H. and Ettouney, H. Plastic compact heat exchangers for single effect desalination systems, *Desalination*, 122, 271–289, 1999.
42. Bourouni, K., Martin, R., Tadrist, L., and Tadrist, H., Experimental investigation of evaporation performance of a desalination prototype using aero-evapo-condensation process, *Desalination*, 114, 111–128, 1997.
43. Bums, J. and Jachuck, R., Condensation studies using cross-corrugated polymer film compact heat exchangers, *Applied Thermal Engineering*, 21, 495–510, 2001.
44. Cheng, L. and Van Der Geld, C., Experimental study of heat transfer and pressure drop characteristics of air/water and air-steam/water heat exchange in a polymer compact heat exchanger, *Heat Transfer Engineering*, 26, 18–27, 2005.
45. Van Der Geld, C., Ganzevles, F., Simons, C., and Weitz, F., Geometry adaptations to improve the performance of compact polymer heat exchangers, *Transactions IChemE: Part A*, 79, 357–362, 2001.
46. Harris, C., Kelly, K., Wang, T., McCandless, A., and Motakef, S., Fabrication, modeling and testing of micro-cross flow heat exchangers, *Journal of Microelectromechanical Systems*, 11, 726–735, 2002.
47. Rousse, D., Martin, D., Theriault, R., Leveillee, F., and Boily, R., Heat recovery in greenhouses: a practical solution, *Applied Thermal Engineering*, 20, 687–706, 2000.
48. Jia, J., Peng, X., Sun, J., and Chen, T., An experimental study on vapor condensation of wet flue gas in a plastic heat exchanger, *Heat Transfer - Asian Research*, 30, 571–580, 2001.
49. Saman, W. and Alizadeh, S., Modeling and performance analysis of a cross-flow type plate heat exchanger for dehumidification/cooling, *Solar Energy*, 70(4), 361–372, 2001.
50. Saman, W. and Alizadeh, S., An experimental study of a cross flow type plate heat exchanger for dehumidification/cooling, *Official Journal of AIRAH*, 1(8), 26–30, 2002.
51. Bahadur, R. and Cohen, A., Thermal design and optimization of staggered polymer pin fin natural convection heat sinks, *The Ninth Intersociety Conference on Thermal and Thermomechanical Phenomena in Electronic Systems*, 1, 268–275, 2004.
52. Norley, J., Tzeng, J. W., and Klug, J., Graphite-based heat sink, *US Patent 6503626*, 2003.
53. Norley, J. and Chen, G.G., High performance, lightweight graphite heat sinks/ spreaders, *Proceedings of PCIM 2002*, Nürnberg, 2002.

54. Marotta, E. E., Ellsworth Jr., M. J., Norley, J., and Getz, G., The development of a bonded fin graphite/epoxy heat sink for high performance, *Advances in Electronic Packaging*, 2, 139–146, 2003.
55. Shives, G., Norley, J., Smalc, M., Chen, G., and Capp, J., Comparative thermal performance evaluation of graphite/ epoxy fin heat sinks, *Thermomechanical Phenomena in Electronic Systems—Proceedings of the Intersociety Conference*, 1, 410–417, 2004.
56. Chen, L., Li, Z., and Guo, Y., Experimental investigation of plastic finned- tube heat exchangers, with emphasis on material thermal conductivity, *Experimental Thermal Fluid and Science*, 33(5), 922–928, 2009.
57. Dobbs, G. M., Plate type heat exchanger, *US Patent Application No. 20020185266*, 2002.
58. Reay, D. A., *Polymer Heat Exchangers: Exploitation of the Technology—A Discussion of Some Options*, David Reay & Associates, Newcastle, 2000.
59. Sterem, F., Polymer heat exchangers, available from <*http://www.faviersterem.com*>, 19 November 2003.
60. Ametek, AMETEK Brochure, Heat exchangers of Teflon, Case History, US, 1990.
61. Fischer, G., Compact all plastic heat exchanger, available from <*http://www.globalspec.com*> 19 November 2003.
62. Ferreira, C. A. I., Compact Air-Refrigerant Evaporator, 2000, available from <*http://www-pe.wbmt.tudelft.nl/kk/comair.htm*> 11 October 1999.
63. Morcos, V. H. and Shafey, H. M., Performance analysis of a plastic shell and tube heat exchanger, *Journal of Elastomers and Plastics*, 27(2), 200–213, 1995.
64. Perry, C. R., Dietz, L. H. S., and Shannon, R. L., Heat exchange apparatus having thin film flexible Sheets, The Boeing Company, *US Patent No. 4411310*, 1983.
65. Shuster, J. and Cesaroni, A. J., Heat exchanger fabricated from polymer compositions, *US Patent No. 4955435*, 1990.
66. Metwally, M. N., AbouZiyan, H. Z., and ElLeathy, A. M., Performance of advanced corrugated duct solar air collector compared with five conventional designs, *Renewable Energy*, 10(4), 519–537, 1997.
67. Smith, W. F., Solar energy absorber, *US Patent No. 4763641*, 1988.
68. Gush, H. P., A compact low thermal conductivity support for cryogenic use, *Review of Scientific Instruments*, 62(4), 1106–1108, 1991.
69. Patel, A. B. and Brisson, J. G., Experimental performance of a single stage superfluid Stirling refrigerator using a small plastic recuperator, *Journal of Low Temperature Physics*, 111(1–2), 210–212, 1998.
70. Welles, C. G., Portable heat generating device, *US Patent No. 6062210*, 2000.
71. Moore, D. A., Molded heat exchanger structure for portable computer, Compaq Computer Corporation, *US Patent No. 6026888*, 2000.
72. Martin, G. L., Johnson, J. E., and Sparrow, E. E. M., Air-to-air heat and moisture heat exchanger, *US Patent No. 6145588*, 2000.
73. Georg, P., Dittman, B., Golbis, K., Stastna, J., Hohmann, M., Oberbeck, S., and Schwalbe, T., Miniaturised reaction apparatus, Cellular Process Chemistry, *European Patent No. EP1123734*, 2001.
74. Piao, C. C., Sakamoto, R., Watanabe, Y., Yoshimi, M., and Yanemoto, K., Air-conditioner, *US Patent No. 6484525*, 2002.
75. Haogland, L. C., Desiccant-free heat and moisture exchange wheel, *US Patent No. 6565999*, 2003.

Appendix A: Physical Properties of Metals and Nonmetals

Nomenclature

b	normal boiling point
c	critical point
c_p	specific heat at constant pressure, kJ/(kg · K)
c_v	specific heat at constant volume, kJ/(kg · K)
c_{pf}	specific heat at constant pressure of the saturated liquid, kJ/(kg · K)
c_{pg}	specific heat at constant pressure of the saturated vapor, kJ/(kg · K)
h	specific enthalpy, kJ/(kg · K)
h_f	specific enthalpy of the saturated liquid, kJ/(kg · K)
h_g	specific enthalpy of the saturated vapor, kJ/(kg · K)
k	thermal conductivity, W/(m · K)
k_f	thermal conductivity of the saturated liquid, W/(m · K)
k_g	thermal conductivity of the saturated vapor, W/(m · K)
M	molecular weight, kg/k mol
P	pressure, bar
P_f	saturation vapor pressure, liquid, bar
P_g	saturation vapor pressure, gas, bar
Pr	Prandtl number, $c_p \mu / k$
Pr_f	Prandtl number of the saturated liquid, $(c_p \mu)_f / k_f$
Pr_g	Prandtl number of the saturated vapor, $(c_p \mu)_g / k_g$
s	specific entropy, kJ/(kg · K)
s_f	specific entropy of the saturated liquid, kJ/(kg · K)
s_g	specific entropy of the saturated vapor, kJ/(kg · K)
T	temperature, °C, K
T_b	normal boiling point temperature, °C, K
T_m	normal melting point temperature, °C, K
v	specific volume, m³/kg
v_f	specific volume of the saturated liquid, m³/kg
v_g	specific volume of the saturated vapor, m³/kg
v_s	velocity of sound, m/s
Z	compressibility factor

Greek Symbols

α	thermal expansion coefficient
β	isothermal compressibility coefficient
γ	ratio of specific heat, c_p/c_v
μ	viscosity, Pa · s
μf	viscosity of the saturated liquid, Pa · s
μg	viscosity of the saturated vapor, Pa · s
σ	surface tension, N/m
ρ	density, Kg/m^3

TABLE A.1
Thermophysical Properties of Metals

Metal	Temperature Range T (°C)	Density ρ (g/cm³)	Specific Heat c (kJ/(kg·K))	Thermal Conductivity k (W/m·K)	Emissivity ε
Aluminum	0–400	2.72	0.895	202–250	0.04–0.06 (Polished) 0.07–0.09 (Commercial) 0.2–0.3 (Oxidized)
Brass (75% Cu, 30% Zn)	100–300	8.52	0.38	104–147	0.03–0.07 (Polished) 0.2–0.25 (Commercial) 0.45–0.55 (Oxidized)
Bronze (75% Cu, 25% Sn)	0–100	8.67	0.34	26	0.03–0.07 (Polished) 0.4–0.5 (Oxidized)
Constantan (60% Cu, 40% Ni)	0–100	8.92	0.42	22–26	0.03–0.06 (Polished) 0.2–0.4 (Oxidized)
Copper	0–600	8.95	0.38	385–350	0.02–0.04 (Polished) 0.1–0.2 (Commercial)
Iron (C ≈ 0.4%, cast)	0–1000	7.26	0.42	52–35	0.2–0.25 (Polished) 0.55–0.65 (Oxidized) 0.6–0.8 (Rusted)
Iron (C ≈ 0.5%, wrought)	0–1000	7.85	0.46	59–35	0.3–0.35 (Polished) 0.9–0.95 (Oxidized)
Lead	0–300	11.37	0.13	35–30	0.05–0.08 (Polished) 0.3–0.6 (Oxidized)
Magnesium	0–300	1.75	1.01	171–157	0.07–0.13 (Polished)
Mercury	0–300	13.4	0.125	8–10	0.1–0.12
Molybdenum	0–1000	10.22	0.251	125–99	0.06–0.10 (Polished)
Nickel	0–400	8.9	0.45	93–59	0.05–0.07 (Polished) 0.35–0.49 (Oxidized)

(Continued)

TABLE A.1 (CONTINUED)
Thermophysical Properties of Metals

Metal	Temperature Range T (°C)	Density ρ (g/cm³)	Specific Heat c (kJ/(kg·K))	Thermal Conductivity k (W/m·K)	Emissivity ε
Platinum	0–1000	21.4	0.24	70–75	0.05–0.03 (Polished)
					0.07–0.11 (Oxidized)
Silver	0–400	10.52	0.23	410–360	0.01–0.03 (Polished)
					0.02–0.04 (Oxidized)
Steel (C. 1%)	0–1000	7.80	0.47	43–28	0.07–0.17 (Polished)
Steel (Cr. 1%)	0–1000	7.86	0.46	62–33	0.07–0.17 (Polished)
Steel (Cr 18%, Ni 8%)	0–1000	7.81	0.46	16–26	0.07–0.17 (Polished)
Tin	0–200	7.3	0.23	65–57	0.04–0.06 (Polished)
Tungsten	0–1000	19.35	0.13	166–76	0.04–0.08 (Polished)
					0.1–0.2 (Filament)
Zinc	0–400	7.14	0.38	112–93	0.02–0.03 (Polished)
					0.10–0.11 (Oxidized)
					0.2–0.3 (Galvanized)

Source: From Kakaç, S. and Yener, Y., *Convective Heat Transfer*, 2nd ed., CRC Press, Boca Raton, FL, 1995. With permission.

TABLE A.2
Thermophysical Properties of Nonmetals

Metal	Temperature Range T (°C)	Density ρ (g/cm^3)	Specific Heat c (kJ/(kg·K))	Thermal Conductivity k (W/m·K)	Emissivity ε
Asbestos	100–1000	0.47–0.57	0.816	0.15–0.22	0.93–0.97
Brick, rough red	100–1000	1.76	0.84	0.38–0.43	0.90–0.95
Clay	0–200	1.46	0.88	1.3	0.91
Concrete	0–200	2.1	0.88	0.81–1.4	0.94
Glass, window	0–600	2.2	0.84	0.78	0.94–0.66
Glass wool	23	0.024	0.7	0.038	
Ice	0	0.91	1.9	2.2	0.97–0.99
Limestone	100–400	2.5	0.92	1.3	0.95–0.80
Marble	0–100	2.60	0.79	2.07–2.94	0.93–0.95
Plasterboard	0–100	1.25	0.84	0.43	0.92
Rubber (hard)	0–100	1.2	1.42	0.15	0.94
Sandstone	0–300	2.24	0.71	1.83	0.83–0.9
Wood (oak)	0–100	0.6–0.8	2.4	0.17–0.21	0.90

Source: From Kakaç, S. and Yener, Y., *Convective Heat Transfer*, 2nd ed., CRC Press, Boca Raton, FL, 1995. With permission.

Appendix B: Physical Properties of Air, Water, Liquid Metals, and Refrigerants

TABLE B.1
Properties of Dry Air at Atmospheric Pressure

Temperature (°C)	ρ (kg/m³)	c_p (kJ/ kg·K)	k (W/m·K)	$\beta \times 10^3$ (1/K)	$\mu \times 10^5$ (kg/m·s)	$\nu \times 10^6$ (m²/s)	$\alpha \times 10^6$ (m²/s)	Pr
−150	2.793	1.026	0.0120	8.21	0.870	3.11	4.19	0.74
−100	1.980	1.009	0.0165	5.82	1.18	5.96	8.28	0.72
−50	1.534	1.005	0.0206	4.51	1.47	9.55	13.4	0.715
0	1.2930	1.005	0.0242	3.67	1.72	13.30	18.7	0.711
20	1.2045	1.005	0.0257	3.43	1.82	15.11	21.4	0.713
40	1.1267	1.009	0.0271	3.20	1.91	16.97	23.9	0.711
60	1.0595	1.009	0.0285	3.00	2.00	18.90	26.7	0.709
80	0.9908	1.009	0.0299	2.83	2.10	20.94	29.6	0.708
100	0.9458	1.013	0.0314	2.68	2.18	23.06	32.8	0.704
120	0.8980	1.013	0.0328	2.55	2.27	25.23	36.1	0.70
140	0.8535	1.013	0.0343	2.43	2.35	27.55	39.7	0.694
160	0.8150	1.017	0.0358	2.32	2.43	29.85	43.0	0.693
180	0.7785	1.022	0.0372	2.21	2.51	32.29	46.7	0.69
200	0.7475	1.026	0.0386	2.11	2.58	34.63	50.5	0.685
250	0.6745	1.034	0.0421	1.91	2.78	41.17	60.3	0.68
300	0.6157	1.047	0.0453	1.75	2.95	47.85	70.3	0.68
350	0.5662	1.055	0.0485	1.61	3.12	55.05	81.1	0.68
400	0.5242	1.068	0.0516	1.49	3.28	62.53	91.9	0.68
450	0.4875	1.080	0.0543	—	3.44	70.54	103.1	0.685
500	0.4564	1.092	0.0570	—	3.86	70.48	114.2	0.69
600	0.4041	1.114	0.0621	—	3.58	95.57	138.2	0.69
700	0.3625	1.135	0.0667	—	4.12	113.7	162.2	0.70
800	0.3287	1.156	0.0706	—	4.37	132.8	185.8	0.715
900	0.321	1.172	0.0741	—	4.59	152.5	210	0.725
1000	0.277	1.185	0.0770	—	4.80	175	235	0.735

Source: From Kakaç, S. and Yener, Y., *Convective Heat Transfer*, 2nd ed., CRC Press, Boca Raton, FL, 1995. With permission.

TABLE B.2
Thermophysical Properties of Saturated Ice-Water-Steam

P (bar)	T (K)	v_f^* (10^{-3} m³/kg)	v_g (m³/kg)	h^* (kJ/kg)	h_g (kJ/kg)	μ_g (10^{-4} Pa·s)	k_f^* (W/m·K)	k_g (W/m·K)	Pr_f^*	Pr_g
0.001	252.84	1.0010	1167	−374.9	2464.1	0.0723	2.40	0.0169		
0.002	260.21	1.0010	600	−360.1	2477.4	0.0751	2.35	0.0174		
0.003	265.11	1.0010	408.5	−350.9	2486.0	0.0771	2.31	0.0177		
0.004	267.95	1.0010	309.1	−344.4	2491.9	0.0780	2.29	0.0179		
0.005	270.74	1.0010	249.6	−337.9	2497.3	0.0789	2.27	0.0180		
0.006	273.06	1.0010	209.7	−333.6	2502	0.0798	2.26	0.0182		
0.0061	273.15	1.0010	206.0	−333.5	2502	0.0802	2.26	0.0182		
0.0061	273.15	1.0002	206.0	0.0	2502	0.0802	0.566	0.0182	13.04	0.817
0.008	276.73	1.0001	159.4	21.9	2508	0.0816	0.568	0.0184	11.66	0.823
0.010	280.13	1.0001	129.2	29.4	2513.4	0.0829	0.578	0.0186	10.39	0.828
0.02	290.66	1.0013	67.00	73.5	2532.7	0.0872	0.595	0.0193	7.51	0.841
0.03	297.24	1.0028	45.66	101.1	2544.8	0.0898	0.605	0.0195	6.29	0.854
0.04	302.13	1.0041	34.80	121.4	2553.6	0.0918	0.612	0.0198	5.57	0.865
0.05	306.04	1.0053	28.19	137.8	2560.6	0.0933	0.618	0.0201	5.08	0.871
0.06	309.33	1.0065	23.74	151.5	2566.6	0.0946	0.622	0.0203	4.62	0.877
0.08	314.68	1.0085	18.10	173.9	2576.2	0.0968	0.629	0.0207	4.22	0.883
0.10	318.98	1.0103	14.67	191.9	2583.9	0.0985	0.635	0.0209	3.87	0.893
0.20	333.23	1.0172	7.65	251.5	2608.9	0.1042	0.651	0.0219	3.00	0.913
0.30	342.27	1.0222	5.23	289.3	2624.6	0.1078	0.660	0.0224	2.60	0.929
0.40	349.04	1.0264	3.99	317.7	2636.2	0.1105	0.666	0.0229	2.36	0.941
0.5	354.50	1.0299	3.24	340.6	2645.5	0.1127	0.669	0.0233	2.19	0.951
0.6	359.11	1.0331	2.73	359.9	2653.0	0.1147	0.673	0.0236	2.06	0.961
0.8	366.66	1.0385	2.09	391.7	2665.3	0.1176	0.677	0.0242	1.88	0.979

Appendix B: Physical Properties

1.0	372.78	1.0434	1.6937	417.5	2675.4	0.1202	0.6805	0.0244	1.735	1.009
1.5	384.52	1.0530	1.1590	467.1	2693.4	0.1247	0.6847	0.0259	1.538	1.000
2.0	393.38	1.0608	0.8854	504.7	2706.3	0.1280	0.6866	0.0268	1.419	1.013
2.5	400.58	1.0676	0.7184	535.3	2716.4	0.1307	0.6876	0.0275	1.335	1.027
3.0	406.69	1.0735	0.6056	561.4	2724.7	0.1329	0.6879	0.0281	1.273	1.040
3.5	412.02	1.0789	0.5240	584.3	2731.6	0.1349	0.6878	0.0287	1.224	1.050
4.0	416.77	1.0839	0.4622	604.7	2737.6	0.1367	0.6875	0.0293	1.185	1.057
4.5	421.07	1.0885	0.4138	623.2	2742.9	0.1382	0.6869	0.0298	1.152	1.066
5	424.99	1.0928	0.3747	640.1	2747.5	0.1396	0.6863	0.0303	1.124	1.073
6	432.00	1.1009	0.3155	670.4	2755.5	0.1421	0.6847	0.0311	1.079	1.091
7	438.11	1.1082	0.2727	697.1	2762.0	0.1443	0.6828	0.0319	1.044	1.105
8	445.57	1.1150	0.2403	720.9	2767.5	0.1462	0.6809	0.0327	1.016	1.115
9	448.51	1.1214	0.2148	742.6	2772.1	0.1479	0.6788	0.0334	0.992	1.127
10	453.03	1.1274	0.1943	762.6	2776.1	0.1495	0.6767	0.0341	0.973	1.137
12	461.11	1.1386	0.1632	798.4	2782.7	0.1523	0.6723	0.0354	0.943	1.156
14	468.19	1.1489	0.1407	830.1	2787.8	0.1548	0.6680	0.0366	0.920	1.175
16	474.52	1.1586	0.1237	858.6	2791.8	0.1569	0.6636	0.0377	0.902	1.191
18	480.26	1.1678	0.1103	884.6	2794.8	0.1589	0.6593	0.0388	0.889	1.206
20	485.53	1.1766	0.0995	908.6	2797.2	0.1608	0.6550	0.0399	0.877	1.229
25	497.09	1.1972	0.0799	962.0	2800.9	0.1648	0.6447	0.0424	0.859	1.251
30	506.99	1.2163	0.0666	1008.4	2802.3	0.1684	0.6347	0.0449	0.849	1.278
35	515.69	1.2345	0.0570	1049.8	2802.0	0.1716	0.6250	0.0472	0.845	1.306
40	523.48	1.2521	0.0497	1087.4	2800.3	0.1746	0.6158	0.0496	0.845	1.331
45	530.56	1.2691	0.0440	1122.1	2797.7	0.1775	0.6068	0.0519	0.849	1.358
50	537.06	1.2858	0.0394	1154.5	2794.2	0.1802	0.5981	0.0542	0.855	1.386
60	548.70	1.3187	0.0324	1213.7	2785.0	0.1854	0.5813	0.0589	0.874	1.442

(Continued)

TABLE B.2 (CONTINUED)

Thermophysical Properties of Saturated Ice-Water-Steam

P (bar)	T (K)	v_f^* (10^{-3} m³/kg)	v_g (m³/kg)	h_f^* (kJ/kg)	h_g (kJ/kg)	μ_g (10^{-4} Pa·s)	k_f^* (W/m·K)	k_g (W/m·K)	Pr_f^*	Pr_g
70	558.94	1.3515	0.0274	1267.4	2773.5	0.1904	0.5653	0.0638	0.901	1.503
80	568.12	1.3843	0.0235	1317.1	2759.9	0.1954	0.5499	0.0688	0.936	1.573
90	576.46	1.4179	0.0205	1363.7	2744.6	0.2005	0.5352	0.0741	0.978	1.651
100	584.11	1.4526	0.0180	1408.0	2727.7	0.2057	0.5209	0.0798	1.029	1.737
110	591.20	1.4887	0.0160	1450.6	2709.3	0.2110	0.5071	0.0159	1.090	1.837
120	597.80	1.5268	0.0143	1491.8	2689.2	0.2166	0.4936	0.0925	1.163	1.963
130	603.98	1.5672	0.0128	1532.0	2667.0	0.2224	0.4806	0.0998	1.252	2.126
140	609.79	1.6106	0.0115	1571.6	2642.4	0.2286	0.4678	0.1080	1.362	2.343
150	615.28	1.6579	0.0103	1611.0	2615.0	0.2373	0.4554	0.1307	1.502	2.571
160	620.48	1.7103	0.0093	1650.5	2584.9	0.2497	0.4433	0.1280	1.688	3.041
170	625.41	1.7696	0.0084	1691.7	2551.6	0.2627	0.4315	0.1404	2.098	3.344
180	630.11	1.8399	0.0075	1734.8	2513.9	0.2766	0.4200	0.1557	2.360	3.807
190	634.58	1.9260	0.0067	1778.7	2470.6	0.2920	0.4087	0.1749	2.951	8.021
200	638.85	2.0370	0.0059	1826.5	2410.4	0.3094	0.3976	0.2007	4.202	12.16

P (bar)	s_f^* (kJ/kg·K)	s_g (kJ/kg·K)	c_{pf}^* (kJ/kg·K)	c_{pg} (kJ/kg·K)	μ_l (10^{-4} Pa·s)	γ_f	γ_g	V_{sf} (m/s)	V_{sg} (m/s)	σ^* (N/m)
0.001	-1.378	9.848	1.957							
0.002	-1.321	9.585	2.015							
0.003	-1.280	9.456	2.053							
0.004	-1.260	9.339	2.075							
0.005	-1.240	9.250	2.097	1.851						
0.006	-1.222	9.160	2.106	1.854						
0.0061	-1.221	9.159		1.854						

Appendix B: Physical Properties

0.0061	0.0000	9.159	4.217	1.854	17.50				0.0756
0.008	0.0543	9.0379	4.206	1.856	15.75				0.0751
0.010	0.1059	8.9732	4.198	1.858	14.30				0.0747
0.02	0.2605	8.7212	4.183	1.865	10.67				0.0731
0.03	0.3543	8.5756	4.180	1.870	9.09				0.0721
0.04	0.4222	8.4724	4.179	1.874	8.15				0.0714
0.05	0.4761	8.3928	4.178	1.878	7.51				0.0707
0.06	0.5208	8.3283	4.178	1.881	7.03				0.0702
0.08	0.5925	8.2266	4.179	1.887	6.35				0.0693
0.10	0.6493	8.1482	4.180	1.894	5.88				0.0686
0.20	0.8321	7.9065	4.184	1.917	4.66				0.0661
0.30	0.9441	7.7670	4.189	1.935	4.09				0.0646
0.40	1.0261	7.6686	4.194	1.953	3.74				0.0634
0.5	1.0912	7.5928	4.198	1.967	3.49				0.0624
0.6	1.1454	7.5309	4.201	1.978	3.30				0.0616
0.8	1.2330	7.4338	4.209	2.015	3.03				0.0605
1.0	1.3027	7.3598	4.222	2.048	2.801	1.136	1.321	438.74	472.98	0.0589
1.5	1.4336	7.2234	4.231	2.077	2.490	1.139	1.318	445.05	478.73	0.0566
2.0	1.5301	7.1268	4.245	2.121	2.295	1.141	1.316	449.51	482.78	0.0548
2.5	1.6071	7.0520	4.258	2.161	2.156	1.142	1.314	452.92	485.88	0.0534
3.0	1.6716	6.9909	4.271	2.198	2.051	1.143	1.313	455.65	488.36	0.0521
3.5	1.7273	6.9392	4.282	2.233	1.966	1.143	1.311	457.91	490.43	0.0510
4.0	1.7764	6.8943	4.294	2.266	1.897	1.144	1.310	459.82	492.18	0.0500
4.5	1.8204	6.8547	4.305	2.298	1.838	1.144	1.309	461.46	493.69	0.0491
5	1.8604	6.8192	4.315	2.329	1.787	1.144	1.308	462.88	495.01	0.0483
6	1.9308	6.7575	4.335	2.387	1.704	1.144	1.306	465.23	497.22	0.0468

(Continued)

TABLE B.2 (CONTINUED)
Thermophysical Properties of Saturated Ice-Water-Steam

P (bar)	s_f^* (kJ/kg·K)	s_g (kJ/kg·K)	c_{pf}^* (kJ/kg·K)	c_{pg} (kJ/kg·K)	μ_l (10^{-4} Pa·s)	γ_f	γ_g	V_{sf} (m/s)	V_{sg} (m/s)	σ^* (N/m)
7	1.9918	6.7052	4.354	2.442	1.637	1.143	1.304	467.08	498.99	0.0455
8	2.0457	6.6596	4.372	2.495	1.581	1.142	1.303	468.57	500.55	0.0444
9	2.0941	6.6192	4.390	2.546	1.534	1.142	1.302	469.78	501.64	0.0433
10	2.1382	6.5821	4.407	2.594	1.494	1.141	1.300	470.76	502.64	0.0423
12	2.2161	6.5194	4.440	2.688	1.427	1.139	1.298	472.23	504.21	0.0405
14	2.2837	6.4651	4.472	2.777	1.373	1.137	1.296	473.18	505.33	0.0389
16	2.3436	6.4175	4.504	2.862	1.329	1.134	1.294	473.78	506.12	0.0375
18	2.3976	6.3751	4.534	2.944	1.291	1.132	1.293	474.09	506.65	0.0362
20	2.4469	6.3367	4.564	3.025	1.259	1.129	1.291	474.18	506.98	0.0350
25	2.5543	6.2536	4.640	3.219	1.193	1.123	1.288	473.71	507.16	0.0323
30	2.6455	6.1837	4.716	3.407	1.143	1.117	1.284	472.51	506.65	0.0300
35	2.7253	6.1229	4.792	3.593	1.102	1.111	1.281	470.80	505.66	0.0200
40	2.7965	6.0685	4.870	3.781	1.069	1.104	1.278	468.72	504.29	0.0261
45	2.8612	6.0191	4.951	3.972	1.040	1.097	1.275	466.31	502.68	0.0244
50	2.9206	5.9735	5.034	4.168	1.016	1.091	1.272	463.67	500.73	0.0229
60	3.0273	5.8908	5.211	4.582	0.975	1.077	1.266	457.77	496.33	0.0201
70	3.1219	5.8162	5.405	5.035	0.942	1.063	1.260	451.21	491.31	0.0177
80	3.2076	5.7471	5.621	5.588	0.915	1.048	1.254	444.12	485.80	0.0156
90	3.2867	5.6820	5.865	6.100	0.892	1.033	1.249	436.50	479.90	0.0136
100	3.3606	5.6198	6.142	6.738	0.872	1.016	1.244	428.24	473.67	0.0119
110	3.4304	5.5595	6.463	7.480	0.855	0.998	1.239	419.20	467.13	0.0103
120	3.4972	5.5002	6.838	8.384	0.840	0.978	1.236	409.38	460.25	0.0089

Appendix B: Physical Properties

130	3.5616	5.4408	7.286	9.539	0.826	0.956	1.234	398.90	453.00	0.0076
140	3.6243	5.3803	7.834	11.07	0.813	0.935	1.232	388.00	445.34	0.0064
150	3.6859	5.3178	8.529	13.06	0.802	0.916	1.233	377.00	437.29	0.0053
160	3.7471	5.2531	9.456	15.59	0.792	0.901	1.235	366.24	428.89	0.0043
170	3.8197	5.1855	11.30	17.87	0.782	0.867	1.240	351.19	420.07	0.0034
180	3.8765	5.1128	12.82	21.43	0.773	0.838	1.248	336.35	410.39	0.0026
190	3.9429	5.0332	15.76	27.47	0.765	0.808	1.260	320.20	399.87	0.0018
200	4.0149	4.9412	22.05	39.31	0.758	0.756	1.280	298.10	387.81	0.0011

Source: From Liley, P. E., Thermophysical properties, *Boilers, Evaporators, and Condensers*, S. Kakaç (Ed.), Wiley, New York, 1987.

* Above the solid line, solid phase; below the line, liquid.

TABLE B.3
Thermophysical Properties of Steam at 1-bar Pressure

T (K)	v (m³/kg)	h (kJ/kg)	s (kJ/ kg·K)	c_p (kJ/ kg·K)	c_v (kJ/ kg·K)	γ	Z	v_s (m/s)	μ (10^{-5} Pa·s)	k (W/m·K)	Pr
373.15	1.679	2676.2	7.356	2.029	1.510	1.344	0.9750	472.8	1.20	0.0248	0.982
400	1.827	2730.2	7.502	1.996	1.496	1.334	0.9897	490.4	1.32	0.0268	0.980
450	2.063	2829.7	7.741	1.981	1.498	1.322	0.9934	520.6	1.52	0.0311	0.968
500	2.298	2928.7	7.944	1.983	1.510	1.313	0.9959	540.3	1.73	0.0358	0.958
550	2.531	3028	8.134	2.000	1.531	1.306	0.9971	574.2	1.94	0.0410	0.946
600	2.763	3129	8.309	2.024	1.557	1.300	0.9978	598.6	2.15	0.0464	0.938
650	2.995	3231	8.472	2.054	1.589	1.293	0.9988	621.8	2.36	0.0521	0.930
700	3.227	3334	8.625	2.085	1.620	1.287	0.9989	643.9	2.57	0.0581	0.922
750	3.459	3439	8.770	2.118	1.653	1.281	0.9992	665.1	2.77	0.0646	0.913
800	3.690	3546	8.908	2.151	1.687	1.275	0.9995	685.4	2.98	0.0710	0.903
850	3.921	3654	9.039	2.185	1.722	1.269	0.9996	705.1	3.18	0.0776	0.897
900	4.152	3764	9.165	2.219	1.756	1.264	0.9996	723.9	3.39	0.0843	0.892
950	4.383	3876	9.286	2.253	1.791	1.258	0.9997	742.2	3.59	0.0912	0.886
1000	4.614	3990	9.402	2.286	1.823	1.254	0.9998	760.1	3.78	0.0981	0.881
1100	5.076	4223	9.625	2.36			0.9999	794.3	4.13	0.113	0.858
1200	5.538	4463	9.384	2.43			1.0000	826.8	4.48	0.130	0.837
1300	5.999	4711	10.032	2.51			1.0000	857.9	4.77	0.144	0.826
1400	6.461	4965	10.221	2.58			1.0000	887.9	5.06	0.160	0.816
1500	6.924	5227	10.402	2.65			1.0002	916.9	5.35	0.18	0.788
1600	7.386	5497	10.576	2.73			1.0004	945.0	5.65	0.21	0.735
1800	8.316	6068	10.912	3.02			1.0011	999.4	6.19	0.33	0.567
2000	9.263	6706	11.248	3.79			1.0036	1051.0	6.70	0.57	0.445

Source: From Kakaç, S. and Yener, Y. *Convective Heat Transfer*, 2nd ed., CRC Press, Boca Raton, FL, 1995. With permission.

TABLE B.4
Thermophysical Properties of Water-Steam at High Pressures

T (K)	v (m³/kg)	h (kJ/kg)	s (kJ/kg·K)	c_p (kJ/kg·K)	c_v (kJ/kg·K)	γ	Z	v_s (m/s)	μ (Pa·s)	k (W/m·K)	Pr
P = 10 bar											
300	1.003.–3[a]	113.4	0.392	4.18	4.13	1.01	0.0072	1500	8.57.–4[b]	0.615	5.82
350	1.027.–3	322.5	1.037	4.19	3.89	1.08	0.0064	1552	3.70.–4	0.668	2.32
400	1.067.–3	533.4	1.600	4.25	3.65	1.17	0.0058	1509	2.17.–4	0.689	1.34
450	1.123.–3	749.0	2.109	4.39	3.44	1.28	0.0054	1399	1.51.–4	0.677	0.981
500	0.221	2891	6.823	2.29	1.68	1.36	0.957	535.7	1.71.–5[c]	0.038	1.028
600	0.271	3109	7.223	2.13	1.61	1.32	0.987	592.5	2.15.–5	0.047	0.963
800	0.367	3537	7.837	2.18	1.70	1.28	0.994	686.2	2.99.–5	0.072	0.908
1000	0.460	3984	8.336	2.30	1.83	1.26	0.997	759.4	3.78.–5	0.099	0.881
1500	0.692	5224	9.337	2.66			1.000	917.2	5.35.–5	0.18	0.80
2000	0.925	6649	10.154	3.29			1.002	1050	6.70.–5	0.39	0.57
P = 50 bar											
300	1.001.–3	117.1	0.391	4.16	4.11	1.01	0.0362	1508	8.55.–4	0.618	5.76
350	1.025.–3	325.6	1.034	4.18	3.88	1.08	0.0317	1561	3.71.–4	0.671	2.31
400	1.064.–3	536.0	1.596	4.24	3.64	1.16	0.0288	1519	2.18.–4	0.691	1.34
450	1.120.–3	751.4	2.103	4.37	3.43	1.27	0.0270	1437	1.52.–4	0.681	0.975
500	1.200.–3	976.1	2.575	4.64	3.24	1.43	0.0260	1246	1.19.–4	0.645	0.856
600	0.0490	3013	6.350	2.85	1.94	1.47	0.885	560.5	2.14.–5	0.054	1.129
800	0.0713	3496	7.049	2.31	1.74	1.32	0.966	674.5	3.03.–5	0.075	0.929
1000	0.0911	3961	7.575	2.35	1.85	1.27	0.987	756.5	3.81.–5	0.102	0.880

(Continued)

TABLE B.4 (CONTINUED)

Thermophysical Properties of Water-Steam at High Pressures

T (K)	ν (m³/kg)	h (kJ/kg)	s (kJ/kg·K)	c_p (kJ/kg·K)	c_v (kJ/kg·K)	γ	Z	ν_s (m/s)	μ (Pa·s)	k (W/m·K)	Pr
1500	0.1384	5214	8.589	2.66			1.000	918.8	5.37,−5	0.18	0.81
2000	0.1850	6626	9.398	3.12			1.002	1053	6.70,−5	0.33	0.64
$P = 100\ bar$											
300	9.99,−4	121.8	0.390	4.15	4.09	1.01	0.0722	1516	8.52,−4	0.622	5.69
350	1.022,−3	329.6	1.031	4.17	3.87	1.08	0.0633	1571	3.73,−4	0.675	2.31
400	1.061,−3	539.6	1.590	4.23	3.64	1.16	0.0575	1532	2.20,−4	0.694	1.34
450	1.116,−3	754.1	2.097	4.35	3.43	1.27	0.0537	1452	1.53,−4	0.685	0.975
500	1.193,−3	977.3	2.567	4.60	3.24	1.42	0.0517	1269	1.21,−4	0.651	0.853
600	0.0201	2820	5.775	5.22	2.64	1.97	0.726	502.3	2.14,−5	0.073	1.74
800	0.0343	3442	6.685	2.52	1.82	1.38	0.929	662.4	3.08,−5	0.081	0.960
1000	0.0449	3935	7.233	2.44	1.88	1.30	0.973	753.3	3.85,−5	0.107	0.876
1500	0.0692	5203	8.262	2.68			1.000	921.1	5.37,−5	0.18	0.82
2000	0.0926	6616	9.073	3.08			1.003	1057	6.70,−5	0.31	0.67
$P = 250\ bar$											
300	9.93,−4	135.3	0.385	4.12	4.06	1.02	0.1792	1542	8.48,−4	0.634	5.50
350	1.016,−3	341.7	1.022	4.14	3.84	1.08	0.1572	1599	3.78,−4	0.686	2.28
400	1.053,−3	550.1	1.578	4.20	3.62	1.16	0.1426	1568	2.24,−4	0.704	1.33
450	1.105,−3	762.4	2.078	4.30	3.41	1.26	0.1330	1496	1.57,−4	0.696	0.969
500	1.175,−3	981.9	2.541	4.50			0.1273	1331	1.24,−4	0.666	0.838
600	1.454,−3	1479	3.443	5.88	4.22	1.40	0.1313	896.9	8.63,−5	0.532	0.952
800	0.0120	3261	6.086	3.41			0.813	627.3	3.29,−5	0.109	1.03

Appendix B: Physical Properties

1000	0.0173	3845	6.741	2.69	1.97	1.36	0.935	745.9	3.98.–5	0.125	0.856
1500	0.0277	5186	7.827	2.73			1.000	929.1	5.40.–5	0.18	0.819
2000	0.0372	6608	8.642	3.04			1.008	1068			
$P = 500$ bar											
300	9.83.–4	157.7	0.378	4.06	3.98	1.02	0.3549	1583	8.45.–4	0.650	5.28
350	1.005.–3	361.8	1.007	4.10	3.81	1.08	0.3112	1644	3.87.–4	0.700	2.27
400	1.041.–3	567.8	1.557	4.14	3.59	1.15	0.2820	1623	2.31.–4	0.719	1.33
450	1.088.–3	776.9	2.050	4.23	3.39	1.25	0.2618	1561	1.62.–4	0.714	0.960
500	1.151.–3	991.5	2.502	4.37			0.2493	1418	1.29.–4	0.689	0.822
600	1.362.–3	1456	3.346	5.08	3.72	1.37	0.2459	1080	9.34.–5	0.588	0.808
800	4.576.–3	2895	5.937	5.84	2.79	2.10	0.620	597.8	4.04.–5	0.178	1.33
1000	8.102.–3	3697	6.302	3.17	1.81	1.76	0.878	742.1	4.28.–5	0.150	0.905
1500	0.0139	5157	7.484	2.82			1.004	943.6			
2000	0.0188	6595	8.310	3.04			1.018	1086			

[a] Notation –3 signifies $\times 10^{-3}$.
[b] Notation –4 signifies $\times 10^{-4}$.
[c] Notation –5 signifies $\times 10^{-5}$.

TABLE B.5

Properties of Liquid Metals

Liquid Metal and Melting Point	Temperature, °C	k W/(m · K)	ρ kg/m^3	c kJ(kg · K)	$\mu \times 10^4$ kg/(m · s)
Bismuth (288°C)	315	16.40	10000	0.1444	16.22
	538	15.70	9730	0.1545	10.97
	760	15.70	9450	0.1645	7.89
Lead (327°C)	371	18.26	10500	0.159	24.0
	482	19.77	10400	0.155	19.25
	704	—	10130	—	13.69
Mercury (−39°C)	10	8.14	13550	0.138	15.92
	149	11.63	13200	0.138	10.97
	315	14.07	12800	0.134	8.64
Potassium (64°C)	315	45.0	804	0.80	3.72
	427	39.5	740	0.75	1.78
	704	33.1	674	0.75	1.28
Sodium (98°C)	93	86.0	930	1.38	7.0
	371	72.5	860	1.29	2.81
	704	59.8	776	1.26	1.78
56% Na, 44% K (19°C)	93	25.6	885	1.13	5.78
	371	27.6	820	1.06	2.36
	704	28.8	723	1.04	5.94
22% Na, 78% K (−11°C)	93	24.4	850	0.95	4.94
	400	26.7	775	0.88	2.07
	760	—	690	0.88	1.46
44.5% Pb, 55.5% Bi (125°C)	315	9.0	10500	0.147	—
	371	11.9	10200	0.47	15.36
	650	—	9820	—	11.47

Source: From Kakaç, S. and Yener, Y. *Convective Heat Transfer*, 2nd ed., CRC Press, Boca Raton, FL, 1995. With permission.

TABLE B.6
Thermophysical Properties of Saturated Refrigerant 12

P (bar)	T (K)	v_f (10^{-4} m³/kg)	v_g (m³/kg)	h_f (kJ/kg)	h_g (kJ/kg)	s_f (kJ/ kg·K)	s_g (kJ/ kg·K)
0.10	200.1	6.217	1.365	334.8	518.1	3.724	4.640
0.15	206.3	6.282	0.936	340.1	521.0	3.750	4.627
0.20	211.1	6.332	0.716	344.1	523.2	3.769	4.618
0.25	214.9	6.374	0.582	347.4	525.0	3.785	4.611
0.30	218.2	6.411	0.491	350.2	526.5	3.798	4.606
0.4	223.5	6.437	0.376	354.9	529.1	3.819	4.598
0.5	227.9	6.525	0.306	358.8	531.2	3.836	4.592
0.6	231.7	6.570	0.254	362.1	532.9	3.850	4.588
0.8	237.9	6.648	0.198	367.6	535.8	3.874	4.581
1.0	243.0	6.719	0.160	372.1	538.2	3.893	4.576
1.5	253.0	6.859	0.110	381.2	542.9	3.929	4.568
2.0	260.6	6.970	0.0840	388.2	546.4	3.956	4.563
2.5	266.9	7.067	0.0681	394.0	549.2	3.978	4.560
3.0	272.3	7.183	0.0573	399.1	551.6	3.997	4.557
4.0	281.3	7.307	0.0435	407.6	555.6	4.027	4.553
5.0	288.8	7.444	0.0351	414.8	558.8	4.052	4.551
6.0	295.2	7.571	0.0294	421.1	561.5	4.073	4.549
8.0	306.0	7.804	0.0221	431.8	565.7	4.108	4.546
10	314.9	8.022	0.0176	440.8	569.0	4.137	4.544
15	332.6	8.548	0.0114	459.3	574.5	4.193	4.539
20	346.3	9.096	0.0082	474.8	577.5	4.237	4.534
25	357.5	9.715	0.0062	488.7	578.5	4.275	4.527

(Continued)

TABLE B.6 (CONTINUED)

Thermophysical Properties of Saturated Refrigerant 12

P (bar)	T (K)	v_f (10^{-4} m³/kg)	v_g (m³/kg)	h_f (kJ/kg)	h_g (kJ/kg)	s_f (kJ/ kg·K)	s_g (kJ/ kg·K)
30	367.2	10.47	0.0048	502.0	577.6	4.311	4.517
35	375.7	11.49	0.0036	515.9	574.1	4.347	4.502
40	383.3	13.45	0.0025	532.7	564.1	4.389	4.471
41.2[a]	385.0	17.92	0.0018	548.3	548.3	4.429	4.429

P (bar)	c_{pf} (kJ/ kg·K)	c_{pg} (kJ/ kg·K)	μ_f (10^{-4} Pa·s)	μ_g (10^{-5} Pa·s)	k_f (W/m·K)	k_g (W/m·K)	Pr_f	Pr_g	σ (N/m)
0.10	0.855		6.16		0.105	0.0050	5.01		
0.15	0.861		5.61		0.103	0.0053	4.69		
0.20	0.865		5.28		0.101	0.0055	4.52		
0.25	0.868		4.99		0.099	0.0056	4.38		
0.30	0.872		4.79		0.098	0.0057	4.26		
0.4	0.876		4.48		0.097	0.0060	4.05		0.0189
0.5	0.880	0.545	4.25	1.00	0.095	0.0062	3.94	0.89	0.0182
0.6	0.884	0.552	4.08	1.02	0.094	0.0063	3.84	0.88	0.0176
0.8	0.889	0.564	3.81	1.04	0.091	0.0066	3.72	0.88	0.0167
1.0	0.894	0.574	3.59	1.06	0.089	0.0069	3.61	0.88	0.0159
1.5	0.905	0.600	3.23	1.10	0.086	0.0074	3.40	0.89	0.0145
2.0	0.914	0.613	2.95	1.13	0.083	0.0077	3.25	0.90	0.0134

Appendix B: Physical Properties

2.5	0.922	0.626	2.78	1.15	0.081	0.0081	3.16	0.91	0.0125
3.0	0.930	0.640	2.62	1.18	0.079	0.0083	3.08	0.91	0.0118
4.0	0.944	0.663	2.40	1.22	0.075	0.0088	3.02	0.92	0.0106
5.0	0.957	0.683	2.24	1.25	0.073	0.0092	2.94	0.93	0.0096
6.0	0.969	0.702	2.13	1.28	0.070	0.0095	2.95	0.95	0.0087
8.0	0.995	0.737	1.96	1.33	0.066	0.0101	2.95	0.97	0.0074
10	1.023	0.769	1.88	1.38	0.063	0.0107	3.05	1.01	0.0063
15	1.102	0.865	1.67	1.50	0.057	0.0117	3.23	1.11	0.0042
20	1.234	0.969	1.49	1.69	0.053	0.0126	3.47	1.30	0.0029
25	1.36	1.19	1.33		0.047	0.0134	3.84		0.0019
30	1.52	1.60	1.16		0.042	0.014	4.2		0.0009
35	1.73	2.5			0.037	0.016			0.0005
40									0.0001
41.2									0.0000

Source: From Kakaç, S. and Yener, Y. *Convective Heat Transfer*, 2nd ed., CRC Press, Boca Raton, FL, 1995. With permission.
[a] Critical point.

TABLE B.7

Thermophysical Properties of Refrigerant 12 at 1-bar Pressure

T (K)	ν (m³/kg)	h (kJ/kg)	s (kJ/kg · K)	μ (10⁻⁵ Pa · s)	c_p (kJ/kg · K)	k (W/m · K)	Pr
300	0.2024	572.1	4.701	1.26	0.614	0.0097	0.798
320	0.2167	584.5	4.741	1.34	0.631	0.0107	0.788
340	0.2309	597.3	4.780	1.42	0.647	0.0118	0.775
360	0.2450	610.3	4.817	1.49	0.661	0.0129	0.760
380	0.2590	623.7	4.853	1.56	0.674	0.0140	0.745
400	0.2730	637.3	4.890	1.62	0.684	0.0151	0.730
420	0.2870	651.2	4.924	1.67	0.694	0.0162	0.715
440	0.3009	665.3	4.956	1.72	0.705	0.0173	0.703
460	0.3148	697.7	4.987	1.78	0.716	0.0184	0.693
480	0.3288	694.3	5.018	1.84	0.727	0.0196	0.683
500	0.3427	709.0	5.048	1.90	0.739	0.0208	0.674

Source: From Kakaç, S. and Yener, Y. *Convective Heat Transfer*, 2nd ed., CRC Press, Boca Raton, FL, 1995. With permission.

Appendix B: Physical Properties

TABLE B.8
Thermophysical Properties of Saturated Refrigerant 22

T (K)	P (bar)	v_f (m³/kg)	v_g (m³/kg)	h_f (kJ/kg)	h_g (kJ/kg)	s_f (kJ/kg·K)	s_g (kJ/kg·K)	c_{pf} (kJ/kg·K)	c_{pg} (kJ/kg·K)
150	0.0017	6.209.−4[a]	83.40	268.2	547.3	3.355	5.215	1.059	
160	0.0054	6.293.−4	28.20	278.2	552.1	3.430	5.141	1.058	
170	0.0150	6.381.−4	10.85	288.3	557.0	3.494	5.075	1.057	
180	0.0369	6.474.−4	4.673	298.7	561.9	3.551	5.013	1.058	
190	0.0821	6.573.−4	2.225	308.6	566.8	3.605	4.963	1.060	
200	0.1662	6.680.−4	1.145	318.8	571.6	3.675	4.921	1.065	0.502
210	0.3116	6.794.−4	0.6370	329.1	576.5	3.707	4.885	1.071	0.544
220	0.5470	6.917.−4	0.3772	339.7	581.2	3.756	4.854	1.080	0.577
230	0.9076	7.050.−4	0.2352	350.6	585.9	3.804	4.828	1.091	0.603
240	1.4346	7.195.−4	0.1532	361.7	590.5	3.852	4.805	1.105	0.626
250	2.174	7.351.−4	0.1037	373.0	594.9	3.898	4.785	1.122	0.648
260	3.177	7.523.−4	0.07237	384.5	599.0	3.942	4.768	1.143	0.673
270	4.497	7.733.−4	0.05187	396.3	603.0	3.986	4.752	1.169	0.703
280	6.192	7.923.−4	0.03803	408.2	606.6	4.029	4.738	1.193	0.741
290	8.324	8.158.−4	0.02838	420.4	610.0	4.071	4.725	1.220	0.791
300	10.956	8.426.−4	0.02148	432.7	612.8	4.113	4.713	1.257	0.854
310	14.17	8.734.−4	0.01643	445.5	615.1	4.153	4.701	1.305	0.935
320	18.02	9.096.−4	0.01265	458.6	616.7	4.194	4.688	1.372	1.036
330	22.61	9.535.−4	9.753.−3[b]	472.4	617.3	4.235	4.674	1.460	1.159
340	28.03	1.010.−3	7.479.−3	487.2	616.5	4.278	4.658	1.573	1.308
350	34.41	1.086.−3	5.613.−3	503.7	613.3	4.324	4.637	1.718	1.486
360	41.86	1.212.−3	4.036.−3	523.7	605.5	4.378	4.605	1.897	
369.3[c]	49.89	2.015.−3	2.015.−3	570.0	570.0	4.501	4.501	∞	∞

(Continued)

TABLE B.8 (CONTINUED)
Thermophysical Properties of Saturated Refrigerant 22

T (K)	μ_f (10⁻⁴ Pa·s)	μ_g (10⁻⁴ Pa·s)	k_f (W/m·K)	k_g (W/m·K)	ν_{sf} (m/s)	ν_{fg} (m/s)	Pr_f	Pr_g	σ (N/m)
150			0.161						
160			0.156						
170	7.70		0.151			142.6	5.39		
180	6.47		0.146			146.1	4.69		
190	5.54		0.141			149.4	4.16		
200	4.81		0.136		1007	152.6	3.77		0.024
210	4.24		0.131		957	155.2	3.47		0.022
220	3.78		0.126		909	157.6	3.24		0.021
230	3.40	0.100	0.121	0.0067	862	159.7	3.07	0.89	0.019
240	3.09	0.104	0.117	0.0073	814	161.3	2.92	0.89	0.017
250	2.82	0.109	0.112	0.0080	766	162.5	2.83	0.89	0.0155
260	2.60	0.114	0.107	0.0086	716	163.1	2.78	0.89	0.0138
270	2.41	0.118	0.102	0.0092	668	163.4	2.76	0.90	0.0121
280	2.25	0.123	0.097	0.0098	622	162.1	2.77	0.93	0.0104
290	2.11	0.129	0.092	0.0105	578	161.1	2.80	0.97	0.0087
300	1.98	0.135	0.087	0.0111	536	160.1	2.86	1.04	0.0071
310	1.86	0.141	0.082	0.0117	496	157.2	2.96	1.13	0.0055
320	1.76	0.148	0.077	0.0123	458	153.4	3.14	1.25	0.0040

330	1.67	0.157	0.072	0.0130	408	148.5	3.39	1.42	0.0026
340	1.51	0.171	0.067	0.0140	355	142.7	3.55	1.60	0.0014
350	1.30		0.060		290	135.9	3.72		0.0008
360	1.06								
369.3									

Source: From Kakaç, S. and Yener, Y. *Convective Heat Transfer*, 2nd ed., CRC Press, Boca Raton, FL, 1995. With permission.

[a] Notation –4 signifies $\times 10^{-4}$.
[b] Notation –3 signifies $\times 10^{-3}$.
[c] Critical point.

TABLE B.9
Thermophysical Properties of Refrigerant R22 at Atmospheric Pressure

T (K)	v (m³/kg)	h (kJ/kg)	s (kJ/kg·K)	c_p (kJ/kg·K)	Z	v_s (m/s)	μ (10^{-6} Pa·s)	k (W/m·K)	Pr
232.3	0.2126	586.9	4.8230	0.608	0.9644	160.1	10.1	0.0067	0.893
240	0.2205	591.5	4.8673	0.6117	0.9682	163.0	10.4	0.0074	0.860
260	0.2408	604.0	4.8919	0.6255	0.9760	169.9	11.2	0.0084	0.838
280	0.2608	616.8	4.9389	0.6431	0.9815	176.2	12.0	0.0094	0.820
300	0.2806	630.0	4.9840	0.6619	0.9857	182.3	12.8	0.0106	0.804
320	0.3001	643.4	5.0274	0.6816	0.9883	188.0	13.7	0.0118	0.790
340	0.3196	657.3	5.0699	0.7017	0.9906	193.5	14.4	0.0130	0.777
360	0.3390	671.7	5.1111	0.7213	0.9923	198.9	15.1	0.0142	0.767
380	0.3583	686.5	5.1506	0.7406	0.9936	204.1	15.8	0.0154	0.760
400	0.3775	701.5	5.1892	0.7598	0.9945	209.1	16.5	0.0166	0.755
420	0.3967	717.0	5.2267	0.7786	0.9953	214.0	17.2	0.0178	0.753
440	0.4159	732.8	5.2635	0.7971	0.9961	218.8	17.9	0.0190	0.752
460				0.8150		223.5	18.6	0.0202	0.751
480				0.8326		227.9	19.3	0.0214	0.751
500				0.8502			19.9	0.0225	0.750

Source: From Kakaç, S. and Yener, Y. *Convective Heat Transfer*, 2nd ed., CRC Press, Boca Raton, FL, 1995. With permission.

Appendix B: Physical Properties

TABLE B.10
Thermophysical Properties of Saturated Refrigerant R134a

T (K)	p (bar)	v_f (m³/kg)	v_g (m³/kg)	h_f (kJ/kg)	h_g (kJ/kg)	s_f (kJ/kg·K)	s_g (kJ/kg·K)	c_{pf} (kJ/kg·K)	c_{pg} (kJ/kg·K)	μ_f (10⁻⁴ Pa·s)	μ_g (10⁻⁴ Pa·s)	k_f (W/m·K)	k_g (W/m·K)	Pr_f	Pr_g	σ (N/m)
200	0.070	0.000661	2.32	−36.0	201.0	−0.1691	1.0153									
210	0.187	0.000674	0.906	−26.5	208.1	−0.1175	0.9941									
220	0.252	0.000687	0.698	−15.3	214.5	−0.0664	0.9758									
230	0.438	0.000701	0.416	−3.7	220.8	−0.0158	0.9602	1.113	0.732							
240	0.728	0.000716	0.258	8.1	227.1	0.0343	0.9471	1.162	0.764	4.25	0.095	0.099	0.008	4.99	0.90	0.0145
250	1.159	0.000731	0.167	20.3	233.3	0.0840	0.9363	1.212	0.798	3.70	0.099	0.095	0.008	4.72	0.96	0.0131
260	1.765	0.000748	0.112	32.9	239.4	0.1331	0.9276	1.259	0.835	3.25	0.104	0.091	0.008	4.49	1.02	0.0117
270	2.607	0.000766	0.077	45.4	244.8	0.1817	0.9211	1.306	0.876	2.88	0.108	0.087	0.009	4.31	1.08	0.0103
280	3.721	0.000786	0.055	59.2	251.1	0.2299	0.9155	1.351	0.921	2.56	0.112	0.083	0.009	4.17	1.14	0.0090
290	5.175	0.000806	0.040	72.9	256.6	0.2775	0.9114	1.397	0.972	2.30	0.117	0.079	0.010	4.07	1.20	
300	7.02	0.000821	0.029	87.0	261.9	0.3248	0.9080	1.446	1.030	2.08	0.121	0.075	0.010	4.00	1.27	
310	9.33	0.000865	0.022	101.5	266.8	0.3718	0.9050	1.497	1.104	1.89	0.125	0.071	0.010	3.98	1.34	
320	12.16	0.000895	0.016	116.6	271.2	0.4189	0.9021	1.559	1.198	1.72	0.129	0.068	0.011	3.98	1.57	
330	15.59	0.000935	0.012	132.3	275.0	0.4663	0.8986	1.638	1.324	1.58	0.133	0.064	0.011	3.94	1.44	
340	19.71	0.000984	0.0094	148.9	277.8	0.5146	0.8937	1.750	1.520	1.45	0.137	0.060	0.012	4.23	1.74	
350	24.60	0.00105	0.0071	166.6	279.1	0.5649	0.8861	1.931	1.795	1.34	0.14	0.056	0.012	4.62	2.09	
360	30.40	0.00115	0.0051	186.5	277.7	0.6194	0.8721	2.304	2.610	1.20	0.16	0.054	0.013	5.16	3.21	
370	37.31	0.00139	0.0035	216.0	270.0	0.6910	0.8370			0.95	0.26					
374.3[a]	40.67	0.00195	0.0020	248.0	248.0	0.7714	0.7714									

Source: From Liley, P. E. Thermophysical properties. *Boilers, Evaporators, and Condensers*, S. Kakaç (Ed.), Wiley, New York, 1987. With permission.

[a] Critical point.

TABLE B.11

Properties of Refrigerant 134a at Atmospheric Pressure

T (K)	v (m³/kg)	h (kJ/kg)	s (kJ/kg · K)	c_p (kJ/kg · K)	Z	v_s (m/s)
247	0.1901	231.5	0.940	0.787	0.957	145.9
260	0.2107	241.8	0.980	0.801	0.965	150.0
280	0.2193	258.1	1.041	0.827	0.974	156.3
300	0.2365	274.9	1.099	0.856	0.980	162.1
320	0.2532	292.3	1.155	0.885	0.984	167.6
340	0.2699	310.3	1.209	0.915	0.987	172.8
360	0.2866	238.8	1.263	0.945	0.990	177.6
380	0.3032	347.8	1.313	0.976	0.992	182.0
400	0.3198	367.2	1.361	1.006	0.994	186.0

Source: From Liley, P. E. Thermophysical properties. *Boilers, Evaporators, and Condensers*, S. Kakaç (Ed.), Wiley, New York, 1987. With permission.

TABLE B.12

Thermophysical Properties of Unused Engine Oil

T (K)	v_f (m³/kg)	c_{pf} (kJ/kg·K)	μ_f (Pa · s)	k_f (W/m·K)	Pr_f	α_f (m²/s)
250	1.093–3[a]	1.72	32.20	0.151	367,000	9.60–8[b]
260	1.101–3	1.76	12.23	0.149	144,500	9.32–8
270	1.109–3	1.79	4.99	0.148	60,400	9.17–8
280	1.116–3	1.83	2.17	0.146	27,200	8.90–8
290	1.124–3	1.87	1.00	0.145	12,900	8.72–8
300	1.131–3	1.91	0.486	0.144	6,450	8.53–8
310	1.139–3	1.95	0.253	0.143	3,450	8.35–8
320	1.147–3	1.99	0.141	0.141	1,990	8.13–8
330	1.155–3	2.04	0.084	0.140	1,225	7.93–8
340	1.163–3	2.08	0.053	0.139	795	7.77–8
350	1.171–3	2.12	0.036	0.138	550	7.62–8
360	1.179–3	2.16	0.025	0.137	395	7.48–8
370	1.188–3	2.20	0.019	0.136	305	7.34–8
380	1.196–3	2.25	0.014	0.136	230	7.23–8
390	1.205–3	2.29	0.011	0.135	185	7.10–8
400	1.214–3	2.34	0.009	0.134	155	6.95–8

Source: From Kakaç, S. and Yener, Y. *Convective Heat Transfer*, 2nd ed., CRC Press, Boca Raton, FL, 1995. With permission.

[a] Notation –3 signifies $\times 10^{-3}$.
[b] Notation –8 signifies $\times 10^{-8}$.

TABLE B.13

Conversion Factors

Area: 1 m^2 = 1550.0 in.^{2} = 10.7639 ft^2 = 1.19599 yd^2 = 2.47104 × 10^{-4} acre 1 × 10^{-4} ha = 10^{-6} km^2 = 3.8610 × 10^{-7} mi^2

Density: 1 kg/m^3 = 0.06243 lbm/ft^3 = 0.01002 lbm/U.K. gal = 8.3454 × 10^{-3} lbm/U.S. gal = 1.9403 × 10^{-3} slug/ft^3 = 10^{-3} g/cm^3

Energy: 1kJ = 737.56 ft-lbf = 238.85 cal = 0.94783 Btu = 3.7251 × 10^{-4} hp · h = 2.7778 × 10^{-4} kW · h

Heat transfer coefficient: 1 W/(m^2·K) = 0.8598 kcal/(m^2 · h · °C) = 0.1761 Btu/(ft^2 · h · °F) = 10^{-4} W/(cm^2 · K) = 0.2388 × 10^{-4} cal/(cm^2 · s · °C)

Inertia: 1 kg · m^2 = 3.41717 × 10^3 lb · in.^{2} = 0.73756 slug · ft^2

Length: 1 m = 10^{10} Å = 39.370 in. = 3.28084 ft = 4.971 links = 1.0936 yd = 0.54681 fathoms = 0.04971 chain = 4.97097 × 10^{-3} furlong = 10^{-3} km = 5.3961 × 10^{-4} U.K. nautical miles = 5.3996 × 10^{-4} U.S. nautical miles = 6.2137 × 10^{-4} mi

Mass: 1 kg = 2.20462 lbm = 0.06852 slug = 1.1023 × 10^{-4} U.S. ton = 10^{-3} t = 9.8421 × 10^{-4} U.K. ton

Mass flow rate: 1 kg/s = 2.20462 lb/s = 132.28 lb/min = 7936.64 lb/h = 3.54314 long ton/h = 3.96832 short ton/h

Power: 1 W = 44.2537 ft · lbf /min = 3.41214 Btu/h = 1 J/s = 0.73756 ft · lbf/s = 0.23885 cal/s = 0.8598 kcal/h

Pressure: 1 bar = 10^5 N/m^2 = 10^5 Pa = 750.06 mm Hg at 0°C = 401.47 in. H$_2$O at 32°F = 29.530 in. Hg at 0°C = 14.504 lbf /in.^{2} = 14.504 psia = 1.01972 kg/cm^2 = 0.98692 atm = 0.1 MPa

Specific energy: 1 kJ/kg = 334.55 ft · lbf/lbm = 0.4299 Btu/lbm = 0.2388 cal/g

Specific energy per degree: 1 kJ/(kg · K) = 0.23885 Btu/(lbm · °F) = 0.23855 cal/(g · °C)

Surface tension: 1 N/m = 5.71015 × 10^{-3} lbf /in.

Temperature: T (K) = T (°C) + 273.15 = [T (°F) + 459.67]/1.8 = T (°R)/1.8

Temperature difference: ΔT (K) = ΔT (°C) = ΔT (°F)/1.8 = ΔT (°R)/1.8

Thermal conductivity: 1 W/(m · K) = 0.8604 kcal/(m · h · °C) = 0.5782 Btu/(ft · h · °F) = 0.01 W/(cm · K) = 2.390 × 10^{-3} cal/(cm · s · °C)

Thermal diffusivity: 1 m^2/s = 38,750 ft^2/h = 3600 m^2/h = 10.764 ft^2/s

Torque: 1 N·m = 1.41.61 oz · in. = 8.85073 lbf · in. = 0.73756 lbf · ft = 0.10197 kgf · m

Velocity: 1 m/s = 100 cm/s = 196.85 ft/min = 3.28084 ft/s = 2.23694 mi/h = 2.23694 mph = 3.6 km/h = 1.94260 U.K. knot = 1.94384 Int. knot

Viscosity, dynamic: 1 (N · s)/m^2 = 1 Pa · s = 10^7 µP = 2419.1 lbm/(ft · h) = 10^3 cP = 75.188 slug/(ft · h) = 10 P = 0.6720 lbm/(ft · s) = 0.02089 (lbf · s)/ft^2

Viscosity, kinematic: (see thermal diffusivity)

Volume: 1 m^3 = 61,024 in.^{3} = 1000 l = 219.97 U.K. gal = 264.17 U.S. gal = 35.3147 ft^3 = 1.30795 yd^3 = 1 stere = 0.81071 × 10^{-3} acre-foot

Volume flow rate: 1 m^3/s = 35.3147 ft^3/s = 2118.9 ft^3/min = 13198 U.K. gal/min = 791,891 U.K. gal/h = 15,850 U.S. gal/min = 951,019 U.S. gal/h

Source: From Liley, P. E. Thermophysical properties, *Boilers, Evaporators, and Condensers*, Kakaç, S. (Ed.), Wiley, New York, 1987.

Index

A

Additives, see Chemical additives to minimize fouling
Aging (fouling), 248
Air conditioning, condensers for, 516–522
Air conditioning industry, heat exchanger types used in, 517
Air-conditioning system, dry expansion coil-finned tube bank for, 525
Air-cooled condenser, 502–503, 519
 design of a compact, 538–539
Air-cooling evaporators (air coolers), 523–525
Air preheaters, 3–6
Air-side conductance, 438
Annulus, thermal and hydraulic analysis of, 278–291
Attachment (fouling), 247

B

Baffles, radial, 396
Baffle types, 371, 377–378
 plate, 377
Base fluid and thermal conductivity, 203
Bell-Delaware method, 395–396
 shell-side heat transfer coefficient, 396–407
 shell-side pressure drop, 407–418
Bends
 heat transfer in, 118
  90° bends, 118–119
  180° bends, 119–120
 pressure drop in, 140–142, 144–145
Biological fouling (biofouling), 245, 266
Blood heat exchanger, 487
 polymer, 572
Boiler with water-cooled furnace and heat recovery units, 26, 27
Boiling flow coefficient, 340
Boiling number, 338
Brinkman number, 162–163

C

Carbon fiber-reinforced polymers (CFRPs), 550
Carbon nanotube reinforced polymers (CNTRPs), 552
Cardiac surgery, heat exchangers for, 487–488, 572
Cavallini-Zechhin correlation, 325, 534
Channel hydraulic diameter, 468
Channel pressure drop, 474
Chemical additives to minimize fouling, 265–266
Chemical processing streams, fouling resistances for, 252
Chemical reaction fouling, 246
Chevron angle reversed on adjacent plates, 457
Chevron-type heat exchanger plate, 452, 456
Chillers, see Evaporator(s)
Circular duct
 with constant properties, 95, 96
 turbulent forced convection through, 95, 96, 105
Clay nanocomposites, inorganic, 551–552
Cleaning plates, 458
Cleaning techniques, see Surface cleaning techniques to control fouling
Cleanliness factor (CF), 256–257, 260, 261
Clustering and thermal conductivity, 204
Coil-finned tube bank, dry expansion, 525
Compact heat exchangers, 427
 defined, 427
 heat transfer and pressure drop, 433–445
 heat transfer enhancement, 427–433
Condensate flow, schematic representation of, 313
Condensate inundation, effect of, 313–322

Condensation, 307–308; *see also* Film condensation
  in downflow over horizontal tubes, 310–311
  film, 308–322
  inside horizontal tubes, 322–327
    idealized condensate profile inside horizontal tube, 323
  inside vertical tubes, 327–329
Condensation modes, 515
Condensation regimes, 515
Condensers, 491
  air-cooled, 502–503, 519, 538–539
  classification, 517
  design and operational considerations, 515–516
    construction considerations, 515
  direct contact, 503–504
  plate, 501–502
  for refrigeration and air conditioning, 516–522
  shell-and-tube, 492–500, 504–514, 518–519
  standards for, 531–536
  steam turbine exhaust, 500–501
  thermal analysis, 525–531
Condenser tube data, 369–370
Convection boiling factor, 338–339
Convection boiling heat transfer coefficient, 339
Convection number, 338
Coolant velocity, 514
Cooling system, closed-circuit, 462
Corrosion, 462–465
Corrosion fouling, 245, 266
Counterflow arrangement, single-pass, 453
Counterflow heat exchangers, 43–47
  temperature variation, 44
Crossflow condenser, features of, 494, 495
Crossflow heat exchangers, 25
  LMTD correction factor for, 52, 53
  temperature distribution in, 51
Cross-linked polyethylene (PEX), 553
Crossover heat exchangers, 47–56
Cross-sectional ducts, noncircular, 132–135

Cross-sectional tubes, circular, 129–132
Crystallization fouling, 245, 265

D

Desuperheating, 515
Direct-contact-type heat exchangers, 6–7
Direct (U-type) expansion (DX) evaporator, 523
Direct simulation Monte Carlo (DSMC), 174
Divided-flow shell-type heat exchanger, 50
Double-pipe geometries with one finned tube, typical, 296
Double-pipe hairpin heat exchanger, 9, 273, 274
Double-pipe heat exchangers, 3, 8–9, 273, 275, 277
  design and operational features, 295, 297
  finned-tube, 303–304
  vs. gasketed-plate heat exchangers, 487
  parallel-series arrangements of hairpins, 291–293
  thermal and hydraulic analysis of annulus, 278–291
  thermal and hydraulic design of inner tube, 276–278
  total pressure drop, 294–295
Dry expansion coil, 526
Dry expansion coil-finned tube bank, 525

E

Ellipse subject to slip-flow, Nusselt number for, 170
Engine oil, thermophysical properties of, 604
Enhancement factor, 338, 339
E-shell condenser, vertical, 495, 497
Ethylene tetrafluoroethylene (ETFE), 545
Evaporative condensers, 519–522
  operation, 520

Index 609

Evaporator(s), 491
　design of an air-conditioning, 539
　for refrigeration and air conditioning, 522–525
　standards for, 531–536
　thermal analysis, 525–531
Extended surface heat exchangers, 17
　classification, 17–23

F

Fanning friction factor and Nusselt number correlations, 469–471
Fiber-filled composites, 548, 550
Film condensation
　on single horizontal tube, 308–312
　in tube bundles, 312–322
Finned flat-tube matrices, 437, 440
Finned-tube double-pipe heat exchanger, 303–304
Finned-tube exchangers, pressure drop for, 441
Finned-tube geometries, 296
　used with circular tubes, 428
Finned-tube matrix
　circular, 435
　friction factor for flow across, 435, 436
Finned tubes, fouling in, 261, 262
Finned wall, 39
Fins, types of, 18–20
Fin-tube air heater, 19, 20
Fittings
　equivalent length in pipe diameters of various, 143
　pressure drop in, 142–147
Flattened tube-platefin compact surfaces, 438
Flat-tube matrices, finned, 437, 440
Flow arrangements, *see* Flow paths in heat exchangers
Flow boiling, 329–334
Flow boiling correlations, 334–352
Flow channel gap, mean, 467
Flow paths in heat exchangers, arrangement of, 24–25, 33, 34
　multipass and multipass crossflow, 25, 33, 34
Flow pattern, 331–334
Fluid-dependent parameter, 528, 529

Fluoropolymers, 545
Force balance of cylindrical fluid element within a pipe, 131, 132
Forced convection, 309–312
　in straight ducts, 107–110
Forced-draft, air-cooled exchanger used as condenser, 22
Fouling, 237–239, 244
　biological, 245, 266
　categories of, 244–246
　cost of, 243–244
　defined, 237
　design of heat exchangers subject to, 250–262
　effects of, 239
　　on heat transfer, 240, 241
　　on pressure drop, 241–243
　　on surface area, 239, 240
　effects of parameters on, 249, 250
　fundamental processes of, 246–248
　operations of heat exchangers subject to, 262–264
　in plain and finned tubes, 261, 262
　prediction of, 248–249
　techniques to control, 264–266
Fouling factors for plate heat exchangers, 465
Fouling resistance, 240, 250–256
　cleanliness factor, percent over surface, and, 261
Fouling time-factor curve, typical, 248, 249
Freon condenser, water-cooled shell-and-tube type, 539
Friction factor(s)
　for circular tube continuous fin heat exchanger, 434
　constricted, 185, 186
　fanning, 469–471
　for flow across finned flat-tube matrix, 437
　for flow across finned-tube matrix, 435, 436
　for gas flows in microchannel, 169–173
　experimental results on, 174
　for hydrodynamically and thermally developed laminar flow in ducts of various cross sections, 110

for plain plate-fin heat transfer matrices, 439
roughness effect on, 185–186
Froude number, 338

G

Gas flow
 engineering applications for, 164–177
 heat transfer in, 165–169
Gasketed-plate heat exchanger assembly, 454
Gasketed-plate heat exchangers, 12–13, 451, 452
 applications, 461–465
  maintenance, 465–466
 vs. double-pipe (hairpin) heat exchangers, 487
 flow regime between plates, 464
 heat transfer and pressure drop calculations, 466–476
  constants for, 472
  performance analysis, 476–481
 materials selection, 464
 mechanical features, 451–453
  gasket arrangements, 455
  plate materials, 455
  plate pack and frame, 453–455
  plate types, 455–457
 operational characteristics, 457
  main advantages, 457–459
  performance limits, 459
 passes and flow arrangements, 460–461
 vs. shell-and-tube heat exchangers, 487
 thermal performance, 481–483
Gas-to-gas heat exchangers, 559–561
Glass fiber-reinforced polymers (GFRPs), 550
Güngör and Winterton correlation, 341, 342, 344, 345, 526, 529–531

H

Hairpin heat exchangers
 with bare inner tube, 278–283
 double-pipe, 9, 273, 274
 with multiple finner inner tubes, 283–291
 multitube, 276

Hairpins, parallel-series arrangements of, 291–293
Hairpin sections arranged in series, 276
Heart surgery, heat exchangers for, 487–488, 572
Heat duty, 281, 285, 384, 475–476, 504
Heat exchanger analysis
 LMTD method, 43–56, 63
 ε-NTU-method, 56–66
Heat exchanger design calculation, 66–67
Heat exchanger design methodology, 70–73
Heat exchanger geometry, 557
Heat exchangers; *see also specific topics*
 applications, 25–26, 29
 approaches used in designing, 504
 basic design procedure, 378–379
  logic structure for process heat exchanger design, 378, 379
  preliminary estimation of unit size, 380–386
  rating of preliminary design, 386–387
 basic equations in design, 35–37
 classification, 1, 2, 6–7, 24, 34
 geometry of construction, 8–23
 selection of, 26–28
Heat sinks, 562
 hybrid vs. laminated graphite/epoxy, 561, 562
Heat transfer coefficient (h), 346, 468–469; *see also* Overall heat transfer coefficient; Shell-side heat transfer coefficient
 average condensation, 327, 328
 average evaporation, 345
 local condensation, 326, 327
 local evaporation, 344
 order of magnitude and range of, 41
Heat transfer mechanisms, 23
Helical coil(s)
 heat transfer in, 114–118
 with laminar flow, 138
 pressure drop in, 137–139
 schematic of, 115
 with turbulent flow, 139
High temperature Nylon (HTN), 553–554

Index 611

Hydraulic diameter
  channel, 468
  defined, 435
Hydraulic drag coefficient
  for inline bundles, 135–137
  for staggered bundles, 135–137

I

Ice-water-steam, saturated
    thermophysical properties of,
    584–589
Immersed heat exchanger, 555
Impingement plate, 496
Indirect-contact-type heat exchangers,
    3, 6–7
Industrial fluids, TEMA design fouling
    resistances for, 251
Initiation (fouling), 246
Inline array, 111, 112
Inline tube arrangement, square, 319
In-tube condenser, horizontal, 492,
    497–500
In-tube downflow condenser, vertical,
    498
In-tube enhancement techniques for
    evaporating and condensing
    refrigerants, 21, 22

J

J-shell, 493
  schematic of, 493, 494

K

Kandlikar correlation, 342–345, 526,
    528–529
Kern method, 402–403, 410, 413, 414
Kern relationship, 314–316
Kevlar fiber-reinforced polymers
    (KFRPs), 550–551
Knudsen number, 159–160, 167–169
  flow regimes based on, 159

L

Lamella heat exchangers, 15–16
Laminar film condensation, 308–309

Laminar flow
  in ducts of various cross sections,
    hydrodynamically and
    thermally developed, 110
  of gases, 92–95
  helical coils with, 138
  of liquids, 90–92
  Nusselt numbers and, 116–117
Laminar forced convection, 84–88
  through circular ducts, 93, 95
Laminar to turbulent flow transition in
    microchannels, experimental
    data on, 175
Laminar to turbulent transition
    regime, 173–177
Liquid crystal polymers (LCPs), 545
Liquid-to-gas heat exchangers, 556–559
Liquid-to-liquid heat exchangers, 552–556
LMTD (log mean temperature
    difference), 293
LMTD method for heat exchanger
    analysis, 43–56, 63
Lockhart–Martinelli parameter, 318, 323
Log mean temperature difference,
    see LMTD
Longitudinally finned bundle
    exchanger, 293
Longitudinally finned inner tube heat
    exchanger, 275
Longitudinally finned tubes, 20
Lumped analysis, 504–505

M

Maxwell–Garnett (M–G) model, 194
Maxwell model, 189–192, 194, 196, 203
Metals and nonmetals, thermophysical
    properties of, 577–581, 594
Microchannel heat sink,
    schematic of, 178
Microchannels
  engineering applications for single-
    phase liquid flow in, 177–186
  gaseous flow in, 158–163, *see also* Gas
    flow
  heat transfer in, 157–158
Multipass heat exchangers, 47–56;
    see also Flow paths in heat
    exchangers

N

Nanocomposites, 551–552
Nanofluid preparation methods, 188
Nanofluids
　convective heat transfer of,
　　186–187, 207–212
　　constant wall heat flux boundary
　　　condition, 212–214
　　constant wall temperature
　　　boundary condition, 214–216
　　experimental correlations of,
　　　216–223
　　particle materials and base
　　　fluids, 187
　　particle size and shape, 187–188
　thermal conductivity, 188–202
　　classical models, 189–191
　thermal conductivity experimental
　　studies of, 203–206
Nanoparticles
　Brownian motion of, 191–193
　clustering of, 193–196
　liquid layering around, 196–202
Natural gas-gas processing
　　streams, fouling
　　resistances for, 252
NTU (number of transfer units),
　　482–483
　defined, 482
　vs. effectiveness for various types of
　　heat exchangers, 59, 61–63
ε-NTU expressions, 59, 60
ε-NTU-method for heat exchanger
　　analysis, 56–66
Nuclear boiling, onset of, 329, 330
Nucleate boiling factor, 339
Number of transfer units, *see* NTU
Nusselt-Graetz correlation, 91–92
Nusselt number, 83, 88, 92, 100, 118, 120,
　　163, 165–170, 212–214, 222–223,
　　346, 416
　average/mean, 84–85, 110, 112, 114,
　　165, 184, 213–214, 217
　for developed laminar flow,
　　162–163
　for different geometries subject to
　　slip-flow, 168
　for helical coils, 116–118
　for hydrodynamically and
　　thermally developed laminar
　　flow in ducts of various cross
　　sections, 110
　for spiral coils, 117
Nusselt number correlation, 165, 216,
　　469–471
　Nusselt number and friction factor
　　correlations for single-phase
　　liquid flow, 179–185
Nusselt number enhancement ratio,
　　213–214
Nusselt theory, 309, 311, 313–314,
　　323, 327
Nylon, high temperature, 553–554

O

Oil cooler, 303–304
　with bare inner tube, 304
　for marine applications, 424
Oil refinery streams, fouling resistances
　　for, 253–255
Onset of nuclear boiling (ONB), 329, 330
Overall heat transfer coefficient (U),
　　37–42, 281, 284, 285, 289, 347,
　　349, 351, 393, 417, 436–437, 505,
　　535, 559
　for clean vs. fouled surfaces,
　　238–240, 256–260, 404–405,
　　475, 479
　estimating, 380, 382
　at inlet and outlet of power
　　condenser, 510–511
　variable, 67–70
　　typical cases of heat exchanger
　　　with, 68

P

Parallel-flow heat exchangers, 43–47
Parallel-series combinations, 277
Particle materials and thermal
　　conductivity, 203
Particle-reinforced composites, 547–548
Particle shape and thermal
　　conductivity, 204
Particle size and thermal
　　conductivity, 203

Index 613

Particle volume fraction and thermal conductivity, 203
Particulate fouling, 244–245, 266
PEEK (polyetheretherketone), 546
PEEK films, corrugated, 557
Percent over surface (OS), 257–262
Plate condensers, 501–502
    types of, 501
Plate exchanger used for dehumidification and cooling, 560, 561
Plate-fin exchangers, pressure drop for, 441–445
Plate-fin heat exchangers, 17–18, 431, 432
    construction, 17
Plate-fin heat transfer matrices, plain, 439, 440
Plate heat exchangers, 12
    classification, 12–16
    fouling factors for, 465
    useful data on, 463
Polycarbonate (PC), 545, 563
Polyetheretherketone, *see* PEEK
Polyethylene (PE), 546, 563; *see also* Cross-linked polyethylene
Polyethylene (PE) tubes, corrugated, 560
Polymer blood heat exchanger, 572
Polymer compact heat exchangers, 563–567
Polymer film compact heat exchanger (PFCHE), 564, 566
    potential applications, 567–570
Polymer heat exchangers, 543, 545–547
    thermal design, 570–571
Polymer matrix composite materials (PMCs), 547–548, 550–551, 562
    properties, 549
Polymers
    application in heat exchangers, 552–562
    thermal and mechanical properties of common, 544
Polyphenylene oxide (PPO), 546, 564
Polyphenylene sulfide (PPS), 546, 563
Polypropylene (PP), 545, 553, 563
Polysulfone (PSU), 546
Polytetrafluoroethylene (PTFE), 545, 563, 570
Polyvinyl chloride (PVC), 548
Polyvinyl chloride (PVC) shell-and-tube heat exchanger, 552, 553
Polyvinylidene fluoride (PVDF), 545, 563, 564
Port pressure drop, 474–475
Power condensers, *see* Condensers
Prandtl number (Pr), 346
Prandtl's correlation, 279–280
Plate condensers, 501–502
Pressure drop; *see also* Gasketed-plate heat exchangers
    from abrupt contraction, expansion, and momentum change, 147–148
    in bends, 140–142, 144–145
    in double-pipe heat exchangers, 294–295
    due to friction in fittings, valves, bends, contraction, and enlargement, 142, 144–145
    for finned-tube exchangers, 441
    fouling and, 241–243
    for plate-fin exchangers, 441–445
    port, 474–475
    shell-side, 387, 389–390, 407–418
    in spiral coils, 137–139
    in tube bundles in crossflow, 135–137
    tube-side, 129–135
Pressured water reactor (PWA), design of steam generator for, 540
Pressurized water reactors (PWRs), 23
Pumping power, heat transfer and, 148–150
Pumping power expenditure for various fluid conditions, 149–150

R

Ramen heat exchangers, *see* Lamella heat exchangers
Rectangular ducts, 133, 134
Recuperators, 1–3
Reflux condenser, 499
Refrigerant condensers, air-cooled, *see* Air-cooled condenser
Refrigerants
    in-tube enhancement techniques for evaporating and condensing, 21, 22
    thermophysical properties, 595–604
Refrigeration, condensers for, 516–518
Refrigeration cycle, two-stage vapor compression, 517, 518

Refrigeration industry, heat exchanger types used in, 517
Regenerators, 1–6
　classification, 3
Removal (fouling), 247
Return bend housing and cover plate, 295
Rotary regenerators, 3, 4
　disk- vs. drum-type, 3, 4
Rotary storage-type heat exchanger, 3, 4
Rotating-plate regenerative air preheaters, 5

S

Shah correlation, 212, 326–328, 338, 339, 341, 344, 345, 526–528
Shah method, 337, 338, 341, 345, 347, 348, 351
Shell-and-tube condensers, 492–500, 518–519
　thermal design of, 504–514
Shell-and-tube evaporator, flooded, 522, 524
Shell-and-tube heat exchangers, 3, 9–11, 48, 49, 361
　basic components, 361
　　allocation of streams, 376
　　baffle type and geometry, 371, 377–378
　　constructional parts and connections, 362
　　front and rear end head types, 363
　　shell types, 361–364
　　tube bundle types, 364–365
　　tube layout, 368, 370–376
　　tubes and tube passes, 366–369
　corrugated PE tubes used in, 560
　film heat transfer coefficients for, 381
　polymer, 552–554
　PVC, 552, 553
Shell-side condensers
　horizontal, 492–495
　vertical, 495
Shell-side heat transfer coefficient, 387–389, 396–407
Shell-side pressure drop, 387, 389–390, 407–418

Shell-type heat exchanger, divided-flow, 361–364
Shell types, 361–364
　schematic sketches of common TEMA, 364
Smooth circular ducts
　hydrodynamically developed and thermally developing laminar flow in, 84–85
　turbulent flow isothermal fanning friction factor correlations for, 101, 102, 131
Smooth ducts
　laminar flow through concentric annular, 86–88
　simultaneously developing laminar flow in, 85–86
Smooth straight circular ducts, laminar forced convection correlations in, 93, 94
Smooth straight noncircular ducts, turbulent flow in, 99–103
Smooth-tube bundles, heat transfer from, 111–114
Specific heat, 208
Spiral coil(s)
　heat transfer in, 114–115, 117
　with laminar flow, 138–139
　pressure drop in, 137–139
　schematic of, 115
　with turbulent flow, 139
Spiral heat exchanger for condenser applications, 502
Spiral plate heat exchangers, 14–15
Spiral-tube-type heat exchangers, 12
Split-flow shell-type heat exchanger, 50
Staggered array, 111, 112
Stationary-plate regenerative air preheaters, 5, 6
Steam, thermophysical properties of, 590; *see also* Ice-water-steam; Water-steam at high pressures
Steam generator for pressured water reactor (PWA), design of, 540
Steam generators, inverted U-tube type, 26, 28
Steam turbine exhaust condensers, 500–501
Subcooled boiling, 329–331

Index

Subcooling, 515
Surface area, heat transfer, 239, 240, 466–467, 475–476
Surface cleaning techniques to control fouling, 264–265
   continuous cleaning, 264
   periodic cleaning, 264–265

T

Taborek method, 403
Teflon, 545
Temperature and thermal conductivity, 204
Temperature jump, 161
Themoplastics, 545
Thermal conductivities of various solids and liquids, 187
Thermal resistances at inlet of power condenser, 511, 512
Thermophoresis, 247
Thermosets, 545
Transfer processes, 6–8
Transport (fouling), 246–247
Tube bank
   dry expansion coil-finned, 525
   ideal, 398–400
Tube bundle arrangements, 111
Tube bundle layout, 315
Tube bundles
   correlation factor for number of rows and average heat transfer from, 112
   in crossflow, pressure drop in, 135–137
Tube evaporator
   flow patterns in horizontal, 333
   temperature profiles in vertical, 332
Tube-fin heat exchangers, 431, 433
Tube layout angles, 370
Tube pitches parallel and normal to flow, 409–411
Tube-platefin compact surfaces, flattened, 438
Tube-shell layouts (tube counts), 372–376
Tube-side condensers, vertical, 495–497
Tube-side pressure drop, 129–135, 390–395
Tube support plate, 496
Tubing, dimensional data for commercial, 367–368
Tubular-fin heat exchangers, 18–23
Tubular heat exchangers, 8
   classification, 8–12
Turbulent flow
   helical coils with, 139
   Nusselt numbers and, 117–118
Turbulent forced convection, 93, 95–99
   effect of variable physical properties in, 103–107
   through circular ducts, 95, 96, 105
Turbulent forced convection correlations in circular ducts
   for gases with variable properties, 109–110
   for liquids with variable properties, 108
Turbulent gas flow in ducts, 104–107
Turbulent liquid flow in ducts, 103–104

U

U-tube configuration, 11, 364, 365
U-tube type steam generators, inverted, 26, 28
U-tube waste heat boiler, horizontal, 22
U-type expansion (DX) evaporator, 523

V

Valves, equivalent length in pipe diameters of various, 143
Vapor belt, 496
Vapor shear, effect of, 317–322
Velocity slip, 160
Viscosity models for nanofluids, 210
Viscosity of nanofluids, 209–212

W

Water; *see also* Ice-water-steam
   fouling resistances for, 256
Water-cooled condensers, 518–519
Water-cooling evaporators (chillers), 522–523
Water-steam at high pressures, thermophysical properties of, 591–593